The Microbial Challenge
Human-Microbe Interactions

The Microbial Challenge
Human-Microbe Interactions

Robert I. Krasner

Department of Biology
Providence College
Providence, Rhode Island

ASM
PRESS

Washington, D.C.

Address editorial correspondence to ASM Press, 1752 N St. NW, Washington, DC
20036-2904, USA

Send orders to ASM Press, P.O. Box 605, Herndon, VA 20172, USA
Phone: (800) 546-2416 or (703) 661-1593
Fax: (703) 661-1501
E-mail: books@asmusa.org
Online: www.asmpress.org

Library of Congress Cataloging-in-Publication Data
Krasner, Robert I.
  The microbial challenge : human-microbe interactions / Robert I. Krasner.
    p. cm.
  Includes index.
  ISBN 1-55581-241-4
  1. Medical microbiology. I. Title.

QR46.K734 2002
616'.01—dc21

                                                                2002020199

10 9 8 7 6 5 4 3 2 1

Cover and title page design: Jon Krasner
Interior design: Susan Brown Schmidler

To Lee
With gratitude for our life of love, happiness, and adventure
and
to our children and their spouses
and
their children

Contents

Acknowledgments

Writing a textbook is a difficult task—more so than I had thought. It is an ongoing collaboration between the author and the publisher requiring trust and mutual respect. I thank those staff members of ASM Press who have helped to transform my manuscript into a finished text of which we are all proud. In particular, I thank Jeff Holtmeier (director, ASM Press) and Ken April (senior production editor). In addition, I thank other ASM staff members who played roles in the publication process, as well as Mary McKenney, the copy editor; Susan Schmidler, who designed the interior of the book; and Patrick Lane and his staff at J/B Woolsey Associates, who prepared the illustrations and In the News pieces. I greatly appreciate their efforts.

The outstanding book cover was designed by my son, Jon Krasner, who continues to amaze me with his creative talents. Thank you, Jon. Several students have helped me during the course of this project in typing the manuscript from endless tapes that seemed to drone on and on. George Dekki has worked with me for the longest time, Lindsay Eberhard is always there to help me meet the never-ending deadlines, and Randy Ingham took on the tedious task of assembling the glossary. I appreciate their efforts. I also express my thanks to the students in my course Contemporary Biology for allowing me the opportunity to field-test the book.

Several professional biologists have generously given their time to review the manuscript; some have reviewed all of the chapters, while others have reviewed particular areas according to their expertise. Whatever the case, I express my gratitude to them for their constructive critiques, which have resulted in a meaningful text. Any errors that remain in the text are my own. These reviewers are as follows: Paul Brown (National Institutes of Health), Toby Glicken, Judith S. Heelan (Memorial Hospital of Rhode Island), Karen Klyczek (University of Wisconsin—River Falls), James H. Maguire (Centers for Disease Control and Prevention), Gary B. Ogden (St. Mary's University), Dennis J. Opheim (Quinnipiac College), Mark L. Petersen (Blue Mountain Community College), Paul Rega (ProMedica Health System and University of Findlay), Moselio Schaechter (formerly of Tufts University), James Snyder (University of Louisville School of Medicine), Charles Toth (Providence College), and Yinsheng Wan (Providence College).

Introduction

New, emerging, and reemerging infections are a major public health problem today, yet 35 years ago, it was predicted that infectious diseases would soon be conquered. Accounts of infectious diseases currently appear almost daily in newspapers and in popular magazines; flesh-eating streptococci, *Escherichia coli*-contaminated strawberries, *Salmonella* disease, Ebola virus, mad cow disease, and AIDS top recent headlines and conjure up frightful images. Several excellent books, including *The Hot Zone* and *The Coming Plague*, have appeared in bookstores. Several years ago the movie *Outbreak* captured the public's interest. News stories warn of the continued threat of the use of biological weapons. In the days after the terrorist attacks on the World Trade Center and the Pentagon, anthrax spores were strategically disseminated to individuals in the government and the media, resulting in a large number of cases of anthrax, some of which proved to be fatal.

Had newspapers been available in ancient times, hieroglyphics would have revealed infectious diseases. Descriptions of what are probably leprosy and tuberculosis appear in the Bible. This is not surprising; after all, microbes were the first inhabitants of the earth, as evidenced by the fact that fossils dating back 3 billion years are those of primitive bacteria.

TEXT OVERVIEW

This text is divided into three parts in a logical sequence. Part I, "The Challenge," comprises chapters 1 to 10; part II, "Meeting the Challenge," comprises chapters 11 to 13; and part III, "Current Challenges," comprises chapters 14 to 16). Chapter 1 introduces the challenge posed by new, emerging, and reemerging microbial diseases, a challenge made more difficult by several factors, including increased urbanization, jet travel, the misuse of antibiotics, and cultural and societal changes. In chapter 2, the array and the biology of microbes are presented in the context of basic biological principles. Chapter 3 emphasizes that disease-producing microbes represent only a very small piece of the microbial pie. Microbes are not "out to get us"; it is not in their best interest. We are just fortuitous hosts who manage to get in the way, and we become part of their food chain. Many microbes are beneficial to humankind. Harnessing them for even greater gain is a challenge. Chapters 4 and 5 present a more in-depth discussion of the biology of bacteria and viruses. The idea of parasitism as a biological association, under the umbrella of symbiosis, is introduced in chapter 6, which also focuses on the mechanisms of microbial virulence. Chapter 7 presents the cycle of microbial disease, with an emphasis on mechanisms of transmission. The major diseases caused by bacteria, viruses, and protozoa and helminths, arranged by modes of transmission, are presented in chapters 8, 9, and 10, respectively. At the conclusion of part I, the student will have the necessary background to meaningfully study chapters in the remaining two sections.

Part II deals with meeting the microbial challenge, a challenge which is present from birth (and before, in some cases). At the level of the individual, the challenge is met by the ability of the immune system to recognize and eliminate invading microbes; this is the subject of chapter 11. Chapter 12 focuses on public health achievements in the control of infectious diseases. The late 1800s and early 1900s ushered in the role of public health and sanitation; immunization and antibiotics became available during the 20th century. The watchwords of public health are surveillance and protection. The successes of the past century were the results of collaborative partnerships ranging from the local to the national and international levels; these partnerships are described in chapter 13.

Part III looks to the past in order to place current challenges in historical perspective. The coevolution and coexistence of microbes and humans from antiquity to current times have led to misery in the form of plagues that have decimated large populations and have affected the development of civilization, as discussed in part III. Chapter 14 is about biological weapons, a current challenge with an ancient history. This chapter covers the terrorist attacks of September 11, 2001, and the anthrax scare which followed in their wake. The fact that plagues cannot be relegated to the past but continue into the present is the basis for chapter 15, which presents mad cow disease, its human counterpart Creutzfeldt-Jakob disease, tuberculosis, and AIDS as examples of current plagues.

TEXT FEATURES

A text needs to be simulating and user-friendly in order to engage the target audience. To this end, I have incorporated the following:

1. *In the News:* Imagine tomorrow's newspaper headlines if a tremendous breakthrough in the fight against AIDS, possibly a cure, were announced today. Newspaper articles and headlines reflect events of the day, including those concerned with public health. Over the past few years, I have clipped newspaper articles and headlines dealing with Ebola virus, AIDS, biological weapons, antibiotics, immunization, and a range of other public health topics. The material presented in "In the News" is adapted from actual headlines and news articles that I have collected.

2. *Photographs:* During my long tenure as a professor of microbiology, I have taken advantage of sabbaticals and leaves of absence to study infectious diseases. I have spent time in Central America, Israel, the

Pasteur Institute in Paris, and most recently Brazil (a trip highlighted by a field experience in the Amazon Basin). During my travels, I took many of the photos that are used to illustrate the text.

3. *Boxes:* Most chapters have one or more boxes. The intent is to present more information on subjects introduced in the narrative. Some boxes are of historical interest, whereas others are based on more contemporary issues.

4. *Layout:* The layout of the text is designed to help keep students focused and organized. Each chapter begins with a preview and ends with an overview, and an outline of the chapter is on the opening page. Section headings and subheadings are extensively used.

5. *Self-Evaluations:* A series of questions follows each chapter. The objective is to assist students in evaluating and reviewing the subject matter.

COURSE CONTENT AND STRUCTURE

The text is designed to allow maximum flexibility in course design. If the plan is to offer a one-semester course, the instructor can choose appropriate chapters (or parts of chapters) to meet his or her needs. Chapters 1 through 7 constitute a core and are preparatory for the remaining chapters. Some instructors may wish to stress pathogens and their diseases and may choose to cover all of the microbes presented in chapters 8, 9, and 10, while others may choose to sample from each mode of transmission in these chapters. If necessary because of time constraints, chapters 3, 11, and 14 could be skipped or assigned for self-study without compromising the flow of information. Certainly, other strategies can be developed, limited only by the instructor's ingenuity. The text can be used for a two-semester course as well, allowing time to supplement each chapter with additional readings, student presentations, guest lecturers, videos, and laboratory demonstrations and exercises that can be done in the classroom. This point is further addressed below.

LABORATORY COMPONENT

It is my expectation that in most cases, this book will be adopted for a course that does not include an assigned laboratory experience. However, there are a number of demonstrations and exercises that can be done in the classroom at nominal cost. Certainly, if a laboratory facility is available, it would be a good added feature. Examples include exercises to demonstrate the effectiveness of hand washing, the use of disks to evaluate antibiotics and household products, culturing normal body floras and floras in the environment, and demonstrations or exercises on fermentation, conjugation, DNA spooling, and person-to-person transmission of microbes. A number of exercises involving yeast (purchased in a supermarket) can be easily implemented in a classroom setting. Agar plates may be available from a microbiology course at the instructor's institution or from a hospital laboratory or may be purchased. Standard microbiology laboratory manuals, the *Journal of College Science Teaching* (published by the National Science Teachers Association), and *American Biology Teacher* (published by the National Association of Biology Teachers) are excellent sources of laboratory exercises that could be adapted to the classroom.

RESOURCES

The American Society for Microbiology maintains the MicrobeLibrary website (http://www.microbelibrary.org) with many resources for microbiology teaching and learning. You can find searchable visual and curriculum resource collections, the *Microbiology Education Journal,* the *Focus On Microbiology Education* newsletter, and a compilation of reviews of books, websites, videos, and software. The Department of Biological Sciences at Kent State University sponsors the Microbiology Learning Center (http://dept .kent.edu/microbiology/default.htm), which has an abundance of resources. The National Science Teachers Association and the National Association of Biology Teachers are also excellent sources. Lastly, I will be pleased to communicate with instructors about course content and structure and classroom (laboratory) demonstrations and exercises.

PART **I**

THE CHALLENGE

IDENTIFYING THE CHALLENGE

PREVIEW This book is about a challenge—a worldwide challenge—posed by microbes, invisible marauders that inhabit the earth and may cause illness and death. They are not "out to get us," and, in fact, their survival is linked to our survival. There are five distinct kinds of microbes: **bacteria**, **viruses**, **protozoans**, **fungi**, and **unicellular algae**. It should be emphasized at the outset that the potential for infection is posed by relatively few members of each of the groups. The vast majority of microbes are often beneficial and are essential to the cycles of nature without which higher life forms could not exist. In many cases, microbes have been harnessed for the benefit of humankind.

But this book, by intent, has a bias because its theme relates to the relatively few microbes that are disease producers; in the language of medical microbiology, they are referred to as **pathogens** or virulent microbes. Why some of these microbes now represent an increased challenge is the subject matter of this chapter and sets the stage for the remaining chapters.

THE CHALLENGE

A few decades ago, there was less need for this book than there is today. The public is currently both fascinated with and terrified by accounts of acquired immune deficiency syndrome (AIDS), Ebola virus, *Escherichia coli*, hantavirus, West Nile virus, *Salmonella*, and flesh-eating "strep," to name only a few. Popular news magazine programs, including *Dateline, 60 Minutes*, and *20/20*, frequently air segments relating to dangerous microbes; newspaper articles and news broadcasts further alert the public. The movie *Outbreak* reached epidemic (a term referring to a large outbreak) proportions in terms of public interest; *Ebola, The Coming Plague, Epidemic, The Hot Zone*, and *The Cobra Event* dominate bookshelves and continue to intrigue readers.

This wave of interest is not based on science fiction; it is the real thing. Consider these examples. In 1992 in New York City, almost 4,000 cases of tuberculosis (TB), a bacterial disease almost relegated to oblivion, were reported. To make matters worse, many of the cases were resistant to a variety

TABLE 1.1 Examples of new, emerging, and reemerging infections

Bacterial diseases
Lyme disease
Ehrlichiosis
Escherichia coli O157:H7 disease
Legionnaires' disease
Tuberculosis
Viral diseases
Hantavirus pulmonary syndrome
Ebola virus hemorrhagic disease
Dengue fever
Rabies
Protozoan diseases
Cryptosporidiosis
Malaria
Fungal diseases
Coccidioidomycosis
Pneumocystis pneumonia

of antibiotics that, in earlier times, stopped the bacteria dead in their tracks. In 1993, an outbreak of cryptosporidiosis, a waterborne protozoan disease characterized by diarrhea, swept through Milwaukee, Wis., causing illness in about 400,000 people, approximately 25% of that city's population; it was the largest reported waterborne illness in U.S. history. In 1993, for the first time in the United States, hantavirus, the causal agent of a potentially fatal influenzalike respiratory illness, was reported to cause fatalities in New Mexico, Utah, Colorado, and Arizona. Ebola hemorrhagic fever, caused by one of the deadliest known viruses, ignited a near worldwide panic in 1995 when an outbreak occurred in Kikwit, Zaire (now the Democratic Republic of Congo), causing 240 people to bleed to death. (These diseases are discussed in chapters 8, 9, and 10.) No wonder people were terrified when Ebola-virus-infected monkeys intended for research were introduced into primate quarantine facilities in the United States. The episode prompted the film *Outbreak* and the best-selling novel *The Hot Zone*. Reports of the bacterium *E. coli* O157:H7 frequent the news, as in 1997 when the Colorado Department of Public Health and Environment detected the organisms in hamburger patties, resulting in a cluster of *E. coli*-infected people and the recall of 25 million pounds of ground beef to avert a nationwide outbreak. Can you imagine a mound of hamburger meat of that size? Lyme disease and the resurgence of malaria are two more examples from a long list of emerging and reemerging infectious diseases (Table 1.1); no nation can afford to be complacent regarding its vulnerability (Fig. 1.1).

In 1992, the Institute of Medicine, a branch of the National Academy of Sciences that advises the government on policy matters pertaining to the health of the public, in its report *Emerging Infections: Microbial Threats to Health in the United States*, defined emerging infections as "new, reemerging or drug-resistant infections whose incidence in humans has increased within the past two decades or whose incidence threatens to increase in the near future."

Despite the tremendous progress in the latter half of the 20th century in controlling infectious diseases, including the eradication of smallpox, the introduction of antibiotics in the 1940s, improvement in sanitation, and an increase in the diseases for which immunization is available, the battle against infectious diseases is uphill and of concern to epidemiologists (Box 1.1). In April 2000, Michael Osterholm, Minnesota's well-respected state epidemiologist for 24 years, stated, "On a good day we hold them at bay. On a bad day they're winning. Our task is a lot like trying to swim against the current of a raging river." David Satcher, former director of the U.S.

FIGURE 1.1 Global microbial threats in the 1990s. (Source: CDC.)

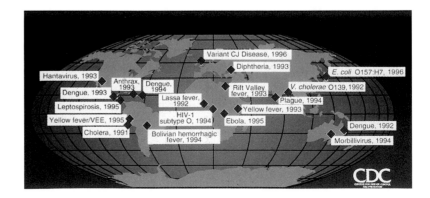

BOX 1.1 Quotations Relating to Health and Infectious Disease

The world may have only a decade or two to make optimal use of the many medicines presently available to stop infectious diseases. We are literally in a race against time to bring levels of infectious disease down worldwide, before the disease wears the drugs down first.

David Heymann, Executive Director, WHO's Communicable Disease Program

Germs come by stealth
And ruin health
So listen, pard,
Just drop a card
To a man who'll clean up your yard
And that will hit the old germs hard.

Sinclair Lewis, *Arrowsmith*

Everyone has the right to a standard of living adequate for the health and well-being of himself and of his family including food, clothing, housing, and medical care. . . .

Article 25, Universal Declaration of Human Rights, adopted by the General Assembly of the United Nations, December 10, 1948

Health is a state of complete physical, mental, and social well-being and not merely the absence of disease or infirmity. . . .

Constitution of WHO, July 22, 1946

Infectious disease is one of the few genuine adventures left in the world. The dragons are all dead and the lance grows rusty in the chimney corner. . . . About the only sporting proposition that remains unimpaired by the relentless domestication of a once free-living human species is the war against those ferocious little fellow creatures, which lurk in the dark corners and stalk us in the bodies of rats, mice and all kinds of domestic animals; which fly and crawl with the insects, and waylay us in our food and drink and even in our love.

Hans Zinsser, *Rats, Lice, and History*, 1935

We stand on the brink of a global crisis in infectious diseases. No country is safe from them. No country can any longer afford to ignore this threat.

Director-General, WHO, World Health Report, 1996

It is time to strengthen our research efforts . . . so that we can unlock the mysteries behind antibiotic resistance and discover new scientific weapons in the battle to detect and control emerging infectious diseases.

Albert Gore, vice president of the United States, June 1996

Ingenuity, knowledge, and organization alter but cannot cancel humanity's vulnerability to invasion by parasitic forms of life. Infectious diseases which antedated the emergence of humankind will last as long as humanity itself and will surely remain, as it has been hitherto, one of the fundamental parameters and determinants of human history.

William H. McNeill, *Plagues and Peoples*, 1976

Our availability to detect, contain, and prevent emerging infectious diseases is in jeopardy.

David Satcher, surgeon general of the United States

Pathogenic microbes can be resilient, dangerous foes. Although it is impossible to predict their individual emergence in time and place, we can be confident that new microbial diseases will emerge.

Emerging Infections: Microbial Threats to Health in the United States (Institute of Medicine, 1992)

The microbe that felled one child in a distant continent yesterday can reach yours today and seed a global pandemic tomorrow.

Pitted against microbial genes, we have mainly our wits.

Joshua Lederberg, 1958 Nobel Prize winner

On a good day, we hold them at bay. On a bad day, they're winning. Our task is a lot like trying to swim against the current of a raging river.

Michael Osterholm, former Minnesota state epidemiologist and founder of an infectious disease control company, 2000

Centers for Disease Control and Prevention (CDC) and later surgeon general of the United States, warned that "our ability to detect, contain, and prevent emerging infectious diseases is in jeopardy."

The International Red Cross, in a report published on June 6, 2000, clearly warns of the danger of infectious diseases. The report speaks of "the silent tragedy" and the death of 13 million people from preventable diseases "primarily infectious in nature" in the previous year; this is symptomatic of deteriorating health services. Further, compared with floods and earthquakes, which grab news headlines and donors' cash, the uncontrolled spread of disease steals far more lives. An estimated 150 million people have died of AIDS, TB, and malaria alone since 1945, compared with 23 million lives lost in wars. In a recent year, 160 times more people died from AIDS, malaria, respiratory diseases, and diarrhea than were killed in natural disasters, including the massive earthquakes in Turkey, floods in Venezuela, and cyclones in India. Malaria kills 2.6 million people a year, of whom 70%

are children. TB has been on the upswing in North Korea, with a report of 40,000 new cases in 1999 and 2000; since the collapse of the Soviet Union, there has been a 40-fold increase in syphilis in Russia.

If the director of the World Health Organization (WHO) were to report on the health status of the world, what might that report state? According to the 1999 WHO report (Fig. 1.2), in 1998 infectious diseases were the second leading cause of death worldwide, resulting in the death of 19.5 million adults and children. Approximately one-third of the 54 million deaths that occur worldwide each year are the result of infectious diseases or their complications. This figure is an underestimate, since surveillance and reporting networks are woefully deficient in many Third World countries. The top 10 killers in WHO's 1999 report (Fig. 1.3) include bacterial, viral, protozoan, and worm diseases (chapters 8, 9, and 10). Initially, one would attribute these devastating statistics to the poverty associated with developing nations, but, as surprising as it may seem, in the United States infectious diseases were ranked as the third leading cause of death (Fig. 1.4). No wonder the director-general of WHO stated in the 1999 report, "We stand on the brink of a global crisis in infectious diseases. No country is safe from them. No country can any longer afford to ignore this threat."

So what's the bottom line? What grade would the world now be awarded in terms of its success in coping with microbial diseases? Most experts in the field would probably give a C–; the grades A and B are considered honor grades. Certainly, under the leadership of the United Nations, WHO, the CDC, and other organizations, the burden of infectious diseases around the world can be lessened and a higher grade can be achieved.

Actually, during the 1950s, 1960s, and 1970s, microbial diseases appeared to be on their way out. It was a span of years heralded by optimism and progress in public health. The first polio vaccine was introduced by Jonas Salk in the 1950s and ushered in a time of successful mass polio vaccination campaigns. Fewer than 1,000 polio cases occurred in 1967 in western Europe and North America as compared with over 75,000 in 1955. In 1948,

AUTHOR'S NOTE *Students are always understandably anxious to know their grades. I hope you will not settle for a C in this or any other course. It's an okay grade but indicates the need for greater effort! Greater effort is needed in the control of infectious diseases too.*

FIGURE 1.2 Leading causes of death, 1998. There were 53.9 million deaths worldwide in 1998. Cancers and cardiovascular, respiratory, and digestive diseases can also be caused by infections, and thus the percentage of deaths due to infectious diseases may be even higher than shown. (Source: WHO report, 1999.)

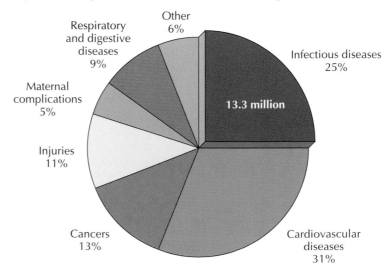

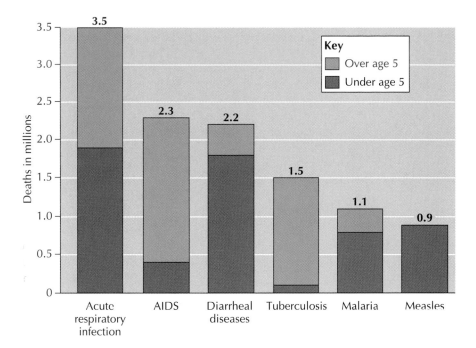

FIGURE 1.3   Leading infectious killers. The graph shows millions of deaths world-wide in 1998 for persons of all ages. Pneumonia and influenza are included in acute respiratory infections. Deaths among HIV-positive people with tuberculosis are included under AIDS. (Source: WHO report, 1999.)

U.S. Secretary of State George C. Marshall announced at an international meeting on tropical diseases that the conquest of microbial diseases was imminent. According to WHO and the Pan-American Sanitary Conference, it appeared that malaria would be taken off the list of diseases "of major importance." The times were good; as the economy of nations improved, poverty decreased, and so did the burden of microbial diseases. The 1966 CDC report, in what might be considered an address on the state of the union's health,

FIGURE 1.4   Leading causes of death in the United States, 1992. (Modified from *JAMA* [*Journal of the American Medical Association*] with permission.)

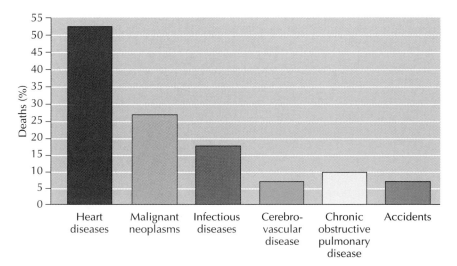

glowed with the promise of the conquest of microbial diseases. William H. Stewart, surgeon general of the United States in 1967, told a gathering of health officers that it was "time to close the book on infectious diseases and shift all national efforts to chronic diseases." Stewart's optimism was echoed by health officials in other developed nations of the world.

But unfortunately, the storm clouds were hovering in developing countries—countries that constitute a large part of the world's population. For the most part, the warnings were not heeded except for a few nagging voices in the background. From 1980 to 1992 alone, the CDC reported a 22% increase in infectious diseases (excluding AIDS). Data presented in a 1996 article published by the *Journal of the American Medical Association* indicated a greater than 50% increase in deaths caused by microbes in the United States since 1980. But the scientific community was slow to acknowledge that the bubble of antisepsis and disease control was about to burst. Life was too good to throw in a monkey wrench!

Why were new diseases emerging and older ones reemerging, as it sometimes appeared, with a vengeance? There were those who claimed it was God's wrath in expressing disapproval of a more liberal society, an explanation without scientific merit. The 1992 Institute of Medicine's report *Emerging Infections: Microbial Threats to Health in the United States* warned that microbes were winning the battle and that our previous complacency and optimism had weakened our ability to counterattack; the report predicted an increase in the burden of infectious diseases. Essentially, it appears that the choreography of adaptation between microbes and humans was beginning to come apart at the seams because of a variety of linked and overlapping factors that are considered in the following sections (Table 1.2): (i) world population growth, (ii) urbanization, (iii) ecological disturbances, (iv) technological advances, (v) microbial evolution and adaptation, and (vi) human behavior.

FACTORS RESPONSIBLE FOR EMERGING INFECTIONS

World Population Growth

On July 4, 2000, at 3:59 p.m., eastern daylight time, the World Population Clock reported that more than 6 billion (6,080,952,933) people inhabited the earth; the growth rate is estimated at 1.25%, which means that by 2050, the population will have soared to over 9 billion (about 9,104,205,830) (Fig. 1.5). To add to the problem of the burgeoning population, 80% of the population is living in less developed countries (of which 60% are tropical and subtropical areas) with a diminished capacity to cope with population increase. Several factors have been cited for the current crisis of new and emerging infectious diseases, but the population explosion is central to the issue (Fig. 1.6).

Thomas Malthus's *Essay on the Principles of Population* warned of the negative influence that unchecked population growth could have on societies, primarily because of inadequate supplies of food. (Charles Darwin's insights into evolution and natural science were strongly influenced by Malthus.) It appears that Malthus was right, as evidenced by famine in Ethiopia and in other parts of the world over the past 100 or more years. But there is another negative consequence of overpopulation, and that is transmission of infectious diseases. The total population of a country or a region, in itself, is not as crucial as its population density—the number of

TABLE 1.2 Factors responsible for emerging diseases

World population growth
Urbanization
Ecological disturbances
Deforestation
Climatic changes
Natural disasters (drought, floods)
Technological advances
Air travel
Unsafe blood supplies
Microbial evolution and adaptation
Antimicrobial resistance
Evasive strategies
Human behavior and attitudes
Complacency
Migration
Societal factors

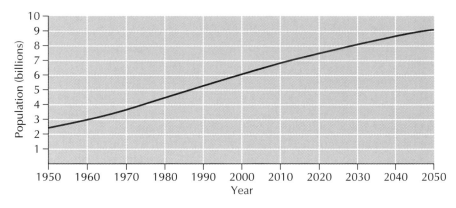

FIGURE 1.5   World population, 1950 to 2050. Projections are based on an estimated annual growth rate of 1.25%. (Source: U.S. Census Bureau, International Data Base, May 10, 2000.)

people per square mile in a defined area. Person-to-person transmission is facilitated as population density increases. Other diseases are carried by **biological vectors**, including mosquitoes, ticks, lice, and flies, from human to human or, in some cases, **zoonotic diseases** are transmitted from animal to human. Whatever the mode of transmission, population density is a significant factor. Consider, for example, a classroom with fixed dimensions and assume that one person in the class has a cold, but there are only 10 other students randomly spaced throughout the room; on the other hand, consider the same classroom with 60 students. Clearly, the chain of transmission is fostered in the larger population, simply because respiratory droplets are able to traverse the shorter distance from contact to contact when the population density is higher.

In considering population size, the distribution of the population is also of considerable significance in terms of the burden of disease. Elderly populations are more susceptible to microbial diseases, presumably because of a decline in their immune status, and serve as a source of infection for family and community members. In the United States in 1980, people over age 65 constituted 11.4% of the population—a figure which is expected to reach almost 28% by 2050. It is difficult to obtain reliable

FIGURE 1.6   Population explosion: the "hub" of the problem.

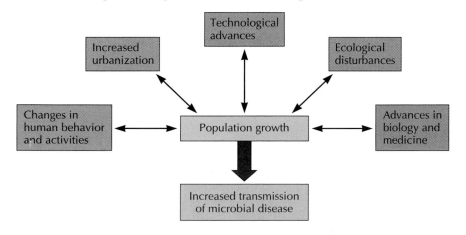

data for developing countries. The point is that predictors of infectious disease need to take into account not only the total population and population density but also the demographics of age distribution in that population.

As the world population increases, there are a number of consequences that foster an increase in infectious diseases (Table 1.3). For example, Dhaka, the capital city of Bangladesh, has many slum areas. Bangladesh is the world's ninth most populous country, with approximately 123 million people. Population control is Bangladesh's most urgent problem, along with the attendant low per capita income. As would be expected, high levels of malnutrition exist, with much of the population getting less than one-third of the normal food intake because agricultural production has not been able to keep pace with population growth. The consequences in terms of infectious diseases, particularly diarrhea, due to poverty and poverty-related conditions are dramatic.

Urbanization

Zaynab Begum lives in Bangladesh in a Dhaka slum, along with her husband and three children in a primitive hut, less than 6 square meters in size, constructed of bamboo and makeshift materials; there is no electricity, running water, or toilet, and an open sewer runs outside the hut. Zaynab's husband is a rickshaw puller; her fate is shared by millions of others who have left their villages and migrated to the cities in search of work. About 25% of the population, over the past 50 years, have left their rural environment for the cities, an increase in urbanization of about 20%, as published in a July 1, 1999, British Broadcasting Corporation news report.

Not only is the world population increasing, but so is urbanization, as people, most notably in developing countries, migrate to the cities in search of work, a higher income, and a better life. The trend to urbanization dates back to early in the 20th century and is projected into the 21st century (Fig. 1.7). By 2030, more than 75% of the world's population will be living in urban areas, according to the 1999 Population Reference Bureau report. The world's 10 largest urban areas are listed in Table 1.4, and

TABLE 1.3 Potential effects of world population growth on variables related to emerging and reemerging infections[a]

Increased potential for person-to-person disease spread
Greater likelihood of global warming
Larger numbers of travelers
More frequent wars
Increased numbers of refugees and internally displaced persons
Increased hunger and malnutrition[b]
More crowding in urban slums
Increased numbers of people living in poverty
Inadequate potable water supply[b]
More large dam construction and irrigation projects

[a]Reprinted from D. B. Louria, in W. M. Scheld, D. Armstrong, and J. M. Hughes (ed.), *Emerging Infections 1* (ASM Press, Washington, D.C., 1998), with permission.
[b]New technologies could prevent or minimize these effects.

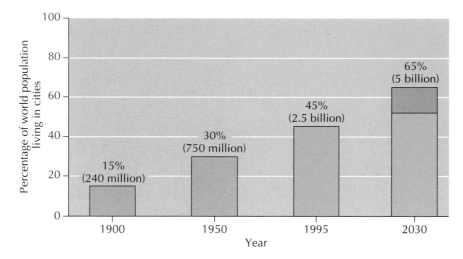

FIGURE 1.7 Progressive urbanization of our planet. The different colors for 2030 indicate the proportions of the urban population that are projected to live in developed countries (20%) and underdeveloped countries (80%). (Redrawn from D. B. Louria, p. 250, *in* W. M. Scheld, D. Armstrong, and J. M. Hughes [ed.], *Emerging Infections 1* [ASM Press, Washington, D.C., 1998], with permission.)

about half of these megacities are in developing countries. The magnitude of the effect of urbanization on communicable diseases varies dramatically in developed and developing countries, as a function of the economy and the public health infrastructure necessary to cope with the stress of increasing population density. The challenge of maintaining acceptable standards of sanitation and hygiene is far more difficult in developing countries, but it is also important to keep in mind that pockets of poverty and despair also exist in the United States and other developed nations. Urbanization frequently leads to poverty; the two go hand in hand and set up a cycle of infectious disease (Fig. 1.8). In July 2000, at a meeting of the G-8 countries (the top seven industrialized countries, plus Russia) in Okinawa, the nations pledged to break the vicious circle of poverty

TABLE 1.4 World's 10 largest urban areas[a]

Rank	Agglomeration	Population	Area (square miles)	Population per square mile
1	Tokyo-Yokohama, Japan	27,245,000	1,089	25,018
2	Mexico City	20,899,000	522	40,036
3	São Paulo, Brazil	18,701,000	451	41,466
4	Seoul, South Korea	16,792,000	885	49,099
5	New York	14,625,000	1,274	11,480
6	Osaka-Kobe-Kyoto, Japan	13,872,000	495	28,024
7	Bombay, India	12,101,000	95	127,379
8	Calcutta, India	11,898,000	209	56,928
9	Rio de Janeiro, Brazil	11,688,000	260	44,954
10	Buenos Aires, Argentina	11,657,000	535	21,789

[a]Source: U.S. Census Bureau, *Statistical Abstract of the United States*, 1992.

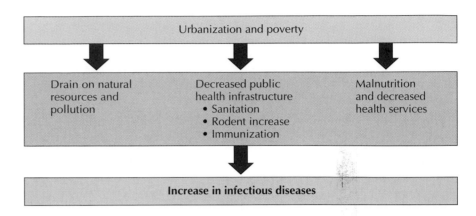

FIGURE 1.8 Relationships among poverty, urbanization, and infectious disease.

I have been to Japan on a few occasions, the first being during my service as a young military officer in the United States Army stationed just outside of Japan. I am always impressed with its cleanliness, despite the ever-present crowds.

suffered by citizens of developing countries. Sub-Saharan Africa is home to 68% of the world's AIDS population because of a variety of factors, an important one of which is poverty-associated urbanization. The drain on natural resources, including safe drinking water, is excessive, while at the same time problems of pollution, including human waste disposal and sanitation, are magnified. Untreated human wastes by the tons are dumped into the rivers, streams, and oceans. Ultimately, slums and shantytowns develop, and people live in filth and squalor (Fig. 1.9); rodent populations increase as sanitation decreases, and the cycle of disease is perpetuated. Rodents may transmit diseases through dry feces by inhalation, as is the case with the hantavirus (chapter 9), and may also harbor fleas, which transmit a variety of diseases, including the Black Death of 14th-century Europe, now known simply as the plague.

In April 2000, the CDC reported on a 5-year study that concluded that city dwellers get sick more often than their rural counterparts and that people living in poverty are sick more often (In the News 1.1). Upton Sinclair's 1906 novel *The Jungle* did not specifically target microbial diseases, but it did portray life in the big city and the risk of infectious diseases. The expression "it's a jungle out there" says it all. Prevention of communicable diseases is a major component of public health and is more problematical in cities than in rural areas and wide open spaces.

On the other hand, it does not necessarily follow that disease runs rampant in the megacities. Consider, for example, the Tokyo-Yokohama area, ranked as the world's largest urban area (Table 1.4), which is hardly poverty stricken or disease ridden. In fact, a recent survey indicates that the Japanese people enjoy the longest life expectancy. The country's economy and public health infrastructure make it possible for them to cope with urbanization. By contrast, in most developing countries, the crush of humanity and the tide of urbanization are overwhelming and beyond the financial resources necessary to construct sewage systems and to develop and maintain a public health infrastructure. Laura Garrett, in *The Coming Plague*, refers to cities as "microbe magnets" and "microbe heavens." "Graveyards of mankind" is a term used by British biologist John Cairns.

Ecological Disturbances

Deforestation

What happens when the village or town in which you live becomes too crowded, whether it be in a poor and developing country or in a poor area

FIGURE 1.9 Slums and shanty towns. Poverty is associated with a lack of sanitary facilities, an increase in rodent populations, a lack of safe drinking water, and other circumstances that contribute to infectious diseases. **(A)** A village in rural Panama. (Author's photo.) **(B)** A slum area in Africa. (Source: WHO.) **(C)** A residential area in Salvador, Brazil. (Author's photo.)

Trash collections to resume after strike
City employees, private companies to help out

Big cities can be unhealthy places

People Packed into City Like Sardines
Some Welcome Influx of Residents, Workers
Others Object to Increased Crowds, Noise, Litter

Life in the Big City

or a wealthy suburb in a developed country? The answer is expansion into the surrounding areas for tracts of land on which to build houses, hospitals, stores, shopping centers, and golf courses, depending on the community. Generally, the first event to give notice that construction is about to take place is the whine of the chain saws signaling deforestation, and then the bulldozers move in to uproot the tree stumps (Fig. 1.10). What a pity to disrupt another ecosystem, particularly when an almost deserted shopping area is across the street. Every time a tree is felled, and every time a bulldozer digs up the soil, microbes and other organisms are displaced. Fungi and bacteria and their spores are released into the environment and may alight and colonize on a human or animal and possibly give rise to a new or reemerging disease. Examples of outbreaks of certain fungal diseases have been reported in construction workers, particularly in the southwest. Almost half of the earth's forests either no longer exist or have been damaged, possibly to the point of no return, as a result of large-scale development projects, logging, and mining over the past 8,000 years. "Save the forests" is a popular battle cry of conservation-minded individuals. The expression "you can't see the forest for the trees" will likely become increasingly meaningless with each passing decade. Perhaps this is what Louis Pasteur was referring to 150 years ago when he advised, "the microbe is nothing; the terrain is everything."

Deforestation is a major factor in the eruption of new, emerging, and reemerging diseases. Wilderness habitats serve as reservoirs for a large variety of insects and other animals that harbor infectious agents. Our intrusion into the environment fosters contact with these species and their microbes and, further, fosters the migration of the displaced species into villages, communities, and backyards in search of food. An example is the rise of rabies in the eastern part of the United States (chapter 9) as a result of raccoons (a rabies vector) foraging for food in the garbage cans of suburban and rural communities. Chagas disease is a protozoan disease (chapter 10) carried by

FIGURE 1.10   Deforestation. As people move into areas that were formerly forests, there is increased contact with animals, including insects, that harbor infectious microbes. Further, the displaced animals return to neighborhoods that were once their lands in search of food. (Author's photo.)

beetles, commonly called kissing bugs, and is particularly prevalent in Brazil and other areas of South America. In the early 1900s, construction of the Central Railroad in Brazil was undertaken through the heavily forested tropical wilderness, a project which necessitated large-scale deforestation. You can guess the outcome—the indigenous mammals were displaced, and so were the beetles that fed upon them for their blood meal. Humans and their domesticated animals took up the slack and became infected, as did rodents; the latter conveyed the disease to species of beetles which inhabit housing in urban populations.

Leishmaniasis, a protozoan disease carried by infected sand flies (chapter 10), is a striking example of the consequence of deforestation and the emergence of urban disease. The disease, once limited to mammals of the forest, is now urban, primarily as a result of deforestation. The circumstances are similar to those described for Chagas disease.

The most contemporary example of the potential consequences of humans' intrusion into the forest is that of AIDS (chapter 9). Most scientists agree that the origin of human AIDS is the result of the propagation of human immunodeficiency virus (HIV) that made the leap from infected monkeys to humans (Fig. 1.13).

Although it can be an attempt to relieve suffering and death, humans' intrusion into ecosystems can backfire as a result of inadequate assessment of the public health impact and inadvertently increase microbial disease. This was the case in the construction of Egypt's $1 billion Aswan High Dam, a 10-year project completed in 1970, which resulted in an increase in the incidence of schistosomiasis, a disease caused by a parasitic worm with a complicated life cycle requiring snails for its completion. There is both "good news" and "bad news" in this story. The good news is that the dam harnessed the uncontrolled Nile River by creating Lake Nasser (Gamal Abdel Nasser was the Egyptian president from 1956 to 1970). The bad news is that the walls of the dam served as a new and convenient habitat for the snails and fostered an increase in their population, resulting in an upswing in schistosomiasis.

Rift valley fever, a viral hemorrhagic disease carried by mosquitoes, is

AUTHOR'S NOTE   *In the spring of 1999, I spent 6 weeks in Salvador, Brazil, at the Institute for Tropical Medicine in completion of the requirements of the M.P.H. degree at the Harvard School of Public Health to study tropical diseases in the natural context of their host. The last week of my stay was in the Amazon and included a visit to a small village with a high incidence of leishmaniasis. It was readily apparent why this was the case; the village bordered a heavy forest (Fig. 1.11). Trees had been felled to allow for the construction of primitive dwellings, and, yet, less than a mile away, leishmaniasis was not prevalent. The reason—sand flies are poor fliers and could not fly far from the forest. As a result of the deforestation, the village habitants and their dogs provided a blood meal for the sand flies. Fig. 1.12 illustrates my Amazon adventure! Somewhere along my travels, I picked up an infection but managed to complete my stay in the Amazon. A few days after returning home, I was hospitalized for severe dehydration.*

FIGURE 1.11   A village with a high incidence of leishmaniasis. Leishmaniasis is a protozoan infection transmitted by infected sand flies. **(A)** The village occupies an area that was formerly a forest and is encircled by a perimeter of trees. Sand flies are poor fliers but can traverse the short distance from their forest habitat. **(B)** Primitive living conditions in the village. (Author's photos.)

another example of the downside of the Aswan High Dam project. An epidemic of this viral disease occurred close to the dam area, resulting in illness in 200,000 people and over 500 deaths. The epidemic was the result of a thriving mosquito population in the flood lands created by the dam.

Climatic Changes

What about the effect of climate on the emergence of microbial diseases? There is ample evidence that global warming and climatic changes are also ecological disturbances which affect the incidence and distribution of infectious diseases (Table 1.5). The 20th century witnessed an increase in average global temperature attributed largely to the burning of fuels and forests, resulting in an increase in carbon dioxide and other heat-trapping greenhouse gases. The effects are seen not only in human health but also in the disruption of ecosystems and the resulting interference with food productivity. A meeting of world leaders was held in Kyoto, Japan, in 1997 to develop countermeasures against the impending threat; a second meeting involving about 150 countries took place in the spring of 2000 to ratify the Kyoto protocol with the goal of implementation by 2002.

FIGURE 1.12   A day's boat trip on the Amazon was beautiful and exciting. The Amazon River is the largest in the world, in terms of the volume of water, and is about 4,000 miles long. (The Nile River in Africa is longer, about 4,160 miles.) The Amazon is rich in vegetation, wildlife, and microbes. New species of life are still being discovered. (Author's photo.)

Nineteen ninety-eight was the warmest year in several centuries and was called the year of El Niño; it was manifested by unseasonably warm temperatures in some areas, increased amounts of rainfall, and increased flooding. The year brought into focus what scientists had long theorized, that global warming could favor outbreaks of a variety of infectious diseases, in particular, those that are insect borne, including malaria, dengue fever, cholera, yellow fever, onchocerciasis, and filariasis. Other weather-related diseases not directly linked to El Niño are hepatitis, *E. coli* infection, cryptosporidiosis, leptospirosis, and Ross River virus.

In the case of **vector-borne** diseases, the vector or the microbe, or both, may be influenced by the temperature. Table 1.6 summarizes WHO data on tropical diseases and indicates the likelihood of alteration in their distribution as a result of climate change. Malaria, a protozoan disease transmitted by mosquitoes, is at the top of the list; an increase in both temperature and

FIGURE 1.13   The interspecies leap. AIDS, which originated in Africa, is presumed to have jumped the species barrier from infected monkeys to humans. Other infectious diseases of humans have made a leap from animals to humans. The photo, taken in the Amazon, shows a wild monkey in the vicinity of a village. (Author's photo.)

TABLE 1.5   Infectious diseases linked to climatic changes

Disease	Biological agent	Transmission
Malaria[a]	Protozoan	Mosquitoes
Rift valley fever[a]	Virus	Mosquitoes
Hantavirus[a]	Virus	Mice
Cholera[a]	Bacterium	Waterborne
E. coli infection	Bacterium	Waterborne
Cryptosporidiosis	Protozoan	Waterborne
Hepatitis	Virus	Waterborne
Leptospirosis	Bacterium	Waterborne
Lyme disease	Bacterium	Ticks
Dengue (breakbone) fever	Virus	Mosquitoes

[a]Directly related to El Niño.

rainfall extends habitats favorable to mosquitoes. (On the other hand, increased temperature and decreased rainfall favor the distribution of sand flies, the vectors responsible for transmission of leishmaniasis, a protozoan disease.) Estimates are that an increase in mean ambient temperature in central Africa by 2°C would extend the range of the vectors of sleeping sickness, filariasis, and leishmaniasis, allowing for these diseases of the tropics to invade marginal temperature zones. Further, higher temperatures may push malaria transmission to higher altitudes, causing epidemics, as has occurred in the highlands of Ethiopia and Madagascar. In Rwanda, in late 1987, malaria incidence increased by 337% over the previous 3-year period as a result of increases in temperature and rainfall. In the last decade, the reported cases of malaria (the form of malaria with the highest fatality rate) in the North-West Frontier Province of Pakistan rose from a few hundred in 1983 to more than 25,000 in 1990. This dramatic rise is attributed to unusually high temperatures at the end of the normal malaria season that extended the season.

Diseases in which infectious agents cycle through invertebrates to complete their development are particularly sensitive to subtle climate variations compared with diseases spread from human to human. Hence, it is imperative that consideration be given to and appropriate measures be enacted regarding the influence of global warming and other climatic changes on microbes and their vectors.

Natural Disasters

Flooding and drought are environmental disturbances at opposite ends of the scale which place populations at risk of an increased burden of infectious diseases. In March and April of 2000, the worst floods in recorded

TABLE 1.6   Status of major vector-borne diseases and predicted sensitivity to climate change[a]

Disease	Population at risk, millions[b]	Prevalence of infection	Present distribution	Possible change of distribution as a result of climatic change
Malaria	2,100	270 million	Tropics, subtropics	Highly likely
Lymphatic filariases	900	90.2 million	Tropics, subtropics	Likely
Onchocerciasis	90	17.8 million	Africa, Latin America	Likely
Schistosomiasis	600	200 million	Tropics, subtropics	Very likely
African trypanosomiasis	50	25,000 new cases per year	Tropical Africa	Likely
Leishmaniasis	350	12 million infected + 400,000 new cases per year	Asia, southern Europe, Africa, South America	Not known
Dracunculiasis	63	1 million	Tropics (Africa, Asia)	Unlikely
Arboviral diseases				
Dengue	NA[c]	NA	Tropics, subtropics	Very likely
Yellow fever	NA	NA	Africa, Latin America	Likely
Japanese encephalitis	NA	NA	East and Southeast Asia	Likely
Other arboviral diseases	NA	NA	Tropical to temperate zones	Likely

[a]Data from WHO.
[b]Based on a world population estimate of 4.8 billion (1989).
[c]NA, not available.

history put the people of Mozambique and other southern African countries at risk for several diseases, particularly for malaria and cholera. Up to 250,000 people in Mozambique alone were endangered by these two diseases. According to a WHO press release, "The threat of a malaria epidemic in the country is increasing and will be at its most dangerous in around three to six weeks time as flood waters gradually subside, the rain stops, and warm temperatures return—ideal breeding conditions for mosquitoes. . . . Before the floods, there were between 6 and 10 cholera cases a week; since the floods it has increased to 120 cases per week."

Drought is related to increased famine and results in an increase in infectious diseases, as has been witnessed in eastern Africa since early 1999, placing 16 million people at risk. In Ethiopia alone, about 8 million people are affected. The country is one of the five poorest countries in the world, with an average life expectancy of 43 years, the fifth lowest in the world.

In the best of times, life is a struggle against poverty and disease, leading to a high child mortality rate. Under usual circumstances, access to safe drinking water is a privilege for less than 20% of the population, and the lack of food and water takes its toll on the maintenance of a healthy immune system, particularly in children.

In 1993, a hantavirus, the cause of a potentially fatal disease (chapter 9), emerged in the Four Corners area of the United States (where New Mexico, Utah, Colorado, and Arizona meet). The disease is transmitted by deer mice, the principal animal hosts of the virus, which feed on pine kernels. Higher than normal humidity favored an abundant crop of the pine kernels, which, in turn, led to a 10-fold increase in the deer mouse population between 1992 and 1993. This is an excellent example of a climatic condition triggering a chain of events that resulted in the emergence of an infectious agent.

Technological Advances

Human activities lead to technological advances which may pose public health risks; jet travel is an example (In the News 1.2). "It's a small world after all" is a theme song in Disneyland and Disney World. President Clinton showed how small the world is in July 1999, when he flew from Washington to New York City to attend the Kennedy-Bessette funeral, back to Washington, on to Cincinnati for a fundraiser, to Denver, to Aspen for a conference, back to Washington, and then across the Atlantic to Morocco for the funeral of King Hassan II (Fig. 1.14)—all in one weekend. (Talk about jet lag, but at least he had plenty of room to stretch his legs, and one would assume that bags of peanuts did not constitute his menu.) Federal Express boasts deliveries to over 200 countries in less than 48 hours.

Table 1.7 lists approximate flying times from New York to distant places. The farthest destination is Sydney, Australia, taking 22 hours, less than the incubation time for many microbial diseases. This means that a passenger harboring an infectious agent could board a jet and arrive at any world destination in less than the time it takes for the agent to establish infection. Such an incident occurred in the spring of 2000: a tourist left Tel Aviv bound for Newark International Airport, an approximately 10-hour nonstop flight, and died of bacterial meningitis approximately 2 hours after landing. Bacterial meningitis has an incubation period of only a few hours to about 2 days. What about the other passengers, particularly those sitting close to him?

In the summer of 1999, I was at an **AUTHOR'S NOTE** *airport in Europe and noted the large number of international flights to destinations the world over. It occurred to me that each aircraft was like a huge winged vector carrying microbes to distant destinations (Fig. 1.15). I admit this is a distorted view.*

IN THE NEWS 1.2

Bloodstream infections can spread to pacemakers

Rochester man dies of meningitis hours after plane trip from Israel

Hepatitis C may have spread via blood transfusions

Modern air travel makes the world smaller

UN Says Most Countries Have Unsafe Blood Supplies

Technological Advances and Health Risks

FIGURE 1.14 "It's a small world after all." In a span of a couple of days, President Bill Clinton demonstrated the truth of this cliché. He flew from Washington to New York and back. Then he flew to Cincinnati, Denver, and Aspen. After briefly returning to Washington again, he flew to Morocco. From there, he returned to Washington one more time. All that travel took place in one weekend.

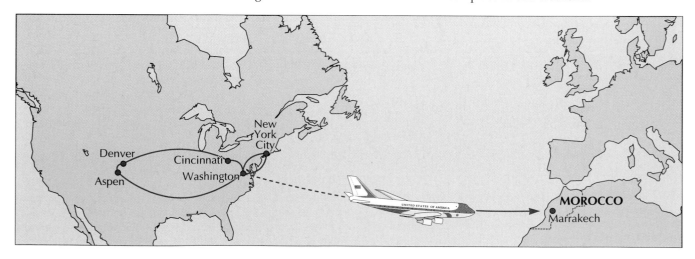

FIGURE 1.15 A flight departure board at an international airport. Jet aircraft, a major technological advance of the 20th century, serve as vectors for microbes around the world. (Author's photo.)

Another case of air travel associated with meningococcal disease occurred on May 24, 2001, when a 62-year-old man en route from Sydney, Australia, became ill prior to arriving at John F. Kennedy International Airport in New York. He was hospitalized and diagnosed with bacterial meningitis. Fortunately, in these cases, the risk of infection was low because the disease is spread by intimate contact. In another scenario, a passenger with TB on a jet from Russia to New York infected 13 other travelers. In these cases, the best that can be done is to notify other passengers to seek medical advice. Microbes can be harbored and transported across borders not only in their human hosts but also in their baggage and personal items. Further, vectors harboring infectious agents can also travel; fleas can be carried in rugs transported by jet cargo from the Middle East and Asia. Public health officials inspect many items being transported from country to country and are authorized to impose quarantine in an effort to minimize the risks.

The use of whole blood and blood products is a life-giving and lifesaving practice. Unfortunately, this statement does not apply to all countries; blood and blood products may be hazardous to your health (In the News 1.2). According to WHO, most of the countries with an unsafe blood supply are developing nations, in which the chances of acquiring infectious diseases are highest. As population pressure increases, so does the demand for blood. In some countries, blood is not screened and may harbor the causative agents of HIV infection, hepatitis, syphilis, malaria, and trypanosomiasis.

To some extent, advances in medical technology that make organ transplantation possible contribute to the burden of infectious diseases. Recipients are at increased risk for infection because they are on a regimen of immunosuppressive drugs to minimize organ rejection. Other conditions leading to immunosuppression include AIDS, certain inherited diseases, and malnutrition.

Microbial Evolution and Adaptation

The 1940s ushered in the dawn of antibiotics—agents which were rightfully called "wonder drugs." Penicillin was the first, and numerous others

TABLE 1.7 Jet travel: microbes without passports

Approximate flying time from New York City
Sydney, Australia: 22 hours (1 stop)
Tokyo, Japan: 14 hours (nonstop)
Tel Aviv, Israel: 10 hours (nonstop)
Nairobi, Kenya: 16 hours (1 stop)
Incubation period for selected diseases
Whooping cough: 7–10 days
Gonorrhea: 2–6 days
Salmonella food poisoning: 8–48 hours
Ebola fever: 4–16 days
Measles: 12–32 days
Chicken pox: 10–23 days

quickly followed; some were tailored to be effective against a broad spectrum of bacteria, while others were more specific. (Antibiotics and other antimicrobial compounds are discussed in chapter 12.) It should be noted that antibiotics are not effective against viruses and should not be prescribed for viral infections. The number of lives saved worldwide over the past 50 years because of antibiotic therapy is beyond estimation. An individual today who is infected with any of a variety of life-threatening bacteria has a fighting chance, assuming antibiotics are administered promptly, whereas an individual infected 50 or 60 years ago had none. The development of antibiotics was a major factor leading to the optimism of the 1970s. Many dread diseases, so it seemed, were about to become vanquished. But it turns out that the tables are turning—antibiotics are losing their punch, and increasing numbers of microbes are resistant. The expression "I'm resistant to such and such an antibiotic" has no meaning; people do not become resistant to antibiotics—their microbes do.

Emblazoned on the cover of the September 12, 1994, issue of *Time* are the headline "Revenge of the Killer Microbes" and the question "Are we losing the war against infectious diseases?" Antibiotic resistance is at a crisis level worldwide (Table 1.8). Vancomycin, an antibiotic considered by many to be the last stronghold in certain situations, is no longer effective against many bacterial strains that responded 10 years ago. Some refer to antibiotic-resistant bacteria as "super bugs." The WHO, the National Institutes of Health, the CDC, and the American Society for Microbiology—national and international health organizations that are charged with the health of the world—are attempting to meet the challenge. What's happening? To put it in a nutshell, antibiotics have been grossly misused and have promoted the emergence of antibiotic-resistant organisms in a Darwinian fashion; the forces of natural selection are in play. The antibiotic-resistant strains are the result of chance mutations, and their survival is favored by the presence of antibiotics. A paradox between developed and developing countries in terms of the meaning of "misuse" of antibiotics is related to the laws of supply and demand (Fig. 1.16). In developing countries, antibiotics are a luxury and cannot be afforded by the majority of the population, even assuming that they are available. For those fortunate enough to procure antibiotics, the expense is major; therefore, it is common for people to take only a few pills and to save the rest. This practice favors the selection of antibiotic resistance, since only the more susceptible strains will be wiped out by only a few

TABLE 1.8 Examples of drug-resistant diseases

Bacterial disease	Viral disease	Protozoan disease
Tuberculosis	HIV infection	Malaria
Gonorrhea	Hepatitis B	Visceral leishmaniasis
Staphylococcal infection		
Shigellosis		
Typhoid fever		
Pneumococcal pneumonia		
Enterococcal infection		

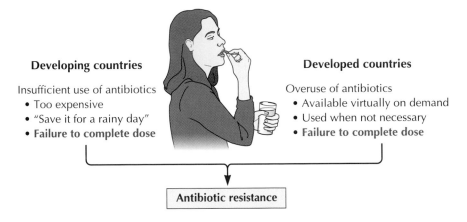

Developing countries

Insufficient use of antibiotics
- Too expensive
- "Save it for a rainy day"
- **Failure to complete dose**

Developed countries

Overuse of antibiotics
- Available virtually on demand
- Used when not necessary
- **Failure to complete dose**

Antibiotic resistance

FIGURE 1.16 The paradox of antibiotic misuse.

doses. In developed countries, where antibiotics are usually readily available, "misuse" is due to antibiotics' being too readily available. Patients demand antibiotics from their physicians and are not satisfied with the explanation that the particular disease is running its normal course, nor are they satisfied with being advised that the infection is caused by a virus and, hence, antibiotics are not indicated. (Sometimes, even though the presumptive diagnosis is a viral infection, antibiotics are prescribed to avoid or minimize secondary bacterial infections; this is particularly true in the case of the elderly.) Too often, as occurs in developing countries for completely different reasons, after taking an antibiotic for 5 or 6 days, the patient begins to feel better and does not take the remainder of the doses that may be prescribed for a 10-day period. Antibiotic-resistant bacteria are thereby selected which may then infect another person or a body of water or may gain access to vectors. Once an antibiotic-resistant organism is selected, its progeny will inherit those genes that code for antibiotic resistance. As stated earlier, the forces of natural selection are at work in the competition for the survival of the fittest; in this situation, the "fittest" refers to those that are most fit to resist a particular antibiotic(s). Continued misuse of antibiotics is dangerous and runs the risk of reversion to a preantibiotic era. Further discussion of antibiotics and their mechanisms of activity, as well as the dilemma of antibiotic resistance, is found in chapter 12.

Another remarkable example of microbial evolution and adaptation is the evasive strategy exhibited by some microbes that allows them to evade the immune system of their host. They are truly "masters of disguise." The influenza virus and the trypanosome protozoan responsible for African sleeping sickness are two well-known examples of microbes that "change their coats" (Fig. 1.17). A major function of the human immune system is to defend the body against microbial infection; certain cells of the immune system recognize components of microbes, called antigens, as foreign and produce antibodies specifically targeted against these antigens, which mark the microbes for destruction. (Antigens and antibodies are discussed in detail in chapter 11.) The influenza virus and the trypanosome adapt by changing their coats (surface antigens), an example of antigenic variation. The upshot is that antibodies fail to recognize and to target these new coats. (Imagine if you recognized an individual only by his or her coat and that person were to switch coats.) Trypanosomes have genes for as many as 1,000 distinct surface antigens and switch from one gene to

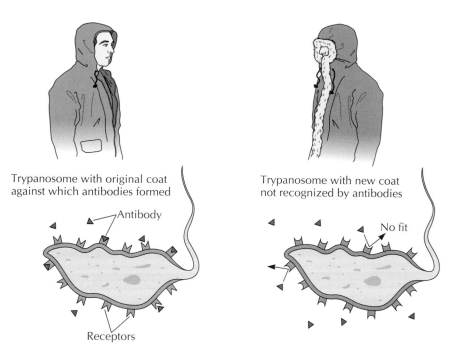

Trypanosome with original coat
against which antibodies formed

Trypanosome with new coat
not recognized by antibodies

Antibody

No fit

Receptors

FIGURE 1.17   Changing one's coat. Trypanosomes can form new surface antigens not recognized by antibodies. This is an important evasion strategy.

another, resulting in disguise—quite a trick! Relapsing fever, a louse-borne bacterial infection, is characterized by one bout of fever after another. As many as five or six or even more episodes may occur because this organism, like the agents of influenza and sleeping sickness, is a master of the art of antigenic variation.

The battle against the natural process of microbial adaptation and change, whether exhibited by resistance against antibiotics or by evasive strategies, is an ever-present and ongoing struggle for survival; failure to meet the challenge affords microbes the upper hand in the continual battle for survival of both humans and microbes.

Another point to keep in mind is that insect vectors—carriers of disease—are also able to adapt to a changing environment. Malaria, a mosquito-borne protozoan disease, was thought to be a disease of the past, thanks to the application of the insecticide dichlorodiphenyltrichloroethane (DDT). Little did scientists realize that the forces of natural selection would again interfere in the battle against malaria as a result of the misuse of the insecticide—a thwarted potential success story. DDT-resistant mosquitoes emerged with a vengeance, and other vector-borne diseases shared their triumph over DDT. West Nile virus, a mosquito-borne agent, now threatens New York and neighboring states, prompting ground and aerial spraying. Can insecticide-resistant mosquitoes carrying West Nile virus emerge?

Dengue fever is caused by a virus (chapter 9) transmitted by the bite of extremely aggressive mosquitoes (*Aedes albopictus*) known as Asian tiger mosquitoes. These mosquitoes made their way to the Americas aboard a shipment of waterlogged tires en route from Japan to Houston for retreading. One biologist described the voracious appetite of tiger mosquitoes for blood as "They will feed on anything—a rat for example—and then turn right around and feed on a human." Again, adaptation to a new environment is the name of the game (abetted by travel) and is responsible for the emergence of urban dengue fever.

Human Behavior and Attitudes

Complacency

How easy it is to cut corners on health-related matters when it appears that progress and improvement have taken place, leading to the false assumption that prevention and control are no longer necessary. Complacency is the belief that "it can't happen to me." The failure of people to complete their full dose of antibiotics because they are feeling better is a prime example.

A dramatic example of complacency is evident in the threatened resurgence of AIDS, particularly among young gay men, because of a return to risky sexual behavior fueled by glowing reports of new drug therapies for the management of AIDS. Estimates are that approximately 5 million Americans have sex and/or drug habits that put them at high risk for acquiring AIDS. The number of cases in the United States has fallen dramatically since the peak of the 1980s; the decrease is primarily attributed to safe sex habits and avoidance of dirty needles by drug abusers, but public health officials worry that this could cause complacency and result in an increase in the number of cases as people return to unsafe sex practices. At the World AIDS Conference in Durban, South Africa, in July 2000, Helen Gayle, AIDS chief at the CDC, said, "I am scared by the trends we are starting to see." In San Francisco a sharp rise in new HIV infections occurred between 1997 and 1999, accompanied by an increase in gonorrhea and other sexually transmitted diseases. High-risk factors include six or more sexual partners per year, failure to use a condom, sex with an HIV-infected person, male homosexual contact, female or male prostitution, and injecting drugs, all of which contribute to the spread of infectious diseases.

Over the years, the developed world has seen a marked decrease in the incidence of infectious diseases due not only to antibiotics but also to the availability of immunization. Unfortunately, the developing world does not have equal access to these medical advances. People have become complacent about receiving immunization shots or keeping their immunization boosters up to date and need to be aware that, with the exception of smallpox, infectious diseases have only been controlled, not eliminated. According to the CDC, only 78% of 2-year-olds in the United States have been given all the recommended immunizations, primarily because of parents' complacency or undue concern about safety. This could mean trouble down the line. A 10% decline in measles vaccination occurred between 1989 and 1991, resulting in an outbreak of 55,000 cases, several thousand hospitalizations, and 120 deaths. Fortunately, children entering school are required to have proof of being up to date on their immunizations; students entering college must also have proof of being fully immunized. Nevertheless, some slip through the cracks. Individuals traveling to foreign countries need to be aware that particular immunizations may be necessary against diseases prevalent in that area. For example, in 1996, tourists traveling to yellow-fever areas neglected to be immunized against the disease and were responsible for importing the disease into the United States and Switzerland. In 1996, approximately 10,000 reported cases of malaria were imported into Europe by travelers; to date there is no effective immunization against malaria.

Human Migration

Populations on the move contribute to the emergence of disease beyond that resulting from voluntary urbanization fueled by a search for a better life. Population movement is frequently not a matter of choice but rather a forced movement because of wars and conflicts resulting from political upheavals. **Internally displaced persons (IDPs)** are refugees who move from one part of a country to another. These migrants, as well as refugees who cross national borders, carry with them their microbes and microbial vectors, resulting in an exchange with intermingling populations. Masses of people are forced to settle in uninhabitable environments without adequate shelter, food, clean drinking water, and latrines. Personal hygiene and sanitation may be virtually nonexistent, and what few facilities are available become quickly overwhelmed. People live in filth and squalor like herds of animals. A safe supply of drinking water and construction of primitive facilities for disposal of human wastes are of prime concern in refugee camps. No wonder these camps are hotbeds for epidemics and their potential spread as refugees continue to flee from one area to another (Fig. 1.18). The World Resources Institute estimates the number of IDPs as 30 million worldwide, mostly concentrated in 35 countries. The Population Reference Bureau estimates that in the mid-1990s about 145 million people lived outside their native countries; this figure is expected to increase by 4 million per year.

Human migration is a major factor in the emergence and reemergence of many communicable diseases. Malaria is an excellent example; refugees migrating through regions where malaria is endemic can acquire the infection and disseminate the disease to other areas. Malaria is a common cause of death among refugees in numerous countries, including Thailand, Somalia, Rwanda, the Democratic Republic of the Congo, and Tanzania.

Wars and civil unrest, in addition to promoting refugees and IDPs, disrupt the public health infrastructure and favor the spread of disease. The destruction of housing leads to increased human-to-human and human-to-

FIGURE 1.18 A refugee camp. Refugee camps are hotbeds of infection. Crowding and lack of hygiene and sanitation favor the incidence and transmission of disease. (Source: WHO.)

vector contact; decline in water management programs and a lack of treatment facilities are contributory factors.

Societal Factors

In many societies, particularly in developed countries, family life and structure have changed as a result of economic growth and increased opportunities for women. In most American families both parents work, leading to an increase in child care centers. More than 13 million U.S. children attend day care centers, which put them at risk for a variety of intestinal parasites, diarrhea, middle ear infections, and meningitis. For example, outbreaks of shigellosis, a diarrheal disease, have caused problems in many day care centers around the country. (To a large extent, the simple act of hand washing on the part of personnel as they move from diaper to diaper would be an effective control measure.) Children convey the microbes to their family members, many of whom, in turn, bring their microbes to the workplace.

As longevity increases, so does the number of elderly citizens requiring nursing homes, day care centers, and assisted living environments. Like child day care centers, these facilities are potential hotbeds for the emergence and spread of communicable diseases within the resident population and the staff, visitors, and their contacts.

Food production and dietary habits also affect the spread of microbial diseases. Globalization of the food supply, centralized processing, fast-food restaurants, dining out, and take-out food are all significant. Food-borne diseases are a major public health problem in the United States. For example, an outbreak of hepatitis A in March and April 1997 resulted from children's eating contaminated strawberries in school lunches and was traced to a food processing company in California. In 1994 and 1995, a salmonella-contaminated snack food manufactured in Israel caused illness in Israel, the United Kingdom, and the United States. In 1997, contaminated raspberries from Guatemala were responsible for food-associated illness in the United States and Canada. In a better economy and a family structure in which both parents work, people rely more on prepared foods to reduce household chores. Fast-food restaurants and take-out restaurants are part of our social structure; images of Mom cooking and baking are disappearing. Food has increasingly become a source of recreation. Consider, for example, a typical Saturday night conversation: "So, what'll we do tonight?" Answer: "Let's eat out!"

This is all well and good, assuming that personal hygiene and sanitary control measures practiced by food handlers are not compromised. Television news shows have aired segments in poultry and meat slaughterhouses and in so-called fancy restaurants that are enough to make you sick! (Food-borne diseases are discussed in chapters 8, 9, and 10.)

Tattooing and body piercing are ancient art forms that have continued through the centuries. In modern times in developed countries, these practices have been largely associated with sailors, the hippie generation of the 1960s, and motorcycle enthusiasts. Young people in the 1990s and the new millennium have resurrected the trend. Tattoo and body piercing parlors are found in many countries; for only a low price, you can get just about any part of your body tattooed or pierced (Fig. 1.19). The risk of infection with a variety of microbes, particularly staphylococci, is a real possibility, and avoidance requires the use of sterilized instruments and trained personnel. Even if the establishment is certified by a local health authority, let the user beware!

FIGURE 1.19   Tattooing and body piercing. Tattooing and body piercing are a risky part of popular culture. The skin is invaded, potentially resulting in serious infections because of the use of unclean instruments. **(A)** A tattoo parlor in Copenhagen; **(B)** tattooed legs, Bergen, Norway. (Author's photos.)

OVERVIEW   This chapter makes the case that, despite the optimism of 30 or 40 years ago, microbial diseases have not been eliminated (with the single exception of smallpox) but rather have been on the upswing around the world. These diseases remain the number 1 cause of death worldwide and the third leading cause of death in the United States. The reasons for this turnabout are the result of world population growth, urbanization, ecological disturbances, technological advances, microbial evolution and adaptation, and human behavior. A quotation from Donald A. Henderson of the U.S. Office of Science and Technology Policy serves as an excellent way to close this chapter.

> The recent emergence of AIDS and dengue hemorrhagic infections, among others, [is] serving usefully to disturb our ill-founded complacency about infectious diseases. Such complacency has prevailed in this country [USA] throughout much of my career. . . . It is evident now, as it should have been then that mutation and change are facts of nature, that the world is increasingly interdependent, and that human health and survival will be challenged, ad infinitum, by new mutant microbes, with unpredictable pathophysiological manifestations. ∎

SELF-EVALUATION

PART I   Choose the *single* best answer.

1. World population is now estimated at
  a. 8 billion   **c.** 4 billion
  b. 6 billion   **d.** more than 8 billion

2. Which one of the following people warned of unchecked population growth?
  a. Satcher   **c.** Malthus
  b. Stewart   **d.** Darwin

3. In the United States, by 2050, estimates are that people over the age of 65 will constitute about what percentage of the population?
  a. 10%   **c.** 30%
  b. 20%   **d.** not predictable

4. The construction of the Central Railroad in Brazil led to an increase in
  a. leishmaniasis   **c.** tuberculosis
  b. malaria   **d.** Chagas disease

5. Which disease has been shown most conclusively to be linked to climate change?
  a. leishmaniasis   **c.** leptospirosis
  b. *E. coli* infection   **d.** malaria

6. Antigenic variation is exhibited by
  a. trypanosomes   **c.** malaria parasite
  b. influenza virus   **d.** more than one of the above

7. Tattooing carries a risk of infection. Which organism (or disease) is most likely to be involved?
  a. staphylococci   **c.** meningitis
  b. *E. coli*   **d.** cryptosporidiosis

PART II   Fill in the blanks.

1. In the United States, infectious diseases are the leading cause of death. (True or false?) _____

2. Name an infectious disease considered to be in the "top 10" worldwide. _____

3. According to the U.S. Census Bureau, the world's largest urban area is _____.

4. In a recent outbreak of encephalitis in New York, it was found that the St. Louis virus was not the culprit as expected, but rather the culprit was a virus referred to as _____.

5. Diseases transferred from animals to humans are called _____ diseases.

6. What does the abbreviation IDP stand for? _____

PART III   Answer the following.

1. List five reasons why infections are emerging and increasing.

2. Choose two of the reasons you gave in question no. 1 and discuss them.

3. Why was the Aswan Dam a "disaster"?

4. Cairns described cities as "graveyards of mankind." Explain.

5. A number of quotations are cited in this chapter. Develop your own quotation that targets the problem of new, emerging, and reemerging infections.

6. The statement was made that a grade of C– should be given to the world for its success in coping with microbial diseases. What grade would you award, and why? Do you think C– is appropriate?

7. In 1967 the surgeon general of the United States declared it was "time to close the book on infectious diseases," but events proved otherwise. What was the basis of the surgeon general's remark?

8. The misuse of antibiotics is a major factor in infectious disease control. Explain the antibiotic misuse paradox that exists in developed and developing countries.

THE MICROBIAL WORLD

PREVIEW In chapter 1, the term microbe, or microorganism, was used extensively because this book is about microbes, particularly those few that are pathogens. The term microbe was not defined or even adequately described, except that the five groups of microbes were named—bacteria, viruses, protozoans, unicellular algae, and fungi. (Worms, biologically known as helminths, are frequently included in microbiology texts even though they are not by any means microbes because many species cause infections resembling microbial infections.)

Each of these five groups is a distinct biological entity with characteristics that distinguish it from the other groups, just as giraffes are distinct from earthworms, and plant life is distinguishable from animal life. The characteristic they have in common is that they all, with the single exception of viruses, exhibit life properties and, hence, are considered organisms—microorganisms. Viruses are not regarded by most scientists as being "alive," but they come close. They are in that gray area between living and nonliving—but more about that later.

SOME BASIC BIOLOGICAL PRINCIPLES

Cell Theory

To further understand microbes, whether pathogens or not, it is necessary to review a few very basic concepts of biology, since all microbes are biological packages with certain unique characteristics. **Cells** are considered the basic unit of life, based on the observations of Robert Hooke in 1665. Hooke used the word "cell" in his examination of cork, which revealed tiny compartments that reminded him of the cells in which monks lived. His studies ultimately gave rise to the cell theory, a fundamental concept in biology, as postulated by Matthias Schleiden and Theodor Schwann (1838) and Rudolf Virchow (1858). The major points of the cell theory are as follows.

1. All **organisms** are composed of fundamental units called cells.
2. All organisms are unicellular (single cells) or multicellular (more than one cell).
3. All cells are fundamentally alike with regard to their structure and their metabolism.
4. Cells arise only from previously existing cells ("life begets life").

An understanding of the cell theory is the basis for an understanding of life, including microbial life. The term organism is a descriptive term and implies life. Viruses are more correctly referred to as biological agents rather than as organisms. Nevertheless, license is taken for the sake of convenience, and viruses are included as microorganisms.

Metabolic Diversity

The term life is elusive and cannot be given an exact definition; at best, it can only be described. Nevertheless, several attributes are associated with living systems that, collectively, establish life. By one strategy or another, all organisms exhibit these characteristics, which are summarized in Table 2.1. A major property of life is the ability to constantly satisfy the requirement for energy. It takes energy for every cell to stay alive; this is translated into the total energy requirement for multicellular organisms. Your body constantly expends energy. It takes energy to breathe even during sleep and for the heart to constantly push blood through an interconnected and tortuous maze of blood vessels. Since you don't fill up at the gas station, it's obvious that your energy is derived from the foods you eat. Through a complex series of biochemical reactions, the body metabolizes the **organic** compounds (proteins, fats, and carbohydrates) of your diet and releases the energy stored in their chemical bonds into a biologically available high-energy compound known as adenosine triphosphate **(ATP)**. Most organisms, including most microbes, are **heterotrophs,** meaning that they require organic compounds as an energy source. Other microorganisms and plant life are **autotrophs** and do not require organic compounds but, rather, are able to directly utilize the energy of the sun (**photosynthetic autotrophs**) or derive energy from the metabolism of **inorganic** compounds (**chemosynthetic autotrophs**). In so doing, autotrophs produce

AUTHOR'S NOTE *Even when you doze in class, it takes energy to keep from slithering out of your chair and onto the floor. I have seen this happen only once in all my years at the lecture podium.*

TABLE 2.1   Characteristics of life

Characteristic	Description
Cellular organization	The cell is the basic unit of life; organisms are unicellular or multicellular.
Energy production	Organisms require energy and a strategy to meet their energy requirement.
Reproduction	Organisms have the capacity to reproduce by asexual or sexual methods and in doing so pass on DNA to their progeny.
Irritability	Organisms respond to internal and external stimuli.
Growth and development	Organisms grow and develop in each new generation; specialization and differentiation occur in multicellular organisms.

organic compounds and oxygen (O_2). Hence, heterotrophs are dependent on autotrophs for energy.

Requirement for Oxygen

In addition to metabolic diversity, organisms exhibit diversity in their O_2 requirement. The "higher" organisms that are more familiar to you are **aerobes,** meaning that they require O_2 for their metabolic activities. Some bacteria are **anaerobes** and do not require oxygen, whereas other anaerobes are actually killed by O_2. **Facultative anaerobes** are bacteria that grow better in the presence of O_2 but can shift their metabolism, allowing them to grow in the absence of O_2. A knowledge of the oxygen requirements of pathogens is important in clinical microbiology. For example, specimens from infections caused by bacteria suspected of being anaerobes must be cultured under anaerobic conditions, allowing the bacteria to grow (Fig. 2.1).

Genetic Information

The genetic information for the structure and functioning of all cells is stored in molecules of **deoxyribonucleic acid (DNA)**, a large and complex organic molecule. **Genes** are segments of the DNA molecule. "Life begets life" is part of the cell theory and emphasizes that life comes only from life; it is a refutation of the doctrine of **spontaneous generation** which was abandoned by the end of the 19th century. Since the establishment of DNA as the hereditary material, the expression "life begets life" can be expanded to explain the mechanism by which a particular life form gives rise to the same life form; that is, tomatoes produce tomatoes, humans produce humans, and *Escherichia coli* produces *E. coli*. Each of these groups has characteristics embedded in its DNA that confer its identity. This is because the DNA is transferred, by a variety of reproductive strategies, from parent to offspring.

WHAT MAKES A MICROBE?

With these basic principles in mind, the term microbe (or microorganism) can now be better described. The question to be considered is "what makes a microbe a microbe?" As will become apparent, this question is not easily answered. Your first response would probably be "they are all too small to

FIGURE 2.1 Culturing anaerobic bacteria. Some bacteria cannot grow in the presence of oxygen. The GasPak jar is a means of culturing anaerobes. (Author's photo.)

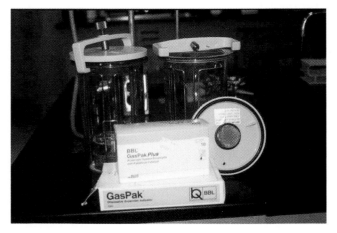

be seen without a microscope." Wrong. At first thought this would appear to be true, but what about the algae and the fungi? Are they **microscopic,** meaning they cannot be seen without a microscope? No doubt you have seen molds, classified as fungi, growing on food left too long in the refrigerator, or perhaps on a pair of old sneakers in the closet that you forgot about. They are **macroscopic**; that is, they can be seen with the naked eye. Hence, "microscopic" is not a distinguishing microbial characteristic. To describe all microbes as being unicellular is also not correct since, as pointed out above, the fungi and many of the algal forms are macroscopic and clearly multicellular. These organisms must be multicellular; if they were unicellular, that one cell would be enormous—a ridiculous idea!

There are exceptions to the rule that all bacteria are unicellular and microscopic. This might seem like an amazing fish story, but in 1985 a large cigar-shaped microorganism was found in the guts of the Red Sea brown surgeonfish. This organism was subsequently identified as a bacterium, approximately a million times larger in volume than *E. coli,* a common bacterial species, and was christened *Epulopiscium fishelsoni.* An even more monstrous bacterium was discovered in African sediment samples residing in the greenish ooze off the coast of Namibia in 1997 (Box 2.1); the organism has the tongue-twisting name *Thiomargarita namibiensis,* and to date is the "mother" of all bacteria and one for the *Guinness Book of World Records.* It is also more affectionately known as the "sulfur pearl of Namibia" because these organisms frequently form strands of dozens of cells and glisten white

---

BOX 2.1 "Monster" Bacteria

If asked to describe bacteria, just about everyone would reply that they are too small to be seen without a microscope. But in 1991, *E. fishelsoni,* a giant bacterium that can be seen without a microscope and was then the largest known bacterium, was discovered in the guts of surgeonfish in the warm waters of the Red Sea and off the coast of Australia. The organism can grow to about 500 micrometers, or about the size of the period at the end of this sentence. To give you some idea of size, one scientist projected that "if ordinary bacteria were mouse sized, *E. fishelsoni* would be equivalent to a lion." These bacteria are referred to as "epulos" and were originally thought to be protozoanlike. However, analysis of their DNA revealed that they are, in fact, bacteria.

In 1997, *T. namibiensis* stole the "prize for size" from *Epulopiscium.* This "monster" bacterium was discovered in samples of sediment in the greenish ooze off the coast of Namibia in Africa.

It is estimated that, if ordinary bacteria were the size of mice and *E. fishelsoni* were the size of a lion, then *T. namibiensis* would be the size of a blue whale, the world's largest animal. Scientists equate the bacterium's actual size to that of the head of a fruit fly. These spherical cells range from 100 to 750 micrometers in size. Dispersed throughout their cytoplasm are globules of sulfur. The bacteria tend to organize into strands of cells that glisten white from light reflected off their sulfur globules. Based on this observation, *T. namibiensis* has been nicknamed the "sulfur pearl of Namibia."

Both epulos and the sulfur pearl are anomalies in the bacterial world. The sizes of cells of all kinds, not only bacterial cells, are limited by the surface area of the membrane, since nutrients and waste are transported in and out of the cells by diffusion across the cell membrane. As cells increase in size, both volume and surface area increase, but surface area increases to a lesser

degree than does volume. At some point, the surface area becomes too limited to allow for sufficient diffusion.

So how did epulos and *T. namibiensis* manage to become so big? What are the physiological adaptations? In the case of epulos, microscopic examination reveals that the cell membrane, rather than being stretched smoothly around the cell, is convoluted (wrinkled), resulting in "hills and valleys," a phenomenon which greatly increases cell surface. (This adaptation is not unique to bacterial cells; the surface of the human brain is highly convoluted, resulting in a greater surface area, a factor which correlates with species intelligence.) The large size of *T. namibiensis* is attributed to the presence of a large fluid-filled sac occupying over 90% of the cell's interior. The sac is packed with nitrate which the cell uses in its metabolism to produce energy, making it less dependent on constant diffusion across the membrane to transport nutrients and waste.

from light reflecting from the sulfur inside them. They are visible to the naked eye and are approximately the size of the head of a fruit fly.

To give you some idea of size relationships, if an ordinary bacterium were the size of a mouse, *E. fishelsoni* would be equivalent to a lion, and *T. namibiensis* would be the size of a blue whale, the world's largest animal. The blue whale measures up to 90 feet (29 meters) and weighs about 120 tons. Here are more amazing facts: the heart of the blue whale weighs 1,000 pounds (450 kilograms) and is about the size of a Volkswagen Beetle, and the aorta (a large artery stemming from the heart) is like a huge blood-filled drain pipe large enough in circumference to allow a human to crawl through. (What a great idea for a wild ride at a water theme park! Can you come up with a name?) How many cells might make up such an enormous creature? The number would be in the trillions. Each of these cells exhibits the same fundamental life characteristics as does the single microbe. Microbes are sometimes described as "simple" because many consist of only a single cell or are less than a cell (viruses). Consider, however, that this single cell must fulfill all the functions of life. On the other hand, in a multicellular organism (like the whale), although each cell must fulfill all the criteria for life, there is a "sharing" of function because of specialization into a variety of cell types (for example, muscle cells, nerve cells, and blood cells). Perhaps that makes life easier. Hence, single-celled organisms, and even those consisting of only a small number of cells without evidence of true specialization, are simple only in the numbers sense and not in a physiological (functional) sense.

So if microbes are not necessarily microscopic and/or unicellular, then what is a microbe? The fact is that there really is no unifying principle; the term microbe, or microorganism, is a term of convenience used to describe biological agents which, in general, are too small to be seen without the aid of a microscope. Based on what has been presented here, it is clear that this description is not always true; from a practical point of view, the term is used for those biological agents for which similar laboratory techniques are used to culture and identify them. Some biologists consider microbes to be organisms which are at less than the tissue level of organization. This requires some explanation and is based on what is referred to as "biological hierarchy," or levels of biological organization. Recall that a cell is the fundamental unit of biological organization and that groups of cells establish multicellularity. Consider the human, or any other multicellular animal or plant, and it is obvious that in addition to an increase in cell numbers, the process of differentiation and specialization has taken place. For example, over 200 cell types make up the human, including red blood cells, five categories of white blood cells, epithelial cells, connective tissue cells, nerve cells, and muscle cells. All of these cells, as stated in the cell theory, share common fundamental characteristics, but superimposed upon their basic structure and function is a specialization of structure and function. Cells of the same type constitute the **tissue** level of organization, as exemplified by nerve tissue, blood tissue, and connective tissue. Tissues in turn constitute **organs,** structures composed of more than one tissue type; the heart, brain, stomach, and kidney are examples. Organs in turn constitute **organ systems,** a collection of organs which contribute to an overall function or functions. The digestive system, nervous system, respiratory system, excretory system, and reproductive system are examples familiar to you. This hierarchy is summarized as follows and is further illustrated in Fig. 2.2: subcellular → cells → tissues → organs → organ systems.

All microbes, whether unicellular, multicellular, microscopic, or macro-

---

AUTHOR'S NOTE

Some years ago I attended the annual meeting of the American Society for Microbiology in Miami Beach, Fla., and overheard two airport baggage handlers commenting that about 12,000 microbiologists were expected to attend. One asked the other, "What's a microbiologist, anyway?," to which the other replied, "Beats me! I suppose it's a small biologist." Several miles from the airport was a huge billboard with the words "Orkin Pest Control welcomes microbiologists." It was a memorable meeting.

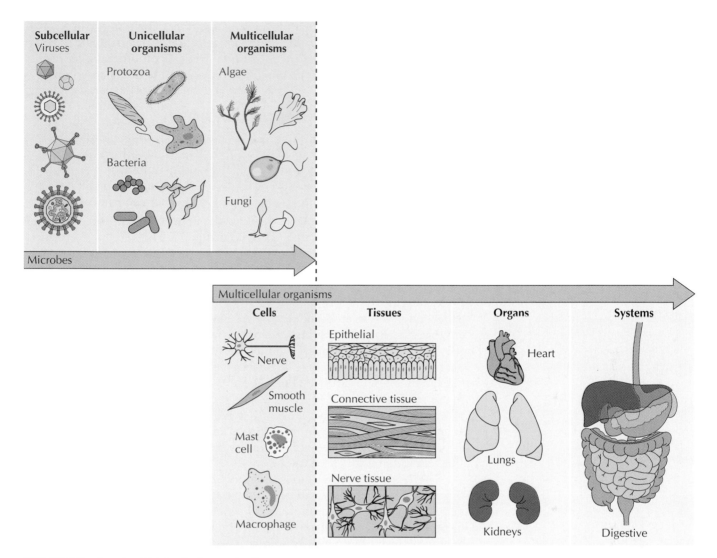

FIGURE 2.2   Levels of biological organization.

scopic, are devoid of tissues. That is, they are all at the cellular level of organization, although fungi and some algae hint at specialization and approach the tissue level of organization. According to this scheme, viruses can now be properly placed. They are at the acellular or subcellular level, which, simply put, means that they are less than cells and are at the threshold of life.

PROCARYOTIC AND EUCARYOTIC CELLS

Biologists recognize the existence of two very distinct types of cells, referred to as **procaryotic** and **eucaryotic** cells (Greek, *pro,* before, + *carion,* nut or kernel, + *eu,* true). Procaryotic cells have a simpler morphology than eucaryotic cells and are primarily distinguished by the fact that there is no membrane around the nucleus. There is a nuclear area rich in DNA which serves as the carrier of genetic information, as in all cells, but that DNA is not enclosed within a nuclear membrane. This DNA-rich area is referred to as a **nucleoid** rather than as a true nucleus. Further, procaryotic cells lack the membrane-bound structures (organelles) which are present in eucaryotic cells. Procaryotic and eucaryotic cells are compared in Table 2.2 and in Fig. 2.3. Bacteria are procaryotic microorganisms; all other microbes (except viruses, which are not cells) and all other forms of life are eucaryotic.

MICROBIAL EVOLUTION AND DIVERSITY

Procaryotes date back 3.5 billion years, and eucaryotes descended from them. Aristotle pondered the relationships among organisms as scientists continue to do today. In the 18th century, the botanist Carolus Linnaeus classified all life forms as belonging to either the plant or the animal kingdom. (Students would be delighted if only this were the case today.) Microbes were largely ignored since little was known about them, but since they had to be placed somewhere, they were considered plants, probably because they possess cell walls. Various schemes of classification have been proposed over the last few centuries, and **taxonomy,** the science of classification, became more and more complex. In 1866, Ernst Haeckel proposed a three-kingdom system—animals, plants, and a new kingdom, Protista, a collection of the microbes. In the light of modern biology, it became apparent that even three kingdoms were not enough. In 1969, a five-kingdom system was proposed by Robert Whittaker, and it has been accepted by most biologists. This classification describes organisms as belonging to the kingdoms Monera, Protista, Fungi, Animalia, and Plantae (Fig. 2.4). Recall that microbes consist of five groups and are accommodated in one of Whittaker's five kingdoms as follows: bacteria are classified as Monera, protozoans and unicellular algae are classified as Protista, and fungi are classified as Fungi; note that viruses are not considered in this scheme of classification, since they are neither procaryotic nor eucaryotic cells but are subcellular.

The 1950s ushered in the tide of molecular biology, and its wake introduced new techniques enabling scientist Carl Woese in 1977 to propose a novel scheme of classification based on analysis of a significant complex molecule known as **ribosomal ribonucleic acid** (rRNA) as a "fingerprint" in gaining insight into the evolutionary history of microbes. The Woese system assigns all organisms to one of three domains or "superkingdoms"— the *Bacteria, Archaea* (*Archaebacteria*), and *Eucarya* (Fig. 2.5), all of which arose from a single ancestral line. (All of Whittaker's five traditional kingdoms can be reassigned among the three domains.) The *Bacteria* and the *Eucarya* first diverged from an ancestral stock, followed by the divergence of

TABLE 2.2   Comparison of procaryotic and eucaryotic cells

Characteristic	Procaryotes	Eucaryotes
Life form	Bacteria	All cells with the exception of bacteria
Nucleus	Nuclear material present but not enveloped by membrane	Nucleus present and enveloped by membrane
Cell size	About 1–10 micrometers	Over 100 micrometers
Chromosomes	Single circular DNA	Multiple paired chromosomes present in nucleus
Cell division	Asexual binary fission; no "true" sexual reproduction	Cell division by mitosis; sexual reproduction by meiosis
Internal compartmentalization	No internal compartments	Organelles bound by membrane
Ribosomes	Smaller than eucaryotic and not membrane bound	Membrane bound

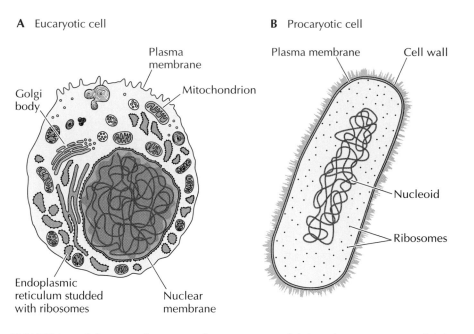

A Eucaryotic cell

Plasma membrane

Golgi body

Mitochondrion

Endoplasmic reticulum studded with ribosomes

Nuclear membrane

B Procaryotic cell

Plasma membrane

Cell wall

Nucleoid

Ribosomes

FIGURE 2.3 Schematic drawings of a eucaryotic cell **(A)** and a procaryotic cell **(B)**.

FIGURE 2.4 Whittaker's five-kingdom system.

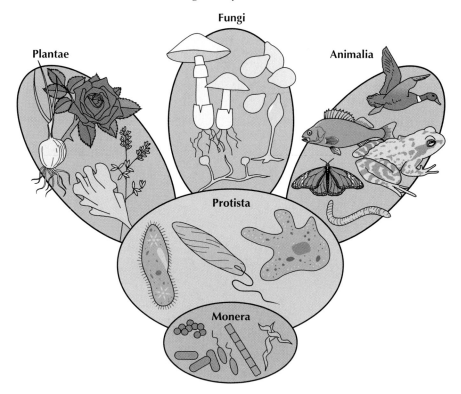

Plantae

Fungi

Animalia

Protista

Monera

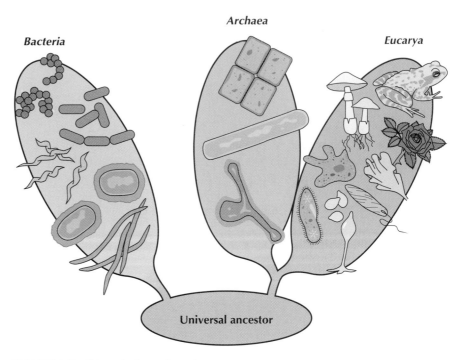

FIGURE 2.5   Woese's three-domain system.

the *Archaea* from the *Eucarya* line. The domains differ remarkably from one another in their chemical composition and in other characteristics, as summarized in Table 2.3. The term bacteria, in a broad sense, includes both the bacteria and the archaea.

It should be apparent that classification, particularly at the level of microorganisms, is not cast in concrete but is constantly under revision as new information becomes available. It is a credit to the scientific process that reevaluation is the name of the game. Admittedly, it is confusing, but to quote William Shakespeare (who probably never even took a course in biology), "What's in a name? That which we call a rose by any other name would smell as sweet." So you need not sweat it too much! No matter what the classification, bacteria were the first forms of life on Earth. Fossilized bacteria have been discovered in **stromatolites,** stratified rocks dating back 3.5 billion to 3.8 billion years, a long time ago in the history of the estimated 4.6-billion-year-old planet Earth. When life arose, the earth's ancient atmosphere contained little or no free oxygen but consisted principally of carbon dioxide and nitrogen with smaller amounts of gases, including hydrogen

TABLE 2.3   Comparison of *Bacteria, Archaea,* and *Eucarya*[a]

Domain	Membrane-bound nucleus	Cell wall	Antibiotic susceptibility	Example
Bacteria	No	Present	Yes	Large number of bacterial species
Archaea	No	Present	No	"Extreme" bacteria growing in high-salt environment and at extreme temperatures
Eucarya	Yes	Variable	No	Algae (most), fungi, protozoans, animals

[a]Major differences are present in the biochemistry of cell walls, cell membranes, genetic material, and structures in the cytoplasm.

(H$_2$), hydrogen sulfide (H$_2$S), and carbon monoxide (CO). This ancient atmosphere, devoid of O$_2$, would not have supported life as we know it. Only microbes that were able to meet their energy requirements by non-oxygen-requiring chemical reactions populated the primordial environment. The early microbes were photosynthetic and utilized water and carbon dioxide (CO$_2$) in photosynthetic reactions, resulting in the production of free oxygen (O$_2$) and **carbohydrates.** This process was responsible for the generation of O$_2$ in the earth's atmosphere approximately 2 billion years ago.

Since their origin on Earth billions of years ago, bacteria have exhibited remarkable diversity and have filled every known ecological niche. Yet, according to some estimates, fewer than 2% of microbes have been identified, and even fewer have been cultured and studied. Bacteria belonging to the domain *Archaea* continue to be found in environments that were only recently considered too extreme or too harsh for life at any level. In most cases, these organisms, **extremophiles,** cannot be grown by existing culture techniques; evidence of their presence has been obtained by molecular biology techniques that allow scientists to examine minute amounts of their deposited ribonucleic acid (RNA). Some like it hot and are called **hyperthermophiles** ("heat lovers"). Some hyperthermophiles have been identified in the hot springs in Yellowstone Park at temperatures exceeding 70°C. Some extremophiles, the **psychrophiles,** like it cold. Psychrophiles are happy in Arctic and Antarctic environments. Some are extreme **halophiles** ("salt lovers") (Fig. 2.6), and some produce methane gas in their metabolism. These bizarre examples indicate that many of the archaea live at the extremes of life zones (Box 2.2). Eubacteria, too, show tremendous diversification. Several years ago, scientists discovered bacteria that use arsenic in meeting their energy requirements in the same sense that humans require oxygen to release energy from foodstuffs. Here is another strange story. *Deinococcus radiodurans* is a bacterium that can withstand 3,000 times more gamma radiation than that which would kill a human because of its unique ability to repair its damaged DNA. It has been dubbed "Conan the bacterium" (Box 2.3).

You may not like to hear this, but the human mouth is considered one of the most diverse ecosystems and rivals the biological diversity of tropical

FIGURE 2.6   The Dead Sea. This sea has a salt concentration well above that found in the Great Salt Lake in Utah; it lies further below sea level than any other spot on Earth. You can lie on your back and float without any effort. Amazingly, this extreme environment is home for a variety of halophilic (salt-loving) bacteria. (Author's photo.)

BOX 2.2 Some Bizarre Bacteria

The TV show *Lifestyles of the Rich and Famous* described the unusual lifestyle of its characters and, in so doing, intrigued the viewers. Well, some microbes, too, exhibit an unusual lifestyle and remarkable characteristics which provide fascinating stories and illustrate the tremendous diversity of the microbial world.

Consider *D. radiodurans,* a bacterium further described in Box 2.3, which can survive a dose of radiation greater than 3,000 times the dose that can kill a human. Further, the Dead Sea, characterized by its extreme salinity, is erroneously named; it is not dead at all but teems with salt-loving (halophilic) bacteria. You will be surprised to learn that microbes can grow in your car's battery acid or that some bacteria thrive on arsenic. How about magnetotactic bacteria? They manufacture minute iron-containing magnetic particles that are employed as compasses by which the organisms align themselves to the earth's geomagnetic field. These curious microbes prefer life in the deeper parts of their aquatic environment where there is less oxygen.

Their magnetic compass points the way down.

And then there are the as yet unnamed bacteria living in symbiotic partnership with giant tube worms, which are as long as 2 meters, living in the hydrothermal vents of the ocean floor. As these worms mature, their entire digestive tract disappears, including their mouth and anal openings. Now that presents a problem, and it's bacteria to the rescue! The tissues of the worm are loaded with bacteria that obtain energy, sufficient for their own needs and for those of their worm hosts, from the surrounding chemical environment. In turn, the worms provide a safe harbor for the bacteria, ensure an adequate environment for energy production, and provide nitrogen-rich waste materials, allowing for synthesis of microbial cellular components.

Here is a strange story about *Serratia marcescens,* a bacterium whose colonies form a deep red pigment when grown on the surface of a nutrient agar plate, on a piece of damp bread, or in other moist environments. In 1264, in the Italian town of Bolsena, the German priest Peter of Prague was celebrating Mass. When he broke the Communion wafers, he found what he thought was blood on them, presumed to be the blood of Christ. This event was regarded as a miracle, but it turns out that it was not a miracle at all but contamination of the wafers with the red pigment-producing *S. marcescens* during their storage in the dampness of the ancient church. Raphael's painting *The Miracle of Bolsena,* depicting this event, hangs on a wall in the Vatican.

"Some like it hot and some like it cold" are the words of an old nursery rhyme but are indicative of the lifestyle of some bacteria. *Thermus aquaticus* is a heat-loving bacterium with an optimum growth temperature of 72°C. *Pyrococcus furiosus* lives in boiling water bubbling from undersea hot vents and freezes to death in temperatures below 70°C; some microbes do best at temperatures above 100°C. In sharp contrast, microbes have been isolated from the frigid environments of the Arctic and the Antarctic that have growth temperatures lower than 20°C.

rainforests. Within the past few years, scientists at Stanford University have discovered 37 new organisms in the mouth, pushing the total to more than 500. These new microbes were found in the common scum (plaque) in the deep gum pockets between teeth. (Your dentist would love this tidbit!) Their presence remained unknown simply because traditional culture methods do not allow their growth. Enterprising microbiologists (sometimes known as plaque pickers) extracted DNA from plaque and mapped out their DNA sequences, revealing bacteria that had not been previously known to inhabit the mouth. In fact, some new bacterial species were identified. This supports the statement that less than 2% of the microbial population has been identified.

A comprehensive global microbial survey to identify the microbes that make up the biosphere is in the early stages as a cooperative effort of the National Science Foundation and the American Society for Microbiology. Biodiversity research sites are being implemented by the establishment of a network of "microbial observatories." Other international efforts are in the works to develop a worldwide microbial inventory of genetic sequences.

The origin of life on Earth is an intriguing and mind-boggling question to which the explanation is purely speculative. It may be that life did not arise on this planet but was seeded by life forms from Mars. A recent

BOX 2.3   Conan the Bacterium

Conan the Barbarian, a 1982 movie starring Arnold Schwarzenegger, was the first Conan movie. In this fantasy story, from the mythical age of sword and sorcery, Arnie portrays Conan as only Arnie can do!

D. radiodurans has been nicknamed "Conan the bacterium"; it is one of nature's "toughest cookies." It can survive the rigors of being completely dried out, have its chromosomes disrupted, and be exposed to 1.5 million rads of radiation, a dose 3,000 times greater than that which would kill a human. Further, it can transform toxic mercury, a component of nuclear waste sites, into a less toxic form. No wonder it has been dubbed "Conan the bacterium." *D. radiodurans* is one of about 2 dozen bacteria whose DNA has been sequenced. According to Owen White of the Institute for Genomic Research in Rockville, Md., "The Department of Energy is very jazzed about *D. radiodurans,* because the agency has a pretty big toxic cleanup problem at its waste development sites." Genes from bacteria that can digest toxic waste but cannot survive radiation have been genetically engineered into *D. radiodurans,* resulting in bacteria which can transform toxic mercury into a nontoxic form and unstable uranium into a stable form. These genetically engineered bacteria are powerful tools in cleaning up the 3,000 waste sites containing millions of cubic yards of contaminated soil and contaminated groundwater estimated to be in the trillions of gallons. The ability of *D. radiodurans* to repair its own DNA is of interest to biologists because the process provides an insight into the mechanisms of aging and into the biology of cancer.

announcement supports this view. Photographs taken from the orbiting Mars global surveyor spacecraft indicated the possibility of water running just below the surface of the Red Planet, adding credence to this idea. If, in fact, Mars has water, it is possible that the planet entertains, or entertained, life. According to the Laboratory for Atmospheric and Space Physics at the University of Colorado at Boulder, "Mars meets all the requirements for life." This is quite a shift, since the National Aeronautic and Space Administration's *Mariner 4* photographs revealed a wasteland inhospitable to life. The possibility that life originated on Mars and was subsequently carried to Earth is plausible. Meteors and meteorites are constantly bombarding the earth and could have transported ancestral procaryotic cells (Fig. 2.7); estimates are that approximately 20 kilograms of material from Mars falls to Earth annually. Bacteria have been cultured out of Siberian and Antarctic permafrosts that have been in the deep freeze for millions of years, leading to speculation that life originated on Mars and seeded Earth. The National Aeronautic and Space Administration is now planning the Mars Sample Return Mission, which will bring Martian rocks back to Earth by 2008, and this will help to resolve the question of the beginnings of life. This may seem like a long time to wait for an answer, but, after all, what's several years considering the few billion years that life has been around?

INTRODUCING THE MICROBES

Although there is no clear definition of microbes, it is time to introduce those biological agents that fall under the microbial umbrella (Fig. 2.8 and Table 2.4). Fungi and algae will not be discussed in detail in this book beyond this chapter, although they are highly significant in terms of food chains and other beneficial aspects. Further, many fungi are human pathogens, and many contribute to the death toll of patients with AIDS (acquired immunodeficiency syndrome). With the exception of viruses, all microbes have both DNA and RNA, as do all cells. Microbes are measured in very small units of the metric system called **micrometers** (equal to 1

FIGURE 2.7   Life on Earth possibly "seeded" from Mars.

millionth of a meter), abbreviated as μm, and **nanometers** (equal to 1 billionth of a meter), abbreviated as nm. These numbers are probably not very meaningful to you, since they do not allow you to appreciate the size of microbes relative to more familiar objects, but there are some points that may help you think in this scale. A meter is equivalent to about 39 inches, so a micrometer is equal to 1 millionth of 39 inches—figure that one out. A spoonful of fertile soil contains trillions of microbes, and the number of microbes that can be accommodated on the period at the end of this sentence is in the millions. Figure 2.9 indicates the relative size of microbes. The monstrous bacteria described earlier are exceptions. Bacteria are many times smaller than eucaryotic cells but are about 50 times (or more) larger than viruses.

Smallness has its advantages. Smallness provides a large surface area per unit volume, allowing for rapid uptake of nutrients from the environment. *E. coli,* for example, has a surface-to-volume ratio about 20 times greater than that of human cells.

Bacteria

Bacteria are the best known of the microbial groups and are further discussed in chapters 4 and 8. They are microscopic, unicellular, and procaryotic and have cell walls (with the exception of a single subgroup), and they

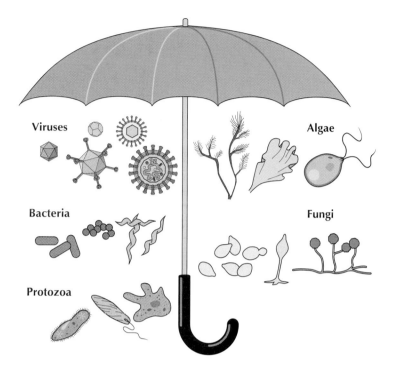

FIGURE 2.8   The microbial umbrella.

reproduce asexually by binary fission. In terms of size, they can be seen with a regular (light) microscope. Many bacteria are heterotrophs and utilize organic compounds as a source of energy. Others are autotrophs and utilize the energy of the sun, whereas some derive energy from the utilization of inorganic substances. Although there are a number of bacteria that are pathogens and are the subject of this text, the vast majority of bacteria are nonpathogenic and play essential roles in the environment without which life would not be possible (chapter 3).

Fungi

Fungi are eucaryotes. Morphologically, they can be divided into two groups, the yeasts and the molds. The yeasts are unicellular and are larger than bacteria; many reproduce by budding. Molds are the most typical fungi and are

TABLE 2.4   Comparison of microbial groups[a]

Characteristic	Bacteria	Protozoans	Fungi	Unicellular algae
Cell type	Procaryotic	Eucaryotic	Eucaryotic	Eucaryotic
Size	Microscopic	Microscopic	Macroscopic	Microscopic
Cell wall	Present	Absent	Present	Present
Reproduction	Mostly asexual (binary fission)	Sexual and asexual	Sexual and asexual	Asexual
Energy process	Mostly heterotrophic	Heterotrophic	Heterotrophic	Autotrophic

[a]Viruses are not cells and, therefore, are not included.

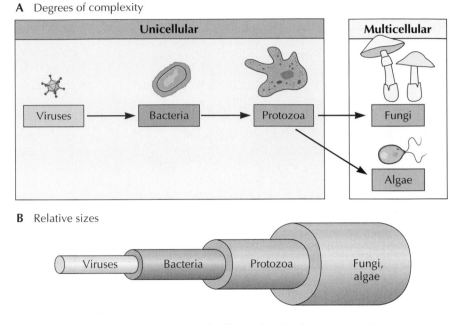

A Degrees of complexity

B Relative sizes

FIGURE 2.9 Comparison of sizes of different kinds of microorganisms.

multicellular, consisting of long, branched, and intertwined filaments called **hyphae.** In the early schemes of classification, fungi were considered plants, primarily because they have cell walls. However, their cell wall composition is quite different from that of plants and from the cell walls of bacteria. A few of the fungi are pathogenic and cause diseases that are difficult to treat; they play a highly significant role as **opportunistic pathogens** (organisms which are not usually considered to be pathogens), because, as in AIDS, when the immune system is depressed, they cause disease.

Algae

Algae are photosynthetic eucaryotes and in the photosynthetic process produce oxygen and carbohydrates which are utilized by forms requiring organic compounds. Hence, they are highly significant in the balance of nature. **Dinoflagellates** and **diatoms** are examples of unicellular algae and fall under the umbrella of microbes. Dinoflagellates are the primary source of food (plankton) in the oceans of the world. There are no algae that are pathogenic for humans directly, but diatoms and dinoflagellates have been implicated in causing neurological disturbances in humans as a result of our consumption of fish and shellfish that had fed on diatoms and dinoflagellates. *Pfiesteria,* a dinoflagellate dubbed the "cell from hell," threatened the fishing industry in the eastern United States in 1997. A bloom of these algae resulted in the release of large amounts of neurotoxin, causing neurological symptoms in fishermen and consumer panic.

Protozoans

Protozoans and the diseases they cause are the subject of chapter 10. These organisms are unicellular and eucaryotic and are classified according to their means of locomotion. Their energy generation requires the utilization of organic compounds. Many diseases, including malaria, sleeping sickness, and amebic dysentery, are caused by protozoans.

Viruses

As stated above, strictly speaking, viruses are not organisms; Fig. 2.2 indicates their subcellular position. Chapters 5 and 9 describe viruses and viral diseases. Two major distinguishing characteristics of viruses are that, in contrast to cells, they contain either RNA or DNA (never both) and they are submicroscopic particles (and thus they can be seen only with an electron microscope). Some have an additional coat or envelope encompassing them. Viruses are described as **obligate intracellular parasites,** meaning that they must be (obligate) inside living cells (intracellular) to replicate; they are not capable of autonomous replication. They take over the metabolic machinery and reap the benefits of energy production by the host cell. Perhaps this is the ultimate in parasitism.

The terms "bugs" and "germs" are part of our popular speech but have no scientific meaning. It should be clear from the above descriptions that each group of microbes is truly distinct from the others. When your physician diagnoses you as having a "bug," you might ask what kind.

OVERVIEW The microbial world is remarkable for its extreme diversity, as is evident in the distinct characteristics of the five microbial groups—bacteria, viruses, protozoans, unicellular algae, and fungi. Further, within each group, there is considerable diversity. Not all microbes are unicellular and microscopic; some are multicellular and macroscopic. In recent times, "monster" bacteria have been found that are unique in being unicellular and macroscopic, a rare combination. Viruses are subcellular and are not considered life forms. There is no clear definition of what makes a microbe a microbe. It is clear that they are all at less than the tissue level of biological organization.

All bacteria are procaryotic, and all of the other microbes are eucaryotic. (Viruses are not cells and, therefore, cannot be classified as either procaryotic or eucaryotic.) Taxonomy evolved from a two-kingdom system (in which bacteria were considered plants) to a five-kingdom system, with various other schemes along the way; the trend has been toward recognizing the uniqueness of microbes. A recent classification of bacteria based on analysis of their rRNA assigns bacteria to one of three domains and reflects their evolutionary history.

Since their origin on Earth, microbes have adapted to extreme ecological diversity and can be isolated from all environments from hot springs to permafrosts. Some live at the extremes.

All organisms must meet a basic requirement for energy, and microbial evolution has fostered a diversity of strategies. Some microbes obtain energy from organic compounds, while others utilize the energy of the sun or derive their energy from the metabolism of inorganic compounds.

The major characteristics of each of the five microbial groups show that each category is distinctive. The popular terms "bugs" and "germs" are used in a collective sense, but there is no basis for lumping these diverse microbial agents together. Further, these terms have a negative connotation, since they are usually used to describe microbial diseases, but it is important to remember that only a handful of microbes are disease producers. ■

SELF-EVALUATION

PART I Choose the *single* best answer.

1. A major distinction between procaryotic and eucaryotic cells is based on the presence of
   a. a cell wall
   b. DNA
   c. a nuclear membrane
   d. a cell membrane

2. Most bacteria are considered
   a. harmful
   b. anaerobes
   c. autotrophs
   d. heterotrophs

3. The smallest of these units of measurement is
   a. millimeter
   b. nanometer
   c. micrometer
   d. centimeter

4. Which of the following are obligate intracellular parasites?
   a. bacteria
   b. viruses
   c. unicellular algae
   d. diatoms

5. Which of the following is the smallest?
   a. rickettsia
   b. amoeba
   c. bread mold
   d. AIDS virus

6. The five-kingdom system of taxonomy is credited to
   a. Haeckel
   b. Woese
   c. Whittaker
   d. Darwin

7. According to Woese,
   a. *Eucarya* arose from *Archaea*.
   b. *Archaea* arose from *Eucarya*.
   c. *Bacteria*, *Archaea*, and *Eucarya* all arose independently.
   d. None of the above are correct.

PART II Fill in the following.

1. Bacteria, viruses, fungi, and protozoans are microbes. Name another group that falls under the microbial umbrella. _____

2. The cell theory is credited to _____ .

3. Compounds of carbon are called _____ compounds.

4. Organisms that do not require organic compounds are called _____ .

5. The "energy compound" is called _____ .

PART III Answer the following.

1. Criticize the terms "bugs" and "germs" as used in a collective sense to describe microbes. List the categories of microbes, and write a one-sentence description of each.

2. What makes a microbe a microbe?

3. What is the relevance to microbiology of Shakespeare's "What's in a name? That which we call a rose by any other name would smell as sweet."?

4. Heterotrophs are dependent on autotrophs. Why is this the case?

THE BENEFICIAL ASPECTS OF MICROBES
The Other Side of the Coin

PREVIEW It would be understandable if the preceding chapters of this book biased you against microbes; at this point, you have more reason to hate them than to love them. In this chapter the take-home message is that few microbes are disease producers, a point emphasized in chapter 1, and many are beneficial in our daily lives. The goal of microbiologists is not to annihilate pathogens but to treat them with respect and to avoid circumstances leading to a collision course. After all, microbes were the first inhabitants of our planet; they are the senior citizens from which evolution to eucaryotic cells and multicellularity proceeded. Microbes constitute the largest component of the earth's biomass and are present in the most extreme habitats of life. They make the planet's ecosystems go around.

Recycling of household wastes is a relatively new practice, and recycling bins are on the increase in homes and in businesses (Fig. 3.1). But the greatest recyclers of all time are microbes, without which life would be a dead end and, ultimately, would cease. Imagine a huge garbage dump into which materials are deposited daily and continue to accumulate year after year and generation after generation. Under these circumstances, and without replenishment, the resources of the planet would soon run out. Microbes are the foundation of the **biosphere;** they are frequently referred to as **decomposers** or **scavengers,** a description dating back to the 1870s when Ferdinand Cohn spoke of the role of microbes in rejuvenating materials. Bacteria are the underpinnings of the carbon, nitrogen, phosphorus, sulfur, iron, and other **biogeochemical cycles.** Their role in these cycles is unseen and taken for granted; the cycles occur without our initiative or our intervention. In fact, there is a danger that our increasing technology could inadvertently interfere with and shut down the cycles of nature as a result of nonbiodegradable products and pollution of the environment.

Since antiquity, societies the world over learned to harness microbes for their beneficial aspects long before there was any awareness of a microbial world. Societies were content with the empirical evidence that certain practices simply "worked." The production of distilled spirits (alcoholic

FIGURE 3.1 Recycling is on the rise. (Author's photo.)

beverages) and a variety of food products, including breads, yogurt, and cheeses, are examples. Yogurt was prescribed centuries ago for "stomach ailments" and, in some cases, seemed to do the trick. Yogurt contains live bacterial cultures, and the bacteria proliferate and replace the normal flora; yogurt is still occasionally used for the same reasons. More will be said about yogurt in a later section. As knowledge of microbes and the enzymes released in their metabolism increased, the manufacture of alcoholic beverages and foodstuffs became increasingly sophisticated, resulting in an increasing array of fermented food products.

Microbes are powerful biological research tools because of the relative ease of culturing and obtaining them in large populations in a short period of time; the microbial growth curve is described in chapter 4. Evidence establishing DNA as the genetic material is the result of experiments utilizing bacterial viruses (bacteriophage) and bacterial cells. Industry, particularly the pharmaceutical industry, has learned to harness microbes for the production of many products including antibiotics, vaccines, genetically engineered medicinals, pesticides, and a large variety of other compounds. Bioremediation, the use of microorganisms to clean up polluted environments, is on the increase; microbes played a role in reducing the impact of the *Exxon Valdez* oil spill that occurred off the coast of Alaska on March 24, 1989. These topics will be discussed in this chapter, and, hopefully, will make you rethink your love-hate relationship with the microbial world.

MICROBES IN THE ENVIRONMENT

Microbes as Decomposers

Perhaps, at some point in your life, you had an aquarium with goldfish or tropical fish. If so, you recall the necessity of paying attention to the fish and their physical and chemical environment. The aquarium simulates an **ecosystem**—a population of organisms in a particular physical and chemical environment. The fish and the plants are the added **biotic** components, while the chemical and physical environment constitutes the **abiotic** component. You will recall paying attention to the light source, temperature, acidity, and cleanliness of the water in order to maintain a healthy and balanced ecosystem. You can assume the presence of microbes as additional biotic components (Fig. 3.2). In the fish tank, the green plants are considered the **primary producers** because of their photosynthetic capabilities that result in the production of organic compounds and oxygen; the fish are the **consumers** and take oxygen from the water and exhale carbon dioxide; the bacteria and fungi are the decomposers and are the link between the producers and the consumers. The microbial population decomposes waste materials of the fish, dead fish, and dead leaves of the plant and, in so doing, are functioning as recyclers. And so it is in nature—witness plant debris, animal wastes, and the bodies of dead animals. If you walk through swampy areas, you may detect the unmistakable odor of methane—marsh gas—resulting from the bacterial action on decomposing materials. The microbes involved as scavengers are nonpathogenic and free-living. The resources on Earth are limited and are recycled through food webs with microbes as the decomposers (Fig. 3.3). Nature has always practiced this, but it has only been in the past 30 or 40 years that society has realized the inextricable link between populations, soil, water, air, and energy, all of which are interdependent and dependent upon microbes.

A

B

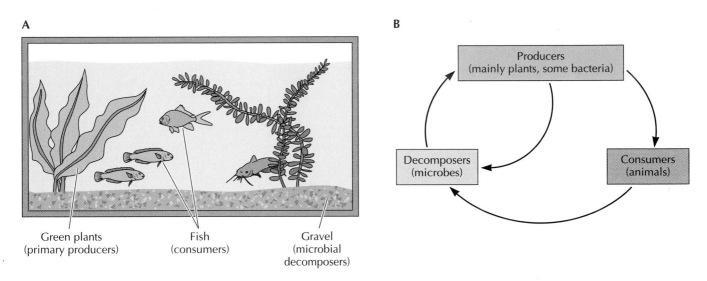

FIGURE 3.2 **(A)** A fish tank: an artificial ecosystem. **(B)** The cycle of life in an ecosystem: producers, consumers, and decomposers.

Microbes and the Biogeochemical Cycles

As stated above, microbes are the basis for the processes involved in the recycling of carbon, nitrogen, sulfur, iron, and phosphorus, resulting in the return of these elements to nature for reuse. These cycles will be discussed separately, but in fact they are linked.

The Carbon Cycle

Carbon atoms are key elements in living systems and are found in proteins, carbohydrates, fats, and DNA. Most of the carbon used by organisms is present in association with carbon dioxide, a simple compound consisting of one carbon atom attached to two oxygen atoms. Photosynthetic organisms are responsible for capturing the sun's energy and using it for the conversion of atmospheric carbon dioxide to make glucose and other energy-

FIGURE 3.3 Part of a food chain. Microbes are the ultimate decomposers. (Author's photo.)

rich carbohydrates. Hydrogen and water are necessary reactants in photosynthesis. An important spin-off of photosynthesis is the release of oxygen from the carbon dioxide back into the atmosphere.

Plants are associated with photosynthesis, but some microbes are also photosynthetic and are the primary producers in the ocean. *Chlorella* is a photosynthetic alga which is found on the surface of ocean water, and **cyanobacteria** are photosynthetic bacteria.

Carbon, captured as carbon dioxide, is ultimately recycled back to the environment through food chains. Cellulose, a polymer (chain) of glucose (sugar) molecules, is an energy-rich carbohydrate product of photosynthesis. Bacteria produce enzymes that are able to break down cellulose into single molecules of glucose and allow **herbivores** (grazers) to feed on plants, although they lack the necessary digestive enzymes to break down the cellulose. But microbes come to the rescue! They are a part of the normal flora residing in the intestinal tract of grazers, and their enzymes digest cellulose, allowing these animals to utilize cellulose as an energy source. Microbes in the intestinal tract of termites produce enzymes that break down cellulose, allowing termites to lunch on your house. Ultimately, the grazers are fed upon by predators, including humans, but one way or another, they and their waste material enter the food web. From there, microbial decomposition takes over, and the cycle is completed (Fig. 3.4).

FIGURE 3.4   The carbon cycle. Microbes are essential in the conversion of atmospheric carbon dioxide to organic compounds and back to the atmosphere.

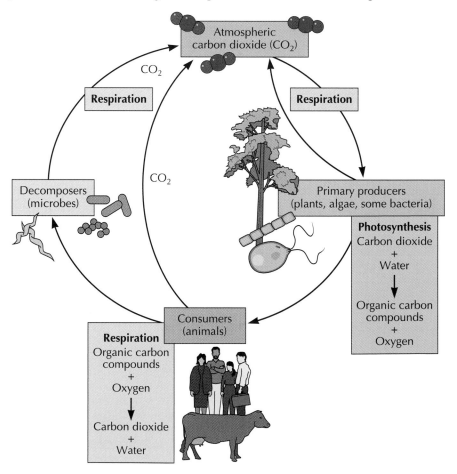

The Nitrogen Cycle

Nitrogen is a constituent of amino acids, the building blocks of proteins and of the nucleic acids of microbes, plants, and animals. It is the most common gas in the atmosphere (about 80%) but cannot be tapped by animals or by most plants. Here again, microbes come to the rescue; only bacteria fix nitrogen and, ultimately, recycle it back to the atmosphere. The **nitrogen cycle** is illustrated in Fig. 3.5. The process begins with the **fixation** of atmospheric nitrogen and its conversion to ammonia by **leguminous** plants. These are plants which have swellings or nodules along their root systems containing *Rhizobium* and other nitrogen-fixing bacteria. Peas, soybeans, alfalfa sprouts, peanuts, and beans are examples of leguminous plants. The association of nitrogen-fixing bacteria and leguminous plants is an example of **symbiosis.**

The next phase of the nitrogen cycle is called **nitrification;** in this process, ammonia is converted into nitrates, the nitrogen form most utilized by plants. Members of the bacterial genera *Nitrobacter* and *Nitrosomonas* carry out these processes. Feeding into the cycle is ammonia resulting from the decay of plants and animals and their waste products, giving meaning to the expression "death yields life." Urine is a waste product particularly rich in nitrogen. Finally, **denitrifying** bacteria are responsible for the return of nitrogen to the atmosphere as nitrogen gas.

Horticulturists and agriculturists have long realized the importance of nitrogen in growing flowers and food crops and use a variety of fertilizers containing nitrogenous compounds.

Other Cycles

In addition to the carbon and nitrogen cycles, the movement of other elements, including sulfur, phosphorus, and iron, through ecosystems in a cyclical manner is dependent upon microbial communities. It is really

AUTHOR'S NOTE *You might be interested in seeing these resident bacteria. All you need to do is to dig up peas, a patch of clover, or some other leguminous plant, being sure to take some of the root system. Wash away the soil and crush a nodule onto a clean slide. Add a drop of water and a dye, such as methylene blue, and spread the preparation with a toothpick or a matchstick onto the slide to establish a thin film. Allow the preparation to dry and examine it under a microscope. You will observe bacilli, probably members of the genus* Rhizobium. *If you perform a Gram stain (chapter 4), the bacterial population will be dominated by gram-negative (pink) bacilli.*

AUTHOR'S NOTE *Look at a variety of fertilizers and you will see three numbers, for example, 22-3-12. The first number pertains to the nitrogen content, the second pertains to the phosphorus content, and the third pertains to the potassium content. Some farmers might not add fertilizer to their soil but instead may include leguminous crops which are plowed under during the off-season as a way of enriching the nitrogen content of the soil.*

FIGURE 3.5 The nitrogen cycle. Microbes are essential in the conversion of atmospheric nitrogen to organic compounds and back to the atmosphere.

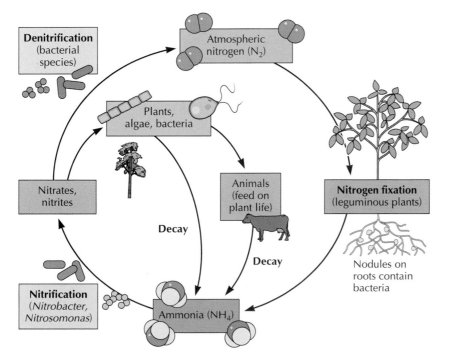

TABLE 3.1   Foods produced by using microbes

Milk products
Cheese
Yogurt
Buttermilk
Kefir
Acidophilus milk
Sour cream
Meats
Bologna
Salami
Country-cured ham
Sausage
Breads
Sourdough bread
Numerous other breads and rolls
Miscellaneous products
Sauerkraut
Pickles
Olives
Vinegar
Tofu
Soy sauce
Kimchi
Alcoholic beverages
Beer
Wine
Distilled spirits (e.g., brandy, whiskey, rum, vodka, gin)

microbes that "make the world go around." Their role in these cycles, which take place in all imaginable ecosystems and at all extremes of temperature, demonstrates the diversity of the microbial world.

MICROBES IN FOOD PRODUCTION

Next time you shop at the market, look around at the shelves of foods for those that are dependent on microorganisms for their production; some products, such as yogurt, may even contain live bacterial cultures. In just about all categories of foodstuffs, you will find examples (Table 3.1). Figure 3.6 presents a microbial feast; all of the items are dependent on microbes for their production. Use your imagination and come up with your own microbial banquet. Chapters 8, 9, and 10 will focus on food-borne epidemics and food-borne diseases, but now it is time to delight in the fanciful images that come to mind as you think about the wonderful foods and beverages whose tastes and aromas are the result of microbial activities that stimulate your taste buds and your sense of smell.

As stated in the introduction of this chapter, microbes have been engaged in the production of fermented foods for thousands of years. **Fermentation** is a series of chemical reactions, mediated by bacterial enzymes, that break down sugars to acids and carbon dioxide. The choice of microorganism to carry out fermentation determines the taste and aroma of the product. The specific microbial strain and the process employed for many products are carefully guarded secrets, and starter cultures are handed down in families from generation to generation. In a survey of the microbial world (chapter 2), fungi were briefly presented. Yeasts are fungi and are efficient fermenters of sugar into alcohol and carbon dioxide, a property exploited in the production of breads and alcoholic beverages. *Saccharomyces cerevisiae* is a commonly employed yeast. A yeast cake containing live yeasts can be bought inexpensively at the supermarket (Fig. 3.7). Wines and other alcoholic beverages have been used in religious ceremonies dating back many centuries, and samples of bread dating back to 2100 B.C. are on display at the British Museum. Cheese production, too, dates back thousands of years, as do fermented milk beverages that have been promulgated for centuries to treat a variety of intestinal tract disorders from constipation to flatulence (gas) (Box 3.1).

FIGURE 3.6   A microbial feast.

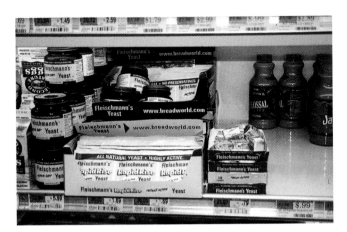

FIGURE 3.7   Yeast. Yeast is used in baking to get the dough to "rise" as a result of carbon dioxide production. Baker's yeast (*S. cerevisiae*) is available in various forms at supermarkets. (Author's photo.)

BOX 3.1   Claimed Medical Benefits of Probiotics

Yogurt and other fermented milk products are centuries-old foods, many of which originate from eastern European countries. We tend to think of the presence of microbes in the intestinal tract, other than the normal flora, as detrimental to health. Frequently, this is the case, as evident in the large number of bacterial food-borne infections, including those caused by strains of *Salmonella*, *Listeria*, and *E. coli* (chapter 8). Also, a number of viruses are transmitted by foods and are responsible for serious gastroenteritis (chapter 9).

Yogurt is a particularly popular food in the United States and many other countries. It is produced commercially, but "do-it-yourself" yogurt-making kits are readily available. A look in the dairy section of a supermarket attests to the popularity of yogurt. The shelves are stacked with no-fat, low-fat, "fruit on the bottom," and flavored yogurts, including mocha latte, vanilla, coffee, peach, strawberry, and apricot. Yogurt comes in a variety of sizes and is produced by a number of companies. Why the yogurt craze? The answer is simple: it is good for you. Many people, in an attempt to lose weight, consume a container of yogurt as a meal. Yogurt is an excellent source of calcium, some vitamins, and protein as indicated on the

label. It can be low in fat, low in calories, or low in both fat and calories. The choice is yours!

What is it about yogurt that, according to some, makes it a beneficial food? Most yogurt, and many other fermented milk products, contains live bacteria. Look at the label; it will state "live, active cultures," or words to that effect, depending upon the brand. *Lactobacillus acidophilus* and other lactobacilli are the predominant live cultures; other bacteria are present as listed on the yogurt container. A gram of yogurt contains about 1 million lactobacilli; a 6-ounce container, intended as a single serving, weighs 170 grams and therefore contains about 170 million live bacteria. Generally speaking, about 1 billion live *L. acidophilus* cells are necessary for effectiveness.

Foods supplemented with live microbes are called probiotics. They are primarily dairy products but are also available as tablets or capsules that can be purchased in pharmacies, health food stores, markets, and other retail outlets. The proposed beneficial effects of probiotics are based on the assumption that consumption of live lactobacilli and certain other bacteria promotes an impressive list of beneficial attributes by complementing or, in

some cases, partially replacing the normal microbial flora. Advocates of probiotics claim that benefits include reduction in blood pressure, regression of tumors, reduction in allergy, decreased duration of diarrhea, and decreased gas production. Some individuals on antibiotic therapy suffer from yeast infections, most commonly manifested in the mouth and in the vagina. Some evidence suggests that consumption of yogurt and other probiotics during antibiotic therapy may be of value in preventing and treating yeast infections. Many yogurt consumers are convinced that regular consumption of probiotics improves their health.

There are skeptics, however, who feel that the claimed benefits associated with probiotics are exaggerated and based on weak science, primarily a poor understanding of the intestinal flora. Some skeptics say "they [probiotics] go in at one end of the digestive tract and come out the other, and hopefully something good happens along the way." Those who are not convinced of the value of yogurt and similar products call for research conducted in a scientific manner.

Meanwhile, if you enjoy your yogurt, continue to eat it!

FIGURE 3.8 Leavened and unleavened breads. Leavened bread utilizes yeast, which causes the bread to rise. Unleavened bread, called matzo, is eaten by Jews during the Passover holiday to commemorate their flight from bondage in Egypt. (Author's photo.)

Bread is a staple in primitive societies. History tells us the 3,000-year-old story of Passover, a Jewish celebration commemorating the time when the pharaoh of Egypt finally relented to Moses' plea to free the Israelites from bondage. In their haste, the Israelites left their homes without time to bake the breads necessary to sustain them during their journey. They took the raw dough and baked it on rocks under the hot sun, producing flat crackers called matzo, an unleavened bread (Fig. 3.8). Hence, it is obvious that leavened bread, resulting from the use of a dough containing yeasts as the leavening agent, was already known. Yeasts in their metabolism produce the gas carbon dioxide, which causes bread to rise and increase in size and acquire a lighter and fluffier texture.

Dairy Products

A variety of fermented milk products, many with centuries-old origins in the Middle East, are marketed (Table 3.1 and Fig. 3.9). They vary in their texture, taste, and aroma, depending on the type of milk, incubation period, and, most significantly, the microbial culture used to carry out the fermentation process. Note that a variety of lactobacillus species and a few species of streptococci are commonly used. Cultured buttermilk is generally made by adding *Streptococcus cremoris* to pasteurized milk. Buttermilk with a variety of flavors is the result of the particular microbes used. Sour cream is made from cream to which certain species of lactobacilli or streptococci are added. *Lactobacillus bulgaricus* and *Streptococcus thermophilus* are commonly used to produce yogurt and yogurt drinks. Kefir is a cultured milk product of increasing popularity in the United States that is made from the milk of cows, sheep, goats, or buffalo. Kefir dates back many centuries to the shepherds of the Caucasus Mountains, who discovered that fresh milk carried in goatskin bags is sometimes fermented into an effervescent beverage.

FIGURE 3.9 Fermented milk and soy products. Yogurt and a variety of other milk products are produced by microbes, which carry out fermentation. Most yogurts contain "live" cultures. **(A)** Soymilk and kefir; **(B)** yogurt. (Author's photos.)

Soft   Hard

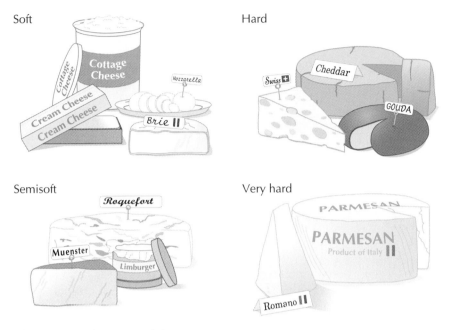

Semisoft   Very hard

AUTHOR'S NOTE *Believe it or not, there is now a new standard allowing grade A Swiss cheese to have smaller holes, or "eyes," to keep the cheese from getting tangled in high-speed slicing machines. The older standard required that the eyes had to be 11/16 to 13/16 inch in diameter, but new regulations reduce minimum eye size to 3/8 (6/16) inch.*

FIGURE 3.10   A variety of cheeses.

(The Caucasus Mountains are between the Black and Caspian Seas and range through Georgia, Armenia, Azerbaijan, and the southwest region of Russia.)

It is not mice, but bubbles of carbon dioxide produced by the fermentative activity of the bacteria, that put the holes in Swiss cheese. Cheeses are classified as soft, semisoft, hard, and very hard (Fig. 3.10 and 3.11), and you have probably sampled many of them. Over a thousand varieties of cheese exist in countries around the world, and cheeses from The Netherlands (Fig. 3.12), Switzerland, Italy, and France are particularly popular.

The texture, aroma, and taste of the cheese are dependent primarily on the process and the microorganisms used. Some varieties of cheese have a wonderful aroma, while others really stink! Cheese is made by adding lactic acid-producing bacteria and the enzyme rennin (or bacterial enzymes) to milk. The lactic acid sours the milk, and the enzymes coagulate casein (a protein in milk) to the solid curd portion and a watery portion known as whey. The curd is pressed to further remove the whey. Cottage cheese and cream cheese are packaged and sold without further ripening. Other cheeses can be ripened without the addition of other microorganisms, while, for some cheeses, additional microbes are added during the ripening process. In the production of blue cheese and Roquefort cheese, spores of the mold *Penicillium roqueforti* are added during the ripening period. In producing Swiss cheese, bacteria known as propionibacteria are added for the desired taste. The length of time allowed for ripening and the microbes involved in the ripening process determine the consistency of the cheese.

Wine, Beer, and Other Alcoholic Beverages

The next time you drink a cold and refreshing beer, enjoy the fragrance and flavor of a fine wine on your taste buds, celebrate an important event (like getting a high grade in this course) by sipping on champagne, or lie on a beach drinking a vodka Collins, remember that none of these alcoholic

FIGURE 3.11 Cheeses of the world. An array of cheeses from many countries is displayed in a supermarket. Their texture, aroma, and taste are the results of the strain of microbe and the fermentation process employed. (Author's photo.)

beverages would have been possible without the fermentation process carried on by yeasts.

Many countries are famous for their alcoholic beverages. French, Italian, and German wines are considered among the best; the wine industry in the United States has, in recent years, produced top wines, and sake, sometimes called rice wine although it is technically not a wine, is produced in Japan. Beers are popular, as attested to by the huge number of varieties available from numerous countries, each with its own distinctive flavor. Wine and beer are products of the fermentation of a variety of sugars and grains by strains of *S. cerevisiae* and other yeasts.

Wine making is a secret carefully guarded by each vineyard; **enology** is the science of wine making. Most wines are derived from the sugary juice extracted from grapes, but other fruits can be used; even dandelions are used to make wine. The extracted juice is usually treated with sulfur dioxide to kill naturally occurring yeasts that would produce uncontrolled and undesirable fermentation products. The yeast strain is added, and fermentation is allowed to proceed for a few days at a temperature between 20 and 25°C, followed by the aging process, which is carried out in wooden casks and takes weeks, months, or even years. During the aging process, the flavor, aroma, and bouquet of the wine develop as a result of the production of a variety of compounds resulting from the metabolism of the yeast. A number of factors are involved in the quality of the wine, including characteristics of the grapes, the strain of yeast, the wooden casks in which the wine is aged, and the duration of aging. Wine connoisseurs pride themselves on knowing a particular year for a fine vintage wine and are prepared to pay hundreds of dollars for this treasure. Their expressions, as they ceremoniously sniff the cork of the opened bottle and roll the first taste of the wine around in their mouths, are almost comical. "Colorful," "crisp," "fruity," "fragrant," and "full bodied" are popular terms in their jargon. Wine-tasting clubs are numerous and, although the actions of the tasters may seem to the uninitiated to be somewhat snobbish and frivolous, the rating of wines is a very serious business. Some wine tasters break the process down to the five basic components of color, swirl, nose, taste, and finish (Box 3.2), while others have different criteria.

A variety of wines to please every taste are available and fill the shelves of liquor stores (Fig. 3.13). All grapes have white juices; red wines are made from red grapes, and the color is due to the pigments of the grape skins.

FIGURE 3.12 A cheese factory in a village near Amsterdam, The Netherlands. (Author's photos.)

A

B

BOX 3.2 Wine Tasting

Some people are true connoisseurs of wines (or put on a good act). Next time you have occasion to dine in a fancy restaurant, watch the antics of a patron who appears to be sophisticated in choosing, tasting, and approving a wine once it is brought to the table. Note, also, the manner of presentation of the bottle of wine to the diner; it is held in a way to prominently display the label on which is clearly stated the year. A fine bottle of wine could cost as much as a few hundred dollars. Wine-tasting clubs for the amateur and for the professional are popular.

Wine has been used for centuries to celebrate religious events, including masses, bar mitzvahs and bat mitzvahs, and weddings. It is, after all, made from grapes, a fruit of the earth.

There is a certain mystique about choosing a good wine. Connoisseurs consider sweetness; wines are characterized as "sweet" when their taste is dominated by sugars and "dry" when other flavors mask the sugar. Acidity is another attribute; the words "tart," "crisp," and "fresh" are part of the jargon. "Astringency" refers to what is known as the "bitterness" of the wine.

One wine taster refers to the following five steps in judging a wine:

Color   Color is a reflection of the type of grape used as the source, the age of the wine, and the aging process. White wines increase in color with aging, while the color of red wine decreases.

Swirl   Gently swirl the glass of wine to oxygenate the wine. This releases those beautiful aromas characteristic of a good wine and complements the taste. By swirling, the wine is allowed to "breathe"; this can also be accomplished by uncorking the bottle and letting it sit open for a while before drinking.

Nose   Swirling the wine releases the aroma, or "bouquet." Here some subjective and quite imaginative terms such as "bountiful," "cherry," "heady," and "nutty" are used. Some people sniff the cork, but the smell can be just as well detected from the wine in the glass. The main point is that wine may have some unpleasant odors.

Taste   Take a small sip of wine from your glass, taking care not to swallow it. Let it caress and bathe the taste buds on your tongue. Your taste buds are sending signals to your brain. Are the sensations evoked pleasant ones?

Finish   The finish is the summation of the previous steps. Is the wine satisfying, mellow, and laced with a pleasant taste and no unpleasant aftertaste? If so, give the waiter a pleasant nod and be prepared to pay the price.

Now you can talk like a wine connoisseur and (if you are of legal drinking age) on the next occasion impress your companions as you delicately inspect the color, gently swirl, and fashionably "nose" the wine and then settle back a few seconds for the "finish." You may then say to your waiter, "Yes, the color is perfect, the bouquet is superb, and it has a perfect crispness."

Sweet wines are those in which fermentation is stopped while a significant amount of sugar is still present; in dry wines, little sugar remains. Champagnes result from continued fermentation that takes place in bottles. In the 1860s, the French wine industry was in a state of chaos and collapse due to poor-quality wines. Emperor Napoleon III called on Louis Pasteur to seek a solution, and within only about 3 years, he determined that the problem was the result of the wines' being contaminated. His solution was simple—heat the wine to 50 to 60°C, a process later applied to milk and other food products that is now referred to as pasteurization. Pasteur's manuscript *Études sur le Vin* (*Studies on Wine*) was published in 1866, and his experience with "sick wines" played a role in his shift to studying diseases in humans.

While wines are produced primarily from grape juices, beers are products of the fermentation of cereal grains, including barley (the most common), wheat, and rice. The grains are "malted" by being moistened and kept warm until partially germinated, which begins the enzymatic breaking down of the starch to simpler carbohydrates. (The malt is then oven dried; the length of the drying process contributes to the flavor and determines the final color of the beer.) The dried, cracked grains are steeped in hot water (mashed) to extract the sugars, starches, and other flavor compounds. The resulting liquid, called wort, is boiled in order to sterilize it, stop enzyme activity, and establish flavor. Hops (dried flowers of the

FIGURE 3.13   A drink for all occasions. Alcoholic beverages, including wines, beers, and distilled spirits, are available for all tastes and occasions. Their production is dependent on the fermentation process carried out by yeasts. (Author's photo.)

female *Humulus lupulus* vine or their extract) are added at various times for flavor, aroma, preservative qualities, and retention of the head (the foam at the top of a glass of beer). After filtration and cooling, yeast (usually a strain of *S. cerevisiae*) is added to ferment the wort, resulting in production of ethyl alcohol, carbon dioxide, and distinctive aroma and flavor compounds.

Brandy, whiskey, rum, vodka, and gin are referred to as **distilled spirits.** Their production resembles that of wine fermentation. A raw product is used as the starting point; it is fermented by yeast species and then aged in casks. After fermentation, distillation is carried out, yielding a product with a higher alcohol content than beer or wine. Scotch whiskey results from the fermentation of barley and rye, brandy results from the fermentation of wine or fruit juice, vodka results from the fermentation of potatoes or grains such as rye or barley, and rum results from the fermentation of molasses.

Alcoholic beverages vary in their content of alcohol. The alcoholic content of beer is usually 4 to 6%, that of wine is about 12 to 13%, and that of spirits ranges from 40 to 50%.

HARNESSING MICROBES AS RESEARCH TOOLS

Microbes (excepting viruses) offer to biologists microscopic packets of life complete with enzymes, energy-generating mechanisms, nucleic acids, structure, and reproductive ability. Even viruses, although not cellular, have some of these properties and are equally important as biological tools. Biologists have capitalized on the fact that microorganisms are easy and inexpensive to grow and reproduce rapidly. As knowledge of the microbial world and techniques to manipulate microbes became available over the past century, many of the major advances in biology resulted from experimentation with microbes. Virtually all fields in biology, and many aspects of physics and chemistry, have been enhanced by exploration with microbes. Genetics and molecular biology, in particular, are beneficiaries of the use of microbes in the laboratory.

One of the greatest achievements of the 20th century was the success of the **Human Genome Project**—the mapping of the 23 pairs of human chromosomes. The announcement of the completion of the project ahead of schedule amazed the world and captured the headlines. Microbes played a major role in this triumph; the proof that DNA is the genetic material was based on research utilizing bacteria and bacterial viruses. Ask most people who discovered DNA and its significance, and the likely answer will be James Watson and Francis Crick. Their brilliant work, announced in 1953, was the determination of the structure of the DNA molecule—the final piece in the puzzle of heredity dating back to Gregor Mendel's experiments in the mid-1800s. The discovery of DNA dates back to the work of Johann Miescher in 1869 when he isolated **"nuclein,"** today's DNA, from the nuclei of pus cells on bandages. There was no hint that nuclein (DNA) would prove to be the repository of genetic information; biologists had begun to speculate on the nature of the genetic material, but all bets were on protein. The establishment of DNA as the genetic material was based on three classical experiments (Fig. 3.14). Frederick Griffith, in 1928, demonstrated that

FIGURE 3.14   The classical proofs establishing DNA as the genetic material were accomplished by using microbes. **(A)** The Griffith experiment; **(B)** the Avery-MacLeod-McCarty experiment; **(C)** the Hershey-Chase experiment.

A Griffith (1928)

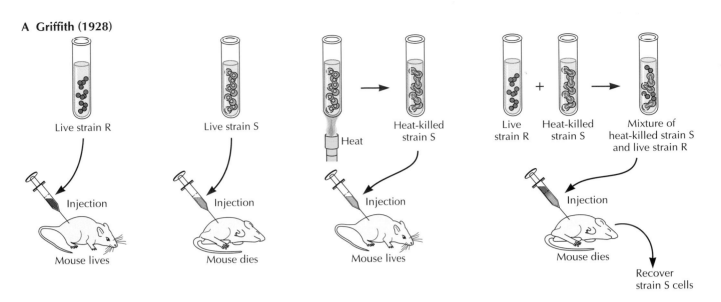

B Avery, MacLeod, McCarty (1944)

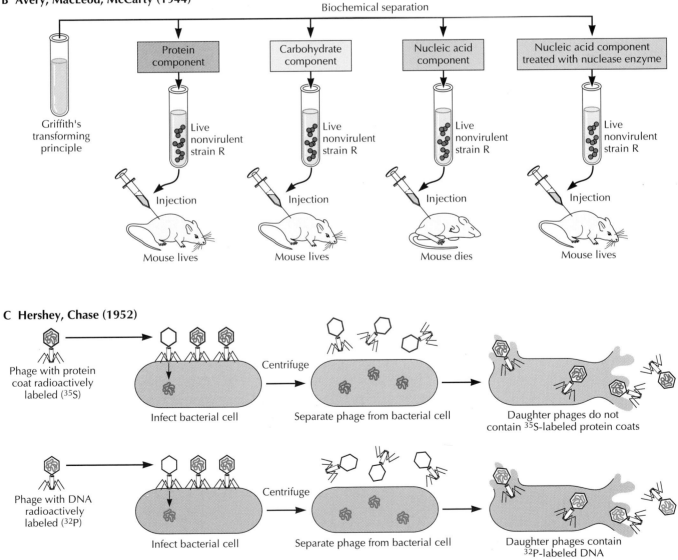

C Hershey, Chase (1952)

strains of pneumococci that were virulent for mice could transform avirulent strains to a state of virulence. He had no way of knowing the nature of what was being transferred to promote virulence and referred to a **"transforming principle"** without realizing that he had accomplished a feat of genetic engineering. Experiments performed with live animals (in vivo) pose variables and need validation in more tightly controlled test tube (in vitro) experiments. In 1944, three biologists at Rockefeller University—Oswald Avery, Colin MacLeod, and Maclyn McCarty—turned their attention to Griffith's transforming principle and demonstrated that the active transformation component was DNA, not a protein as had been anticipated, establishing DNA as the genetic material. The final confirmation was provided in 1952 by two biochemists, Alfred Hershey and Martha Chase. These investigators labeled the DNA in bacteriophages (chapter 5) with radioactive phosphorus (^{32}P) and labeled the viral protein coat with radioactive sulfur (^{35}S); the radioactive viruses were then mixed with bacteria. The radioactive labels demonstrated that the DNA entered the cell and was responsible for supplying the genetic information. Numerous other researchers contributed to these and other studies that led to an understanding of genetic mechanisms at the molecular level. Intensive research led to elucidation of the structure of DNA by Watson and Crick in 1953, rapidly followed by cracking of the genetic code and determination of the mechanisms by which genes function. **Genetic engineering,** also known as **recombinant DNA technology,** is a product of these studies.

The Human Genome Project was initiated in 1990 with the mission of mapping and sequencing the entire human genome—a genetic human blueprint with enormous potential impact on humankind in the coming years. Without microbes, none of this would have happened. The Microbial Genome Program was initiated in 1994 with the goal of sequencing the genomes of medically, environmentally, and industrially significant microbes; this program will lead to further success in harnessing these microbes for the benefit of humans. Genetics has come a long way since Mendel's mid-19th-century observation on the inheritance of color and other characteristics in plants.

An interesting and bizarre idea, utilizing the techniques of genetic engineering, was reported in the *Washington Post* in March 2001. According to the article, researchers interested in developing a cure for cystic fibrosis, a lung disease, hope to deliver new genes, a strategy referred to as gene therapy, by employing viruses for "delivering therapeutic payloads." As described in chapter 2, viruses gain access to cells and, perhaps, can act as shuttle vehicles, an idea that has also been proposed for other applications. Researchers plan to combine pieces of two of the world's deadliest viruses, Ebola virus, which causes hemorrhagic fever (bleeding disease), and human immunodeficiency virus (HIV), the cause of acquired immune deficiency syndrome (AIDS). The choice of these two viruses as partners is based on Ebola virus's predilection to attach to lung cells and on HIV's reputation for persisting in the body. The hybrid virus would be used to deliver new genes into the cells of the lungs in patients with cystic fibrosis.

The idea of this dynamic duo is frightening. Robert Gallo, codiscoverer of HIV, stated, "I wouldn't want this thing put into me." Gallo warns of harmful immune reactions and the possibility that the hybrid virus could combine with HIV to create a new monster.

On the other hand, W. French Anderson, a pioneer and leader in gene

AUTHOR'S NOTE *The logic is there, but what a choice for people suffering from cystic fibrosis to have to make!*

therapy, and other biologists consider the proposed hybrid virus to be safe. "It's not even HIV anymore, it's just pieces. Ebola sounds horrible, this has nothing to do with the Ebola virus that knocks out all your defense mechanisms and kills you," according to Anderson.

HARNESSING MICROBES IN INDUSTRY

In industry, including the pharmaceutical industry, the challenge is to harness microbes as factories and capture their metabolic products. The list of products (Table 3.2) is impressive. Antibiotics, genetically engineered products, proteins, carbohydrates, nucleic acids, and cleaning products are examples. Certain characteristics of microbes promote their use as microbial factories. (i) The high ratio of surface area to volume leads to rapid replication; the product yield is dependent on the number of microbes maintained under optimum conditions. (ii) Microbes are versatile and can be grown in vats on a large scale and under a variety of growth conditions. (iii) Some products can be produced only by microbes. (iv) Microbes produce a large variety of enzymes which can be harvested to obtain desired products. (v) Microbes can be genetically engineered to produce biological products that are used in the prevention and treatment of cardiac disease and other medical problems. (vi) Microbes can be genetically engineered to increase their productivity. (vii) Microbial factories are cost-effective.

In the early years of antibiotic production, only about 5 milligrams of penicillin could be recovered per liter of culture, whereas new strains of *Penicillium* have increased the yield to over 50,000 milligrams per liter. Industrial microbiologists are always on the hunt for microorganisms that will synthesize new products or synthesize known products at a greater yield, for example, mutants that are not able to control synthesis of a particular

TABLE 3.2 Examples of products of genetic engineering

Product and/or microbe	Function
Products used in human medicine	
Alpha interferon (*E. coli*)	Treatment for some viruses
Insulin (*E. coli*)	Treatment for diabetes
Human growth hormone (*E. coli*)	Treatment for pituitary dwarfism
Interleukin-2 (*E. coli*)	Stimulation of immune system
Tumor necrosis factor (*E. coli*)	Treatment of certain cancers
Epidermal growth factor (*E. coli*)	Treatment of skin wounds and burns
Hepatitis B vaccine (*S. cerevisiae*)	Vaccine used in prevention of hepatitis B
Products used in animal husbandry	
Bovine growth hormone (*E. coli*)	Increases weight gain and milk production
Porcine (swine) growth hormone (*E. coli*)	Increases weight gain
Genetically engineered microbes	
Pseudomonas fluorescens	Carries genes from *Bacillus thuringiensis* that produce an insect poison
Pseudomones syringae (ice-minus bacterium)	Engineered to remove protein that initiates ice formation on plants and affords protection from frost

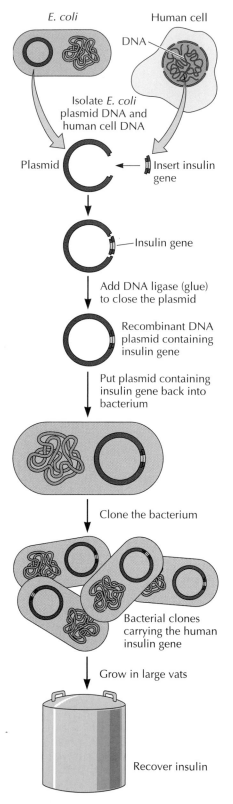

FIGURE 3.15 Production of insulin by genetic engineering. The human insulin gene is cloned and expressed in *E. coli.*

E. coli   Human cell

DNA

Isolate *E. coli* plasmid DNA and human cell DNA

Plasmid   Insert insulin gene

Insulin gene

Add DNA ligase (glue) to close the plasmid

Recombinant DNA plasmid containing insulin gene

Put plasmid containing insulin gene back into bacterium

Clone the bacterium

Bacterial clones carrying the human insulin gene

Grow in large vats

Recover insulin

product and thereby produce "overruns." The search for soil and other antibiotic-producing microorganisms continues, particularly in light of the antibiotic resistance problem.

A field trip to an industrial plant is a worthwhile experience. You cannot help being amazed at the sheer magnitude and complexity of industrial microbiology and the skills of the bioengineers. Vats (fermentors) are two or more stories high and are connected to an array of garden hose-like tubes, dials, gauges, and electronic monitoring devices; the growth medium depends on the particular microbe and the desired product. Many companies, including breweries, offer tours of their facilities (and free samples!).

Genetic engineering has promoted the growth of biotechnology and pharmaceutical companies aimed at harnessing microbes for the production of medicinals, including vaccines, antibiotics, hormones, and immune regulatory factors, a number of which are currently available. **Human insulin** and **human growth hormone** are remarkable examples of genetic engineering products (Fig. 3.15), as are some of the newer recombinant DNA vaccines (chapter 11). Human insulin is now produced in *Escherichia coli* by cloning the human "insulin gene" into its genetic material. Prior to this, insulin was produced from the pancreas of slaughtered cows and pigs; it was not as effective as human insulin and more expensive to produce. Prior to genetic engineering, human growth hormone for the treatment of dwarfism was produced from human cadaver brains obtained at autopsy, and about 16 brains were required to treat one individual. This posed a severe limitation of supply and favored "the rich and the famous." Gene therapy is the wave of the future and has already taken place on a trial basis. Viruses, as obligate intracellular parasites, will play an important role as vectors to deliver replacement genes into defective cells.

Microbes have been tapped as a source of insecticides; a bioinsecticide called Bt (using *Bacillus thuringiensis*) has been marketed for years to control caterpillars and other leaf-eating insects. Other bioinsecticides using microbes or their products are available to control mosquitoes, Japanese beetles, crickets, and grasshoppers. These products, unlike some chemicals, are ecologically safe.

HARNESSING MICROBES FOR BIOREMEDIATION

Some of the coastline in the Prince William Sound off the coast of Alaska is tarnished with an oil slick. This is because on March 24, 1989, the oil tanker *Exxon Valdez* ran aground, spilling 10.8 million gallons, or 257,000 barrels, of oil (Box 3.3). To appreciate the magnitude of the spill, the largest in U.S. history but only about the 34th largest worldwide, consider that the oil would fill about 125 Olympic-sized swimming pools. Feelings of shock and disbelief spread through the world (In the News 3.1) as massive efforts took place to clean up the spill and come to the rescue of the thousands of seabirds, sea otters, seals, bald eagles, killer whales, and salmon and herring eggs affected by the spill. It is impossible to estimate the damage caused to microbial life and to food chains (Fig. 3.16). One of the techniques used to clean up the mess was **bioremediation.** Bioremediation is an "emerging technology based on the enzymatic capacity of microbes to digest or to degrade a variety of materials including oil, paper, concrete, and textiles," as defined in a publication of the American Academy of Microbiology. Bioremediation can be enhanced by the spraying of nutrients on beaches or on other problem sites to foster the growth of the microbes

BOX 3.3 An Account of the *Exxon Valdez* Oil Spill

What actually happened?

The *Exxon Valdez* departed from the Trans-Alaska Pipeline terminal at 9:12 p.m. on March 23, 1989. William Murphy, an expert ship's pilot hired to maneuver the 986-foot vessel through the Valdez Narrows, was in control of the wheelhouse. At his side was the captain of the vessel, Joe Hazelwood. Helmsman Harry Claar was steering. After passing through Valdez Narrows, Pilot Murphy left the vessel and Captain Hazelwood took over the wheelhouse. The *Exxon Valdez* encountered icebergs in the shipping lanes, and Captain Hazelwood ordered Claar to take the *Exxon Valdez* out of the shipping lanes to go around the icebergs. He then handed over control of the wheelhouse to Third Mate Gregory Cousins with precise instructions to turn back into the shipping lanes when the tanker reached a certain point. At that time, Claar was replaced by Helmsman Robert Kagan. For reasons that remain unclear, Cousins and Kagan failed to make the turn back to the shipping lanes, and the ship ran aground on Bligh Reef at 12:04 a.m. on March 24, 1989. Captain Hazelwood was in his quarters at the time.

How did the accident happen?

The National Transportation Safety Board investigated the accident and determined that the probable causes of the grounding were:

1. the failure of the third mate to properly maneuver the vessel, possibly due to fatigue and excessive workload;
2. the failure of the captain to provide a proper navigation watch, possibly due to impairment from alcohol;
3. the failure of the Exxon Shipping Company to supervise the captain and provide a rested and sufficient crew for the *Exxon Valdez;*
4. the failure of the U.S. Coast Guard to provide an effective vessel traffic system; and
5. the lack of effective pilot and escort services.

Was the captain drunk?

The captain was seen in a local bar and admitted to having some alcoholic drinks; a blood test showed alcohol in his blood even several hours after the accident. The captain has always insisted that he was not impaired by alcohol. The state charged him with operating a vessel while under the influence of alcohol. A jury in Alaska, however, found him not guilty of that charge. The jury did find him guilty of negligent discharge of oil, a misdemeanor. Hazelwood was fined $50,000 and sentenced to 1,000 hours of community service in Alaska. He began a 5-year plan to serve his sentence in 1999.

How much oil was spilled?

The amount of oil spilled was 10.8 million gallons (257,000 barrels, or 38,800 metric tons). The amount of spilled oil was roughly equivalent to 125 Olympic-sized swimming pools.

How much oil was the *Exxon Valdez* carrying?

The ship was carrying 53,094,510 gallons, or 1,264,155 barrels.

How does the *Exxon Valdez* spill compare to other spills?

The *Exxon Valdez* spill is the largest ever in the United States but ranks 34th largest worldwide. It is widely considered the number 1 spill worldwide in terms of damage to the environment, however. The timing of the spill, the remote and spectacular location, the thousands of miles of rugged and wild shoreline, and the abundance of wildlife in the region combined to make it an environmental disaster well beyond the scope of other spills.

How many miles of shoreline were affected by oil?

Approximately 1,300 miles. Two hundred miles was heavily or moderately oiled (meaning the impact was obvious); 1,100 miles was lightly or very lightly oiled (meaning light sheen or occasional tar balls). By comparison, there is more than 9,000 miles of shoreline in the spill region.

How large an area did the spill cover?

From Bligh Reef, the spill stretched 460 miles to the tiny village of Chignik on the Alaska Peninsula.

How was the spill cleaned up?

This is a complicated question. It took more than four summers of cleanup efforts before the cleanup was called off. Some beaches remain oiled today. At its peak, the cleanup effort included 10,000 workers, about 1,000 boats, and roughly 100 airplanes and helicopters, known as Exxon's army, navy, and air force. It is widely believed, however, that wave action from winter storms did more to clean the beaches than all the human effort involved.

How much did it cost?

Exxon says it spent about $2.1 billion on the cleanup effort.

What techniques were used?

Check out *National Geographic,* January 1990, pages 18 and 19, for a great illustration of how shoreline cleanup was conducted.

Hot water treatment was popular until it was determined that the treatment could be causing more damage than the oil. Small organisms were being cooked by the hot water.

High-pressure cold water treatment and hot water treatment involved dozens of people holding fire hoses and spraying the beaches. The water, with floating oil, would trickle down to the shore. The oil would be trapped within several layers of boom and either be scooped up, sucked up, or absorbed using special oil-absorbent materials.

Mechanical cleanup was attempted on some beaches. Backhoes and other heavy equipment tilled the beaches to expose oil underneath so that it could be washed out. Many beaches were fertilized to promote growth of microscopic bacteria that eat the hydrocarbons. Known as bioremediation, this method was successful on several beaches where the oil was not too thick. For further technical information,

(continued)

An Account of the *Exxon Valdez* Oil Spill *(continued)*

go to the Environmental Protection Agency website (http://www.epa.gov) and search using the terms "bioremediation" and "Exxon Valdez."

How many animals died outright from the oil spill?

No one knows. The carcasses of more than 35,000 birds and 1,000 sea otters were found after the spill, but since most carcasses sink, this is considered to be a small fraction of the actual death toll. The best estimates are 250,000 seabirds, 2,800 sea otters, 300 harbor seals, 250 bald eagles, up to 22 killer whales, and billions of salmon and herring eggs.

How are they doing over 10 years later?

Lingering injuries continue to plague most species. Of 23 species listed as injured by the spill, only two have been declared "recovered."

How does oil harm birds and mammals?

See *National Geographic,* January 1990, pages 26 and 27, for an excellent illustration of how oil affects the fur and feathers of wildlife.

There are three primary ways that oil injures wildlife:

1. The oil gets on the fur and feathers and destroys the insulation value. Birds and mammals then die of hypothermia (they get too cold).
2. They eat the oil, either while trying to clean the oil off their fur and feathers or while scavenging on dead animals. The oil is a poison that causes death.
3. The oil affects them in ways that do not lead to a quick death, such as damaging the liver or causing blindness. An impaired animal cannot compete for food and avoid predators. Oil also affects animals in non-lethal ways, such as impairing reproduction.

How were the oiled birds and sea otters cleaned?

A professional team and dozens of volunteers, including veterinarians, set up a cleaning facility and recovery facility. Dawn dishwashing detergent was the cleaning agent of choice.

Where is the *Exxon Valdez* today?

The Exxon Shipping Company was renamed the Sea River Shipping Company. The *Exxon Valdez* was repaired and renamed the *Sea River Mediterranean,* and it is used to haul oil across the Atlantic Ocean. The ship is prohibited by law from returning to Prince William Sound.

Source: Exxon Valdez Oil Spill Trustee Council, 2001.

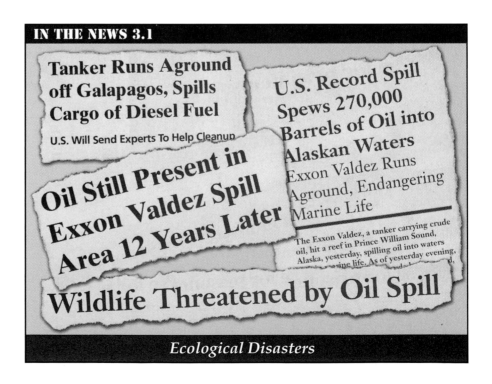

IN THE NEWS 3.1

Ecological Disasters

indigenous to the area to accelerate degradation of the pollutant; this process is called **bioaugmentation.**

It is difficult to believe that microbes can bring about the decomposition of these tough materials, but this is another illustration of microbial diversity. There are several advantages to the use of microbes as recyclers to clean up the environment, including cost-effectiveness, self-destruction once the conditions are improved, and minimal disruption of the environment. Hopefully, in the not too distant future, bioremediation will be further advanced and play an important role in coping with ecological disturbances.

The possibilities of bioremediation are exciting and seemingly unlimited as microbes from diverse habitats are identified, and as new microbes are genetically engineered with the enzymatic capability of breaking down environmental contaminants, including toxic products, petroleum products, landfill wastes, soils, polychlorinated biphenyls, plastics, and even disposable diapers. There is some evidence that global warming, an ecological disturbance with worrisome consequences (chapter 1) may be at least partially slowed down by *Synechococcus,* a bacterium that decreases carbon dioxide released in industrial processes, counteracting the greenhouse effect. *Methylosinus trichosporium* is another naturally occurring microbe which conceivably could be harnessed to reduce the process of global warming; the organism breaks down the chlorofluorocarbon gases that are products of refrigerants, air conditioners, foam packaging, and spray can propellants. Chlorofluorocarbons are "greenhouse" gases that contribute to global warming and deplete the earth's protective ozone layer.

Bioremediation is at the point where it is now moving from the laboratories and computer simulations to field research and to applicability in the real world.

A mall of factory outlets in Wrentham, Mass., is a pioneer in water recycling and recycles 90% of its 35,000 gallons of water a day into clean water that can be reused for toilet flushing. The system eliminates pathogens but allows nonpathogens to "clean" the water—another potential application of bioremediation.

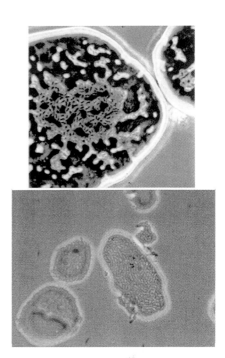

FIGURE 3.16 Oil globules from the *Exxon Valdez* oil spill. The globs are oil droplets in water. The small, black oblong objects in the globules are bacteria that have utilized the oil as a nutrient source and are replacing the oil as the bacterial population increases. (Source: Oppenheimer Biotechnology, Inc.)

OVERVIEW In this chapter are presented positive aspects of our association with the microbial world, a world whose presence is unseen but is manifested in many ways every day. The fact that life exists is, in itself, a manifestation of microbes and their role in the cycles of nature. Many microbes are too small to be seen without a microscope and, therefore, are taken for granted, but the reality is that they constitute a major component of Earth's biomass and are the underpinnings of life itself.

Earlier civilizations learned by experience to harness the metabolic activities of microbes in the production of foods (including fermented milk products) and alcoholic beverages. In more recent years, as the science of microbiology developed, the diversity of microbes has been further exploited. Microbes continue to be used as research tools which have spawned many disciplines in biology, as factories for the production of many useful products, including products of genetic engineering, and as agents of cleaning by means of bioremediation.

Because of the barrage of microbial threats, perhaps there is too much emphasis on strategies to wipe out, destroy, sanitize, or scrub away microbes. In so doing, we may be killing "microbial friends" and thereby "throwing out the baby with the bathwater." *The Microbes' Contribution to Biology* was published in 1956; if that text were updated today, many more pages would be necessary. The following two quotations serve as an appropriate conclusion to this chapter: "If you take care of your microbial friends, they will take care of your future" and "Never underestimate the power of the microbe." ■

SELF-EVALUATION

PART I Choose the *single* best answer.

1. The presence of bacteria in the intestinal tract of grazers and of termites is an example of
   a. commensalism   c. parasitism
   b. mutualism   d. "the good life"

2. Nitrification is characterized by
   a. conversion of nitrates into ammonia
   b. fixation of atmospheric oxygen
   c. conversion of ammonia into nitrates
   d. return of nitrogen to atmosphere

3. Red wines (unlike white) are the result of
   a. red juices of red grapes
   b. aging process
   c. skins of red grapes
   d. white grapes to which red skins have been added

4. A "transforming principle," a start in the proof of DNA as the genetic material, was the work of
   a. Pasteur   c. Hershey and Chase
   b. Griffith   d. Prusiner

5. A sweet wine is characterized by
   a. distillation
   b. little remaining sugar
   c. fermentation in a bottle
   d. a relatively high sugar concentration

PART II Fill in the following.

1. Plants with nodules along their root systems are called _____ plants.

2. What is the significance of *Rhizobium*?
   _____

3. Name the yeast (genus and species) frequently used in making bread and alcoholic beverages.
   _____

4. What causes bread to "rise"? _____

5. Wines are produced from grape juices, while beers are produced from fermentation of _____.

6. What is bioremediation? _____

PART III Answer the following.

1. Distinguish between bioremediation and bioaugmentation.

2. Using specific examples, explain the expression "microbes make the world go around."

3. Cite and briefly explain the proofs that DNA is the genetic material.

BACTERIA

PREVIEW "Dear God, what marvels there are in so small a creature." These were the words of Antonie van Leeuwenhoek, a Dutch merchant by vocation and a lens grinder by avocation, in a report to the Royal Society of London in 1675. Leeuwenhoek's development of a simple microscope allowed him to see and describe **"small animalcules."** Microbes date back to the antiquity of life; perhaps bacteria were the first forms of life. A scientist at California Polytechnic Institute recently reported the isolation of ancient bacteria from a bee in a specimen of amber, a hardened resin from ancient pine trees, dating back 25 million to 40 million years. This account is reminiscent of the novel and movie *Jurassic Park*, in which a scientist extracted dinosaur DNA from entombed mosquitoes that had fed on dinosaurs.

As was emphasized in chapter 2, the notion that bacteria are simple because they are unicellular is far from the truth; they may be small, but they are not simple. They are complex biological entities. The fact is, within a matter of only several hours following the invasion of a human body by some pathogenic bacteria, severe illness and even death may result. This is equally true of infected elephants and whales, whose mass is measurable in tons and yet they fall victim to microscopic microbes. It would take a fantastic number of bacteria to equal 1 ton. The anatomy of the bacterial cell as an organized and functional structure meeting the characteristics associated with life will be described in this chapter. Under appropriate conditions, some bacteria can undergo **binary fission** in as little as 20 minutes, resulting in huge populations in 24 hours. An appreciation of the bacterial growth curve is necessary to understand the dynamics of growth when only a few pathogens are introduced into the body.

Most bacteria are easily grown in the laboratory, facilitating the diagnosis of bacterial infection. Mycoplasmas, chlamydiae, and rickettsiae are atypical, "oddball" bacteria, and their unique characteristics are described later in the chapter. They cause a variety of diseases in humans (chapter 8).

CELL SHAPES AND PATTERNS

Examination of bacterial cells under a microscope reveals the presence of a variety of shapes and patterns of arrangement (Fig. 4.1). The majority are either rod shaped, known as **bacilli** (singular, bacillus); spherical, known as **cocci** (singular, coccus); or spiral shaped, known as **spirilla** (singular, spirillum). For the most part, bacilli tend to occur as single cells, but some species tend to form chains. Characteristic groupings of cocci are more common and are useful in identification. **Streptococci** are chains of cocci, resembling a string of pearls; **staphylococci** look like a bunch grapes; **diplococci** occur in pairs; and **tetrads** are groupings of four. Undoubtedly, the terms "strep" and "staph" are familiar to you. Spirilla are categorized as spiral (wavelike), **spirochetes** (corkscrews), and **vibrios** (comma-shaped curved rods). Other bacteria have been described as star shaped, triangular, flat, or square. Microscopic determination of shape and pattern is often the first step in the identification of bacteria.

NAMING BACTERIA

Young couples spend hours searching through baby-naming books, conferring with family and friends, and otherwise agonizing over their choice of the perfect name for their expected child, particularly the firstborn. Biologists have established names for microbes and have put considerable thought into doing so, because the names frequently identify important characteristics. As you read through this book, you may feel overwhelmed by the long and difficult-to-pronounce names, but at least you can impress your family and friends.

Microbes are named in accordance with the **binomial** system of nomenclature established by Carolus Linnaeus in 1735. This system is not restricted to microbes but applies to all organisms. The names are latinized, and each organism carries two (binomial) names, the first designating the **genus** and the second designating the **species.** For example, humans are *Homo* (genus) *sapiens* (species), and the fruit flies most commonly used in genetic studies are *Drosophila melanogaster*. Note that the first letter of the genus name is always capitalized, but the species name is not. Further, both the genus and species names are italicized.

Unfortunately for the student, microbiologists have not been consistent in assigning names. Some microbes are named in honor of the scientist

FIGURE 4.1   Bacterial cell shapes and patterns.

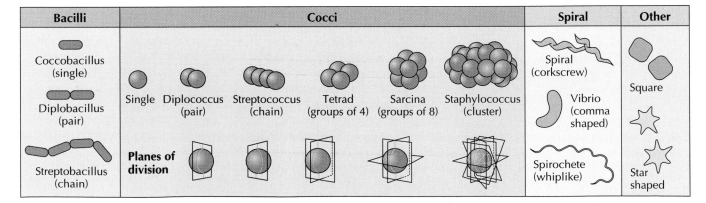

responsible for first describing them, while others indicate the microbe's habitat, shape, associated disease, or combinations of these factors. A few examples will make this clear. *Rickettsia prowazekii* (epidemic typhus) is named in honor of Howard T. Ricketts and Stanislaus von Prowazek, both of whom died of the disease that they were studying. *Legionella pneumophila* is named after an outbreak of pneumonia at an American Legion convention in 1976. *Escherichia coli* is named in commemoration of Theodor Escherich and identifies the coli (large intestine) as the organism's habitat.

The use of descriptive morphological terms (e.g., bacillus) can lead to confusion. *Bacillus anthracis* belongs to the genus *Bacillus* (with a capital "B") and morphologically is a bacillus (with a lowercase "b"). But *Escherichia coli* is also a bacillus ("b") belonging to the genus *Escherichia*. There are many other bacilli that do not belong to the genus *Bacillus*. So all members of the genus *Bacillus* are bacilli, but not all bacilli are classified under *Bacillus*. The same case can be made for streptococci and diplococci. *Diplococcus pneumoniae* strains are all diplococci, but so is *Neisseria gonorrhoeae*.

THE ANATOMY OF THE BACTERIAL CELL

To appreciate the challenge posed by bacteria, it is necessary to be familiar with their structure and properties. The fact that they are procaryotic cells sets them apart from the other microbes. The structures that compose these cells are outlined in Table 4.1 and illustrated in Fig. 4.2; note that some anatomical features are not common to all bacteria.

The Envelope

The bacterial **envelope** consists of the **capsule, cell wall,** and **cell membrane.** A capsule is not present in all species, but the cell wall (with the exception of the mycoplasmas, to be described later in the chapter) and the cell membrane are structures present in all bacteria.

The Capsule

When present, the capsule is not integral to the life of the cell. In fact, the capsule can be easily removed by treating a culture with appropriate enzymes or by manipulating the presence of nutrients in the culture; in either case, the cells grow just as well in the laboratory but lack capsules. The progeny of these noncapsulated cells will have capsules. In some species, the presence of a capsule promotes **virulence** (the capacity to produce disease), as, for example, in the bacillus that causes anthrax (*B. anthracis*). The streptococci, responsible for strep throat and flesh-eating streptococcal disease, are another example of capsulated organisms. The presence of the capsule interferes with **phagocytosis,** an important body defense mechanism by which bacteria are ingested and killed by a variety of phagocytic cells. The presence of the capsule makes it difficult for phagocytosis to occur; it is like trying to grab on to a slippery fish. Phagocytosis is described in more detail in chapter 11. The capsular material may become so thick that it resembles a **slime layer.** Occasionally, hunks of this mucuslike layer can be found in the broth used to grow the culture. A condition known as **"ropy milk"** is a nuisance to the milk industry because of the growth of the nonpathogen *Alcaligenes viscolactis* and the shedding of its slime layer into milk. The fact that this organism is not a disease producer is of little comfort as your tongue and palate unexpectedly encounter this ropy, mucuslike material.

AUTHOR'S NOTE

I would love to have a microbe or disease named after me!

TABLE 4.1   Anatomical features of the bacterial cell

Structure	Function
Cell envelope	
Glycocalyx[a] (capsule, slime layer)	Promotes virulence in some cases
Cell wall (with the exception of mycoplasmas)	Corsetlike structure which confers shape and provides tensile strength
Cell membrane	Controls movement of molecules into and out of cell; referred to as gatekeeper
Cytoplasm[b]	
Nucleoid	DNA-rich area not enclosed by a membrane
Plasmids[a]	Nonchromosomal DNA that confers properties, including antibiotic resistance
Spores[a]	Confers extreme resistance to environmental factors
Ribosome	Protein synthesis
Chromosome	Determinant of genetic traits
Inclusion bodies	Storage and reserve supply of nutrient materials
Appendages	
Flagella[a]	Motility
Pili[a]	Adhesion to surfaces, bridge for transfer of DNA

[a]Structure not found in all bacterial cells.
[b]Area within membrane; contains numerous organelles.

FIGURE 4.2   A "composite" bacterial cell.

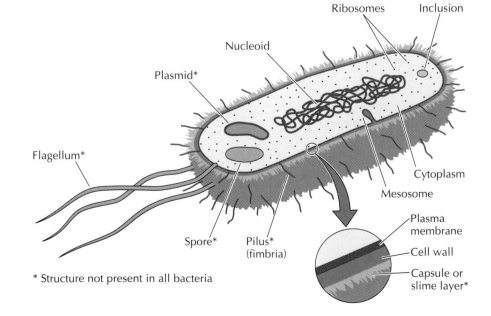

* Structure not present in all bacteria

The Cell Wall

Cell walls are characteristic of all bacteria (again, with the exception of the mycoplasmas) and are a structure shared with plant, algae, and fungal cells. The cell wall is a rigid, corsetlike structure responsible for the characteristic shape of the cell. Further, it confers resistance to the cell from the inward **diffusion** of water. Without a strong cell wall, the membrane would swell, and **lysis** (cell bursting) would occur. In the 1940s, penicillin came into widespread use despite the fact that its mechanism of action was not yet understood. Several years later, scientists discovered that penicillin interferes with the ability of cells to synthesize normal cell walls, making these cells subject to lysis. This is the mechanism by which the "wonder drug" penicillin and related antibiotics indirectly deliver a death blow to certain bacteria. (The story of the discovery of penicillin is one of many examples of serendipity [chance] in scientific discovery and is further described in chapter 12.) A number of other antibiotics target bacterial cell walls, leading to the death of these cells. Since human cells do not have cell walls, humans are able to take antibiotics.

Differences in the chemistry of the cell wall make it possible to divide the many bacteria responsible for human infections into two major groups based on staining properties. The Gram stain, an important early step in bacterial identification, was introduced by Hans Christian Gram in 1884. The two groups are **gram-positive** and **gram-negative** organisms, a characteristic of considerable medical significance. Perhaps you, or someone you know, have been diagnosed with a gram-positive or a gram-negative infection. The distinction is important, because the choice of antibiotics is strongly influenced by the Gram stain reaction of the causative bacteria (Table 4.2). For example, penicillin and a variety of related antibiotics are

TABLE 4.2   Examples of diseases produced by gram-positive and by gram-negative bacteria

Bacterium	Disease
Gram positive	
Bacillus anthracis	Anthrax
Clostridium botulinum	Botulism
Clostridium tetani	Tetanus
Corynebacterium diphtheriae	Diphtheria
Listeria monocytogenes	Listeriosis
Staphylococcus aureus	Skin infections, toxic shock
Streptococcus pyogenes	Streptococcal sore throat; scarlet fever
Gram negative	
Escherichia coli O157:H7	Bloody diarrhea
Campylobacter jejuni	Gastroenteritis
Legionella pneumophila	Legionnaires' disease
Salmonella typhi	Typhoid fever
Neisseria gonorrhoeae	Gonorrhea
Yersinia pestis	Plague
Vibrio cholerae	Cholera

effective against gram-positive but not gram-negative cells; on the other hand, chloramphenicol has a different spectrum of activity.

Cell walls of gram-positive and gram-negative bacteria are composed of **peptidoglycan,** a backbone which gives the cell wall its characteristic rigidity. Some of the molecules associated with the bacterial cell wall are not found elsewhere in nature. Cell walls of gram-positive bacteria have a thick peptidoglycan layer, whereas this layer is relatively thin in gram-negative cells. Additionally, in gram-negative cells, there is an **outer membrane** external to the peptidoglycan layer that is associated with virulence. When these cells undergo lysis, an endotoxin that causes damage to the host is released from the outer membrane (chapter 6).

The Cell Membrane

The passage of molecules between the bacterial cell and its external environment is controlled by the cell membrane. To understand this point, think of chains of streptococci residing on the surface of your pharynx (Fig. 4.3). They are probably quite content, and why shouldn't they be? The surface of your throat is a warm, moist, and nutrient-rich environment that is optimal for bacterial multiplication. This requires the passage of oxygen and nutrients into the "strep" cells through their cell membrane to allow for metabolic processes consistent with life and reproduction. During their growth and reproduction, as a natural part of their metabolism, they secrete a number of products, some of which are toxic and produce the sore throat and fever associated with a "strep throat."

The cell membrane is **selectively permeable,** meaning that not all molecules are able to pass freely into or out of the cell. The structure of the membrane, inherent in its chemistry and physical arrangement, accounts for its "gatekeeper" function; it is double-layered and has pores and transport molecules which are instrumental in the movement of materials between the cells and their environment. The physical processes of diffusion and **osmosis,** plus the size and the charge of the molecules, are factors which govern the selective permeability of the membrane. The bacterial cell membrane is the site for energy-generating reactions of the cell and also contains enzymes which are instrumental in cell wall assembly and

FIGURE 4.3   Streptococci on the surface of the pharynx.

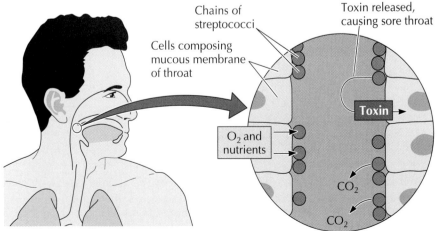

Chains of streptococci

Toxin released, causing sore throat

Cells composing mucous membrane of throat

Toxin

O_2 and nutrients

CO_2

CO_2

membrane synthesis. A cell membrane is not unique to bacterial cells but is common to all cells, procaryotic and eucaryotic, and has a remarkably similar structure in all cells.

The Cytoplasm

The **cytoplasm** is the part of the cell enclosed within the cell membrane, and within it are numerous cellular constituents that function in cell growth and multiplication. Figure 4.2 illustrates the variety of structures found within this region.

The Nucleoid

As established in chapter 2, the bacterial cell lacks a nuclear membrane and, hence, it is a procaryotic cell. Within the cytoplasm, there is a DNA-rich area, demonstrated by electron microscopy, that is referred to as a **nucleoid;** there is no membrane defining this area. It is relatively easy to extract bacterial DNA from *E. coli* and some other bacteria. Most bacterial cells contain a single **chromosome** present as a circular, double-stranded stretch of DNA. (The organism *Vibrio cholerae* has two chromosomes.) This is in contrast to eucaryotic cells, in which the DNA is organized into discrete bodies, the chromosomes. The mechanism of cell division is binary fission, which, simply put, means "splitting in two" (Fig. 4.4).

Plasmids

Plasmids are small molecules of nonchromosomal DNA that are found in some bacteria; they are located in the cytoplasm (Fig. 4.2) and are independent of the chromosomal DNA. Under laboratory conditions, plasmids are not essential to the cell; it is possible to "cure" (eliminate) them by laboratory manipulation without killing the cell. Although plasmids may consist of only a few to a large number of genes, they code for the production of significant properties. Some carry genes which confer antibiotic resistance; these plasmids are referred to as **R (resistance) factors.** Other plasmids carry genes which confer virulence. Plasmids may be transferred from one bacterial cell to another, and because of this, they are referred to as **infectious agents.** Imagine, therefore, a population of bacteria in your intestinal tract in which, initially, only a few cells have plasmids conferring resistance to three different antibiotics. It is possible that within only several hours plasmids may infect other cells, resulting in a majority of the microbial population's becoming multiply drug resistant (Fig. 4.5). Some plasmids allow the exchange of DNA between bacterial cells. Since plasmids are self-replicating DNA segments, the donor cells retain their plasmids while transferring copies to the recipient cell.

Spores

The genera *Bacillus* and *Clostridium,* both of which contain disease-producing species (chapter 8), are spore producers. **Spores** are extremely hardy structures that are highly resistant to heat, drying, radiation, and a variety of chemical compounds, including alcohol. Most bacteria will be killed in boiling water within only a few minutes, but this is not true of bacteria that make spores; they remain viable even after a few hours of exposure to boiling water. Spores can exist for centuries in soil and have been recovered from soil samples dating back several thousand years. They are in a state of dormancy with little or no metabolism. In fact, some scientists maintain that the first life-forms on Earth were bacterial spores that had drifted from

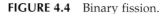

FIGURE 4.4   Binary fission.

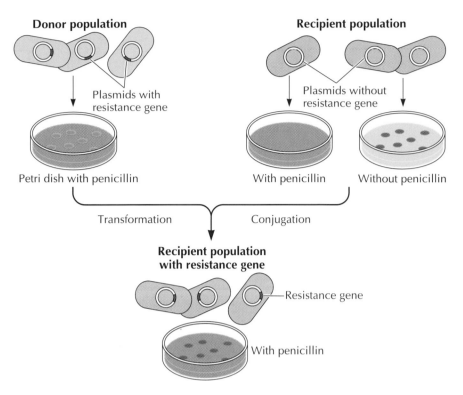

FIGURE 4.5 Infectious nature of plasmids.

some distant planet. The spore is actually a condensed form of the cell, including the cell's DNA, encased within a several-layered, chemically complex coat. A plant seed is somewhat analogous. If you plant a tomato seed, you get a tomato plant because the seed contains the genetic information. Spore-forming bacteria without spores are termed **vegetative cells.** In the laboratory, "hard times" (for example, old cultures, deficient growth media) induce the vegetative cells to produce spores. Providing these sporulating cells with more optimal growth conditions will result in their spores' germinating as vegetative cells (Fig. 4.6). Note that bacterial spores, unlike spores of fungi, are not involved in multiplication. The rule is "one cell, one spore, one cell." The organism *B. anthracis* is a sporeformer. The spore's hardiness and infectivity make it a top candidate for biological warfare and terrorism (chapter 14). You may have seen news accounts of threatened terrorism employing the anthrax bacillus; some have proved to be hoaxes, but others, unfortunately, have been the real thing (chapter 14). The potential for biological warfare and terrorism with the use of this organism certainly exists.

The spore-forming anaerobic genus *Clostridium* includes three species that cause serious and potentially life-threatening diseases of humans. They are tetanus (lockjaw), botulism (a type of food poisoning), and gas gangrene (chapter 8). The stage for tetanus is set when you step on a rusty nail carrying spores and sustain a puncture wound, providing anaerobic conditions for germination of the spores. (It does not have to be a nail, nor does the object need to be rusty. Have you had your tetanus booster shot?) Those who like to do home canning should be aware that they are not protected by boiling the canning jars and other items that are used. Remember that spores are resistant to boiling; botulism is a potential menace.

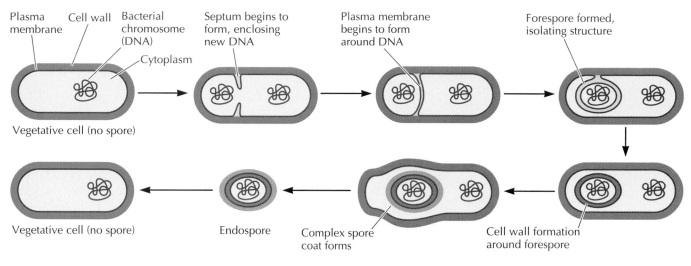

FIGURE 4.6   The spore cycle.

Appendages

Flagella

Flagella, which confer motility, are composed of a protein called **flagellin** and are found in some species of bacilli and cocci. One or more flagella are arranged in clumps or are distributed all over the cell; the particular arrangement of the flagella is constant for those species that have flagella (Fig. 4.7). Each flagellum is like a long helix and measures over 10 micrometers, many times longer than the length of the cell. They are extremely thin and difficult to stain and observe under a light microscope. The anatomy of flagella reveals that they are anchored by their **basal bodies** to the cell wall and membrane from which they extrude through the cell wall. The motility of live and unstained bacteria can be observed by suspending a drop of culture on a slide and examining this **"hanging drop"** under a microscope.

The flagellum rotates like a propeller and is driven by rings in the basal body. Descriptive and colorful terms, including runs, tumbles, twiddles, and tweedles, are used to describe its behavior. Receptor sites are present within the cell, resulting in directed movements—a process called **chemotaxis**—toward or away from a chemical stimulus. Some bacteria will "swim" toward an attractant (glucose) and away from repellents (acids). It's amazing to think they can detect minute changes in their environment. This is remarkable behavior. Motility appears to play a role in the ability of some disease-producing bacteria to spread through the tissues.

Pili

Pili are composed of the protein **pilin** and, like flagella, extrude through the cell wall and are present in many gram-negative species. They are shorter, straighter, and thinner than flagella. There may be only a single pilus or up to several hundred pili per cell; in some bacteria, including those that cause the sexually transmitted disease gonorrhea, pili serve as adhesins and anchor the bacteria to the mucous membrane of the vagina or the penis. This is an early step in the **colonization** and subsequent establishment of disease. Other pili, referred to as **sex pili,** function in genetic exchange by forming a bridge between cells through which DNA passes from

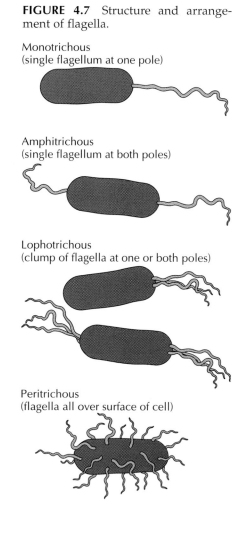

FIGURE 4.7 Structure and arrangement of flagella.

Monotrichous
(single flagellum at one pole)

Amphitrichous
(single flagellum at both poles)

Lophotrichous
(clump of flagella at one or both poles)

Peritrichous
(flagella all over surface of cell)

donor to recipient (Fig. 4.5); this is a major factor in the spread of antibiotic resistance in bacteria among different species.

THE GROWTH CURVE

For unicellular organisms, both procaryotic and eucaryotic, the terms **growth** and **multiplication** are used synonymously. In contrast, for multicellular organisms (consider the human), growth refers to an increase in the size of an individual as a unit and is the result of multiplication at the cellular level (mitosis) leading to an increase in the total number of cells in an individual, but not in the number of individuals. A population, microbial or otherwise, if allowed to grow unchecked, would take over the earth and, ultimately, exterminate all other forms of life. Can you imagine a universe composed entirely of *E. coli*? Abiotic factors, including availability of oxygen (O_2), temperature, and rainfall, and biotic factors, including predator-prey relationships and competition, constitute a system of checks and balances limiting population size. A course in ecology usually considers these factors as they apply to plants and animals, but the point is that these same concepts apply to microbes as well. (Microbial ecology is a specialized course that might be of interest to a microbiology major.) Interaction between bacterial species, as well as between multicellular plants and animals, is an important consideration. As Fleming observed in 1928, the product secreted by the mold *Penicillium chrysogenum* (later termed penicillin), which is inhibitory to a wide spectrum of bacteria, is a defense mechanism for the mold in the same sense that poisonous secretions of some jellyfish and the poisonous venom of certain snakes serve as defense mechanisms. These secretions function in predator-prey relationships.

Bacterial growth curves provide insight into the dynamics of population growth; they yield information as to what might be occurring when a streptococcus happens to land on the surface of your pharynx or when a food-borne pathogenic microbe is swallowed as a component of the mayonnaise in your tuna salad after being incubated at the beach in the cooler in which you neglected to put sufficient ice. Your pharynx, that tuna salad, and your intestinal tract allow multiplication of these disease-producing microbes. What takes place in terms of bacterial population size can be simulated by what occurs in a test tube. For instance, assume that you have a test tube of nutrient broth that meets the growth requirements for the streptococcus. Introduce a small drop of a streptococcal culture into that tube of clear broth, incubate the tube overnight, and the next morning, you will note that the tube is so cloudy that you can no longer see through it (Fig. 4.8). What happened? The answer is that the multiplication of those few cells in the original inoculum occurred at a phenomenal rate. The time that it takes a cell to undergo binary fission, producing two "new cells," is known as the **generation time.** Most of the common bacteria involved in human disease have short generation times. *E. coli,* for example, has a generation time of only 20 minutes, assuming optimal growth conditions (e.g., temperature and nutrients). At the other end of the spectrum is *Mycobacterium tuberculosis,* the causative agent of tuberculosis, with a generation time of about 18 hours; the spirochete *Treponema pallidum,* the causative agent of syphilis, has a generation time of 33 hours. Examples of generation times are given in Table 4.3.

If you started with a single *E. coli* bacillus, how many would you have in a 24-hour period during which 72 generations would have occurred?

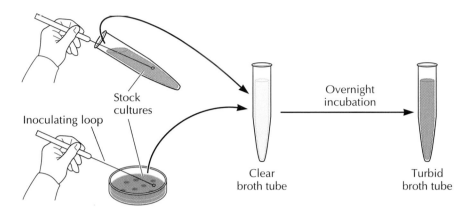

FIGURE 4.8   Overnight growth in a broth tube—clear to turbid.

After the first 20 minutes there would be two cells, 20 minutes later (elapsed time of 40 minutes) there would be four cells, and after another 20 minutes (elapsed time of 60 minutes) there would be eight cells, and so on. Mathematically, this can be expressed as 2^1, 2^2, and 2^3, respectively, with the exponents 1, 2, and 3 representing the number of generations. Therefore, at the end of 24 hours (72 generations) the total population would be 2^{72}, and at 48 hours (144 generations), the mass would be greater than that of the earth! Calculate the actual number, and you will be amazed. However, several factors, including exhaustion of nutrients, space, and accumulation of toxic materials, limit this growth.

Figure 4.9 illustrates a typical growth curve for *E. coli;* a similar curve would result with *M. tuberculosis* but the time scale would be in weeks because of the protracted generation time (three generations in 54 hours versus three generations per hour for *E. coli*). A growth curve of *E. coli* can be demonstrated by introducing a few organisms into a tube of nutrient broth. Samples are taken at frequent intervals, and population counts are performed by allowing the cells to develop into (macroscopic) **colonies** (Fig. 4.10), which can be counted. (One-hour intervals are convenient, except for the student who needs to culture-sit throughout the night and into the next day.) Based on these counts, a growth curve is constructed illustrating the four major phases.

TABLE 4.3   Examples of microbial generation times[a]

Microbe	Generation time[a]
Bacteria	
Escherichia coli	20 minutes
Staphylococcus aureus	30 minutes
Clostridium botulinum	35 minutes
Mycobacterium tuberculosis	18 hours
Treponema pallidum	33 hours
Protozoans	
Leishmania donovani	10 hours
Giardia lamblia	18 hours

[a]Times are approximate.

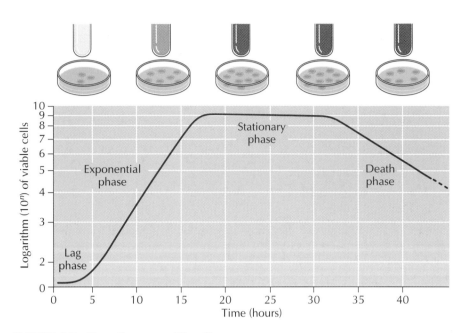

FIGURE 4.9 Growth curve of *E. coli*.

The Lag Phase

The lag phase is a "get ready for growth" stage and represents time of adaptation to the new environment; adaptation also occurs in actual infection. Under both circumstances, there may be an initial drop in the count because of adjustment and survival of only a part of the microbial population. In infection, immune mechanisms of the body more readily kill those cells that are "less fit" for survival, allowing for survival of the fittest, a nice example of Darwinism at the microbial level.

The Exponential (Logarithmic) Phase

During the "log" phase, there is an explosion of growth governed by the generation time. A doubling of the population occurs at the conclusion of each generation time, so that at the end of three generations (2^3) (assuming one cell initially and a generation time of 20 minutes), there will be a population of 8, which will double to 16 after completion of the fourth generation. Although this may not be particularly impressive, consider that at 4 hours (12 generations) the 16 cells which existed at the completion of the fourth generation have now reached a population of 2^{12}, or 4,096 cells. Now that is an impressive jump from 16 to 4,096 in 4 hours! What about after 10 hours (30 generations) or after 24 hours (72 generations)? There is a limit set by the conditions within the test tube, as there is in any functioning ecosystem. Sometime after about 10 or 12 hours, depending on the number of cells started with, adverse conditions set in and binary fission slows down, leading into the stationary phase.

The Stationary Phase

What are the adverse conditions that arrest the growth of a culture? Having started with a single species, there is no species competition, but there is a competition for the depleting supply of nutrients and O_2. Toxic materials, including carbon dioxide (CO_2), accumulate. Binary fission continues to

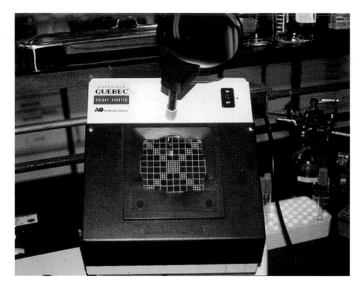

FIGURE 4.10 A colony counter. Agar plates with colonies are placed on the counting grid. A magnifying glass facilitates counting. (Author's photo.)

occur but at a rate about equal to the death rate, and thus the culture has now reached a plateau, so there is no net increase in numbers. The culture has about had it at this time, but as far as the bacteria are concerned, the worst is yet to come—the dreaded death phase!

The Death Phase

The adverse conditions in the test tube become more and more pronounced. The number of cells dying exceeds the rate of binary fission. It still might be possible to save the culture by simply transferring it into a larger tube with fresh medium or, alternatively, by taking a small sample of the culture which should contain some viable cells and inoculating a fresh tube of medium. If the organism is a sporeformer (*Bacillus* or *Clostridium*), spore formation may have resulted, and the cells may remain viable for years in the spore stage.

The dynamics of growth is of more than academic interest. What is the significance of all of this in terms of the natural state of infection? Think of the intestinal tract, the ear, an area of skin, or the urinary bladder as a test tube representing a particular ecological niche, each with its own distinctive environment. Certainly, the "climate" of the intestinal tract differs markedly from that of the middle ear. In most cases, symptoms of infection do not appear until there is a relatively large bacterial population. How do you apply this to the growth curve? Under most circumstances, the initial introduction of bacteria into the body would escape notice, and this would represent the lag phase. Within only several hours, depending on the particular organism and its generation time, a dramatic increase in the bacterial population takes place, and that is when you experience the symptoms associated with that infection; common examples are lethargy, abdominal pain, nausea, vomiting, diarrhea, headaches, and fever. Intervention, by taking an antibiotic or other medicines as a supplement to the immune system, is an attempt to interrupt the logarithmic phase, enter the stationary phase, and, finally, enter the death phase.

In most cases, the body's immune system responds to the presence of the microbial pathogen by calling into play defense mechanisms (chapter

FIGURE 4.11 Primary isolation. The specimen is streaked onto the surface of an agar plate, which is then incubated to allow development of colonies. (Author's photo.)

11) which attempt to destroy the invading microbes, thereby halting the continuation of the logarithmic phase. Should this not occur, the log phase continues, and the person will die of overwhelming infection.

CULTURING BACTERIA: DIAGNOSTICS

What happens to the throat swab that your physician uses to swab your pharynx, possibly when "strep throat" is suspected, or to the urine, fecal, or blood specimens that might be taken when bacterial infection is considered? (You cannot forget the throat swab! It is quite normal to gag, because, in fact, the swab does hit the spot that is responsible for your gag reflex. The idea is to get a sample from your pharynx and not from the roof of your mouth.) The purpose is to **culture,** that is to grow, the suspect organism so that it can be identified. The swab or a sample of the specimen is put into a tube of liquid nutrient broth and/or onto the surface of a **petri dish** containing nutrients and agar. Agar is a semisolid material with a consistency slightly more firm than that of Jell-O. This allows "primary" isolation (Fig. 4.11) and further testing to identify the organism. Growth on **nutrient agar** allows the production of colonies with recognizable and identifiable properties. Colony characteristics, including size and pigmentation (Fig. 4.12), are valuable clues in identification. For example, the streptococcus that causes strep throat is easily identifiable on primary cultures, assuming that the swab has been streaked on the surface of an agar plate enriched with sheep blood. The streptococci in question secrete a product (hemolysin) that destroys the sheep red blood cells surrounding the colonies. This results in clear zones because of the lysis of the red blood cells. This finding confirms the diagnosis of a streptococcal sore throat (Fig. 4.13). However, in most cases following primary isolation, further studies are necessary for identification. There are a wide variety of broth and agar media available for growth and identification, and the particular selection depends on the source of the specimen. After all, the bacteria that are commonly associated with intestinal tract infections are different from those that cause skin, urinary tract, or respiratory tract infections. Additional differential media are

FIGURE 4.12 Colony characteristics. Species of bacteria display characteristic colony morphology, including pigmentation, texture, and size. (Author's photo.)

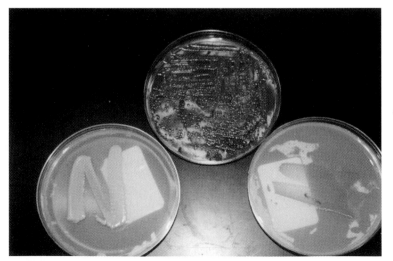

inoculated, and, after incubation at an appropriate temperature, the cultures are examined for characteristic responses (Fig. 4.14). Gram stains are performed to aid in the diagnosis. Further, the isolate is streaked onto the surface of an agar plate, antibiotic disks are added (Fig. 4.15), and the plate is incubated and examined after growth has occurred. Zones of inhibition (no growth) will occur around those disks to which the bacteria are sensitive, since the antibiotic leaches from the disk onto the moist agar surface. This is an important tool for the physician in prescribing an appropriate antibiotic; other factors, including toxicity, cost, and duration of therapy, are taken into account.

The process of identification may take only 24 hours, as in the case of streptococci; organisms with longer generation times may require extended incubation to achieve sufficient growth before moving on to the next stage. In the case of tuberculosis, bacterial identification takes several weeks because of the long generation time characteristic of *M. tuberculosis* (Table 4.3).

It is not always possible to culture the suspect organism, particularly if the infected individual is already receiving antibiotics. Further, as too frequently happens, the urine, throat swab, or blood specimen may have been left at room temperature too long, resulting in its drying out, or the time necessary for the organism to grow out may impose too long a delay. Fortunately, there is another avenue of diagnosis available. Microbes leave evidence of their presence by eliciting the production of specific antibodies by the host; these are telltale "footprints." Antibodies are produced in response to molecules, called antigens, that are a part of the microbe or that are secreted (toxins). Since antibodies are specific and will react only with the antigen that caused their production, the presence of antibodies helps to establish the identity of the bacteria involved. (Antigens and antibodies are covered in chapter 11.)

Medical laboratory microbiologists need to be highly trained to properly identify suspected disease-producing bacteria. The techniques are simple, but interpretation of the laboratory findings can be complicated. The microbiologist must be thoroughly familiar with the normal bacterial inhabitants of the body to distinguish suspected disease producers. Many directors of today's clinical microbiology laboratories have a Ph.D. or M.D. degree.

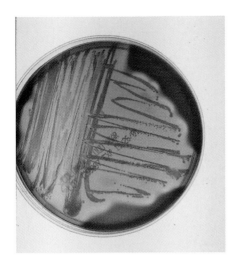

FIGURE 4.13 Diagnosis of strep throat. The streptococci responsible for strep throat are easily identified. A swab of the throat is streaked onto the surface of an agar plate enriched with sheep blood. The presence of clear zones around the areas of growth establishes the diagnosis. (Author's photo.)

FIGURE 4.14 Differential media. A variety of media in test tubes and in petri dishes are used to determine properties of specific bacteria that are useful in identification. (Author's photo.)

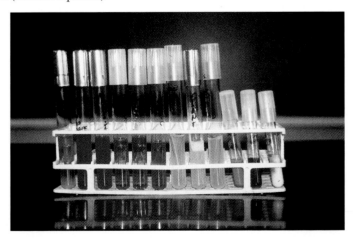

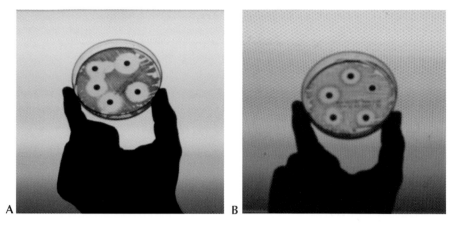

FIGURE 4.15 Determination of antibiotic sensitivity. The specimen is streaked onto the surface of an agar plate, antibiotic disks are added, and the plate is incubated. Zones of inhibition (no growth) around a disk indicate the bacteria are sensitive. Each disk is coded with the name and concentration of the antibiotic. Note that the spectrum of activity varies for each culture. (Author's photos.)

ODDBALL (ATYPICAL) BACTERIA

Included in the "oddball" category are three important groups of bacteria, each of which possesses distinctive properties distinguishing it from the "typical" bacteria and from each other. These groups are significant disease producers in humans (chapter 8). They are (i) **mycoplasmas,** (ii) **chlamydiae,** and (iii) **rickettsiae;** their properties are summarized in Table 4.4.

Mycoplasmas

Mycoplasmas are described as bacteria without cell walls; they are the "smallest of the small." Perhaps you have heard the term "walking pneumonia"; the causative agent is *Mycoplasma pneumoniae.*

Chlamydiae

The chlamydiae are slightly larger than the mycoplasmas and can be seen with a light microscope. These coccoid bacteria are obligate intracellular parasites, a property shared with rickettsiae and viruses. Transmission is from person to person. Chlamydiae are responsible for a variety of diseases, including urethritis (inflammation of the urethra), trachoma (a common cause of blindness), and lymphogranuloma venereum (a sexually transmitted disease).

Rickettsiae

Rickettsiae, which are rod-shaped bacteria, are the largest of the three groups of atypical bacteria and are easy to see with a light microscope. With the exception of Q fever, diseases caused by these organisms are acquired through the bites of arthropods, examples of which are mosquitoes, fleas, ticks, and lice (chapter 7). Like the chlamydiae, they are obligate intracellular parasites and are commonly grown in the yolk sac of (live) chicken embryos for laboratory study. According to some historians, typhus fever (not typhoid) defeated Napoleon's army in the Russian invasion of 1812. Perhaps you have read *The Diary of Anne Frank,* the story of a teenage Jewish girl who hid from the Nazis with her family in a secret attic in Amsterdam between 1942 and

TABLE 4.4   Comparison of mycoplasmas, chlamydiae, and rickettsiae

Bacterial group	Size	Cell wall	Obligate intracellular parasite
Mycoplasmas	0.3–0.8 micrometers	Absent	No
Chlamydiae	0.2–1.5 micrometers	Present	Yes
Rickettsiae	0.8–2.0 micrometers	Present	Yes

1944. The family was betrayed and sent to Bergen-Belsen, a German concentration camp; Anne died there of typhus fever at the age of 15, shortly before the camp was liberated by Allied forces. The Anne Frank House (Fig. 4.16), located in the center of Amsterdam, remains as a memorial.

OVERVIEW   Bacteria are a distinct group of unicellular microbes dating back to antiquity and were, possibly, the beginnings of life. Although they are small, they are not simple. Bacteria display a diversity of shapes and patterns of arrangement. The majority are bacilli, cocci, and spirilla. Cocci are subdivided into streptococci, staphylococci, diplococci, and tetrads.

Bacteria have a complex anatomy consisting of an envelope, cytoplasm, and appendages. The envelope includes the capsule, the cell wall, and the cell membrane. The cytoplasm is the region of the cell within the cell membrane and includes a nucleoid, plasmids, and spores. Appendages are flagella and pili, both of which extrude through the cell wall. The capsule, plasmids, spores, flagella, and pili are not present in all bacteria.

Bacteria exhibit considerable diversity in generation time—the time it takes a cell to undergo binary fission resulting in two new cells. Under optimal laboratory conditions, the generation time for *E. coli* is 20 minutes, whereas for *M. tuberculosis,* the generation time is about 18 hours. The four major phases of the curve are the lag phase, exponential growth phase, stationary phase, and death phase. Bacterial growth curves provide insight into the dynamics of bacterial growth that occur in an infected individual.

The fact that most bacteria can be readily cultured in the laboratory facilitates identification of the causative organism, allowing appropriate treatment.

Mycoplasmas, rickettsiae, and chlamydiae are three groups of atypical bacteria, each of which includes pathogens. ∎

FIGURE 4.16   A memorial statue of Anne Frank in Amsterdam. This young Jewish girl died of typhus fever at the age of 15 in Bergen-Belsen, a German concentration camp. (Author's photo.)

SELF-EVALUATION

PART I Choose the *single* best answer. Questions 1 to 4 are based on the following key:

    a. capsule   **c.** cell membrane
    b. cell wall   **d.** pili

1. Accounts for Gram stain reaction

2. Responsible for gatekeeper function

3. A deterrent to phagocytosis

4. Important in adhesion to host cells

5. Assume that the organism *Robertus krasnerii* has a generation time of 20 minutes and that you are starting with one cell. How many generations will it take to achieve a population of 2,048 cells?
    a. cannot determine   **c.** 11
    b. 10   **d.** 12

6. In examining a petri dish culture with antibiotic sensitivity disks, the greatest sensitivity would be demonstrated by the disk with the
    a. most growth in the vicinity of the disk
    b. smallest zone of no growth
    c. largest zone of no growth
    d. no zone of inhibition

7. The property of "hardiness" is associated with
    a. flagella   **c.** capsule
    b. cell wall   **d.** spores

8. Which group of bacteria is unique in lacking a cell wall?
    a. procaryotes   **c.** rickettsiae
    b. chlamydiae   **d.** mycoplasmas

PART II Fill in the following.

1. Cocci which tend to form chains are called

_____.

2. Corkscrew-shaped spiral bacteria are called

_____.

3. A defense mechanism of the body is the ability of certain cells to engulf and destroy bacteria. This process is known as _____.

4. Name a spore-forming genus. _____

PART III Answer the following.

1. Draw and label a "composite" bacterial cell. Indicate what structures are not found in all bacterial cells.

2. The term "bugs" is conventionally used to describe all microbes. If you are ill and diagnosed with a "bug," what are the limitations of this diagnosis?

3. The chances are that *Clostridium tetani* will not cause tetanus when an individual sustains a cut but might when a puncture wound occurs. Why is this the case?

5

VIRUSES

PREVIEW Viruses are subcellular agents that were responsible for a variety of diseases in humans long before they were recognized as biological agents. Familiar diseases such as smallpox, poliomyelitis, and influenza are but a few examples. It was not until the 1940s that viruses were described in terms of their biological and chemical properties. Bacteriophages are viruses that infect bacterial cells; these viruses have served as models for understanding the dynamics of host-parasite relationships that occur in human viral diseases. Viruses are obligate intracellular parasites, meaning they must gain access to the cells to replicate. Viruses are particularly unique in that they have either DNA or RNA, whereas cells contain both DNA and RNA. Their nucleic acid is wrapped in a protein coat, and some have an envelope around the coat. A generalized cycle of viral replication follows the sequence of adsorption, penetration, replication, assembly, and release, but the strategies differ depending on the specific virus. Because of their obligate intracellular parasitism, cell culture and fertile chicken eggs are used to grow viruses. The diagnosis of a specific viral disease is usually based on the patient's symptoms rather than on isolation of the virus. The use of bacteriophages to treat bacterial infections, particularly since the increase in antibiotic-resistant bacteria, has reemerged as a possible therapy.

VIRUSES AS INFECTIOUS AGENTS

What biological agents are smaller than a cell, not considered to be alive, able to be transmitted from person to person, potentially fatal, and, historically, have caused the largest recorded outbreak of disease on a worldwide scale (pandemic) in the history of humankind? The title of the chapter is a giveaway to the answer to this riddle: viruses. The **pandemic** referred to is the influenza (Spanish flu) catastrophe of 1918, which killed 25 million people around the world in only 1 year. Acquired immune deficiency syndrome (AIDS) is a pandemic viral disease that began in the latter part of the 20th century and is devastating worldwide, particularly in Africa. According to the director of the United Nations Children's Fund (UNICEF), "by any measure the HIV-AIDS pandemic is the most terrible undeclared war in the world with the whole of sub-Saharan Africa a killing field."

As described in chapter 2, viruses are subcellular (simpler than a cell) and, therefore, can replicate (their sole claim to life) only when circumstances result in their gaining access to a cell. They have no life outside the cell; the host cell is taken prisoner and turned into a virus-producing machine. This is the basis for describing viruses as obligate intracellular parasites. They literally need to "get a life." On the positive side, the fact that viruses take up intracellular existence is being exploited in genetic engineering (chapter 3) by employing "tamed" viruses as vehicles for the delivery of "good" genes into cells to replace defective or nonexistent genes.

You probably have experienced numerous bouts of common viral infections, including colds, chicken pox, influenza, and gastroenteritis. Human immunodeficiency virus (HIV), Ebola virus, hantavirus, and West Nile virus cause viral diseases that have been widely publicized (In the News 5.1). In 1997 a potential outbreak of influenza in Hong Kong, dubbed by the media the bird-flu virus, caused a scare around the world and resulted in the slaughtering of over a million chickens as a measure to halt the spread. Fortunately, the scare proved to be unwarranted. In October 1999, an outbreak of 30 cases with at least five deaths caused by a mosquito-borne virus occurred in New York and was, initially, presumed to be due to the St. Louis encephalitis virus. The real culprit, however, turned out to be the West Nile virus, which had never before been seen in the Americas. How the virus came to New York remains an open question at this time.

IN THE NEWS 5.1

Six Die of Viral Pneumonia at Nursing Home

Two Crows Killed by West Nile Virus in Somerset

Health Dept. issues alert urging precautions against mosquitoes

China Says It Has an AIDS "Epidemic"
New Cases Increase by Two-Thirds over Last Year

WHO aims to eradicate polio worldwide

Health Officials Increase Efforts To Fight Rabies

Viruses in the News

Some scientists guess that it was carried on an airplane from elsewhere in the world, possibly from Russia, where the disease is a problem. Thinking back to the material presented in chapter 1, you should recognize this as an example attesting to the occurrence of new and emerging infections.

Smallpox is a terrible viral disease which decimated large segments of the population. The disease was described as early as about 10,000 B.C. Examination of the body of Pharaoh Ramses V, who died in 1157 B.C., revealed pustulelike scars on the face, neck, and shoulders, which appeared to have been caused by smallpox. Shamefully, the smallpox virus was employed in biological warfare long before its identity was known. English troops invading North America, under the guise of being Good Samaritans, gave the Indians blankets. Little did the Indians know that the blankets contained the virus that caused smallpox; the results are too easily imagined. Smallpox was the first disease for which immunization was available, thanks to the pioneering work of Edward Jenner in 1796, and is the first disease to be eradicated from the face of the earth. The word "eradicated" is to be emphasized as compared with the word "controlled." The irony is that this triumph renders the world susceptible to the threat of smallpox, because the virus still exists and could find its way into the hands of terrorists.

The term "virus" can be found in the mid-1800s writing of Pasteur and others who preceded him long before the germ theory of disease was established and the biology of viruses was known. "Virus" means "poison" or "slime" and has a negative connotation. By the 1940s, viruses had been well described, and the term now has a scientific meaning. In more recent years, "virus" has become part of computer jargon and indicates something that is easily spread (infectious) and destroys data.

Based on the work of the Russian scientist Dmitri Ivanovsky in 1892, viruses were described as "filterable." Ivanovsky filtered extracts of tobacco leaves with symptoms of tobacco mosaic disease through filters which were known to retain bacteria. He then injected these extracts into healthy tobacco plants and observed symptoms of tobacco mosaic disease (Fig. 5.1). It was clear that the infectious agent was, indeed, smaller than bacteria. Several years later, the Dutch microbiologist Martinus Beijerinck also demonstrated that tobacco mosaic disease was caused by a filterable agent. The term "filterable" was eventually dropped. A breakthrough occurred in

FIGURE 5.1 Discovery of "filterable" viruses.

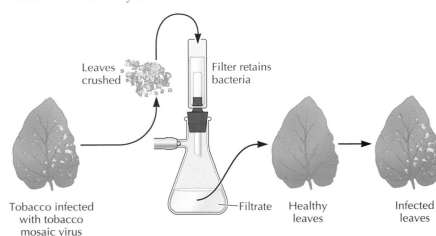

Leaves crushed

Filter retains bacteria

Tobacco infected with tobacco mosaic virus

Filtrate

Healthy leaves

Infected leaves

1935 when Wendell M. Stanley, an American biochemist, purified and crystallized the tobacco mosaic virus, opening the door for more sophisticated examination of a purified virus in an effort to further elucidate its structure. The invention of the electron microscope, at about the same time, allowed viruses to be seen.

All cells are potential hosts for viruses. Even *Escherichia coli* and other bacteria serve as hosts. Bacterial viruses are known as bacteriophages (usually shortened to "phages"). Phages, because of the ease of growing the bacteria that serve as their hosts, have served as models for studying animal and plant viruses. Our current knowledge of these viruses has been largely gained by studying phages.

VIRUS STRUCTURE

Among microbes, viruses are the smallest. Bacteria are described as microscopic, meaning they can be seen only with a light microscope. Viruses, however, are in a different realm, the realm of the electron microscope, and they range in size from about 20 to 350 nm. Well over 1,000 phage particles can fit into an *E. coli* cell.

As viewed through an electron microscope, viruses are simple in structure and may consist only of a nucleic acid wrapped in a protein coat (Fig. 5.2), whereas others may have an envelope around the coat. The larger viruses (e.g., smallpox virus) overlap the size of the smaller bacteria, and, although they contain a few enzymes, they are obligate intracellular parasites. A complete viral particle is referred to as a **virion.**

Nucleic Acids

Viruses are unique in that they have either RNA or DNA genomes, but never both. Cells, whether procaryotic or eucaryotic, contain both DNA and RNA. In the RNA viruses, the RNA serves as the repository for genetic information. For example, the influenza virus is an RNA virus that replicates to produce more influenza viruses. Hence, the genetic information must be encoded in the RNA. Further, some viruses are unique in that their genome is single-stranded DNA (ssDNA); others have double-stranded RNA (dsRNA). Other than in these viruses, DNA is double stranded and

FIGURE 5.2   Basic virus structure.

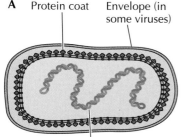

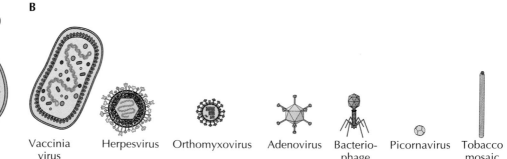

A   Protein coat   Envelope (in some viruses)

Nucleic acid (4 categories):
1. Double-stranded DNA
2. Single-stranded DNA
3. Double-stranded RNA
4. Single-stranded RNA

B

Vaccinia virus   Herpesvirus   Orthomyxovirus   Adenovirus   Bacteriophage   Picornavirus   Tobacco mosaic virus

RNA is single stranded. Hence, in terms of nucleic acid content, there are four categories of viruses: dsDNA, ssDNA, dsRNA, and ssRNA. This fact is important for the classification of viruses (Table 5.1), along with other properties. Small viruses have only a few genes, while larger viruses have a few hundred genes; *E. coli,* by comparison, has approximately 4,000 genes, and human cells are estimated to have about 38,000 genes, possibly many more.

Protein Coat

The protein coat is called the **capsid,** and the term **nucleocapsid** refers to the nucleic acid genome plus the protein coat. The coat, in turn, consists of protein units called **capsomeres.** Three arrangements of these capsomeres have been described: **helical, polyhedral,** and **complex.** Examples are illustrated in Fig. 5.3.

Helical viruses consist of a series of rod-shaped capsomeres which, during assembly, form a continuous helical tube containing the nucleic acid characteristic of the particular virus. The tobacco mosaic virus is an example of a helical virus. Imagine that you are eating a turkey wrap; the shell is like the protein coat, and the turkey is the nucleic acid.

In most polyhedral viruses, the capsomeres form **icosahedrons**—three-dimensional, 20-sided, triangular structures which confer a geodesic-appearing shape to the virus. These viruses, based on their appearance under the electron microscope, are also described as spherical viruses. Triangles are used by engineers to create geodesic structures. If you have been to Epcot Center at Disney World in Florida, you could not escape viewing Spaceship Earth, a giant geodesic dome at the entrance. At the Antarctic research station at the South Pole, a geodesic dome houses one of the facilities. It appears that viruses knew what they were doing when it came to fashioning a protective environment for their nucleic acid.

Some of the commonly studied bacteriophages are complex viruses; they consist of a polyhedral head, a helical tail, and tail fibers which serve for attachment to the host cell. These viruses have an intricate anatomy and resemble a spacecraft. There are still other viruses that are not easily categorized as either polyhedral or helical and are considered atypical or complex viruses.

Viral Envelopes

Some viruses, during the process of emerging from the host cell, acquire a piece of the host cell's plasma membrane, which constitutes an additional layer covering the capsid. These viruses are called **enveloped viruses.** The envelopes are modifications of the host cell membrane in which some of the membrane proteins are replaced with viral proteins. Some of these appear

TABLE 5.1   Virus classification based on nucleic acid composition

Nucleic acid	Virus(es) (disease)
dsDNA	Human papillomavirus, Epstein-Barr virus, smallpox virus, adenovirus, herpes simplex virus, varicella-zoster virus (chicken pox and shingles)
ssDNA	Parvovirus B19 (possibly slapped-cheek disease)
ssRNA	Hepatitis A virus, poliovirus, Norwalk virus, rubella virus, Ebola virus
dsRNA	Rotavirus, Colorado tick fever virus, reovirus

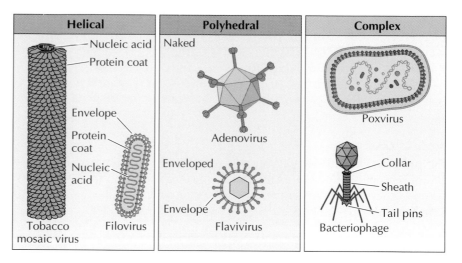

FIGURE 5.3   Representative virus structures.

as **spikes** because they protrude from the membrane and are important for the attachment of viruses to host cells. The AIDS virus (HIV) has spikes which attach to strategic white blood cells (T lymphocytes) of the immune system, leading to serious damage to the individual's ability to fight infection. Viruses lacking an envelope are referred to as **naked** or **nonenveloped** viruses (Fig. 5.3).

VIRAL CLASSIFICATION

Superficially, viruses are classified as plant, animal, or bacterial in accordance with the host organism that they infect. Clinically, they may also be classified according to the organ or organ system involved (Table 5.2), for example, **dermotropic** (skin), **viscerotropic** (internal organs), or **pneumotropic** (respiratory system). This latter classification, however, is limited by the fact that some viruses can infect more than a single organ or organ system. Current classification schemes are based on the biology of the virus, starting with their nucleic acid content. Despite the fact that viral taxonomy (classification) is complex, a workable scheme is based on the type of nucleic acid (ssDNA or dsDNA or ssRNA or dsRNA), capsid structure (helical or polyhedral), number of capsomeres, and other chemical and physical

TABLE 5.2   Clinical classification of viruses

Category	Tropisms (tissue affinities)	Diseases
Dermotropic	Skin and subcutaneous tissues	Chicken pox, shingles, measles, mumps, smallpox, rubella, herpes simplex
Neurotropic	Brain and central nervous system tissues	Rabies, West Nile virus encephalitis, arboviral encephalitides (eastern equine encephalitis, St. Louis encephalitis), polio
Viscerotropic	Internal organs	Yellow fever, AIDS, hepatitis A and B, infectious mononucleosis, dengue fever
Pneumotropic	Lungs and other respiratory structures	Influenza, common cold, respiratory syncytial disease

properties. Well over 1,400 viruses have been described, and each has been assigned to one of 61 recognized families. Viruses that affect humans are found in 21 of these families. Specific viral diseases are described in chapter 9.

VIRAL REPLICATION

The fact that viruses are obligate intracellular parasites makes them difficult to grow and study. While it is easy to grow bacteria in lifeless medium in test tubes or on petri plates, the culture (growth and replication) of viruses requires the presence of living cells; these techniques will be described later in this chapter. This is particularly complicated in the case of animal and plant viruses but, nevertheless, is routinely done in laboratories specialized in these techniques. Bacteriophages, on the other hand, are much easier to study since their host cells are bacteria, which can be readily grown. The system that has been most extensively investigated and that has served as a model for animal viruses is the T4 bacteriophage (a complex dsDNA phage) and *E. coli*. The T4 viruses enter the *E. coli* cell and utilize the cell's metabolic machinery to form a large number of new T4 viruses, which are released by causing the lysis of their bacterial hosts. These new virions are now capable of infecting other *E. coli* cells.

The replication of T4 in *E. coli* is a model for the replication of viruses in general, although there is considerable variation in the strategies demonstrated by specific viruses. The replication cycle is divided into five stages (Table 5.3 and Fig. 5.4): **adsorption, penetration, replication, assembly,** and **release.** Depending upon the particular virus, there is variation in the time required to complete the cycle. For example, the length of the cycle for poliovirus is only 6 hours, whereas it is 36 hours for a herpesvirus. The replication cycle is of more than academic interest. As you study the details, remember that this process occurs in your cells when you have a viral infection.

Adsorption

As an obvious prelude to replication, the virus must penetrate the cell, a process which first requires adsorption onto the host cell surface. Bacterial

TABLE 5.3   Generalized viral replication cycle[a]

Stage	Description
Adsorption (attachment)	Viruses attach to cell surface receptor molecules by spikes, capsids, or envelope.
Penetration	Entire viral particle or only nucleic acid enters via endocytosis or by fusion with cell membrane.
Replication	Process is complex, and details are dependent upon particular viruses and their nucleic acid structure. Replication may occur in the nucleus or cytoplasm; viral nucleic acid replicates, and genes are expressed, leading to production of viral components.
Assembly	Components are assembled into mature viruses.
Release (exit)	Viruses are extruded from host cell by budding (HIV) or lysis of host cell membrane.

[a]Specific strategies vary with particular viruses.

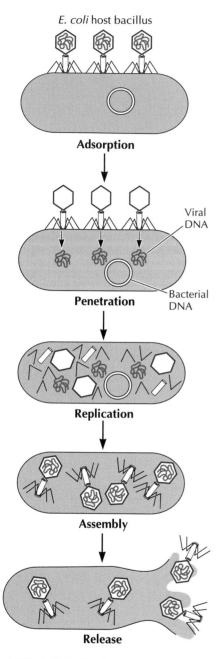

FIGURE 5.4 Replication cycle of a bacterial virus.

cells have receptors on their cell walls which serve as sites of attachment. Animal cells lack cell walls, and receptor molecules are embedded within their cell membranes. Viruses do not seek out a particular host cell; it is a matter of chance encounter. To view it another way, the virus must "dock" with a receptor molecule onto which it fits (Fig. 5.5). Consider a piece from a jigsaw puzzle; it can only fit into an appropriate complementary shape. Specificity is thereby required and establishes host range. For example, *E. coli* strain B serves as the host for a particular phage, called T4, whereas *E. coli* strain C serves as the host for phage φX174. T4 does not infect C, nor does φX174 infect B, despite the fact that these two *E. coli* strains are virtually identical. Yet, there is a difference between these strains in the receptor molecules, which are specific for the tail fibers of either T4 or φX174. Hence, the host for T4 is *E. coli* B, and the host for φX174 is *E. coli* C. Specificity is always the name of the game to some degree, but it may not always be as "fussy" as in the *E. coli*-phage system.

In animal viruses, too, specificity establishes host range. For example, HIV has spikes on its surface which dock with receptor (CD4) molecules on the surface of a particular type of white blood cell (T lymphocyte), allowing, ultimately, the replication cycle of the virus to be completed. During this process the cells are destroyed, rendering the immune system severely impaired, as discussed in chapter 11.

Penetration

At this point, the virus is positioned on the surface of the host cell. This, in itself, is of little consequence since viruses are obligate intracellular parasites. Therefore, one way or another, penetration of the host cell is necessary for replication. Keeping in mind that it is the viral nucleic acid, whether ssDNA, dsDNA, ssRNA, or dsRNA, that carries the genetic message, the critical factor is that the nucleic acid must enter the cell (Fig. 5.6). Perhaps the most bizarre mechanism of penetration is observed in some of the phages, including the T4 phage, of *E. coli* B. This phage attaches to the bacterial cell wall by phage tail fibers and injects the nucleic acid across the cell wall in a fashion resembling that of a miniature syringe because of the contraction of the tail sheath. The capsid remains on the outside of the cell and is of no further consequence.

FIGURE 5.5 Virus "docking" with receptor molecules.

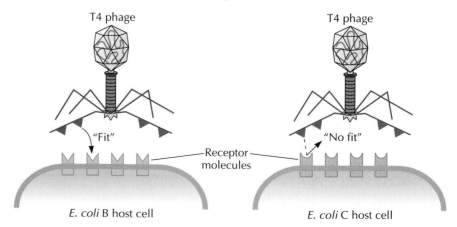

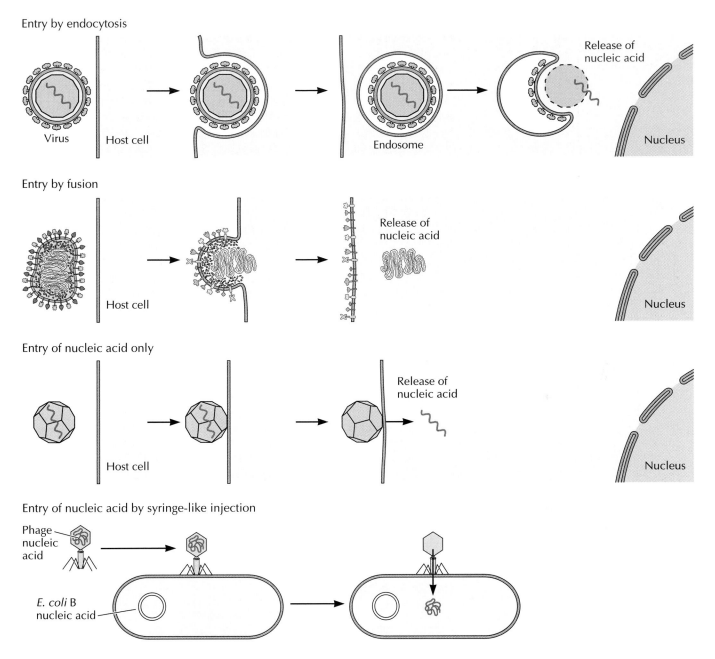

Entry by endocytosis

Virus   Host cell   Endosome   Release of nucleic acid   Nucleus

Entry by fusion

Host cell   Release of nucleic acid   Nucleus

Entry of nucleic acid only

Host cell   Release of nucleic acid   Nucleus

Entry of nucleic acid by syringe-like injection

Phage nucleic acid

E. coli B nucleic acid

FIGURE 5.6 Strategies of viral penetration of cells.

In the case of animal viruses, the nucleic acid is not injected into the host cell. Rather, some enveloped viruses enter the cell by a process called **endocytosis** in which the complete virion is engulfed by the host cell and contained within a **vesicle**. This is the case for smallpox virus and other poxviruses. In other enveloped viruses, **fusion** of the envelope with the cell membrane occurs, and the nucleocapsid enters, for example, in the mumps, cold, and measles viruses. Still a third strategy exists for some of the naked viruses, including the poliovirus, in which only the nucleic acid component enters the cell's cytoplasm. In the first two scenarios (i.e., endocytosis and fusion), the nucleic acid must be uncoated from its site within the protein coat for the cycle to continue. This process of uncoating results from the action of enzymes of the host cell or of the viruses.

Replication

The replication stage is directed by the nucleic acid of the virus. The details are a function of which nucleic acid the virus contains; the process is complicated and involves many steps. The important point is that, ultimately, viral components are synthesized within the host cell, again reflecting the obligate intracellular nature of viruses. Genetic information, whether stored in RNA or DNA, is ultimately expressed (translated) into viral protein molecules specific for the particular virus. Think of the host cell as a factory for viral components analogous to an automobile factory. Instead of making carburetors, dashboards, airbags, and steering mechanisms, the products are tail fibers, capsids, nucleic acids, spikes, and envelopes. The blueprint is the nucleic acid of the virus which penetrated the cell. Depending upon the specific virus, production may occur in the cytoplasm, nucleus, or possibly both production sites. Generally, nucleic acids are synthesized in the nucleus of the host cell, while capsids and other structures are made in the cytoplasm and then move into the nucleus.

Assembly

Continuing with the analogy of the automobile plant, now that the parts have been made, the virus is assembled into a functional structure in a manner similar to that of a production line (Fig. 5.7).

Release

Mission accomplished! At this point, the host cell is teeming with newly formed and complete virions, all potentially infective for host cells. The mechanism of release of virions varies as a function of the specific virus. In some cases, release results in the death of the host cell, whereas in other cases death is not the outcome. For example, with some of the phages (e.g., T4), the release of new phage particles, referred to as the burst, is the result

FIGURE 5.7   Viral assembly.

Assembly of an automobile

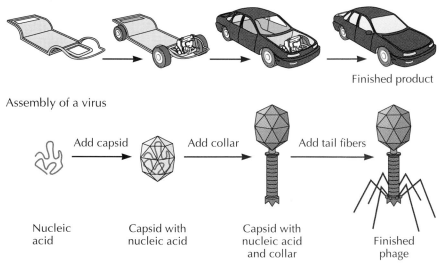

Finished product

Assembly of a virus

Add capsid   Add collar   Add tail fibers

Nucleic   Capsid with   Capsid with   Finished
acid   nucleic acid   nucleic acid   phage
                                      and collar

of the enzymatic splitting open (lysis) of the cell membrane, virtually the "kiss of death" (Fig. 5.8). As many as a few hundred new phage virions may be released. The entire process from adsorption to burst averages about 20 to 40 minutes in phages.

How is release accomplished in animal viruses? Nonenveloped animal viruses are released from the cell by a process similar to the release of some phages involving lysis and death of the host cell.

Release is somewhat more complicated in the case of enveloped viruses (Fig. 5.8). Generally, prior to release, the expression of viral genes brings about the production of viral envelope components, a principal one being spikes, located in the host cell's membrane. **Budding** or **extrusion** is, essentially, the reverse of endocytosis. The virus pushes its way through the membrane and pinches off a piece of the membrane-spike complex, which now envelopes the nucleocapsid. Hence, the envelope consists of elements encoded by viral genes and by host cell membrane genes. Unlike lysis, budding is a gradual and continuous process and does not necessarily kill the cell. Ultimately, cell death is due to the takeover of the cell's machinery and accumulation of viral components. The number of mature virions released is enormous, ranging from a few thousand for poxviruses to over 100,000 for poliovirus. Each of these virions is an infective particle capable of invading a healthy cell and causing damage symptomatic of the particular virus. All it takes is a few viral particles, which, by chance encounter with host cells bearing appropriate receptor molecules, initiate infection and within a short time cause symptoms, possibly death. Selected viral diseases are the subject of chapter 9.

FIGURE 5.8 Mechanisms of release of bacteriophages and animal viruses.

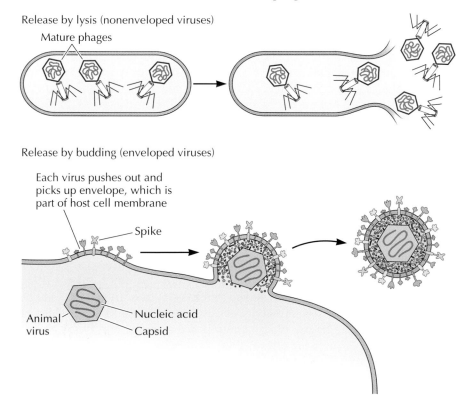

Release by lysis (nonenveloped viruses)

Mature phages

Release by budding (enveloped viruses)

Each virus pushes out and picks up envelope, which is part of host cell membrane

Spike

Animal virus

Nucleic acid

Capsid

As stated earlier, all viruses, including phages, follow a similar sequence of events in their replication cycle (adsorption, penetration, replication, assembly, and release). The strategies differ; this is particularly true in the replication stage and depends on the nucleic acid content. It is remarkable that the minute amount of viral nucleic acid that gains access to the host directs the events leading to new viruses. Whatever the peculiarities of the strategy, the significant point is that viral replication proceeds at a rapid rate; the production of new virions numbers in the thousands. Considering that each virus is infective, allowing for ongoing cycles involving more and more host cells, it is not difficult to understand the overwhelming effects that occur when you acquire a viral infection. Antibacterial drugs (antibiotics) are ineffective against viruses, because, unlike bacteria, viruses lack a cell wall and other structures that are targeted by antibiotics. There are some antiviral drugs, but they are not as effective against viruses as antibiotics are against bacteria. These drugs are described in chapter 12.

HOST CELL DAMAGE

Can you imagine what it must be like in a host cell that is invaded by a virus? It would probably be an awesome experience to witness the devastation. In many cases, the outcome is death or at least cell damage. The virus leaves its specific imprint, known as the **cytopathic effect (CPE),** evidenced by cell deterioration. The particular CPE, viewed under a microscope, is helpful in the identification of many viral infections. In some cases, this is applicable to tissue specimens taken directly from an infected individual or at autopsy. Further, in **cell culture,** a method of growing viruses to be described in the next section, the virus-infected cells may show a characteristic CPE.

Rabies is a viral disease that can be transmitted from certain wild animals, including skunks, raccoons, bats, and foxes, directly to humans or dogs, or from dogs to humans. Consider the situation when a raccoon whose bizarre behavior is suggestive of rabies bites a dog or a human. When the raccoon is shot or captured and killed, brain tissue is examined for the presence of **Negri bodies**, a CPE that is specific to rabies and therefore diagnostic. (An animal suspected to be rabid should not be shot in the head.) Other CPEs are diagnostic for particular viral diseases, including smallpox, herpes, and the common cold.

CULTIVATION (GROWTH) OF VIRUSES

Cultivation of bacteriophages is simple, based on the fact that easily grown bacterial cells serve as convenient host cells and meet the requirements for obligate intracellular parasitism. However, the growth of animal viruses is hindered by the necessity to utilize animal cells as host cells. In earlier years, this presented a serious problem for virologists. Viruses were studied by injecting them into laboratory animals and observing the resulting symptoms; animals were euthanized so that the effects of infection could be observed at the organ and tissue levels, particularly the development of characteristic CPEs. The use of live animals is expensive and poses ethical issues.

In certain circumstances, animals are still employed, but other methods have, to a large extent, replaced them. These methods are the use of **embryonated (fertile) chicken eggs** (Fig. 5.9) and cell culture (Fig. 5.10).

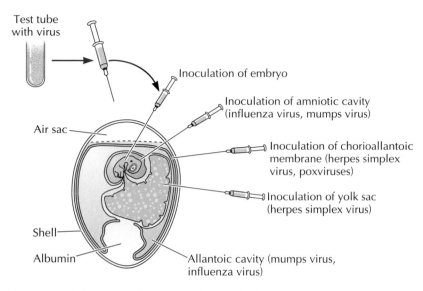

FIGURE 5.9 Cultivation of viruses in fertile chicken eggs.

Embryonated (Fertile) Chicken Eggs

Some viruses can be grown in embryonated chicken (or other) eggs, a discovery made in the 1930s. The fertile chicken egg is a convenient and relatively inexpensive alternative to inoculation of live animals. The embryo is enclosed within the protective eggshell and, further, is covered by the shell membrane, providing a sterile and nutrient-rich environment for the developing embryo. There are several different sites and membranes within the egg, each representing a unique ecological niche which supports the growth of particular viruses. With the use of sterile techniques, a hole is drilled through the shell, and the viral preparation is injected into the selected cavity, onto the appropriate membrane, or into the embryo itself, depending on the particular virus or suspected virus under study. Viral multiplication may be detected by the death of the embryo, pocks or lesions on membranes, or abnormal growth of the embryo. The viruses are recovered by harvesting fluids from the egg.

FIGURE 5.10 Cultivation of viruses in cell culture.

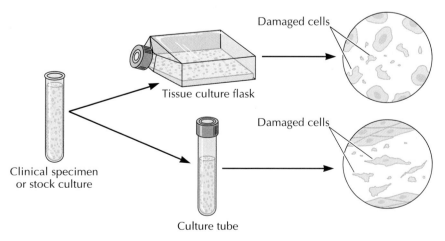

For the manufacture of flu vaccine, influenza virus is grown in embryonated chicken eggs. Hence, before receiving the vaccine, you may be asked whether you are allergic to eggs. Since the vaccine may contain egg proteins, you may be advised not to get a flu shot.

Cell Culture

The application of cell culture techniques to the growth of viruses was implemented in the early 1950s. These techniques opened the door to the advancement of the study of animal viruses since they could now be grown inexpensively, rapidly, and on a large scale. Cells are cultured in a variety of sterile laboratory containers, including flasks, tubes, and petri dishes, containing a complex mixture of nutrients appropriate to the particular cells under culture. The viral suspension is then introduced into the culture vessel; viral growth is detected by microscopic observation to detect CPEs and lysis of cells as evidence of viral infection and replication.

John Enders, Thomas Weller, and Frederick Robbins, in the early 1950s, discovered that poliovirus could be induced to grow in various types of cell culture, making possible the large-scale growth of the poliovirus and leading to the development of the Salk and Sabin polio vaccines. Previously, the virus could be cultured only in live monkeys, a fact which posed severe limits on polio research. Enders, Weller, and Robbins shared the 1954 Nobel Prize for their pioneering work. This is an excellent example of the potentially significant payoff of basic research for medicine and health (Box 5.1).

BOX 5.1   Enders, Weller, and Robbins Awarded 1954 Nobel Prize for Poliovirus Cultivation

There are many examples in biology in which basic research has led to major advances in biology and medicine. The research of John Enders, Thomas Weller, and Frederick Robbins is one such case. Their discovery that poliovirus could be grown in test tube cultures of a variety of nonneural tissue opened the door for research on virus vaccines in addition to polio. Previously, poliovirus could be grown only in culture of human embryonic brain tissue. Nerve tissue is difficult to obtain and to culture, imposing severe limitations. The discoveries of Enders, Weller, and Robbins made possible the development of the research for the Salk vaccine in 1955 and the Sabin vaccine in 1959. Polio is about to be eradicated from the face of the earth. Little more than a generation ago in the United States and across the world, summer was likely to be a season of dread because of the fear that polio might strike. Thank you, John Enders, Thomas Weller, and Frederick Robbins.

The following is excerpted from the text of the presentation speech by S. Gard, member of the Staff of Professors of the Karolinska Institute. Note the formal language of the opening and closing statements.

Your Majesties, Your Royal Highnesses, Ladies and Gentlemen. The principles of cultivation of bacteria were laid down in the late 1870's by Robert Koch. Since that time the bacteriologists could study systematically the diseases caused by bacteria, isolate the causative agents in pure culture and make themselves familiar with their nature. With the aid of the culture technique they were able to trace the routes, along which infection is transmitted, to detect carriers and other sources of infection thereby making a rational combat of epidemics feasible. They could produce therapeutic sera and prophylactic vaccines. Finally, the culture technique was instrumental in the discoveries of the modern wonder drugs, sulfa, penicillin, streptomycin, etc.

. . .

Turning to the virus diseases we meet an entirely different picture. . . . Worst of all, many virus diseases are on the increase, a tendency particularly evident in poliomyelitis. After having been practically unknown at the turn of the century, poliomyelitis in this country is now responsible for almost one fifth of all deaths from acute infections. . . .

It is not difficult to find the reason why the virologists have failed where the bacteriologists were so successful. They have been severely handicapped by the difficulties connected with the cultivation of viruses. Unlike bacteria and other microorganisms, virus is incapable of multiplying in artificial lifeless culture media. In the test tube it appears as an inert chemical substance; only in the interior of a living cell its hidden forces are liberated. Here, it turns the more active and incites a process which may, sometimes within a few minutes, lead to cell destruction and the production of hundreds of new virus particles.

. . .

Then, in 1949 there appeared from a

Enders, Weller, and Robbins Awarded 1954 Nobel Prize *(continued)*

Boston research team a paper, modest in size and wording but with a sensational content. John Enders, director of the Children's Hospital's Research Laboratory, and his associates Thomas Weller and Frederick Robbins reported the successful cultivation of the poliomyelitis virus in test-tube cultures of human tissues. A new epoch in the history of virus research had started.

. . .

Enders' interest in tissue-culture methods dates back to around 1940. . . . In the 1940's together with Weller and a few other associates he studied the viruses of vaccinia, influenza and mumps, thereby gathering valuable experience.

At last, time was ripe for experiments on the poliomyelitis virus. The prospect of a favorable result could not be considered particularly bright. Other scientists had previously attacked the problem with very moderate success. It was generally held that the final word had already been said by Sabin and Olitsky who in 1936 tried to grow the virus in Maitland cultures of various tissues from chick embryos, mice, monkeys, and human embryos. Their results remained completely negative except in cultures of human embryonic brain tissue in which the virus at least seemed to maintain its activity. These findings were taken as a definitive confirmation of the accepted concept of the virus as a strictly neurotropic agent, i.e., capable of multiplying in nerve cells exclusively. Accordingly, the hopes of a practicable method

for the cultivation of the poliomyelitis virus were temporarily shelved. Of all tissues, nerve tissue is the most specialized, the most exacting and consequently the most difficult to cultivate. As, at that, there seemed to be no alternative to the use of human brain tissue, the general resignation is easily understood.

In the 1940's the belief in the neurotropism of the virus began to falter, however, and Enders, Weller and Robbins decided to repeat Sabin and Olitsky's experiment with an improved technique. In their first experiments they used human embryonic tissue. To the great surprise of everybody except perhaps the experimenters themselves they registered a hit in their first attempt. The virus grew not only in brain tissue but equally well in cells derived from skin, muscle, and intestines. Furthermore, in connection with the multiplication of the virus, typical changes appeared in the cellular structure, finally leading to complete destruction, easily recognizable under the microscope. This observation furnished a convenient method of reading the results. . . . Subsequently, Enders, Weller and Robbins found that tissues secured from operations on children as well as adults could be used to advantage; all tissues except bone and cartilage seemed to be equally suitable. Finally they tried to isolate the virus from various specimens directly in tissue cultures. This was likewise achieved. In the latter observation probably the greatest practical importance of their discoveries is to be found. The virologists finally had a tool in the same class

as the culture technique of the bacteriologists.

. . .

We now possess essentially improved technical facilities for the combat of viral diseases. We should be aware, however, of not claiming any victories in advance. It took 75 years to achieve those results in the field of bacteriology to which we now point with pride. Much effort and a considerable time will be required for equivalent achievements in the fight against the virus diseases. However, thanks to Enders, Weller and Robbins' discovery we may look with confidence to the future.

Dr. John Enders, Dr. Frederick Robbins, Dr. Thomas Weller. Karolinska Institutet has decided to award you jointly the Nobel Prize for your discovery of the capacity of the poliomyelitis virus to grow in test-tube cultures of various tissues. Your observations have found immediate practical application on vitally important medical problems, and it has made accessible new fields in the realm of theoretical virus research.

. . .

. . . It is my privilege and pleasant duty on behalf of Karolinska Institutet to extend to you our sincere felicitations.

Dr. Enders, Dr. Robbins, Dr. Weller. May I now request you to receive your Nobel Prize from the hands of His Majesty the King.

Source: The excerpts are reprinted with permission from the Nobel Foundation. © The Nobel Foundation, 1954.

DIAGNOSIS OF VIRAL INFECTION

While cell and embryonic egg techniques are available and routine in the hands of trained laboratory personnel, there is usually no attempt to culture the viruses from those with suspected viral infections, as is routinely done for bacterial infections. There are several reasons for this, including cost and limited viral laboratory facilities, but a prime reason is that results of cell culture take too long to be useful for clinical diagnosis. Nevertheless, definitive tests are available when indicated. The presence of antibody molecules in the patient's blood may be of diagnostic value; these molecules are produced in response to (foreign) antigens, and their presence is an indication of current or past infection. (Antibodies and antigens are discussed in chapter 11.) Accurate identification of a virus may be essential for the protection of the public health, as in the 1999 encephalitis outbreak in New York, when tissue specimens taken at autopsy identified the virus not as the St. Louis en-

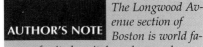

The Longwood Avenue section of Boston is world famous for its hospitals and research centers, including the John F. Enders Pediatric Research Laboratories (Fig. 5.11). While at the Harvard School of Public Health in 1999, I passed the building daily on my way to classes and, on several occasions, went into the lobby to view the magnificent and inspiring pictures of John Enders and his colleagues. Each time, I experienced a sense of awe, admiration, and inspiration for their work, having grown up in the days when the fear of their children's developing polio was uppermost in the minds of parents. Polio eradication is now in sight! The 2000 target was missed, largely because of internal strife in Africa, but eradication of this ravaging disease is anticipated by 2005.

FIGURE 5.11   The John F. Enders Pediatric Research Laboratories. **(A)** Portrait of John Enders in lobby; **(B)** plaque honoring John Enders. (Author's photos.)

cephalitis virus but rather as the West Nile encephalitis virus. Most likely, the diagnosis of viral infection is based on the patient's symptoms, including fever, general aches and pains, weakness, nausea, and muscle fatigue. If there is a telltale rash, as in measles, or other characteristic signs, such as the skin lesions associated with chicken pox, the diagnosis is more apparent and more accurate. Usually, the patient has to settle for "it's a bug, 24-hour grippe, flu, or intestinal virus." Recovery from the more common viral infections usually occurs without complications, thanks to the immune system (chapter 11). Undoubtedly, you have had your share of these infections!

PHAGE THERAPY

You are now familiar with the devastating effect of T4 phage on *E. coli* (Fig. 5.12). The phage can wipe out an entire *E. coli* culture. Hence, the possibility of using bacterial viruses to control bacterial infections may not seem so far-fetched. Hundreds of bacterial species, including many pathogens, are subject to viral infections, frequently resulting in bacterial cell damage or death. Obviously, then, the bacterial growth cycle would be interrupted.

The use of phage has been a model for understanding the dynamics of interactions between animal and plant viruses and their hosts. However,

FIGURE 5.12   The lytic activity of phage. Bacteriophage invade and kill *E. coli*. **(A)** The tube of nutrient broth on the right was inoculated with *E. coli* and incubated for 24 hours, resulting in marked turbidity. The tube on the left was inoculated at the same time, but phage was added; the contents of this tube are clear because the bacteria were lysed. **(B)** A lawn of *E. coli* with plaques (lighter-colored areas). (Author's photos.)

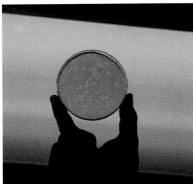

the earlier interest in these viruses was based on their use as therapeutic agents to fight bacterial infections. To put it another way, investigators realized the possible potential of phages to kill pathogenic bacteria (Box 5.2). Phage therapy was controversial in the early part of the 20th century and was essentially abandoned in the Western world when antibiotics became available in the l940s. However, research and implementation of phage therapy have continued over the last 50 years, primarily in eastern Europe. In the past several years there has been a worldwide resurgence of interest in the possibility of phage therapy, sparked by the highly significant problem of antibiotic resistance. The logic is there, and, perhaps, phage therapy has a future. How would you feel about drinking, or being injected with, a culture of bacteriophage?

BOX 5.2   A Return to Phage Therapy

Phage therapy might be the answer to the crisis resulting from the misuse of antibiotics in human medicine and on the farm to promote the growth of farm animals. Antibiotic resistance has been cited as a significant factor in emergence and reemergence of microbial diseases (chapter 1); the mechanisms by which antibiotic resistance is acquired are discussed in chapter 12. Meanwhile, biologists are constantly on the search for new antibiotics and new therapies, including a renewed interest in phage therapy.

Historically, Frederick Twort (1915) and Felix d'Herelle (1917) independently isolated "filterable" agents that could "eat" bacteria. They observed small clear areas, called plaques (Fig. 5.12), which appeared on the surface of bacterial lawns grown on agar plates when exposed to phage. D'Herelle is credited with coining the term "bacteriophage," frequently shortened to "phage." He realized the potential therapeutic use of phage as a natural process to control bacterial infections. Phage therapy has an almost 100-year history. Perhaps, had the antibiotic era not occurred in the 1940s, the early promise of phage therapy might have further developed. However, in eastern Europe, including the former Soviet Union, phage research has continued over the years, particularly in research institutes in Poland and in Tbilisi, Georgia.

Successes in the early years of phage therapy were reported, but so were disappointments and failures. In retrospect, these negative results may well have been due to flawed procedures resulting from the relatively meager knowledge available at the time regarding phage-bacteria interactions.

There are several inherent advantages in the use of phages that warrant continued research in their potential to combat bacterial infections, including the following.

- Phages can be targeted to specific bacteria, minimizing disturbance to the normal bacterial flora of the body.
- Phages are easily grown and purified and are relatively inexpensive.
- Small doses can be used because their numbers increase exponentially as they spread from bacterium to bacterium.
- There is no toxicity because phages invade bacterial cells and not human cells.
- The replication and activity of phages introduced into the body are self-limited. Once their bacterial targets are destroyed, they gradually disappear from the body.

In recent years, researchers have reported on promising new approaches to combat bacterial infections. A group in Canada is studying phage therapy as a means to cure mastitis, an inflammation of the udder in dairy cattle. Researchers at Texas A&M University and at the Rockefeller University have harvested the enzymes that phages use to disintegrate cell walls of their bacterial hosts, allowing for phage release. Their logic is that these same enzymes can be employed to attack the bacteria from the outside by cell wall destruction. Rockefeller University's Vincent Fischetti, who studies the streptococci that cause streptococcal sore throat and other streptococcal infections, states, "It kills the target bacteria instantly. It's amazing—instantly. It does this by punching holes in the cell walls. We can take 10 million organisms in a test tube, add a very small bit of [phage] enzyme, and five seconds later, they are all dead. Nothing other than chemical agents can kill bacteria this quickly." Other researchers are focusing on phages to control bacterial infections in food crops. Scientists at the University of Florida at Gainesville have found that tomato plants treated with phage prior to their inoculation with a microbial pathogen are afforded dramatic protection. Pharmaceutical and biotechnology companies are now expanding research funds for the development of phage therapy.

So, who knows? The time may not be too distant when you will be using live active phage particles as a throat gargle, a nasal or throat spray, a liniment or cream to be rubbed on your skin, eye drops, or a liquid poured directly onto or into a wound. Are you ready to accept this?

OVERVIEW Viruses are subcellular infectious agents. They are obligate intracellular parasites that subvert the host cell's metabolic machinery for their own replication. A unique aspect of viruses is that they contain either DNA or RNA, whereas cells contain both nucleic acids. Further, viral nucleic acid can be ssDNA, dsDNA, ssRNA, or dsRNA.

Some viruses consist of only a protein coat—the capsid—containing nucleic acid, whereas others have an envelope. Viruses are helical, polyhedral, or complex. Bacterial viruses—bacteriophages—are models for understanding viral replication, a process which consists of five stages: adsorption, penetration, replication, assembly, and release. Viruses that cause human diseases exhibit these stages with variations in the strategies employed, depending on the specific virus.

Viruses, as obligate intracellular parasites, cannot be grown on nonliving medium and so are cultured in fertile chicken eggs or in cell culture, both offering a supply of living cells allowing for replication. Identification of the specific virus causing infection is impractical and routinely not done. Diagnosis, in most cases, is based on the symptoms of disease.

Treatment of bacterial diseases with bacteriophages sounds bizarre but is based on sound logic. The idea is attractive because of the growing problem of antibiotic-resistant bacteria.

SELF-EVALUATION

PART I Choose the *single* best answer.

1. Tobacco mosaic virus was crystallized by
   a. Enders
   b. Pasteur
   c. Ivanovsky
   d. Stanley

2. The "turkey wrap" analogy applies to what kind of virus?
   a. complex
   b. polyhedral
   c. icosahedral
   d. helical

3. The units of the viral protein coat are called
   a. capsids
   b. helical
   c. capsomeres
   d. envelopes

4. Viruses are known to infect
   a. plants
   b. bacteria
   c. humans
   d. all organisms

5. The envelope of an animal virus is derived from the _____ of its host cell.
   a. cell membrane
   b. cell wall
   c. cytoplasm
   d. nucleus

6. The general steps in a viral multiplication cycle include (1) release, (2) penetration, (3) adsorption, (4) replication, (5) assembly. The correct sequence of these steps is
   a. 2, 3, 4, 5, 1
   b. 2, 1, 3, 4, 5
   c. 3, 2, 4, 5, 1
   d. 1, 2, 3, 4, 5

PART II Fill in the following.

1. Based on nucleic acid categories, how many viral groups are there?

   _____

2. A complete viral particle is called a

   _____.

3. Viruses can be grown in tissue culture or in

   _____.

4. The abbreviation CPE stands for

   _____.

5. For what disease are Negri bodies diagnostic?

   _____

PART III Answer the following.

1. Describe the five stages of viral replication.

2. The genetic material of viruses is unique. Discuss this statement.

3. Why did the discovery and understanding of viruses lag about 40 years behind knowledge about bacteria?

4. How does the possibility of being treated for a bacterial infection with live bacteriophage strike you? Describe the logic of this strategy.

6

CONCEPTS OF MICROBIAL DISEASE

PREVIEW "No man is an island" is a popular expression having to do with human interactions. Biologically, this expression can be extended to "no organism is an island." Wherever one looks in the biological world, associations between varieties of species are apparent. Even bacterial cells harbor bacteriophages. Humans are part of an invisible ecosystem of microbes and, like it or not, are a part of the food chain and serve as a microbial fast-food restaurant. A variety of microbes live in peace on or in their human hosts as relatively permanent inhabitants that constitute what is called the normal flora or as transient visitors just passing through. Some almost always produce damage and disease to the host upon contact, while others, including the normal flora, may produce disease only when the host immune defense mechanisms are impaired, as in acquired immune deficiency syndrome (AIDS). Whatever the circumstance, the underlying factor is that of a dynamic interplay between the number of microbes and their virulence mechanisms, as described in this chapter, and the immune mechanisms of the host (see chapter 11). Microbial diseases are characterized by a sequential series of five stages.

BIOLOGICAL ASSOCIATIONS

The term **symbiosis** means "living together" and describes an association between two or more species. According to biologist Clark P. Reed,

> Symbiosis, particularly parasitism, is frequently regarded in distasteful terms by the hygiene-conscious citizen. We have a tendency to think of it as a peculiar and abnormal association of some lower organism with a higher one. There is an element of snobbishness in such a view, which must be quickly abandoned when a discerning look is taken of the living world. In nature there is probably no such thing as a symbiote-free organism. The phenomenon of symbiosis is quite as common as life itself.

Symbiosis is an umbrella with three possible characteristics (Fig. 6.1): **mutualism, commensalism,** and **parasitism.** Mutualism may be the "best

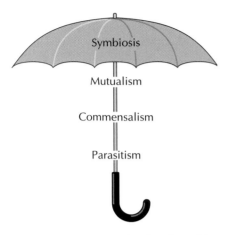

FIGURE 6.1   The umbrella of symbiosis.

of all worlds" and is like a blissful marriage in which both members of the association enjoy benefits. A classic example of mutualism is the association between termites and the protozoans that constitute a part of the normal flora of the termite's intestine. Termites can "eat your house down," a fact exploited by pest control companies with creative slogans such as "Bug off!" and other gimmicks. The termites' ability to digest wood (cellulose, a carbohydrate) is based on the enzymes secreted by the protozoan residents. In turn, the protozoans have it made; they live in a stable environment with plenty of food passing through. Consider *Escherichia coli,* a part of the normal flora and an inhabitant of the human large intestine. These bacteria enjoy the comforts of home while producing vitamin K, needed by its hosts for blood clotting.

Commensalism is described as a relationship between two or more species in which one benefits and the other is indifferent, that is, neither benefited nor harmed. Consider, for example, most of the normal flora of the body, organisms which reside in or on the body and take their nourishment from host secretions and waste products but, in return, make no obvious contribution to the partnership with their host, nor do they damage their host. A humorous poem summarizes the commensal relationship enjoyed by the normal flora and the host (Box 6.1). As pointed out in the poem, the body offers a wide variety of ecological niches (Fig. 6.2), each of which has a particular bacterial species population. Microbes are not the only commensals to enjoy our bodies. If you were to take a scraping from the landscaped furrows of your forehead just below the hairline and examine it under a microscope, you would probably see mites crawling about. Mites are distantly related to spiders and are so small that they take up residence in sebaceous (sweat) glands and hair follicles and live in their luxury condominiums happily ever after. These examples illustrate our role as part of an ecosystem.

While mutualism is at one end of the spectrum, parasitism—a relationship in which the parasite lives at the expense of its host—is at the other end. Parasitism is an association in which one species, the parasite, lives at the expense of the other, the host. There are at least as many definitions of the word "parasitism" as there are letters in the word itself, but they all point to harm to or death of the host. The host (in this book) is the human, and microbes are the parasites. The term parasite is used to encompass all microbes and worms that produce disease. On the other hand, some biologists restrict the term to protozoans (animal-like) and worms, and other biologists restrict it to worms.

**BOX
6.1   Normal Flora: a Poem**

For creatures your size I offer
    a free choice of habitat,
so settle yourselves in the zone
    that suits you best, in the pools
of my pores or the tropical
    forests of arm-pit and crotch,
in the deserts of my fore-arms,
    or the cool woods of my scalp

Build colonies: I will supply
    adequate warmth and moisture,
the sebum and lipids you need,
    on condition you never
do me annoy with your presence,
    but behave as good guests should
not rioting into acne
    or athlete's-foot or a boil.

Source: "A New Year Greeting" by W. H. Auden. Quoted in C. A. Mims, N. J. Dimmock, A. Nash, and J. Stephen, *Mims' Pathogenesis of Infectious Disease,* 4th ed. (Academic Press, San Diego, Calif., 1995).

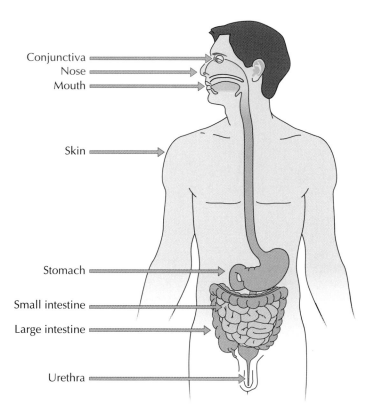

FIGURE 6.2 The normal flora: commensal organisms.

The distinctions between mutualism, commensalism, and parasitism are not airtight. The borders between them are hazy and are frequently crossed. Think of these relationships as being like a "slippery slide" (Fig. 6.3) or like an elevator moving up and down from level to level. For example, in AIDS and other conditions resulting in a depressed immune system, nonpathogenic microbes, including the normal flora, are responsible for potentially fatal infections; these microbes are appropriately referred to as opportunists or opportunistic pathogens. A major cause of death in AIDS is a fungal infection caused by the opportunist *Pneumocystis carinii.* As another example, a major threat to individuals suffering from extensive burns is serious bacterial infection resulting from nonpathogenic microbes, including their own normal flora. Further, organisms that are part of the normal flora can take the slippery slide from commensalism to parasitism. For example, when *E. coli,* a normal resident of the intestine, gains access to another area of the body—as might happen if the intestinal wall is pierced during surgery, resulting in leakage of intestinal fluids containing *E. coli* into the abdomen—a potentially fatal condition called **peritonitis** may result (Fig. 6.4).

FIGURE 6.3 The "slippery slide" of symbiosis from mutualism to parasitism.

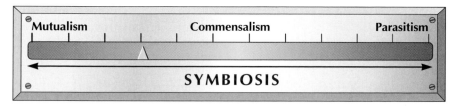

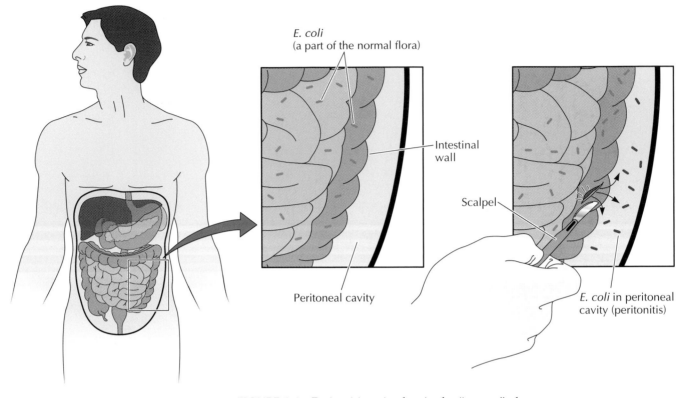

FIGURE 6.4   Peritonitis: microbes in the "wrong" place.

PARASITISM: A WAY OF LIFE

The term parasite has a negative connotation and is occasionally used to describe human interactions. To be called a parasite is hardly a compliment. (You may have had the unpleasant experience of being part of a group project, perhaps a student presentation, and one individual in the group "parasitizes" the others, meaning that he or she makes little contribution and lives off everybody else's effort.)

Parasitism is the basis for microbial disease and for the challenge between microbes and humans. The parasite lives at the expense of the host and, in so doing, causes damage to the host. According to Lewis Thomas (1913–1993), an American physician, author, educator, and administrator known for his popular essays in biology and medicine, "Disease usually results from inconclusive negotiations for symbiosis, an overstepping of the line by one side or the other, a biological misinterpretation of borders." Thomas refers to the "slippery slide" previously described. Parasitism is a constant and delicate dance of biological adaptation in the Darwinian survival of the fittest as both host and parasite coevolve.

Rats, Mice, and History, Hans Zinsser's intriguingly titled 1933 classic text, speaks of parasitism in a somewhat philosophical sense:

> Nature seems to have intended that her creatures feed upon one another. At any rate, she has so designed her cycles that the only forms of life that are parasitic directly upon Mother Earth herself are a proportion of the vegetable kingdom that dig their roots into the sod for its nitrogenous juices and spread their broad chlorophyllic leaves to the sun and air. But these—unless too unpalatable or poisonous—are devoured by the beasts and by man; and

the latter, in their turn, by other beasts and by bacteria. . . . The important point is that infectious disease is merely a disagreeable instance of a widely prevalent tendency of all living creatures to save themselves the bother of building, by their own efforts, the things they require. Whenever they find it possible to take advantage of the constructive labors of others, this is the direction of the least resistance, the plant does the work with its roots and green leaves. The cow eats the plant. Man eats both of them; and bacteria (or investment bankers) eat the man. . . . That form of parasitism which we call infection is as old as animal and vegetable life.

AUTHOR'S NOTE *My wife may well take issue with Zinsser's inclusion of investment bankers. She is the founder and chief executive officer of an investment business. What could Zinsser have been thinking of?*

Zinsser's comments made over 60 years ago are remarkably true to this day.

MICROBES AS AGENTS OF DISEASE

The bubonic plague, or Black Death (chapter 8), a bacterial disease, swept through Europe in the 14th century and killed millions of people. Smallpox, a viral infection, took a terrible toll for century after century. Added to the misery produced by these microbial diseases was the fact that there was little that society could do as preventive measures. In those days the existence of bacteria was not even hinted at; Leeuwenhoek's observations of "animalcules" occurred in 1638. What, then, was the early thinking regarding the cause of the plague, smallpox, and many other infectious diseases? People considered disease and death a part of life and nature's way; the cause was often thought to be "**miasmas**," a term meaning "bad air" or "swamp air." By the 19th century a link had been made between filth and the lack of sanitation and what is now referred to as infectious disease. This observation led to the recognition that nonhygienic conditions foster the prevalence of disease and disease vectors, including rodents, mice, fleas, ticks, lice, and mosquitoes (chapter 12). Bacteria and other microbes had not yet been discovered. It was another 150 years before Robert Koch, Louis Pasteur, and other early microbe hunters established the role of bacteria as causative agents of specific diseases. The French microbiologist Pasteur spent a good part of his life investigating diseases of wine and of silk moths, both of which were important to the economy of France. Later in his career, Pasteur reasoned that if microbes could cause these maladies, perhaps they were responsible for certain diseases of higher animals and humans. In short order, he proved that microbes were responsible for fowl cholera and anthrax in sheep. His crowning achievement was immunization against the dread disease hydrophobia, now known as rabies (chapter 8). It was Koch, a German bacteriologist, who experimentally proved that specific microbes were the causative agents for specific diseases. Koch's work initially centered on tuberculosis (TB), a disease that was rampant in his time. Koch established that this disease was caused by the bacterium *Mycobacterium tuberculosis*; this was the first proven causal relationship between a microbe and a particular disease. Koch's work extended well beyond TB and led to the development of a series of four postulates, **Koch's postulates**, that are still used to prove that a particular organism is the cause of a particular disease (Fig. 6.5):

1. Association: the causative agent must be present in every case of specific disease.
2. Isolation: the causative agent must be isolated in every case of the disease and grown in pure culture (a laboratory procedure).

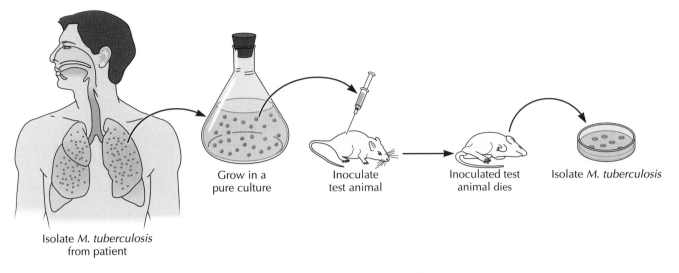

Isolate *M. tuberculosis*
from patient

Grow in a
pure culture

Inoculate
test animal

Inoculated test
animal dies

Isolate *M. tuberculosis*

FIGURE 6.5 Testing Koch's postulates.

3. Causation: the causative agent in the pure culture must cause the disease when inoculated into a healthy and susceptible laboratory animal.
4. Reisolation: the causative agent must be reisolated from the laboratory animal and be identical to the original causative agent.

These postulates have been invaluable to medical microbiologists and physicians throughout the 20th century and into the 21st and continue to play an important role in the identification of new and reemerging infections. They are not perfect and sometimes have to be "taken with a grain of salt." For example, the causative agent for the ancient bacterial disease leprosy (chapter 8) was established long before an animal model was found. (The model turned out to be the armadillo! Who would have thought of that creature as a likely candidate?) Syphilis has long been known to be caused by a corkscrew-shaped (spirochete) bacterium (chapter 8) which still has not been grown in culture medium. What about viruses? From the discussions in chapters 2 and 5, it is clear that these obligate intracellular parasites cannot be grown on lifeless culture media, although there are countless examples of specific viral agents associated with specific diseases. Remember that viruses were not described until the fourth and fifth decades of the 20th century. On the basis of the advent of the techniques of molecular biology, it is now possible to identify bacteria through their DNA fingerprints without growth in culture; this modern approach does not necessitate the fulfillment of Koch's postulates.

MICROBIAL MECHANISMS OF DISEASE

Attention now needs to be focused on the mechanisms by which microbes cause disease. Only infectious diseases are caused by microbes. There are many nonmicrobial, noninfectious diseases, including some cardiac diseases, most cancers, metabolic diseases, mental disorders, nutritional deficiency disorders, and structural disorders, some of which are hereditary. (A growing body of evidence indicates that some of these diseases may well be caused by microbes—a subject discussed in chapter 16.) A distinction needs to be made between the terms **infection** and **disease**. Some microbiologists use these two terms interchangeably; most use the term infection to describe the multiplication of microbes on or in the body without producing definitive symptoms, whereas disease is a possible outcome of infection that results in an impairment of health to some degree. Again, the slippery slide principle exists, as it frequently does in biology. Early symptoms of many infectious diseases include lethargy and a sense of "I don't feel so great." As your experience indicates, infectious diseases run the gamut of severity from mild to moderate to severe to lethal (Fig. 6.6). Some diseases are **subclinical** (asymptomatic), as evidenced by the presence of telltale antibodies against diseases that were never diagnosed.

The chance of acquiring a particular infection and the severity of the accompanying symptoms are dependent upon three major factors: (i) **dose** (*n*)—the number of microorganisms to which the potential host has been exposed, (ii) **virulence** (*V*)—microbial mechanisms or weapons, and (iii) **resistance** (*R*)—the host immune system, as described in chapter 11. There is a dynamic battle going on between disease-producing microbes and the resistance of the host which can be viewed as a tug-of-war (Fig. 6.7) or a seesaw effect. The equation $D = nV/R$ summarizes this struggle. D represents the severity of infectious disease, the numerator refers to the microbes (n = number of organisms, V = virulence factors), and the R of the denominator refers to resistance (immunity). The outcome of the battle determines whether disease occurs and, if so, how severe it is. Particularly significant, in most cases, are the early hours of challenge; if the microbes gain the upper hand over resistance, then they continue to multiply and damage the host; a state of disease occurs.

Pathogenicity and Virulence

The number of organisms (*n*) gaining access to (in or on) the host plays a crucial role in determining whether or not disease will take place. For most bacteria there is a minimal dose, the **infectious dose (ID)**, necessary to establish infection. Highly virulent organisms have smaller IDs and, in some cases, fewer than 100 bacteria are needed, while those that are of low virulence can establish infection only when the number of organisms is large. All it takes is about 10 TB bacilli to establish infection, but as many as 1,000 cholera bacteria or 10 million salmonellae are necessary for infection to result. The 50% lethal dose (**LD$_{50}$**) is a laboratory measurement of virulence

FIGURE 6.6   The gamut of severity of infectious disease.

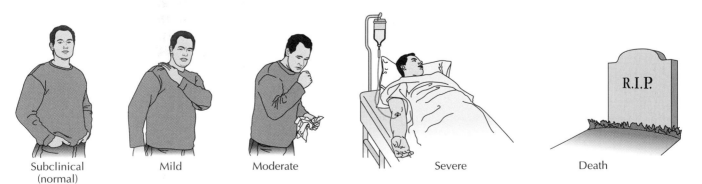

Subclinical (normal)   Mild   Moderate   Severe   Death

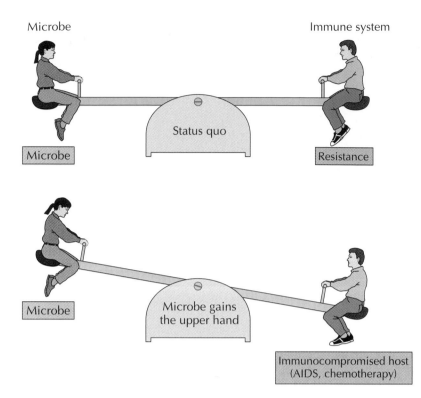

FIGURE 6.7   Host-parasite dynamics.

determined by injecting animals with graded doses of microbes; the dose that kills 50% of the animals in a determined time is the LD_{50}.

The term "pathogen" has been previously introduced and refers to the ability of microbes to produce disease. Some groups of microbes are considered to be highly pathogenic, including those that cause tetanus, botulinum, food poisoning, plague, and Ebola fever. They almost always cause disease. (It should be kept in mind that the vast majority of bacteria are nonpathogens.) Virulence is a measure of pathogenicity and encompasses those specific factors (for example, toxins) that enable pathogens to overcome host defense mechanisms and to multiply and cause damage. Even within the same species of microbes, there is variation in virulence as demonstrated by LD_{50} determinations.

Microbes gain entry into the body through portals of entry which include the skin, mucous membranes of the mouth, nose, and genital tract, and blood. This subject is covered in chapter 7 as part of the cycle of disease.

Usually, only those microbes with adaptive **defensive strategies** that allow them to escape destruction by the host immune system and **offensive strategies** (Table 6.1) that result in damage to the host are able to establish disease. The distinction between these two strategies is not always clear, since some factors act both defensively and offensively. Think of it as being like a basketball team—to win the game (analogous to establishing infection), the team (microbes) must have effective defensive strategies to counter its opponent (the host immune system), but that is not enough. The team must also have an effective offensive strategy to shoot baskets in order to score points.

Each group of pathogens has evolved unique defensive and offensive

TABLE 6.1 Bacterial mechanisms of virulence

Structure or secretion	Function	Examples
Defensive strategies		
Adhesins	Fixation to cell surfaces and linings	*Neisseria gonorrhoeae* (attachment to urethral lining); *Streptococcus mutans* (causes tooth decay and sticks to surface of teeth)
Capsules	Interfere with uptake of bacteria (phagocytosis)	Anthrax, plague, streptococcal diseases (e.g., scarlet fever, "strep" throat, pneumococcal infection)
Miscellaneous structures	Interfere with uptake of bacteria (phagocytosis)	"Waxy coat" of tubercle bacilli; M protein in streptococcal cell walls
Offensive strategies		
Enzymes	Destroy integrity of tissue structure	Hyaluronidase (breaks down hyaluronic acid of connective tissue), hemolysins (bring about lysis of red blood cells), collagenases (break down collagen in connective tissue), leukocidins (destroy white blood cells)
Exotoxins	Specific activities that interfere with vital host functions	Botulinum toxin (one of the most potent toxins known; interferes with transmission of nerve impulse, resulting in flaccid paralysis); tetanus toxin (interferes with transmission of nerve impulses, resulting in irreversible muscle contraction)

virulence factors, and some factors are present in more than one group. (Virulence factors of other microbes will be discussed later in this chapter.)

Defensive Strategies

Adhesins

Many pathogens possess cell surface molecules called adhesins (Fig. 6.8) that allow them to stick onto receptor molecules at the portal of entry in a Velcro-like manner. A clear example of this is the causative agent of gonorrhea, *Neisseria gonorrhoeae*, which produces adhesins that allow the bacteria to hold fast and colonize on the warm, moist membranes of the urogenital tract despite the periodic flow of urine. The bacteria that are primarily responsible for dental decay remain attached to the surface of teeth by adhesins despite the constant irrigation of the teeth by saliva. Bacteria lacking adhesins are subject to being washed or swept away and are not able to colonize the surface of the host cells lining the portal of entry.

Capsules and Other Structural Factors

Recall from chapter 3 that some bacteria have capsules around their cell walls. In many cases the presence of the capsule material interferes with the process of phagocytosis, an important resistance mechanism of the host by which bacteria are engulfed by scavenger cells of the body (chapter 11). It can be experimentally demonstrated that removal of the capsule from encapsulated bacteria renders them more susceptible to phagocytosis (Fig. 6.9). A number of bacteria, including those that cause plague, anthrax,

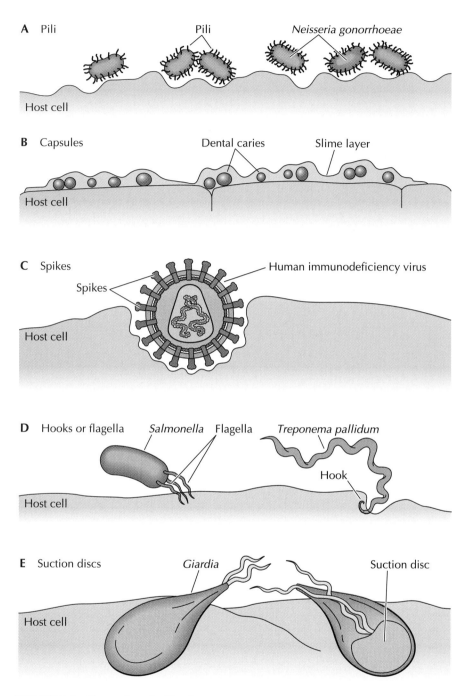

A Pili

Pili

Neisseria gonorrhoeae

Host cell

B Capsules

Dental caries

Slime layer

Host cell

C Spikes

Spikes

Human immunodeficiency virus

Host cell

D Hooks or flagella

Salmonella Flagella

Treponema pallidum

Hook

Host cell

E Suction discs

Giardia

Suction disc

Host cell

FIGURE 6.8   Examples of adhesins.

pneumococcal pneumonia, and strep throat, have capsules. The presence of a capsule does not necessarily confer virulence, nor does the lack of a capsule exclude virulence. It is to be emphasized that bacterial virulence is due to the combination of virulence factors. The streptococci responsible for streptococcal sore throat and a variety of other diseases have a protein in their cell walls termed the **M protein**. Those streptococci containing the M protein display resistance to phagocytosis, whereas mutant strains not producing this factor and strains in which the M protein has been experimen-

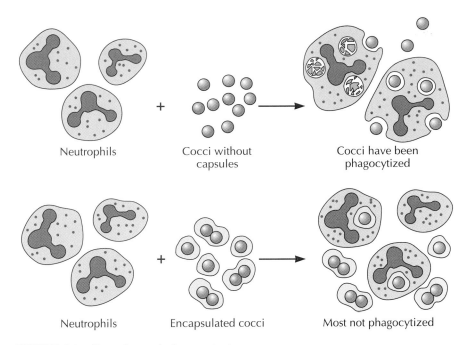

FIGURE 6.9   Capsules and phagocytosis.

tally removed are readily engulfed by phagocytic cells. Note that the streptococci possess both capsules and M protein—a double whammy against phagocytosis. The causative microbe of TB is characterized by a distinctive "waxy" cell wall that allows the bacterium not only to resist digestion within phagocytic cells but, further, to multiply inside them. These are notable examples of structural components that act as virulence factors.

Antigenic Variation

Antigenic variation is discussed in chapter 1 as an example of microbial adaptation. The particular microbes described are influenza virus and the protozoan parasite that causes African sleeping sickness. The causative agents of relapsing fever and gonorrhea, both bacterial diseases, are also masters of antigenic disguise. Antigenic variation is an effective strategy that allows microbes to evade the antibody defense system, an important component of the immune system.

Offensive Strategies: Extracellular Products

A variety of products are produced and released by bacterial cells as they multiply; these factors are referred to as extracellular substances. Some are classified as toxins, and some are classified as enzymes. Some exhibit both enzymatic and toxic activities. The distinction between toxins and enzymes is not always clear, so it is best to think of them collectively as extracellular products.

Exoenzymes

Some pathogens secrete **enzymes** which foster the spread of invading bacteria throughout the tissues by causing damage to host cells in their immediate vicinity and, in so doing, breaking down tissue barriers; they are called spreading factors. Think of this action as being like that of a bulldozer pushing its way through a field of trees and rocks. **Hyaluronidase** is an example; it breaks down the ground substance, the hyaluronic acid content of connective tissue, reducing its viscosity and allowing microbes to spread deeper into the tissues. **Collagenase** is another example of a spreading factor; it breaks down the structural framework of **collagen,** a vital part of various connective tissues, resulting in gas gangrene and massive areas of **necrotic** (dead) tissue. Extensive tissue **debridement** (cutting away of necrotic tissue) or amputation may be necessary to halt the infection and avoid death.

Hemolysins destroy red blood cells through destruction of cell membranes. **Kinases** break down clots, allowing entrapped bacteria to spread. (One kinase—**streptokinase**, a product of streptococci—is now an important therapeutic tool for dissolving blood clots in patients suffering from heart attacks and has proved to be quite effective in preventing strokes and deaths.) **Coagulase** forms a network of threads around bacteria, affording protection against phagocytosis, and **leukocidins** destroy white blood cells.

Exotoxins

The word **toxin** conjures up thoughts of substances that are poisonous to the body. Toxins are major virulence factors for many pathogenic microbes; they are classified as either **exotoxins** or **endotoxins**. These products differ from each other in their chemical composition, modes of action, and nature of their release.

Exotoxins are protein molecules which are synthesized within the microorganism and released ("exo") into the host tissues during the growth and metabolism of the microbes. Exotoxin production is primarily associated with gram-positive organisms (chapter 4); the ability to produce toxins is called **toxigenicity**. Exotoxins are readily soluble in body fluids, are rapidly transported throughout the body, and may act at sites quite distant from the site of colonization of the organism. They are highly toxic and specific in their activity. They are grouped into three principal types: (i) **cytotoxins**, which kill or damage host cells, (ii) **neurotoxins**, which interfere with transmission of nerve impulses, and (iii) **enterotoxins**, which affect the cells lining the gastrointestinal tract, leading to diarrhea. Table 6.2 lists a number of bacterial toxins and their effects on the host. Exotoxins are among the deadliest poisons. It is estimated that 1 milligram of the botulinum toxin (which causes botulism, a type of food poisoning [chapter 7]) is enough to kill 1 million guinea pigs. Others estimate that 1 milligram of botulinum toxin can kill half the population of a major city. This may be an exaggeration, so let's say it will kill only one-fourth of the population. What is the difference? Whatever the numbers, the potency of this toxin is amazing! Toxoids (chapter 12) are toxins that have been detoxified but retain their antigenicity. They are used in immunization against those diseases in which the disease is primarily a result of exotoxin secretion. A number of potentially fatal bacterial diseases are characterized by the production of specific exotoxins, including diphtheria, tetanus, botulism, cholera, and pathogenic *E. coli* infections. These diseases are largely a manifestation of the toxin and not the presence of the bacteria; injection of toxin by itself into laboratory animals mimics the symptoms of the disease. Although each toxin has a specific mechanism of activity (Table 6.2), the **AB model** (Fig. 6.10) has been proposed to explain the general mode of activity of some toxins. According to this model, exotoxins are composed of two subunits referred to as the A (active) fragment

TABLE 6.2   Examples of diseases caused by exotoxins and their activity

Disease	Exotoxin activity and result
Botulism	Neurotoxin: prevents transmission of nerve impulse, resulting in (flaccid) limp paralysis
Tetanus	Neurotoxin: prevents transmission of nerve impulse, resulting in rigid contractions of skeletal muscles (lockjaw)
Clostridial food poisoning	Enterotoxin: diarrhea
Cholera	Enterotoxin: diarrhea
E. coli (enterotoxigenic) traveler's diarrhea	Enterotoxin: diarrhea
Scarlet fever	Cytotoxin: causes damage to blood capillaries, resulting in a red rash
Diphtheria	Cytotoxin: kills cells in throat, resulting in a buildup of dead cells; causes damage to heart
Whooping cough	Cytotoxin: kills cells in throat and inhibits cilia
E. coli O157:H7 disease	Cytotoxin: bloody diarrhea; kidney damage

and the B (binding) fragment. Isolated A fragments have been shown to be enzymatically active but lack the ability to bind and to enter cells; isolated B fragments are able to bind to target cells but lack toxicity. Hence, the AB complex is necessary for exotoxin activity; the activity of the A subunit follows the activity of the B subunit. The specific nature of the A fragment activity is a function of the particular exotoxin.

In the earlier years of microbiology, the effect of toxins on human patients and on laboratory animals could be superficially observed, but little was known about how they worked. With the advent of molecular biology, toxins have been studied in great detail to determine their mechanism of activity. Figure 6.11 illustrates the activity of the botulinum and tetanus toxins.

As discussed in chapter 4, bacteria may suffer from infection by bacteriophages. In some cases, the phage DNA becomes incorporated into the bacterial chromosome in a latent or dormant state and does not bring about lysis of the bacterium. The phage is termed a **prophage**, and the bacterium is said to be **lysogenized**. During cell division, the prophage is replicated as a part of the chromosome, and the resulting daughter cells continue to

FIGURE 6.10   The AB model of exotoxin activity.

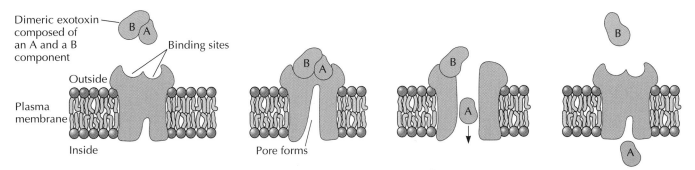

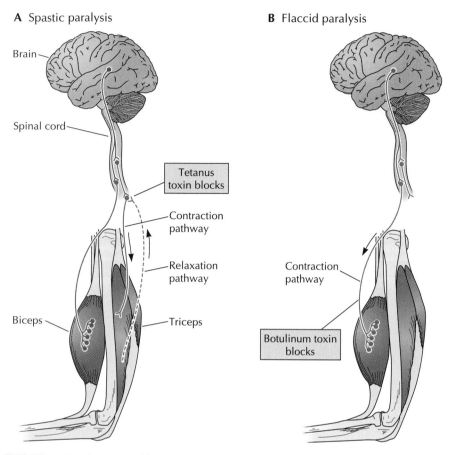

A Spastic paralysis

Brain

Spinal cord

Tetanus
toxin blocks

Contraction
pathway

Relaxation
pathway

Biceps

Triceps

B Flaccid paralysis

Contraction
pathway

Botulinum toxin
blocks

FIGURE 6.11   Activity of botulinum and tetanus toxins.

harbor the phage DNA. The presence of the phage DNA may confer new properties on the lysogenized bacteria, including toxin production. This phenomenon is referred to as **phage conversion** (Fig. 6.12). For example, diphtheria is a **toxemic** disease resulting from the production of an exotoxin, a property conferred by lysogeny with a phage carrying the *tox* gene. When the bacteria are "cured" (the phage is removed), toxin production no longer takes place. Other exotoxins, including the erythrogenic, botulinum, and cholera toxins, are the result of prophage genes.

Endotoxins

Endotoxins are quite different from exotoxins, as summarized in Table 6.3. Unlike exotoxins, endotoxins are not usually released during microbial growth and metabolism; they are structural components of the outer membrane of gram-negative cells (Fig. 4.2). They are released as these cells undergo disintegration, although some endotoxin material may be released during cell multiplication. Endotoxins are not proteins like the exotoxins but are molecules known as **lipopolysaccharides**. As indicated in Table 6.2, each exotoxin has a specific action; in contrast, all endotoxins, regardless of their source, produce the same host response, characterized by shocklike symptoms, chills, fever, weakness, formation of small blood clots, and possibly death. They are considerably less toxic than exotoxins but cause symptoms during overwhelming infections caused by gram-negative organisms.

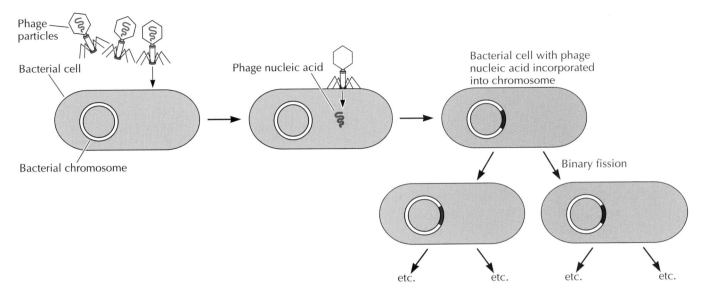

FIGURE 6.12 Mechanism of phage conversion.

Ironically, individuals with these infections may experience exaggerated symptoms during antibiotic treatment because of the release of endotoxins from dead cells.

Mechanisms of bacterial virulence have been studied over the years, and in more recent years they have been elucidated through molecular biology techniques.

Virulence Mechanisms of Nonbacterial Pathogens

Viruses

Recall from chapter 5 that viruses are subcellular obligate intracellular parasites and are not capable of producing enzymes and toxins. They are completely dependent on their host cells for survival and replication. On the defensive side, the fact that viruses are intracellular parasites allows them to hide from components of the immune system. The ability of some viruses to change their antigen coats affords a measure of defense against antibodies.

TABLE 6.3   Characteristics of bacterial exotoxins and endotoxins

Property	Exotoxins	Endotoxins
Site	Released from cell during growth and metabolism	Retained (for the most part) within outer membrane and released when cell disintegrates
Cell source	Primarily gram-positive cells	Gram-negative cells
Activity	Specific for each toxin	Essentially similar for all endotoxins
Chemical nature	Protein	Lipopolysaccharide
Toxicity	High toxicity	Minimal toxicity
Heat stability	Unstable; usually destroyed at about 60°C	Stable; can withstand temperatures of 100°C
Examples of diseases	Tetanus, scarlet fever, diphtheria, gangrene, botulism	Meningococcal meningitis, typhoid fever, salmonellosis

On the offensive side, death of the host cell may result from the lysis of the host cell membrane caused by the production of large numbers of replicating viruses, by shutting down the host cell's protein synthesis, by damaging the cell membrane of the host, and by inhibiting host cell metabolism. Some viruses have attachment molecules enabling them to dock with specific target cells, allowing for adsorption and penetration. For example, consider that the virus which causes AIDS docks with strategic cells (T lymphocytes) of the immune system, resulting in the infected individual's becoming severely immunocompromised. Other viruses produce cytopathic effects that kill the host cell. Some viruses cause adjacent cells to fuse and form a syncytium (a network).

Some viruses leave telltale fingerprints, called inclusion bodies, that are important in diagnosis (chapter 5) because they can be seen microscopically. These bodies are "viral debris" consisting of viral parts (nucleic acids, coats, envelopes) in the process of assembly.

Eucaryotic Microbes (Including Helminths)

Many protozoans, fungi, and helminths are pathogens, but their mechanisms of virulence are not well defined. Many fungi secrete toxins and enzymes that cause damage to host cell tissues and aid in their invasion. *Giardia*, an important water-transmitted protozoan parasite that causes severe diarrhea, attaches to cells lining the small intestine by a virulence factor called an adhesive or sucking disk. The *Plasmodium* protozoan parasite responsible for malaria infects and reproduces within host red blood cells, causing their rupture. The trypanosome protozoan that causes sleeping sickness exhibits antigenic variation, a defensive strategy of virulence, as does influenza virus. Helminths are large extracellular parasites that can obstruct lymph circulation, leading to grotesque swellings in the legs and other body structures, a condition called elephantiasis. *Ascaris* worms can "ball up," obstruct the intestinal tract, and migrate into the liver. Waste products of helminths can cause toxic and immunologic reactions. Diseases caused by protozoa and helminths are described in chapter 10.

STAGES OF A MICROBIAL DISEASE

As explained earlier, the outcome of contact between a pathogen and a host (human) is dependent upon the number of pathogens and their virulence factors versus the immune system of the host (Fig. 6.7). Everyone has experienced the miseries of a common cold and of gastrointestinal disturbances manifested by vomiting and diarrhea or (perish the thought!) both. In many cases, the disease runs its course and you get over it. Treatment may help to alleviate the symptoms and to shorten the duration of illness, but, nevertheless, five general stages (Table 6.4) take place, and their characteristics are significant in diagnosis and treatment.

The Incubation Stage

The **incubation stage** is the time between the pathogen's access to the body through a portal of entry and the display of signs and symptoms. During this time, although the infected individual is not necessarily aware of the presence of a pathogen, the microbes may be spread to others. The incubation time is quite consistent in some diseases but variable in others and is dependent upon the specific microbes, the numbers involved, and the host's resistance. The incubation times for several diseases are as follows:

TABLE 6.4 Stages of microbial disease

Stage	Description
Incubation	Period between initial infection and appearance of symptoms; considerable variation among diseases
Prodromal	Period in which early symptoms appear; usually short and not always well characterized
Illness	Period during which the disease is most acute and is accompanied by characteristic symptoms
Decline	Period during which the symptoms gradually subside
Convalescence	Period during which symptoms disappear and recovery ensues

botulism, 12 to 36 hours; chicken pox, 2 to 3 weeks; staphylococcal food poisoning, 2 to 4 hours; (serum) hepatitis, 45 to 160 days; and AIDS, 6 months to about 12 years. (Chapters 8, 9, and 10 cover a spectrum of bacterial, viral, protozoan, and helminthic diseases; incubation times are noted in those discussions.)

The Prodromal Stage

The **prodromal stage** is relatively short and not always obvious. The symptoms are vague and mild and are frequently characterized by tiredness, headache, muscle aches, and "feeling lousy." (These symptoms may not be indicative of a disease at all, but the result of too much partying or the stress associated with exams, papers, and the routines of student life.) Depending upon the disease, the individual may be contagious during this stage.

The Illness Stage

During the **illness stage,** you might feel like you have been hit in the head with a hammer and wish you could die but don't have the luxury. In this period the disease develops to the most severe stage, accompanied by typical signs and symptoms that may include fever, nausea, vomiting, chills, headache, muscle pain, fatigue, swollen lymph glands, and a rash. (What an impressive and depressing list!) This is the invasive time during which the tug-of-war between the pathogen's virulence factors and the host's immune system is taking place. It is a critical time in that, as in a tug-of-war, one side wins. Either recovery will be complete, or impairment or death of the host will result.

The Stage of Decline

What a relief! You won the battle and are in the **stage of decline**. The signs and symptoms begin to disappear, the body returns to normal, and life is worth living again.

The Convalescence Stage

During the **convalescence stage**, recovery takes place, strength is regained, repair of damaged tissues takes place, and rashes disappear. In some cases, healthy and chronic carrier states develop, as might happen with cholera and typhoid fever; the carrier state can exist for years.

OVERVIEW This chapter considered microbial pathogens and their mechanisms of causing disease in individuals. The next chapter focuses on microbial diseases from an epidemiological and a population point of view. "No organism is an island" is an expression symbolizing that all organisms live in symbiotic associations with other organisms. These associations can take the form of mutualism, commensalism, or parasitism and can slide from one to the other. Parasitism is the basis for microbial disease; the parasite lives at the expense of the host and causes damage to the host.

Throughout history, microbes have caused anguish and death to individuals on a large scale. The bubonic plague, smallpox, and influenza were notorious for decimating large segments of the population and, thereby, influencing the course of civilization. Societies seemed to accept these plagues as a part of life. The curse of these diseases was ascribed to vague notions of "miasmas; swamp air and noxious odors." By the 19th century, a link had been made between filth and a lack of sanitation to what is now referred to as infectious disease, but it wasn't until the end of the century that Koch, Pasteur, and other microbe hunters established the role of bacteria as agents of specific diseases.

The mechanisms of microbial disease have been described as a tug-of-war between (i) the number, or dose, of microorganisms to which a potential host has been exposed and their virulence factors and (ii) the immune mechanisms of the host. The outcome of the host-parasite dynamics determines the severity of infection, varying from subclinical to death. Microbes vary considerably, even within the species, in their pathogenicity, largely as a function of their virulence factors. Pathogenic microbes have evolved defensive strategies by which they are able to withstand the immune system of the host and offensive strategies that cause damage to the host. Although there is considerable variation in the manifestation of illness, depending on the specific microbe, there is a general pattern of infectious diseases illustrated by an incubation stage, a prodromal stage, an illness stage, a stage of decline, and a convalescence stage.

SELF-EVALUATION

PART I Choose the *single* best answer.

1. Endotoxins are associated primarily with
   a. viruses
   b. gram-positive bacteria
   c. gram-negative bacteria
   d. none of the above

2. Capsules
   a. are adhesins
   b. are exotoxins
   c. are associated with viruses
   d. break down T cells

3. Which is *not* an example of a virulence factor?
   a. hyaluronidase
   b. collagenase
   c. hemolysins
   d. All of the above are virulence factors.

4. Microbial diseases are the result of a biological association, which can best be described as

   a. commensalism
   b. symbiosis
   c. mutualism
   d. parasitism

5. Which term describes intensity of disease-producing ability?
   a. virulence
   b. pathogenicity
   c. infectivity
   d. toxicity

PART II Fill in the following.

1. *E. coli,* a resident of the intestinal tract, produces vitamin K. What type of symbiotic association does this demonstrate? _____

2. What does the abbreviation ID stand for, and what does it mean? _____

(continues)

3. What does the term LD_{50} mean?

_____

4. Name a "spreading factor" and its action.

_____

5. Antigenic variation is a significant microbial defense mechanism. Name a disease for which this is demonstrated. _____

6. What is a toxoid?

_____

PART III Answer the following.

1. Discuss Koch's postulates. Illustrate with a diagram.

2. The expression $D = nV/R$ expresses host-parasite relationships. Elaborate on this; identify each term (D, n, V, and R) in the equation.

3. Regarding symbiotic associations, a "slippery slide" exists. What does this mean? Give examples.

4. Does infection always result in disease? Discuss.

7

THE EPIDEMIOLOGY AND CYCLE OF MICROBIAL DISEASE

PREVIEW The preceding chapters have focused on microbes as a potential threat to the world population and on the biology of microbes and the mechanisms by which they cause disease. Chapters 8, 9, and 10 will consider specific microbial diseases and their effects on individuals. This chapter considers microbial diseases in relation to public health and focuses on the factors which are responsible for infectious diseases in populations. Basic concepts of epidemiology are presented so that the occurrence and prevalence of disease in a particular environment at a particular time can be understood. The existence of microbial disease requires a chain of linked factors which constitute the cycle of disease. These factors are reservoir, transmission, portal of entry, and portal of exit.

In hospitals and in long-term health care facilities, all of the factors involved in the cycle of infectious diseases are present in a concentrated way. These facilities are hotbeds for the transmission of microbes among hospital personnel and patients.

CONCEPTS OF EPIDEMIOLOGY

Epidemiology is an investigative methodology designed to determine the source and the cause of diseases and disorders that produce illness, disability, and death in human populations. Epidemiologists have been dubbed "disease detectives"; they are among the first group to be dispatched by the Centers for Disease Control and Prevention (CDC) when the threat of an outbreak occurs anywhere in the world. Their sleuthing is directed at understanding why an outbreak of a disease is triggered at a particular time and in a particular place. Factors considered are age distribution of the population, sex, race, personal habits, geographic location, seasonal changes, modes of transmission, and others. These factors are used to design public health strategies for control and prevention of future outbreaks. Historically, epidemiology is based on an understanding of the causes and distribution of infectious diseases, but modern epidemiology has branched out to other public health problems, including alcohol and drug abuse, cancer, mental conditions, "road rage" and other acts of violence, and exposure to lead paint.

Epidemiology dates back to the time of Hippocrates (460–377 B.C.), who questioned the role of eating and drinking habits, the source of drinking water, lifestyle, and seasons of the year as factors related to causality of disease. He was astute enough to realize that the diseases now identified as yellow fever and malaria were associated with swampy (mosquito breeding) environments. The epidemiological studies of Edward Jenner, a country physician in England, proved that cowpox and smallpox were related and led to smallpox immunization in the late 1700s. In the mid-1800s another early epidemiologist-physician, Ignaz Semmelweis, showed that childbed fever resulted from physicians' proceeding directly from autopsy room dissections to the delivery room without washing their hands, thereby transmitting bacteria from cadavers into the birth canals of pregnant females while doing pelvic examinations or delivering babies (Box 7.1 and Fig. 7.1). Childbed fever was so rampant that women begged to deliver their babies at home rather than in hospitals because of the high death rate associated with hospital deliveries. Semmelweis's attempts to convince his colleagues to wash their hands were ridiculed. He ended up in a mental institution, some say from frustration, and died from a wound infection. A tragic ending for a scientist ahead of his time!

John Snow was another early epidemiologist who laid the groundwork

FIGURE 7.1 Ignaz Semmelweis, defender of motherhood. Before the establishment of microbes as causative agents of disease, Semmelweis realized that childbed fever was transferred from physicians to mothers during delivery. (Source: Semmelweis University Central Library.)

BOX 7.1 Semmelweis and Childbed Fever

Ignaz Semmelweis was a man ahead of his time. This Hungarian-born physician recognized the cause of childbed (puerperal) fever, a disease in which many healthy women developed fatal fevers several days after giving birth. Semmelweis observed that doctors proceeded from autopsy rooms to delivery rooms without washing their hands. Frequently, they wiped them on their aprons, which were already contaminated with body fluids. He postulated that their dirty hands were transferring unknown agents from corpses into the vaginas of pregnant females during examination and delivery. His insistence that medical personnel wash their hands in a chlorinated lime solution was not heeded. Semmelweis's observations came years before the germ theory of disease was established. "Contagion" was a term yet to take on meaning; "bad blood" and mysterious forces were thought to be the cause of childbed fever and other diseases. Semmelweis was ridiculed by his colleagues, his hospital privileges were limited, and his academic rank was reduced. Despondency set in, and his last days were spent in an insane asylum.

Following is a commentary by Irvine Loudon, medical historian, previously a Wellcome Research Fellow and now an Honorary Research Associate at the Wellcome Unit for the History of Medicine at the University of Oxford. He is an honorary Fellow of Green College, University of Oxford.

In 1846, Ignaz Philipp Semmelweis (1818–1865), who was born in Hungary, was appointed what was then by far the largest maternity hospital in the world: the Vienna Maternity Hospital, which was divided into two clinics. Doctors and medical students were taught in the first clinic, midwives in the second, and patients were allocated to the clinics on alternate days. There was no clinical selection of cases for either clinic. From 1840 to 1846, the maternal mortality rate in the first clinic was 98.4 per 1,000 births, while the rate in the second clinic, the midwives clinic, was only

36.2 per 1,000 births. Almost all the maternal deaths were due to puerperal fever. The alarmingly high mortality in the first clinic had defied explanation until Semmelweis was appointed and postulated that the excess deaths in the first clinic were due to the routine procedures carried out in the courses attended by doctors and medical students. Each day started with the carrying out of post-mortems on women who had died of puerperal fever. Then, without washing their hands, the pupils went straight to the maternity wards where they were required, as part of their training, to undertake vaginal examinations on all the women. The pupil midwives in the second clinic did not, of course, carry out post-mortem examinations, and did not undertake routine vaginal examinations.

This was many years before the role of bacteria in diseases was discovered, and Semmelweis suggested that the training procedures of the first clinic resulted in the transfer from the corpses of what he first called "morbid matter," and later "decomposing animal organic matter," on the hands of the students. In 1847, he

(continued)

Semmelweis and Childbed Fever (continued)

therefore introduced a system whereby the students were required to wash their hands in chloride of lime before entering the maternity ward. The result was dramatic. In 1848, the maternal mortality rate fell to 12.7 in the first clinic compared with 13.3 in the second clinic. The large number of deliveries (from 1840 through 1846 there were 42,795 births and 2,977 maternal deaths in the two clinics) ensure that the figures cited above were statistically significant, and the process of admission to the two clinics on alternate days produced, by accident rather than design, a controlled trial.

Semmelweis' observations were clinically astute and potentially of great practical importance. But Semmelweis was a complex, difficult, and dogmatic man, intolerant to the point of paranoia of the slightest criticism, and capable of distorting the views of others when it suited him to do so. Although urged by his friends to publish, he waited for thirteen years before he published his treatise, *The Etiology, Concept, and Prophylaxis of Childbed Fever*, which is dated 1861 but was actually published in 1860. The treatise of over 500 pages contains passages of great clarity interspersed with lengthy,

muddled, repetitive, and bellicose passages in which he attacks his critics. No wonder that it has often been referred to as "the often quoted but seldom-read treatise of Semmelweis." When he wrote the treatise, Semmelweis was probably in the early stages of mental illness that led to his admission to a lunatic asylum in the summer of 1865 where he died a fortnight later. The nature of his illness and cause of death [are] still debated.

Source: I. Loudon, *The Tragedy of Childbed Fever*, p. 88–110 (Oxford University Press, Oxford, United Kingdom, 2000).

AUTHOR'S NOTE *In 1990, I visited the site of the old Broad Street pump. At the site is now a pub called the John Snow Pub (Fig. 7.2) in commemoration of the pump. I had a few beers and reminisced about history.*

FIGURE 7.2 The John Snow Pub and the "ghost of Broadwick Street" (formerly Broad Street). In the foreground is a replica of the "ghost," the Broad Street pump that was the source of contaminated water causing a major outbreak of cholera. (Source: The Wellcome Trust, London.)

for modern methodologies of epidemiology. In 1849 a major epidemic of cholera occurred in the Soho district of London, causing about 500 deaths in the span of only 10 days. Cholera is a bacterial disease (chapter 8) manifested by diarrhea so pronounced that life-threatening amounts of water are lost from the body in a short time, causing death due to dehydration. Snow's epidemiological detective work showed that most of the cholera victims lived in the Broad Street area and drew their water from the Broad Street pump. Further investigation revealed that the pump was contaminated with raw sewage, and when the pump handle was removed, the cholera epidemic was halted. A second outbreak of cholera occurred in London in 1854. Snow's sleuthing revealed that most of the cholera victims purchased their drinking water from the Southwark and Vauxhall Company; the company's source of water was the Thames River downstream from the site where raw sewage was discharged into the river. The Lambeth Company, another water supplier, obtained its water further upstream; the incidence of cholera in the population using Lambeth water was much lower. The actual bacterial contaminant in the water in both outbreaks was *Vibrio cholerae*.

Epidemiologists focus on the frequency and distribution of diseases in populations and classify diseases as **sporadic**, **endemic**, **epidemic**, and **pandemic**. Sporadic diseases are those which occur only occasionally and at irregular intervals in a random and unpredictable fashion. Typhoid fever, eastern equine encephalitis, and tetanus are examples. Diseases which are continually present at a steady level in a population and pose little threat to the public health are endemic diseases. The common cold, mumps, and chicken pox are endemic across the United States, whereas Lyme disease is endemic in certain seasons of the year, primarily in some New England states. A disease is said to be epidemic when there is a sudden increase in the **morbidity** (illness rate) and in the **mortality** (death rate) above the usual levels and when it causes a potential public health problem. Throughout history, epidemics have resulted in more deaths than have wars, and they have influenced the course of history. Plague has bedeviled humankind at least since the reign of Emperor Justinian in the 6th century; the 14th-century epidemic was particularly devastating. Smallpox, cholera, and typhus fever are other examples of epidemics (chapters 8 and 9). Epidemics may arise from an explosion of sporadic or endemic diseases or, it would seem, from "out of nowhere." Pandemic diseases are those which spread across continents and may be worldwide; acquired immune deficiency syndrome (AIDS) is a pandemic. Cholera has been responsible for pandemics on several occasions over time. In 1918, swine influenza was perhaps the greatest pandemic of all times. Epidemiologists use a variety of graphs, charts, and maps as tools to illustrate the frequency and distribution of diseases (Fig. 7.3). The term epidemic has been borrowed to indicate a variety of conditions unrelated to infectious diseases that are present beyond the norm. For example, college officials talk of grade inflation as a problem of epidemic proportions, as are school violence, obesity, and cancer.

With reference to the source and the spread of epidemics caused by microbes, epidemiologists have described **common-source epidemics** and

FIGURE 7.3 Graphs, charts, and maps used by epidemiologists to illustrate disease frequency and distribution. **(A)** Geographic distribution of human ehrlichioses in the United States. HGE, human granulocytic ehrlichiosis; HME, human monocytic ehrlichiosis. **(B)** *Salmonella enterica* serovar Enteritidis isolation rates by region, United States, 1974 to 1994. **(C)** Primary causes of chronic liver diseases in a selected area (Jefferson County, Alabama). **(D)** Causes of death worldwide, 1993. (Adapted from illustrations provided by the CDC; data in panel D originally from WHO [*Global Health Situations and Projections, Estimates 1993*, published 1995].)

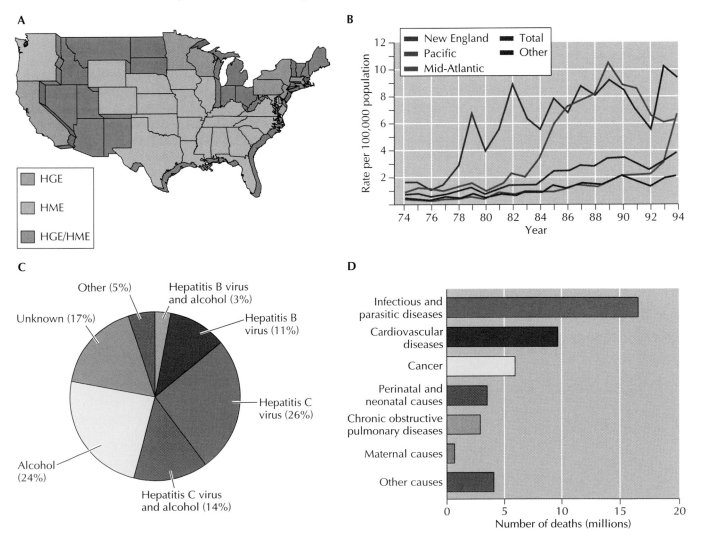

propagated epidemics. Common-source epidemics arise from contact with a single contaminated source and are usually associated with fecally contaminated foods and water. Typically, a large number of people become ill quite suddenly, and the disease peaks rapidly in the population. A propagated epidemic is the result of direct person-to-person (horizontal) transmission; the microbe is spread from infected individuals to noninfected susceptible individuals. As compared with common-source epidemics, the number of infected individuals rises more slowly and decreases gradually. Chicken pox, measles, and mumps are examples of propagated epidemics. Figure 7.4 illustrates the courses of common-source and propagated epidemics.

The number of individuals in a population who are immune (nonsusceptible) to a particular disease as compared with those who are nonimmune (susceptible) is an important factor in the occurrence of epidemics. Immunity can be the result of having had a particular infection or of having been immunized. The term **herd immunity** (group immunity) refers to the proportion of immunized individuals in a population. Disease can only be spread to susceptible individuals, and, therefore, the smaller the number of susceptible individuals, the less opportunity for contact between them and infected individuals. Public health officials strive to maintain high levels of herd immunity against communicable diseases in order to minimize the chances that they will progress to epidemic status. Hence, immunization is required against a variety of diseases beginning in the elementary grades; proof of an up-to-date immunization history is required for college admission.

A decrease in herd immunity can lead to reemergence of a disease. A case in point is the epidemic of diphtheria which occurred in the newly independent states of the former Soviet Union in the early 1990s. A decline in the public health infrastructure resulted in fewer children receiving diphtheria vaccination and a decline in herd immunity. When the disease was introduced into the population, possibly by returning military personnel, diphtheria reached epidemic proportions.

Smallpox immunization is no longer practiced because of the eradication of this disease, as proclaimed by the World Health Organization (WHO) in 1980. Only children born before the war against smallpox was won were vaccinated against smallpox, resulting in little herd immunity and a susceptible world population. This is a potential nightmare, since

FIGURE 7.4 A comparison of the courses of common-source and propagated epidemics.

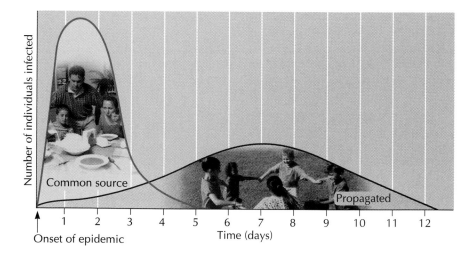

smallpox virus is at or near the top of the list of potential biological weapons (chapter 14).

Surveillance of disease outbreaks and of factors that could trigger outbreaks is an important aspect of the mission of public health organizations throughout the world, including WHO, the CDC, and agencies at the state and local levels. To keep track of these diseases in the United States, physicians are required to report cases of certain diseases, termed notifiable diseases, to their local health departments; these are then reported to the CDC. In 1994, 49 diseases were listed as notifiable; by 1999, the number of notifiable diseases had grown to 55 (Table 7.1). Some of these diseases are discussed in chapters 8, 9, and 10. To further assist public health and medical personnel, the CDC publishes the journal *Emerging Infectious Diseases* as

TABLE 7.1 Infectious diseases designated as notifiable at the national level, effective May 1999[a]

AIDS	Meningococcal disease
Anthrax	Mumps
Botulism	Pertussis
Brucellosis	Plague
Chancroid	Poliomyelitis, paralytic
Chlamydia trachomatis, genital infection	Psittacosis
Cholera	Q fever
Coccidiodomycosis	Rabies, animal
Cryptosporidiosis	Rabies, human
Cyclosporiasis	Rocky Mountain spotted fever
Diphtheria	Rubella (German measles)
Ehrlichiosis	Salmonellosis
Encephalitis, arboviral	Shigellosis
Enterohemorrhagic *Escherichia coli*	Streptococcal disease, invasive, group A
Gonorrhea	Streptococcal toxic shock syndrome
Haemophilus influenzae, invasive disease	*Streptococcus pneumoniae*, drug-resistant invasive disease
Hansen disease (leprosy)	
Hantavirus pulmonary syndrome	*Streptococcus pneumoniae*, invasive disease (<5 years)
Hemolytic uremic syndrome, postdiarrheal	
Hepatitis, viral, acute	Syphilis
Hepatitis, viral, perinatal hepatitis B virus infection	Syphilis, congenital
	Tetanus
HIV infection, adult (>13 years)	Toxic shock syndrome
HIV infection, pediatric	Trichinellosis
Legionellosis	Tuberculosis
Listeriosis	Tularemia
Lyme disease	Typhoid fever
Malaria	Varicella (deaths only)
Measles	Yellow fever

[a]Source: CDC.

well as the *Morbidity and Mortality Weekly Report* (*MMWR*), which contains data organized by states on morbidity and mortality of particular diseases in the United States and throughout the world (Fig. 7.5).

THE CYCLE OF MICROBIAL DISEASE

For infectious diseases to exist at the community level, a chain of linked factors needs to be present, somewhat reminiscent of a parade of circus elephants linked trunk to tail. These factors are reservoirs, modes of transmission, portals of entry, and portals of exit (Fig. 7.6). An understanding of these factors is imperative to attempt to break the cycle somewhere along the path. For example, if insects are involved in transmission, then controlling their population is a target; for those microbes that are transmitted by drinking water, providing safe drinking water is a goal. Shrinking the reservoir (where the microbes exist in nature) is a potential target for other diseases. In some instances, a combination of targets is preferable.

For a particular microbial disease to exist, there has to be a pathogen as the causative agent and a host in which the pathogen takes up residence. The potential for disease to occur and its outcome are a result of the complex interaction between the number of invading microbes and their virulence (chapter 6) and the host immune system (chapter 11). Communicable diseases are infectious diseases in which the pathogen can be transmitted from its reservoir to the host portal of entry.

Reservoirs of Infection

A **reservoir** is a site in nature in which microbes survive (and possibly multiply) and from which they may be transmitted. All pathogens have one or more reservoirs, without which they could not exist. Knowledge and identification of these reservoirs are important, since the reservoirs are prime targets for preventing, minimizing, and eliminating existing and potential

FIGURE 7.5   (A) *Emerging Infectious Diseases;* **(B)** *Morbidity and Mortality Weekly Report.* The CDC publishes these materials as references for physicians and other public health personnel. (Author's photos.)

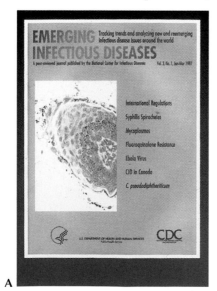

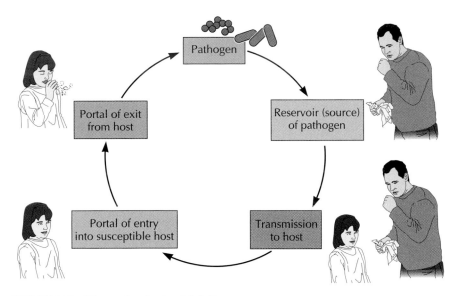

FIGURE 7.6   The cycle of microbial disease.

epidemics. The facts that humans are the only reservoir of smallpox and that person-to-person transmission of smallpox takes place were key factors in the eradication of this disease. Additionally, humans are the only known reservoir for gonorrhea, measles, and polio. Animals, as well as plants and nonliving environments, also serve as reservoirs. In some cases, the source of the pathogen is distinct from the reservoir and is the immediate location from which the pathogen is transmitted. For example, in typhoid fever, the reservoir may be an individual with an active case of the disease who sheds typhoid bacilli in feces; the immediate source would be water or food contaminated with fecal material. On the other hand, in most sexually transmitted diseases, the human body serves as both reservoir and source.

Active carriers are those individuals who have a microbial disease, whereas **healthy carriers** have no symptoms and unwittingly pass the disease on to others. **Typhoid Mary**, a cook and healthy lifetime carrier of typhoid fever, was responsible for about 10 outbreaks, 53 cases, and three deaths due to typhoid fever during her lifetime; her dilemma is revealed in chapter 8. **Chronic carriers** are those who harbor a pathogen for long periods after recovery, possibly throughout their lives, without ever again becoming ill with the disease. In the case of chronic (and healthy) carriers of typhoid fever, removal of the gallbladder may be effective in eliminating the carrier state; intensive therapy with antibiotics works in other cases. Tuberculosis is another disease in which carriers play a significant role. Depending on the particular infection, carriers discharge microbes via portals of exit, including respiratory secretions, feces, urine, and vaginal and penile discharges.

Domestic and wild animals serve as reservoirs for about 150 species of pathogenic microbes that can affect humans. These diseases are referred to as **zoonoses** (Table 7.2). Microbes of animals that are most closely related to humans have the greatest chance of making the "species leap" to humans. Consider, for example, the AIDS virus, which is thought to have a reservoir in green monkeys and is now a human pathogen. Monkeys are also reser-

TABLE 7.2   Selected zoonotic diseases

Transmission by arthropod bites
Bacteria
Ehrlichiosis
Relapsing fever
Lyme disease
Rocky Mountain spotted fever
Plague
Typhus fever
Viruses
Yellow fever
Eastern equine encephalitis
West Nile virus disease
Rift Valley fever
Dengue fever
Protozoans
Babesiosis
Sleeping sickness
Malaria
American trypanosomiasis
Leishmaniasis
Transmission via food, water, or animal bites
Bacteria
Undulant fever
Leptospirosis
Anthrax
Cat scratch fever
Tularemia
Viruses
Rabies
Hantavirus disease
Viral gastroenteritis
Protozoans
Giardiasis
Cyclospora infection
Toxoplasmosis

voirs for the microbes that cause malaria, yellow fever, and numerous other diseases. The reservoirs for the bacteria that cause Lyme disease, a major problem in the northeastern United States, are wild deer and mice. Hantavirus pulmonary syndrome, a relatively new disease in the United States, utilizes a variety of rodent species, particularly the deer mouse, as reservoirs. Many mammals, including dogs, cats, raccoons, skunks, foxes, and bats, serve as reservoirs for rabies.

Eradication of zoonotic diseases is particularly challenging and difficult since it is, ultimately, dependent on eradicating the reservoirs. Malaria is a protozoan disease transmitted by mosquitoes. Intensive spraying with the pesticide DDT (dichlorodiphenyltrichloroethane) in the 1940s markedly reduced the mosquito population and the number of malaria cases. Unfortunately, the mosquitoes developed resistance to DDT, leading to a reemergence of malaria.

Nonliving Reservoirs

Some organisms are able to survive and multiply in nonliving environments. Soil and water are the major nonliving reservoirs of infectious diseases. The tetanus bacillus and the botulinum bacillus, both members of the same bacterial group, are sporeformers and thus can survive for many years in soil. These organisms are part of the normal flora of horses and cattle and are deposited in their feces onto the soil. The use of animal fertilizers contributes to their distribution. Certain helminth (worm) parasites, e.g., hookworms, deposit their eggs onto the soil, establishing a reservoir for human infection (Fig. 7.7).

Contaminated drinking water and foods are major reservoirs for many microbes that cause gastrointestinal tract disease, ranging from mild to severe to fatal. The list includes bacteria, viruses, and protozoans (Table 7.3). Many of these diseases are discussed in chapters 8, 9, and 10. Because of the potential for an outbreak of waterborne and food-borne illnesses, local departments of public health devote considerable attention to sanitary measures designed to minimize risks; their activities include monitoring food service establishments, beaches, swimming pools, and certification of food handlers.

FIGURE 7.7   Soil can be a reservoir for microbes and helminth eggs. This child is at increased risk for infection. (Source: WHO.)

TABLE 7.3   Water and food as reservoirs of infection

Type of microbe	Examples of water- and food-borne infections
Bacteria	Salmonellosis, cholera, *E. coli* infection
Viruses	Hepatitis A, poliomyelitis, viral gastroenteritis
Protozoans	Giardiasis, amebiasis, cryptosporidiosis
Worms	Ascariasis, trichinellosis, *Trichuris* infection

Transmission

The next link in the cycle of disease is **transmission**, the bridge between reservoir and portal of entry (Fig. 7.6). Transmission is the mechanism by which an infectious agent is spread through the environment to another person. More simply put, transmission answers the question "How do you get the disease?" (In chapters 8, 9, and 10, the diseases are grouped by mode of transmission.) There are several modes of transmission, and they can be grouped into two major pathways, direct and indirect. Each of these, in turn, can be subgrouped into three categories (Table 7.4).

Direct Transmission

The most common type of **direct transmission** is person-to-person contact, in which the infectious agent is directly and immediately transferred from a portal of exit to a portal of entry. Sexual contact, kissing, and touching are the most common examples. Transmission is facilitated by contact of the warm, moist mucous membranes of one individual with the warm, moist mucous membranes of another, as occurs in sexually transmitted diseases. This is an example of **horizontal transmission** (i.e., transmission from one person to another). Droplet transmission is also direct and horizontal and involves the projection of infected spray from coughing, sneezing, talking, and laughing onto the conjunctivae of the eyes or onto the mucous membranes of the nose or mouth. Influenza, whooping cough, and measles are spread by droplets. Droplets have a size limit of about 1 to 4 micrometers in diameter and travel less than 1 meter; as many as 20,000 droplets may be produced in a sneeze (Fig. 7.8). (Think about that the next time you cough or sneeze directly into the crowded environment of the classroom!)

A second type of direct transmission involves animal bites, rabies being the most common example. The virus is directly transmitted from the saliva of the rabid animal onto the skin and underlying tissues. Finally, transplacental transmission is an example of **vertical transmission**, in which the pathogens are passed from mother to offspring across the placenta (AIDS, measles, and chicken pox), in breast milk, or in the birth canal (syphilis and

TABLE 7.4   Modes of transmission

Direct	Indirect
Contact (e.g., kissing, sneezing, coughing, singing, sexual contact)	Vehicles (fomites, e.g., doorknobs, eating utensils, toys)
Animal bites	Airborne (via aerosols created by, e.g., shaking bedsheets, sweeping, mopping)
Transplacental	Vectors (e.g., mosquitoes, ticks, flies)

FIGURE 7.8 Droplet transmission. As many as 20,000 droplets may be produced during a sneeze. It is important to carry a handkerchief or tissue and to cover your nose and mouth when sneezing. (Source: Armed Forces Institute of Pathology.)

gonorrhea). Note that in all these categories of direct transmission, there are no intermediaries. Microbes are transferred by the contact of portals of exit with portals of entry. These portals are described below.

Indirect Transmission

Indirect transmission involves the passage of infectious material from a reservoir or source to an intermediate agent and then to a host. The intermediate agent can be living or nonliving. Vehicle-borne transmission is accomplished by food, water, biological products (organs, blood, blood products), and **fomites** (inanimate objects).

Waterborne transmission is a serious problem throughout the world and is a major cause of death in many developing countries as a result of fecal-oral passage, in which pathogens are transmitted from the feces of one individual to another by hand-to-mouth transfer. Public, semiprivate, and private water supplies must all be carefully monitored for the presence of fecal pathogens. Water can serve as both a reservoir and a transmitter of infectious agents. Food-borne infections are an increasing problem; like water, food is both a reservoir and an agent of transmission.

Fomites play a significant role in the transmission of infectious agents. The list of fomites is seemingly endless and includes objects in common use, such as doorknobs, toilet seats, faucets, computer keyboards, and exercise equipment. Toys are fomites and contribute to illness in children at day care centers. Surgical instruments, medical equipment (e.g., catheters, intravenous equipment, and syringes), bedding, and soiled clothing are also fomites. An interesting study involving soiled saris as fomites, conducted in 51 slum areas in Dhaka, Bangladesh, revealed a positive correlation between the number of misuses of dirty saris and episodes of childhood diarrhea.

The list of fomites and their role may cause you some concern, but there is at least a partial solution to the problem. The simple act of frequent hand washing has been shown to markedly reduce hand-to-mouth (and nose and eye) infection. Frequent wiping of tabletops and counters with disinfectants is effective and a sign of good hygiene in a sanitation-conscious restaurant. In health and exercise clubs, it has become a widespread practice to wipe down exercise equipment after use; it is a sign of bad manners not to do so (Fig. 7.9).

Airborne transmission by **aerosols** is the second type of indirect transmission. Aerosols are suspensions of tiny water particles and fine dust in the air; they are distinct from droplet nuclei, as they are larger than 4 micrometers, travel more than 1 meter, and are small enough to remain airborne for extended periods. Outbreaks of Q fever, Legionnaires' disease, and psittacosis (from infected birds) are caused by aerosols. The microbes in aerosols may not come directly from humans or animals but may be present in dust particles and are disseminated by changing bed linens, sweeping, mopping, and other activities; bacteria and viruses can survive for months in dust particles. Hospital personnel are keenly aware of this, as reflected in the practice of using wet mops and damp cloths to wipe surfaces.

The third type of indirect transmission is by vectors (living organisms that transmit microbes from one host to another). The term "vector" is sometimes more broadly used to cover any object that transfers microbes, but this is incorrect usage. Ticks, flies, mosquitoes, lice, and fleas are the most common vectors, and they belong to the same biological phylum, the **Arthropoda,** along with lobsters and crabs. (It may be difficult to understand what flies, fleas, ticks, and lobsters have in common, but the edibility of lobsters certainly sets them apart!) Arthropods are invertebrate animals with jointed appendages ("arthro" means joint, as in "arthritis" [inflammation of joints], and "pod" means foot, as in "podiatrist" [foot doctor]). Further, they all have segmented bodies and a hardened exoskeleton. The arthropods are members of the largest phylum and consist of many diverse species that are divided into three groups (Table 7.5). They are considered to be the most successful of all living animals in terms of the huge number of species and their distribution. (Think about their success when you swat mosquitoes!)

Spiders, ticks, and mites hatch from eggs as six-legged larvae and undergo metamorphosis to eight-legged adults with two body segments and mouth parts adapted for the sucking of blood. Ticks transmit a variety of infectious diseases, including Lyme disease, Rocky Mountain spotted fever, babesiosis, and ehrlichiosis. These diseases are described in chapter 8. In

TABLE 7.5 The phylum Arthropoda

Subphylum Chelicerata
Scorpions
Chiggers
Spiders
Daddy longlegs
Mites[a]
Horseshoe crabs
Ticks[a]
Subphylum Crustacea
Water fleas
Isopods
Fairy shrimp
Crabs
Copepods[a]
Lobsters
Barnacles
Shrimp
Subphylum Uniramia
Insects (many subgroups)
Flies[a]
Fleas[a]
Mosquitoes[a]
Lice[a]
Centipedes
Millipedes

[a]Vectors of human disease.

FIGURE 7.9 Exercise machines can be reservoirs for microbes. You may be exercising to maintain good health, but poor personal hygiene may place others at risk. Wipe down exercise machines after use. (Author's photo.)

addition to their role as vectors, some ticks are important reservoirs because they exhibit **transovarial transmission** (the passage of microbes into their eggs).

Insects are an extremely large group of organisms with well over 500,000 species. They are familiar to you as pests because (depending on the species) they bite, are "ugly" and "disgusting," and are associated with uncleanliness. Insects have three body segments (the head, the thorax, and the abdomen) and six legs; some have two pairs of wings. Some have mouth parts adapted for puncturing the skin and sucking blood. "Kissing bugs" suck blood from their hosts and are vectors of a significant disease in Central and South America known as Chagas disease (chapter 10).

Apart from their nuisance value, arthropods play a significant role in the cycle of infectious diseases and are targets for public health efforts in disease control. Mosquito abatement programs have been carried out on numerous occasions to control malaria. More recently, New York, Connecticut, Massachusetts, and Rhode Island have been threatened by West Nile virus, a mosquito-borne virus, resulting in insecticide spraying to control the mosquito population. Several viruses that cause encephalitis (brain swelling and other neurological damage) belong to a group called arboviruses and are so named because they are arthropod borne (shortened to "arbo").

Arthropods can be either **mechanical vectors** or **biological vectors**. Mechanical vectors transmit microbes passively on their feet and other body parts; the microbes do not invade, multiply, or develop in the vector. House flies, for example, feed on exposed human and animal fecal material and then transfer microbes on their feet to food and eating utensils. Typhoid fever and other gastrointestinal diseases characterized by diarrhea or dysentery may be spread in this way. Covering of human and animal waste to avoid exposure to flies is an obvious answer, but this is not always possible in poverty-stricken areas, under wartime conditions, in refugee camps, and in other circumstances involving large groups of people when it is difficult to maintain good sanitation. Even in the best of circumstances, flies have access to dog feces and can mechanically transmit microbes to kitchen areas. It is disturbing to think that a fly that has just lunched on dog feces in your backyard or in a neighboring park may walk across the chicken salad that you prepare for a picnic. Cockroaches (you probably shudder at the term) also serve as mechanical vectors; remember this when you see them marching across a kitchen counter. An article in a Chinese newspaper in December 2000 stated that Beijing was in the grip of a roach menace. Roaches were invading restaurants, hotels and motels, and even hospitals. Roaches carry more than 40 kinds of bacteria, some of which are pathogens.

Biological vectors, unlike mechanical vectors, are necessary components in the life cycles of many infectious disease agents and are required for the multiplication and development of the pathogen; transmission by biological vectors is an active process. As an example, when a mosquito picks up the malaria parasite while having a blood meal on an infected person, the parasites are not at a stage that is infective. Further development within the body of the mosquito results in parasites that are now infective for hosts. Depending on the particular vector, parasites may be carried in the saliva and injected into the tissue while biting; other vectors have the nasty habit of regurgitating infectious secretions into and around the bite, and others defecate infectious material onto the bite area. Itching usually results, and scratching facilitates entry of the parasite. Mosquitoes, fleas, lice, and ticks are common biological vectors and are responsible for the transmission of many microbial and helminthic diseases (Table 7.6).

In 1994, an outbreak of plague occurred in India and caused widespread panic. Plague is a bacterial disease in which rats are the reservoirs and fleas are the biological vectors. (The disease is discussed in chapter 8.) Public health officials around the world worried that a pandemic might occur. Fortunately, the rapid implementation of public health measures, including rodent control, halted a potential disaster. Rats are important reservoirs for other microbial diseases, including Lassa fever and hantavirus disease. Rats have been known to stow away on ships and carry microbes to distant shores.

Vector-borne infectious diseases are emerging and reemerging throughout the world. Factors responsible include genetic changes in both vectors and pathogens, resulting in resistance to insecticides and drugs; public health policy; funding directed toward emergency response rather than prevention; and societal changes. In the first half of the 20th century, considerable progress was made in the fight against vector-borne diseases.

TABLE 7.6   Diseases transmitted by arthropod bites[a]

Disease	Distribution	Vector
Protozoan and helminthic diseases		
Filariasis	Central and South America, Africa, Indian subcontinent and other parts of Asia	Mosquitoes
Babesiosis	United States, Europe	Ticks
American trypanosomiasis	South and Central America	Kissing bug
Onchocerciasis	Central America, tropical South America, Africa	Blackflies
African trypanosomiasis (sleeping sickness)	West, Central, and East Africa	Tsetse flies
Leishmaniasis	Central and South America, Africa, Indian subcontinent and other parts of Asia, Europe	Sand flies
Viral diseases		
Yellow fever	Tropical South America, Africa	Mosquitoes
Colorado tick fever	United States (Rocky Mountains)	Mosquitoes
Rift Valley fever	Eastern and southern Africa, sub-Saharan Africa, Madagascar	Mosquitoes
West Nile virus disease	Asia, Africa, United States	Mosquitoes
Dengue fever	India, Southeast Asia, Pacific, South America, Caribbean	Mosquitoes
Bacterial diseases		
Plague	Southeast Asia, Central Asia, South America, western North America	Fleas
Relapsing fever	South America, Africa, Asia, western North America	Lice or ticks
Lyme disease	Europe, United States, Australia, Japan	Ticks
Typhus fever (endemic)	Worldwide	Fleas
Typhus fever (epidemic)	Eastern Europe, Asia, Africa, South America	Lice

[a]Adapted from CDC's *Travel Info—1996*.

Most of these diseases were brought under control, and by the 1960s their threat, except in Africa, was greatly diminished (Table 7.7). In fact, malaria was eliminated from many countries of the world. However, no country is immune to the potential threat and spread of microbial diseases. A case in point is West Nile virus, which emerged in 1999 in the state of New York. This was the first appearance of the virus in the United States; it had been previously reported only in Africa and Asia.

In 1989, in response to the growing problem of vector-borne diseases, the CDC established the Division of Vector-Borne Infectious Diseases in Fort Collins, Colo., which is responsible for information, surveillance, prevention, and control of vector-borne diseases. The division is charged with the investigation of national and international epidemics of bacterial and viral diseases transmitted to humans by arthropods, primarily mosquitoes, ticks, and fleas. To prevent and control these diseases, biologists in the division work with the three populations involved: the pathogen, the host, and the vector.

Portals of Entry

The next step in the cycle of disease involves access into (or onto) the body. Some microbes have a single preferred portal, but others have more than one. To some extent, human behavior influences the **portals of entry.** The most common site of entry for sexually transmitted diseases is the urethra in males and the vagina in females, but the throat and the rectum may also serve for entry as a result of oral and anal sex, respectively. The portal of entry is an important consideration in the outcome of host-parasite interactions. Bubonic plague results from the bite of a plague-infected flea, but pneumonic plague, a different manifestation caused by the same bacterium, results when the respiratory tract is the portal of entry into the lungs. Anthrax, also a bacterial disease, is another example. There are three varieties

TABLE 7.7 Successful vector-borne disease control or elimination programs[a]

Disease	Location	Year(s)
Yellow fever	Cuba	1900–1901
Yellow fever	Panama	1904
Yellow fever	Brazil	1932
Anopheles gambiae infestation	Brazil	1938
Anopheles gambiae infestation	Egypt	1942
Louse-borne typhus	Italy	1942
Malaria	Sardinia	1946
Yellow fever	Americas	1947–1970
Yellow fever	West Africa	1950–1970
Malaria	Americas	1954–1975
Malaria	Global	1955–1975
Onchocerciasis	West Africa	1974–present
Bancroftian filariasis	South Pacific	1970s
Chagas disease	South America	1991–present

[a]From *Emerging Infectious Diseases*, vol. 4, no. 3, 1998.

of anthrax. Cutaneous anthrax results when the skin is the portal of entry; gastrointestinal anthrax occurs as the result of oral ingestion of the bacteria, while inhalation anthrax is the result of the organisms' entering through the respiratory tract. Table 7.8 and Fig. 7.10 summarize the portals of entry by anatomical sites. Body orifices (openings to the outside), including the mouth, nose, ears, eyes, anus, urethra, and vagina, and penetration of the skin make it possible for microbes to gain access into the body.

Portals of Exit

Once microbes have gained access into the body, whether or not disease results is determined by the interaction between the number of pathogens and their virulence (chapter 6) and the immune system of the host (chapter 11). To complete the cycle of infectious disease (Fig. 7.6) and to allow the spread of disease into the community, pathogens require a **portal of exit** (Table 7.8 and Fig. 7.10). In some cases, the portal of exit is related to the area of the body that is infected. This is particularly true for organisms that cause diseases of the respiratory tract (such as colds and influenza). On the other hand, this is not always the case. For example, the spirochete that causes syphilis utilizes the urogenital tract as the portal of exit but can invade the skin and nervous system. The eggs of some disease-producing

TABLE 7.8 The infectious disease cycle: portals of entry and exit

Portal	Examples of disease or microbe
Portals of entry	
Mucous membranes	
Respiratory tract	*Streptococcus pneumoniae*, tuberculosis, Legionnaires' disease, influenza, hantavirus, common cold
Gastrointestinal tract	Cholera, salmonellosis, *E. coli*, rotavirus, poliomyelitis, guinea worm disease, giardiasis
Urogenital tract	Gonorrhea, chlamydia, AIDS, genital warts
Skin (hair follicles, sebaceous glands, wounds, arthropod bites)	Boils, abscesses, cutaneous anthrax, rabies, warts, hookworm, schistosomiasis, malaria
Blood (transfusion, blood products, arthropod bites, placental transfer)	Congenital syphilis, AIDS, German measles, toxoplasmosis
Portals of exit	
Respiratory tract	Tuberculosis, Legionnaires' disease, influenza, common cold
Gastrointestinal tract	Cholera, salmonellosis, rotavirus, poliomyelitis, hookworm, guinea worm disease
Urogenital tract	Gonorrhea, chlamydia, HIV, schistosomiasis, genital herpes
Skin	Impetigo, boils, abscesses, warts, cold sores, fever blisters, guinea worm disease
Blood (transfusion, blood products, arthropod bites, placental transfer)	Congenital syphilis, toxoplasmosis, HIV, German measles, malaria

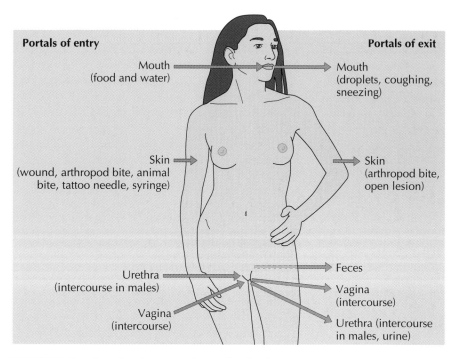

FIGURE 7.10   Portals of entry and portals of exit.

worms exit the body in fecal material, survive in soil, and remain infectious for long periods of time. The AIDS virus exits the body through semen and vaginal discharges as well as through the blood. Arthropod-borne diseases enter the body through the bites of insects, and these insects also serve as avenues of exit. A mosquito biting an individual with malaria will pick up the parasite.

NOSOCOMIAL (HOSPITAL-ACQUIRED) INFECTIONS

When you enter a hospital, not only do you need to worry about the notoriously poor hospital food, but there is a 5 to 15% chance that you will acquire an infection during your stay. People are hospitalized because they are ill and require treatment beyond what home care can provide. Ironically, while in the hospital they are at an increased risk for contracting an infectious disease. **Nosocomial infections** are infections acquired by patients during their hospital stay or during their confinement in other long-term health care facilities; infections acquired by hospital personnel are also considered nosocomial. Based on the number of hospital admissions, estimates are that 2 million to 4 million cases, resulting in 20,000 to 40,000 deaths, occur each year because of nosocomial infections in the United States, accounting for about 50% of all the major complications of hospitalization.

The Hospital Environment as a Source of Nosocomial Infections

What are the factors unique to the hospital environment that place patients and hospital staff at an increased risk for acquiring infections? To begin with, the patient population consists of individuals who are ill and may have an immune system that is compromised (weakened), increasing their susceptibility to pathogens and opportunistic microbes, including their own normal flora. Surgical and other invasive procedures, drugs to purposely suppress the immune system (as in cancer therapy and organ transplantation), prolonged bed rest, and restrictive diets are necessary components of treatment but are traumatic to the body and counterproductive to the maintenance of a healthy immune system.

Antibiotics are heavily misused in hospitals to treat or to prevent infections, fostering the development of antibiotic-resistant strains of bacteria. Diagnostic and treatment protocols frequently involve extensive surgery and the use of invasive procedures, including the insertion of catheters into the urethra, swallowing of tubes, insertion of needles into veins for intravenous therapy, and insertion of nasal tubes. Thermometers, bedpans, urinals, eating utensils, and night table surfaces are only a few of the many fomites that pose potential risk. Hence, the equipment and devices involved in patient care contribute to transmission. Despite the risk of nosocomial infection, be assured that the advances in medicine far outweigh the risks of hospitalization.

The hospital staff, including physicians, nurses, laboratory technicians, and maintenance workers, may unwittingly (and carelessly) transmit microbes from patient to patient; some may be healthy carriers.

All of the factors involved in the cycle of infectious diseases are present in a concentrated way in hospitals and in long-term health care facilities, establishing these environments as reservoirs of pathogens. A relatively small number of bacterial species are responsible for the majority of nosocomial infections; these are organisms that are common to the environment. Some sites in the body are more prone to nosocomial infections than others; the urinary tract is the most susceptible, followed by surgical sites, the respiratory tract, and other sites (Table 7.9).

Control Measures

Nosocomial infections are a serious and growing problem in hospitals and in other medical facilities in terms of mortality, morbidity, and financial burden, and every hospital has strategies of prevention and control. The fact that the frequency and spectrum of antibiotic-resistant organisms are on the rise poses a serious problem in intensive care units (Fig. 7.11). All hospitals are required to have an infection control officer and an infection control committee in order to maintain accreditation by the American Hospital Association. Hospitals spend considerable time and money in an effort to minimize the possibility of microbial contamination in all aspects of

TABLE 7.9   Body site distribution of nosocomial infections

Site	% of all nosocomial infections
Urinary tract (urinary catheterization)	~50
Surgical site (intestinal surgery, joint replacement surgery)	~25
Lower respiratory tract (respirators and other breathing aids)	~12
Bacteremia (blood infection)	~6
Other (including skin)	~7

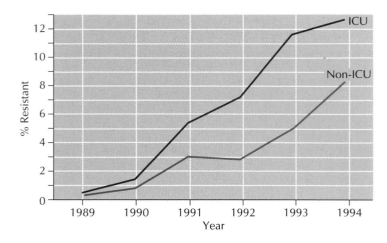

FIGURE 7.11   Nosocomial infections on the rise. The graph shows the percentages of nosocomial enterococci reported to be resistant to the antibiotic vancomycin in intensive care units (ICU) and nonintensive care units (non-ICU) from 1989 to 1994. (Adapted from a graph provided by the National Nosocomial Infections Surveillance System, CDC.)

the hospital environment. Training of hospital personnel in basic infection control procedures is paramount and centers on isolation procedures, proper techniques of disinfection and sterilization, and the surveillance and reporting of cases of infectious diseases in both patients and staff. Insect control, good housekeeping, and safe practices for the disposal of feces, urine, bandages, dressings, and other potentially contaminated materials are also the responsibility of the infection control committee. Education emphasizing the importance of the simple act of hand washing is vital; numerous studies have demonstrated that this single simple procedure is the most important practice in minimizing nosocomial infections. In some studies, shockingly low rates (well under 50%) of hand washing by health care workers, including physicians and nurses, have been revealed.

EPIDEMIOLOGY OF FEAR

The public is both fascinated and terrified by outbreaks of disease. Ebola virus, West Nile virus, bird flu, and *E. coli* have captured attention and led to an explosion of television programs and popular books. The fear of epidemics can reach epidemic proportions. The threat of infectious disease, according to some experts, is out of proportion. West Nile virus, a mosquito-borne virus which infects birds, emerged in the Western Hemisphere for the first time in the summer of 1999 in New York State and caused illness in more than 60 people and the death of 7. By the summer of 2000, infected birds were detected in New York, Massachusetts, Connecticut, and Rhode Island, posing a potential threat in these areas and generating concern among public health officials about an epidemic of fear. As a Massachusetts Department of Health spokeswoman stated, "The message we're trying to get out is to stop people from panicking. West Nile virus is not a major public health threat. It's something people should be aware of, and take precautions, but not let it interrupt their summer."

In his first inaugural address, on March 4, 1933, President Franklin Delano Roosevelt spoke eloquently of the danger of fear. His often quoted words were, "So first of all let me assert my firm belief that the only thing we have to fear is fear itself—nameless, unreasoning, unjustified terror which paralyzes needed efforts to convert retreat into advance."

The take-home message is that awareness, surveillance, common sense precautions, and calmness are paramount. Fear can be paralyzing.

OVERVIEW Epidemiologists classify disease as sporadic, endemic, epidemic, or pandemic, depending on its frequency and distribution. These categories are not absolute; a particular disease can slide from one classification to another. Common-source epidemics arise from contact with a single contaminant, resulting in a large number of people becoming ill suddenly; the disease peaks rapidly. Propagated epidemics are characterized by direct person-to-person (horizontal) transmission, a gradual rise in the number of infected individuals, and a slow decline.

A chain of linked factors is required for infectious diseases to exist and to spread through a population. These factors are reservoirs of disease, transmission, portals of entry, and portals of exit. Understanding the characteristics of microbes and the diseases they cause is necessary to break the cycle somewhere along its path. The reservoir is the site where microbes exist in nature and from which they can be spread. Active carriers, healthy carriers, and chronic carriers are reservoirs, as are wild and domestic animals. Nonliving reservoirs include contaminated water, food, and soil.

Transmission is the bridge between reservoir and portal of entry. Person-to-person contact is the most common type of horizontal direct transmission and allows for the immediate transfer of microbes. Vertical transmission is another type of direct transmission and is categorized by the passage of pathogens from mother to offspring across the placenta, in the birth canal during delivery, or in breast milk. In direct transmission there are no intermediaries. Indirect transmission involves the passage of materials from a reservoir or source to an intermediate agent and then to a host. The intermediate agent can be nonliving or living. Water, food, fomites, and aerosols are significant nonliving vehicles of indirect transmission. Vectors are living organisms (arthropods) that transmit microbes from one host to another. Mechanical vectors passively transfer microbes on their feet or other body parts, while biological vectors are required for the multiplication and development of the pathogen within the vector.

Portals of entry are the next consideration in the cycle of disease. Some microbes have a single preferred portal of entry into the body, while others have more than one. Body orifices, including the mouth, nose, ears, eyes, anus, urethra, and vagina, are portals of entry; the skin can be penetrated and is another portal of entry.

For the cycle of disease to continue in a population, microbes must exit from the body. In many cases, the portals of entry and the portals of exit are the same.

Hospitals and long-term health care facilities are hotbeds of infection for patients. Hospital-acquired infections are called nosocomial infections and account for about 50% of all the major complications of hospitalization.

The public should not be paralyzed by the fear of infection. Awareness, surveillance, common sense precautions, and calmness are the best preventive measures.

SELF-EVALUATION

PART I Choose the *single* best answer.

1. The breakup of the Soviet Union ushered in an unusually high number of cases of diphtheria from about 1990 to 1995. Which term best characterized the situation?
   a. endemic   **c.** pandemic
   b. epidemic   **d.** herd

2. A common-source outbreak would most likely be attributed to
   a. airborne source   **c.** person-to-person contact
   b. change in vector   **d.** water supply
       distribution

3. Chicken pox in the United States is best described as
   a. sporadic   **c.** zoonotic
   b. endemic   **d.** pandemic

4. Which one of the following is not an insect?
   a. fly   **c.** flea
   b. tick   **d.** mosquito

5. Vertical transmission is possible except in the case of
   a. gonorrhea   **c.** influenza
   b. AIDS   **d.** chicken pox

PART II Fill in the following.

1. A worldwide outbreak of a disease is called a
   _____ .

2. Name a disease transmitted by an arthropod vector, and name the vector.
   _____

3. Only one disease has been eradicated. Name this disease.
   _____

4. Give an example of indirect contact transmission (fomites) of disease.
   _____

5. Which body site is most susceptible to nosocomial infection?
   _____

PART III Answer the following.

1. Distinguish between vertical and horizontal transmission. Give examples of each.

2. What is meant by zoonoses? What is a common zoonosis in the northeastern United States? Give three examples of zoonoses.

3. Distinguish between biological vectors and mechanical vectors. Give two examples of each.

BACTERIAL DISEASES

I am not born for one corner; the whole world is my native land.

SENECA (THE YOUNGER)

PREVIEW The list of bacterial diseases presented in this section may appear long, but it represents only a drop in the bucket; only the major bacterial diseases of personal and public health are presented. They are organized according to their route of transmission: food borne and waterborne, airborne, sexually transmitted, contact, soilborne, and arthropod borne. It should be noted that, in some cases, transmission may be accomplished by more than one route. For example, listeria may be food borne or airborne, and anthrax may be airborne or acquired by direct contact. (Viral and protozoan diseases will be similarly presented in chapters 9 and 10.) The practical questions that need to be answered and that will guide this presentation (and that of viruses and protozoans) are as follows.

- *Transmission:* How does an individual acquire the infection?
- *Incubation time:* What is the time from contact to onset of symptoms?
- *Pathogenesis:* What does the disease do?
- *Prevention:* How can chances of getting the disease be minimized? Is immunization available?
- *Treatment:* What measures are used to treat the disease?

FOOD-BORNE AND WATERBORNE BACTERIAL DISEASES

After studying this section, you may be overly anxious about the safety of the foods you eat and the source of the water you drink. You may have pathogens lurking in the kitchen, and it is certainly wise to take certain precautions and to use common sense; it is not necessary, however, to become paranoid! You have protection at two levels: (i) your immune system (chapter 11) and (ii) a system of surveillance practiced by national, state, and city health departments throughout the country. The U.S. Department of Agriculture, through its Food Safety Inspection Service, is charged with ensuring the safety of the nation's commercial supply of meat, poultry, and egg products. In 1997 President Clinton announced the National Food Safety Initiative to deal with the new challenges presented by changes in food supply and distribution reflective of changes in eating habits. Figure 8.1 illustrates the anatomy of the human digestive system, the usual target for food-borne and waterborne microbes resulting in gastroenteritis.

Food-borne and waterborne infections have a significant impact on health in the United States and around the world; increasingly, more of the pathogens responsible for these infections are becoming resistant to antibiotics. Estimates are that in the United States as many as 9,000 people die annually from food-borne illnesses and that more than 70 million cases occur per year. Many of the cases are undiagnosed and are sometimes passed off as "it's just a stomach virus." Further, millions of dollars are spent per year on the recall and disposal of contaminated foods, representing "dollars down the drain." It is not uncommon for a major food-canning or food-freezing company or meat, chicken, or fish processing plant to issue an urgent warning regarding consumption of a particular food item. Even the Gerber baby food company was faced with a gigantic crisis. Suddenly the company's image plummeted as a result of bacterial contamination of some of its food products. The example that truly captured the attention of the public was the Jack-in-the-Box episode that occurred in Washington State in 1993 from the consumption of undercooked hamburgers that were

FIGURE 8.1   Anatomy of the human digestive system.

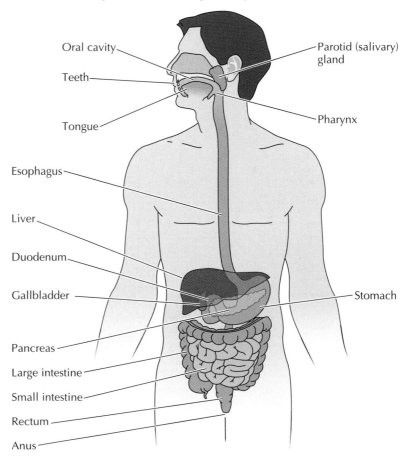

contaminated with *Escherichia coli* O157:H7. In 1999, the Velvet Ice Cream Company voluntarily recalled 244 gallons of chocolate ice cream because of possible contamination with *Listeria monocytogenes*. Examples of food-borne illness due to ingestion of contaminated foods are too common; note that the problem is by no means confined to *E. coli* O157:H7 (In the News 8.1). Popular TV shows such as *Dateline* and *20/20* have presented segments relating to the food industry, some of which are shocking.

Travelers to foreign countries need to be particularly careful about ingestion of food and water, since the many problems of infrastructure, particularly in developing countries, do not ensure an optimal level of safety. It's easy to forget that ice cubes come from water or that thoroughly washing fruits and vegetables is to no avail if the water is contaminated. People who live in an area develop a level of immunity that travelers do not have and, hence, are less prone to get these diseases. Tourists' plans are frequently curtailed, and sight-seeing becomes limited to the hotel room with an adjoining bathroom. A summary list of major food-borne and waterborne infections is presented in Table 8.1. Note that there are common denominators, including the types of food and the symptoms. "Boil it, cook it, or forget it" is an excellent rule, particularly for travelers to foreign countries. Table 8.2 gives a few simple tips for minimizing the likelihood of infection.

Food Intoxication (Food Poisoning)

Botulism

A distinction needs to be made between **food intoxication** (also called poisoning) and **food infection.** In the popular jargon, all food-related illnesses are falsely lumped together as food poisoning. Intoxication refers to the ingestion of already-produced bacterial toxins; the actual organisms may, in fact, no longer be present. Food-borne infection is the result of the ingestion of the bacteria in contaminated foods and their subsequent growth in the

IN THE NEWS 8.1

Frankfurters Recalled after Deaths

E. coli Outbreak at Fair Kills Girl, Sickens Dozens More

Over 100 People Sick in Illinois E. coli Outbreak

Drug-Resistant Salmonella Becoming More Common

A Sampling of Food-Borne Illnesses

TABLE 8.1　Food-borne and waterborne bacterial diseases[a]

Disease	Incubation period	Causative microbe	Transmission	Symptoms
Botulism	8–24 hours	*Clostridium botulinum*	Contaminated foods, home-canned products	Nausea, vomiting, diarrhea, possibly bloody stools
Clostridial food poisoning	8–14 hours	*Clostridium perfringens*	Meat, poultry, beans	Severe cramping, abdominal pain, watery diarrhea
Staphylococcal food poisoning	1–6 hours	*Staphylococcus aureus*	Coleslaw, potato salad, fish, dairy products, cream-filled pastries, spoiled meats	Abdominal cramps, nausea, vomiting, diarrhea
Salmonellosis	12–48 hours	*Salmonella enteritidis* and other species	Undercooked or raw eggs, turtles, chicks, iguanas, reptiles, unpasteurized orange juice, raw alfalfa sprouts	Nausea, vomiting, abdominal cramps
Typhoid fever	7–14 days	*Salmonella typhi*	Fecal-oral; flies, fomites	Blood in feces, fever, delirium, rose-colored spots on the abdomen
Shigellosis	12–72 hours	*Shigella* species	Eggs, shellfish, dairy products, vegetables, water	Diarrhea, dysentery, abdominal cramps, blood in feces
Cholera	24–72 hours	*Vibrio cholerae*	Raw shellfish, water contaminated with fecal material of infected persons	Severe diarrhea and dehydration, muscular cramps, wrinkling of the skin
Campylobacter infection	2–10 days	*Campylobacter jejuni*	Poultry, cattle, water, unpasteurized milk, anal and oral sex	Diarrhea, high fever, bloody stools
Listeriosis	3–70 days (usually within 1 month)	*Listeria monocytogenes*	Vegetables, meats, cheese, ice cream, unpasteurized milk	Sore throat, fever, diarrhea, miscarriage, death
E. coli O157:H7 infection	18–72 hours	*Escherichia coli*	Undercooked hamburgers, radish sprouts	Diarrhea, nausea, abdominal cramps

[a]These diseases are treatable with antibiotic therapy; drug-resistant strains pose a problem in some cases.

intestinal tract accompanied by secretion of a toxin or by invasion of the intestinal lining. The toxin is referred to as an **enterotoxin** (a toxin which affects the intestinal tract) and causes the common symptoms of nausea, vomiting, diarrhea, and, possibly, bloody stools and fever—a very unpleasant menu over which you have no choice.

Botulism, caused by *Clostridium botulinum,* is the most dangerous food intoxication. The neurotoxin produced by this organism is extremely deadly; it interferes with the passage of nerve impulses, resulting in muscle paralysis, including the diaphragm and the muscles of the ribs, leading to respiratory paralysis, that is, the inability to breathe; death occurs within a day or two (Fig. 6.11). Antibiotics are of no value in treatment, since the problem is not one of infection, but intoxication. The number of cases in the United States is low, but the associated death rate in untreated cases is 70%. There are, unfortunately, no telltale signs of *C. botulinum*-contaminated cans, such as bulging of the can, odor, or taste. The best defense is to heat foods for 10 minutes at 90°C, a process which will destroy the toxin. Home-canned string beans, peppers, asparagus, sausage, cured pork and ham, smoked fish, and canned salmon are potential sources of intoxication.

Ironically, minute doses of botulinum toxin called **botox** have been used to treat a variety of common disorders associated with muscle over-activity, including strabismus (crossed eyes), stuttering, and uncontrolled blinking. Botox is used cosmetically by injection into the skin to relieve wrinkling! In May 2001, the journal *Neurology* reported on a study of a small number of patients, in which it was claimed that treatment with botox helped sufferers of chronic low back pain. Botox is advertised as "from toxin to therapeutic agent."

Clostridium perfringens is usually associated with **gas gangrene,** a condition which devastates soldiers wounded in battle because of wound contamination with particles of soil containing spores. This organism is an important cause of food poisoning (some consider it a food infection). It is commonly associated with spore contamination of meat, poultry, and beans. Recovery usually occurs within 24 hours, and there is no specific therapy. In 1993, "wearing of the green" celebrations ended with outbreaks of gastroenteritis associated with contaminated corned beef served at St. Patrick's Day meals in Virginia and Ohio.

Recall from chapter 4 that the genus *Clostridium* is a sporeformer and that spores survive the high temperature used in cooking and germinate as multiplying cells during cooling.

Staphylococcal Food Poisoning

Staphylococcal food poisoning is the most common type of food poisoning, and many cases remain undiagnosed. You probably have heard of this condition as **"staph food poisoning"** or by an older and incorrect term, "ptomaine poisoning." *Staphylococcus aureus* secretes an enterotoxin that produces intestinal tract symptoms, including abdominal cramps, nausea, vomiting, and diarrhea. The symptoms may be severe to the point where you might wish you were dead! There is hardly anyone who has not suffered from this, so all you can do is to wait it out; you will feel better within several hours. The list of potential food sources is large and includes coleslaw, potato salad, fish, dairy products, cream-filled pastries, and spoiled meats. A family day at the beach can turn into a nightmare when, on the way home, all of the occupants of the car complain of nausea and impending diarrhea and vomiting. The problem is compounded when you are in the middle lane of a three-lane highway and cannot pull over to run for the woods and some of the passengers are young children! The lunch, which was brought from home in a cooler, consisted of peanut butter-and-jelly sandwiches, coleslaw, and cream-filled pastries. The suspect food is the coleslaw or the pastries, but the probability of obtaining a bacteriological diagnosis is very low, because all of the lunch has been consumed or thrown away. The symptoms make the diagnosis. How did contamination occur? The main reservoir for *S. aureus* is the nose, but the organism also causes boils, abscesses, or pimples on the skin, and these lesions are the usual source of seeding the staphylococci into food. A few hours under the sun in an improperly iced cooler is all it takes to allow for multiplication of the bacteria and enterotoxin production.

Food-Borne Infection

Salmonellosis

How do you like your eggs? If you answered "over easy" or "sunny side up," you are at risk for **salmonellosis.** If your Caesar salad dressing is made the "old-fashioned" way with raw eggs, or if you have purchased an iguana,

TABLE 8.2 Tips on reducing the risk of food-borne and waterborne infections

Wash hands frequently, particularly after using the bathroom and after changing a baby's diapers.
Cook raw beef and poultry products thoroughly.
Avoid unpasteurized milk and juices and any food made from unpasteurized milk.
Properly wash food utensils, cutting boards, and raw fruits and vegetables.
Beware of "double-dippers" (those who dip their cracker, celery, carrot stick, etc., into a spread, take a bite, and then dip the same item into the spread again).
Avoid eating at "sleazy" joints.
Avoid street vendors, particularly when traveling abroad.
Follow common sense and good personal hygiene.
Frequently wash kitchen sponges in the dishwasher.
Don't overreact!

At about the age of 2, my daughter became ill and required hospitalization. At the time I was on a sabbatical leave at Georgetown University School of Medicine in Washington, D.C. Several days elapsed before a definitive laboratory diagnosis was made. As crazy as it may sound, my wife and I were relieved when the diagnosis was salmonella infection and, therefore, the prognosis was excellent. The likely culprit was a small pet turtle that we had purchased a few weeks earlier. The turtle was flushed down the toilet before I had the opportunity to culture from it—a foiled epidemiological study!)

snake, turtle, or chick, you may be asking for trouble. These are all potential carriers for the variety of salmonella species that cause gastroenteritis, manifested by nausea, vomiting, abdominal cramps, diarrhea, and possibly fever. The Centers for Disease Control and Prevention (CDC) reported on four cases of *Salmonella* infections from 1996 to 1998 in persons who had contact with reptiles (lizards, snakes, or turtles). The patients were as follows: (i) a 3-week-old boy in Arizona (the source of infection was a pet iguana), (ii) a 6-year-old boy and his 3-year-old brother in Kansas (the source of infection was corn snakes), (iii) an 8-year-old boy in Massachusetts (the source of infection was two iguanas), and (iv) a 5-month-old boy in Wisconsin who died (the source of infection was an iguana). One recommendation of the CDC is that pet store owners advise potential reptile owners of the risk of salmonellosis, particularly in children. Concern has been voiced that crows may be more than just a nuisance. These big, black birds have increased in population in recent years, and it is now known that crow feces carry salmonellae. (Think about that when a flock flies overhead, and wear a cap!) Remember that salmonellosis is a food infection and requires the presence of bacteria which multiply in the intestinal tract, whereas staphylococcal and clostridial food poisoning results from ingestion of toxins.

In July 1999, the Food and Drug Administration issued a warning to consumers against drinking unpasteurized orange juice distributed by Sun Orchard of Tempe, Ariz., because of contamination with a species of *Salmonella* which can cause serious and potentially fatal infections; several individuals became ill after drinking the product. Sun Orchard cooperated by stopping the production of unpasteurized juice and producing only pasteurized products. Further, eating raw alfalfa sprouts is commonplace, particularly among diet-conscious individuals, but may be hazardous to your health because of contamination with *Salmonella* or *E. coli* O157:H7.

Salmonellosis associated with the ingestion of eggs infected with *Salmonella enteritidis* is a significant public health problem; by 1994 this microbe had become the most frequently reported cause of *Salmonella* infection. Between 1985 and 1998, the CDC traced 794 outbreaks with 79 associated deaths. Eggs may appear normal, but if they are eaten raw or undercooked, infection may result. Cracked eggs were formerly believed to be the culprits, but it is now known that *S. enteritidis* infects the ovaries of seemingly healthy hens, and eggs may be contaminated before shells are formed. Hence, intact and disinfected grade A eggs can be a source of salmonellosis. Hollandaise sauce, homemade mayonnaise, and Caesar salad dressing may be made with raw eggs. This is an unsafe procedure, but, if it is practiced, pasteurized eggs should be used. The Food and Drug Administration has recently proposed a warning on egg cartons that would read: "Eggs may contain harmful bacteria known to cause serious illness, especially in children, the elderly, and persons with weakened immune systems. For your protection, keep eggs refrigerated; cook eggs until yolks are firm, and cook foods containing eggs thoroughly." The current labeling on egg cartons (Fig. 8.2) is inadequate. *NBC News* (October 22, 1999) reported on the results of an investigation citing shocking violations of the storage and handling of eggs in supermarkets, including their finding that eggs from expired cartons were put into new cartons. Further, despite the fact that eggs are supposed to be refrigerated, eggs were kept at market (room) temperature for well over a week. Beware of those incredible edible eggs!

Typhoid fever is a salmonellosis caused by *Salmonella typhi*. The

FIGURE 8.2 An egg carton with a label containing instructions for safe handling of eggs. (Author's photo.)

organism is able to survive in sewage, water, and certain foods and gains access into the body with the ingestion of fecally contaminated food products, including some imported foods. It is transmitted by flies and fomites. Once an important disease worldwide, its prevalence has markedly decreased in developed countries as a triumph of sanitary engineering leading to sewage treatment facilities, a luxury not available in still-developing countries, as will be further discussed in chapter 12. In the early 1900s, typhoid fever was a major killer, with over 20,000 cases occurring annually. Approximately 400 to 500 cases are now reported annually to the CDC; some of these cases are acquired during overseas travel. After ingestion, the organism invades the lining of the small intestines, causing ulcers and the passage of blood in the feces, accompanied by fever, possibly delirium, and the presence of rose-colored spots on the abdomen due to hemorrhaging in the skin. About 5% of those infected remain as healthy carriers and present a public health problem, although antibiotic therapy usually renders the individual carrier free. In some cases, surgery to remove the gallbladder, a site where the bacilli sometimes take up residence, may be performed to eliminate the carrier state.

The story of **Typhoid Mary** is a classic and significant tale which points out the interface between biology and society. An interesting question is whether Mary's human rights were violated. After reading Box 8.1 and thinking about this, you can decide how you feel about this question. In 1909 the British tabloid *Punch* published a satiric poem about Typhoid Mary attesting to the public's aversion to her (Box 8.2).

The CDC documented the first sexually transmitted outbreak of typhoid fever in 2000. A Cincinnati man infected seven other men with whom he had sexual relations, presumably by highly risky oral-anal contact.

Shigellosis
Shigellosis is manifested by the usual gastrointestinal symptoms with the possible addition of dysentery. Eggs, shellfish, dairy products, vegetables, and water are the sources. It is thought that in the Battle of Crecy in 1346,

BOX 8.1   Typhoid Mary: a Public Health Dilemma

Had you accepted an invitation to dinner in the summer of 1906 at the rented vacation home of Charles Henry Warren, a wealthy New York banker, in Oyster Bay, N.Y., you might have become seriously ill or died from typhoid fever. The culprit would have been the bacterial organism *Salmonella typhi*, harbored unknowingly (at least initially) in the digestive tract of Mary Mallon, the Warrens' cook. Over the next several years, Mary became stigmatized as "Typhoid Mary"; during that time, as a healthy (asymptomatic) salmonella carrier, she was responsible for approximately 50 cases of typhoid fever, at least two of which ended in death. For her crime of being a healthy carrier, she was forced to spend 26 years of her life in relative isolation, shunned by her fellow humans, in a tiny cottage on North Brother Island, a small island in the East River in New York City. Mary, not *Salmonella*, became the culprit. This was hardly the life she envisioned when, as a teenager, she emigrated from Ireland. In a legal sense, she was never tried for her "crime."

Mary Mallon was born in 1869 in Ireland and emigrated to the United States in 1883. The first suspicion that she was a healthy typhoid carrier was during her brief stint as a cook for the Warrens during which time, in a 1-week period, 6 of 11 people in the household (Mrs. Warren, two daughters, two maids, and the gardener) developed typhoid fever. Typhoid and other diseases associated with "filth" were simply "not nice" for the elite of Oyster Bay; hence, George Soper, a sanitary engineer, was hired to determine the cause of the outbreak. Soper quickly ruled out soft clams as the source (sushi was not yet popular!) and revealed a pattern of typhoid in families where Mary had worked as a cook, dating back to 1900 and totaling 26 cases. (She had an excellent reputation as a cook and was famous for her Irish stew.) Soper was unsuccessful in convincing Mary to submit specimens of blood, urine, and feces for microbiology-based laboratory examination to determine whether she was, in fact, a salmonella carrier. She proclaimed that her health was excellent and that she had no history of typhoid and, hence, saw no reason for compliance. The case was referred to Herman Biggs, medical officer of the New York City Health Department. The upshot was that in 1907, police, under Biggs' authority, took Mary by force and against her will, without a trial, to a contagious disease unit in a New York hospital. Examination of her feces revealed a high content of typhoid bacilli, and, for this, she was removed to North Brother Island. Three years later, she was released by the health commissioner on her promise that she would not cook again, but after about 2 years she returned to cooking, her only employment skill. In 1915 she was sent back to the island for the rest of her life. During her 4 years of freedom, she was responsible for another 25 cases of typhoid fever, including two deaths. She died on November 11, 1938, not as Mary Mallon, but as Typhoid Mary, a public menace blamed for three deaths and about 47 other cases of typhoid fever and with the dubious distinction of being the first healthy typhoid carrier in the United States (none of these credits appeared on her tombstone).

The account of Mary Mallon exposes the dilemma of the public's health and safety versus the rights of the individual. The circumstances of this case and its outcome (Mary Mallon's confinement) can be viewed as an application of objective utilitarianism, a consequence-based moral philosophical position succinctly stated as "the greatest good for the greatest number." Proponents of utilitarianism are clear that all individuals count and count equally, but, nevertheless, the sacrifice of some (Mary, in this case) is justified for the sake of the many (the public). Some individuals become a means to an end, rather than an end in themselves. Did Mary get a "square deal"? At the time, those charged with the public's health exhibited social prejudice (e.g., sex and social class prejudice) in considering a broad spectrum of factors and in arbitrarily deciding that liberty was an impossible privilege to allow Mary Mallon. In the case of other disease carriers, particularly males, other options were explored.

Do you think Mary was justly treated? This case serves as a paradigm for today's "Typhoid Mary"—"AIDS Sam" or "Resistant TB Joe." Health officials have not only the right but the obligation to develop public health policies, but, in so doing, they incur the responsibility to provide realistic and appropriate alternatives, including health care, housing, and reemployment training, to those individuals in society who are afflicted. AIDS, in some respects, parallels the Typhoid Mary dilemma.

BOX 8.2   A Satiric Poem

In U.S.A. (across the brook)
There lives, unless the papers err,
A very curious Irish cook
In whom the strangest things occur;

Beneath her outside's healthy glaze
Masses of microbes seethe and wallow
And everywhere that MARY goes
Infernal epidemics follow.

Source: This poem appeared in *Punch*, a British magazine, in 1909. Reprinted with permission of Punch Ltd.

the defeat of the English army by the French was due to the English soldiers' suffering from dysentery; as stated in a current microbiology text, "they were truly caught with their pants down!" In January 1991, a shigellosis scare occurred in Lexington County, Ky., and was traced to several day care centers. A community-based approach, coupled with maximizing the availability of hand soap and water and in-house education, solved a potentially serious situation. The tourists aboard a cruise ship in 1994 on which there was an outbreak of shigellosis headed for the "head." As in all diseases involving loss of large volumes of water, dehydration is the major cause for concern and can lead to death via circulatory collapse. Treatment involves rehydration, either oral or intravenous, and possibly antibiotics.

Cholera

Vibrio cholerae is the cause of **cholera.** This organism, like so many other intestinal tract pathogens, causes diarrhea due to an enterotoxin. The diarrhea can be best described as the "mother of all diarrheas"; as much as 1 liter of fluid can be lost per hour for several hours. A liter is approximately equal to 1 quart, and a human has a total of 5 to 6 liters of blood, which is about 80% water. Unless rehydration is instituted quickly, death results in only a few hours; the mortality rate reaches 70% in untreated cases. The massive dehydration results in the eyes' taking on a sunken appearance, the skin becomes dry and wrinkled, and muscular cramps in the legs and arms are a common occurrence. Cholera reaches epidemic proportions in developing countries (Fig. 8.3). However, outbreaks of the disease also occur in the United States and in other developed countries. Infection is acquired by the ingestion of water or food, including raw shellfish, contaminated by fecal material from cholera-infected persons. This is an unpopular fact with New Englanders who consume raw clams at a high rate.

FIGURE 8.3 Cholera. Poverty, overcrowding, and primitive sanitation are a dangerous trio favoring the emergence and dissemination of this waterborne disease. Note the close contact of people with dirty water in the gutter. (Source: National Library of Medicine.)

Occasionally, notices are posted in newspapers, citing the closing of shellfish beds and beaches because of fecal pollution and the danger of *V. cholerae* and other fecal pathogens.

Pandemics of cholera have been documented for hundreds of years, and it seems that no continent has escaped. Outbreaks have occurred in the United States as the result of tourists returning from Africa, Southeast Asia, or South America; in some cases the source has been contaminated seafood brought back to the United States. In the United States and in other industrialized countries, because of advances in water and sanitary engineering, cholera is rare.

Fortunately, the terrible loss of life from cholera throughout the world can be prevented through **oral rehydration therapy (ORT),** which is available at a cost of pennies, is readily transported to remote areas, and negates the use of intravenous injection and the associated costs that make such therapy prohibitive in countries where it is most needed.

Hopefully, good news regarding the prevention of cholera may not be too far away. The genes for enterotoxin production by the cholera vibrios were identified in 1992 by scientists at the University of Maryland. The removal of these genes renders the organisms nonpathogenic but still alive, and these cells are candidates to be used as a vaccine—a work in progress.

E. coli *Infection*

E. coli is frequently a news item because of an enterotoxin-producing strain called *E. coli* O157:H7, first recognized in the United States in 1982. It has been featured on television news shows and in newspaper stories and magazine articles and is considered an emerging cause of food-borne illness. Its rise to fame is due to an outbreak of disease in 1993, when more than 500 customers of a Jack-in-the-Box fast-food chain restaurant in the state of Washington became ill from eating undercooked meat contaminated with *E. coli* O157:H7. The outbreak resulted in several deaths of children under the age of 5. Numerous other outbreaks have occurred in recent years, including one in Kyoto, Japan, in July 1996 that was traced to ingestion of contaminated radish sprout salad served during lunch at a factory cafeteria; 47 workers were infected. The sprouts were traced to four farms in the area, one of which had been associated with an outbreak of *E. coli* O157:H7 infection in 6,000 children earlier that month. More than half the states in the United States require that isolation of *E. coli* O157:H7 be reported to the state health department for monitoring purposes in an effort to prevent outbreaks. A number of known outbreaks have occurred; a sampling is indicated in Table 8.3. Note that ground beef and other meats are not the only culprits. Ground beef is particularly dangerous, since the bacteria are mixed throughout large batches of meat obtained from many cattle when it is ground, in contrast to steaks, which are from a single source. Contamination of the meat with fecal material may occur during the slaughtering process. With a little imagination, the expression "One bad apple spoils the barrel" can be applied. Are you sure you want that hamburger rare? The best advice is to avoid rare meat, particularly hamburgers. Many of the fast-food chain restaurants now specify the cooking time and temperature as a part of their policy. In most cases, recovery occurs within 5 to 10 days without specific treatment. In children under 5 years of age, there is a greater risk of death due to kidney damage resulting from toxin-produced hemorrhages in the kidney, a condition known as **hemolytic uremic syndrome.**

TABLE 8.3   Sampling of articles on *E. coli* O157:H7 outbreaks

Year	Article title
1987	A severe outbreak of *Escherichia coli* O157:H7-associated hemorrhagic colitis in a nursing home
1993	An outbreak of diarrhea and hemolytic uremic syndrome from *Escherichia coli* O157:H7 in fresh-pressed apple cider
1993	A two-restaurant outbreak of *Escherichia coli* O157:H7 enteritis associated with consumption of mayonnaise
1993	Yet another outbreak of hemorrhagic colitis
1994	A multistate outbreak of *Escherichia coli* O157:H7-associated bloody diarrhea and hemolytic uremic syndrome from hamburgers: the Washington experience
1994	A severe outbreak of hemorrhagic colitis and hemolytic uremic syndrome associated with *Escherichia coli* O157:H7 in Japan
1995	A university outbreak of *Escherichia coli* O157:H7 infections associated with roast beef and an unusually benign chemical course
1997	Outbreaks of *Escherichia coli* O157:H7 infection associated with eating alfalfa sprouts—Michigan and Virginia
1997	A multistate outbreak of *Escherichia coli* O157:H7 infections associated with eating mesclun mix lettuce
2000	Outbreak of *Escherichia coli* O157:H7 infection associated with eating fresh cheese curds—Wisconsin

What is *E. coli?* The genus is named for Theodor Escherich, who described the organism in the 1880s, and the species (*coli*) indicates the location: the organisms are part of the normal flora that live in the large intestines (colons) of humans and animals. Bacteria, particularly *E. coli,* account for much of the weight of fecal material; these residents are considered beneficial because they produce vitamins which are absorbed into the body. So why the bad rap for *E. coli* O157:H7, one of the more than 100 strains of *E. coli* normally found in the intestines of animals? Somewhere and somehow, *E. coli* picked up the genes that are responsible for the production of an enterotoxin which produces severe damage, possibly to the point of death.

In addition to O157:H7, there are several other strains of *E. coli* that exhibit variation in the mechanism by which they cause diarrhea. The **enterotoxigenic *E. coli* (ETEC) strains** are one example. This group is the most common cause of **traveler's diarrhea.**

Too bad about O157:H7 damaging the reputation of the usually nonpathogenic *E. coli.* An *E. coli* culture (nonpathogenic) can be purchased for only about $7 from a variety of biological supply houses and is widely used in laboratories in high schools and colleges. *E. coli* is probably the most widely investigated species of life. (Harnessing of *E. coli* and other microbes for the benefit of humankind is discussed in chapter 3.) When you hear that a particular beach or shellfish bed is closed because of fecal pollution, you now know that it is probably because *E. coli* is present.

Campylobacter *Infection*

Campylobacter is a relatively new cause of intestinal disease in humans, although it has a long history of association with disease in animals, causing abortion and enteritis in sheep and cattle. It is a major cause of bacterial diarrhea in the United States and causes more cases of diarrhea than

Salmonella and *Shigella* combined; the *Campylobacter* species associated with this disease is *Campylobacter jejuni.* Poultry, cattle, drinking water, and unpasteurized milk have been identified as potential sources; sexual activity involving oral-anal contact is also a mechanism of transmission. The disease is usually self-limiting and lasts about 1 week. Fewer than 500 organisms are enough to cause illness. The October 20, 1998, issue of the *New York Times* shocked the public with its report that 70% of chicken meat sampled from supermarkets was contaminated with *Campylobacter.* To make matters worse, many isolates were antibiotic resistant because of the widespread use of antibiotics in feed.

Listeriosis

Listeriosis, caused by *L. monocytogenes,* is an increasingly significant cause of food infection. The organism is distributed worldwide and has been isolated from soil, water, manure, plant materials, and the intestines of healthy animals (including humans, birds, and fish). The disease is associated with the ingestion of *Listeria*-contaminated foods, including vegetables, meats, cheeses, ice cream, and raw (unpasteurized) milk. Processed foods, such as cold cuts, hot dogs, and soft cheeses, may become contaminated after processing and serve as the source of epidemics or threatened epidemics. *Listeria* can grow in refrigerators, contributing to the problem. Table 8.4 lists episodes occurring in 1999. In many cases, voluntary recall

TABLE 8.4   Sampling of *Listeria* recalls and outbreaks in 1999[a]

Food source	Company	Description
Hot dogs, lunch meat	Thorn Apple Valley	Voluntarily recalled products
Hot dogs, cold cuts	Sara Lee subsidiary	75 cases, including 15 deaths in 16 states; voluntarily recalled product
Chocolate ice cream	Velvet Ice Cream	Voluntarily recalled 244 gallons of ice cream; possible contamination; no reports of illness
Headcheese (a seasoned loaf made of the head meat of a calf or pig in its natural aspic)	Ba Le Meat Processing & Wholesale, Inc.	Voluntarily recalled 2,600 pounds of headcheese; no reports of illness
Beef wieners, hot dogs	Valley Meat Supply	Voluntarily recalled 150 pounds of wieners and frankfurters; no reports of illness
Weisswurst sausage	Alpine Wurst & Meat House	Voluntarily recalled 60 pounds of sausage; no reports of illness
"Hawaiian Winners" sausage	Redondo's, Inc.	Voluntarily recalled 9,620 pounds of sausage
Sliced ham, salami, luncheon meats	White Packing Co.	Voluntarily recalled 16,392 pounds of sliced ham, salami, and luncheon meats
Deli meats	Oscar Mayer	Voluntarily recalled 28,313 pounds of deli meat
Hot dogs	Marathon Enterprises	Voluntarily recalled 2.1 million pounds of hot dogs; 2 deaths

[a]In some cases, bacteria were presumed to be *Listeria.*

of the contaminated foods prevented illness, although there is the likelihood of unreported cases.

A wrongful death suit against the Sara Lee Corporation was filed on February 2, 1999, after the death of a woman who ate *Listeria*-contaminated hot dogs. The U.S. Department of Agriculture reported that on July 4, maintenance of the company's air-conditioning system and the resulting dust "could have caused wide-spread contamination of meat products through the plant." It was reported that 15 adults died and three miscarriages occurred in 16 states.

Listeriosis is usually mild or subclinical in healthy adults and children and is manifested by nonspecific symptoms, including sore throat, fever, and diarrhea. Infants, the elderly, immunocompromised individuals (e.g., cancer patients receiving chemotherapy and persons with acquired immune deficiency syndrome [AIDS]), and pregnant women are at high risk. Persons with AIDS are approximately 300 times more susceptible than healthy adults, and pregnant women are approximately 20 times more susceptible than nonpregnant adults. Newborns can be infected as a result of their mothers' eating contaminated foods containing only a few *Listeria* bacteria and may become acutely ill with meningitis (nervous system involvement); the mortality rate is about 60%, and antibiotics given promptly to pregnant women can often prevent infection of the fetus or newborn.

AIRBORNE BACTERIAL DISEASES

Airborne bacterial diseases are respiratory tract infections that are transmitted by air droplets or by contact with contaminated inanimate objects; in some cases, the disease spreads beyond the respiratory tract to other areas of the body. The humidity and temperature in air allow bacteria to remain viable for extended periods of time; hence, air serves as an excellent vehicle for transmission. Air serves as a passive transmitter, since there is no multiplication of the microbes, as is the case with food-borne bacterial infections. Respiratory infections generally respond well to antibiotic therapy, although the problem of antibiotic resistance is a recurring theme. Immunization is available for some of these infections but may not be practiced in many countries because of problems of availability, cost, and public health infrastructure. The upper respiratory tract includes the tonsils and pharynx, while the lower respiratory tract includes the trachea, larynx (voice box), bronchi, bronchioles, and lungs; Fig. 8.4 illustrates the respiratory system. Table 8.5 provides a list of major airborne bacterial diseases categorized into **upper** and **lower respiratory tract infections.**

Before the advent of immunization and antibiotic therapy, there were so-called communicable disease hospitals in all communities of the United States and other developed countries of the world. Diphtheria and whooping cough, along with other airborne bacterial and viral diseases, are "infectious," a term that is now preferred to "contagious." Communicable-disease hospitals were better equipped to manage the constant threat of bacteria being introduced into the air by their victims through coughing and nasal secretions; young children were especially susceptible to these diseases. High fever, delirium, racking coughs, labored breathing, and horrible rashes were associated symptoms. Little could be done other than supportive therapy. Antibiotics were not available, nor was immunization. It was a "wait, see, and pray" attitude; death rates were very high. Frequently, the young and hapless victims were not hospitalized but were confined to

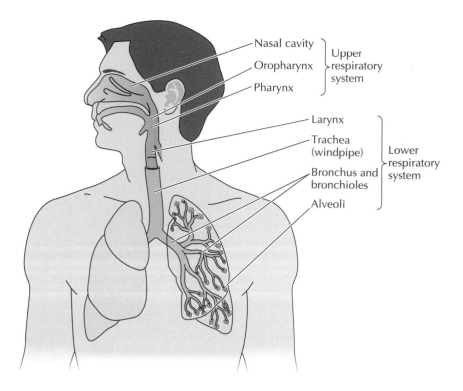

FIGURE 8.4   Anatomy of the human respiratory system.

their homes; quarantine signs were placed on the doors by local public health officials to discourage visitors. These were the days of the horse-and-buggy doctors, and their selflessness was more than admirable while they sat by the bedsides and could do little more than apply cold towels to the patients who were burning with raging fevers.

Upper Respiratory Tract Infections
Diphtheria

Diphtheria is caused by *Corynebacterium diphtheriae*; infection occurs in the upper part of the respiratory tract, close to the tonsils (Fig. 8.5). The pathology is the result of the production of a powerful exotoxin which destroys the cells constituting the epithelial lining. These dead cells, along with mucus and scavenger cells, pile up and form a pseudomembrane (false membrane). Its leathery consistency may cause respiratory blockage and death by suffocation, particularly in young children. Many a **tracheotomy** (cutting a hole in the throat) was performed by a physician with penknife in hand on the kitchen table in a desperate attempt to save a child's life. Further, the toxin diffuses into the bloodstream, causing widespread damage, particularly to the heart.

The incidence of diphtheria in the United States is now fewer than 50 cases per year. In fact, many modern physicians, having never seen a case, would have difficulty in making a diagnosis. Worldwide, however, the disease remains a significant problem. In the early to mid-1990s, a massive reemergence of diphtheria occurred in the newly independent states of the former Soviet Union as a result of numerous factors, primarily a collapse of

TABLE 8.5 Airborne bacterial diseases[a]

Disease	Incubation period	Causative microbe	Symptoms
Upper respiratory tract			
Diphtheria	2–4 days	*Corynebacterium diphtheriae*	Respiratory blockage causing suffocation and possibly death
Pertussis (whooping cough)	7–10 days	*Bordetella pertussis*	Respiratory blockage causing violent and persistent coughing and possibly death
Streptococcal infections (group A) (pharyngitis, tonsillitis, strep throat, scarlet fever, flesh-eating strep, toxic shock syndrome, childbed fever)	2–5 days	Streptococci (group A)	Red and sore throat, fever, headache, red rash, strawberry tongue, multi-organ failure, destruction or "eating" of the flesh, infant and newborn death
Meningitis	1–3 days	*Neisseria meningitidis*	Coldlike symptoms, fever, delirium, stiffness of the back and neck
Meningitis	5–6 days	*Haemophilus influenzae* type b	
Lower respiratory tract			
Legionellosis	2–10 days	*Legionella pneumophila*	Fever, muscle pain, cough, pneumonia, death
Tuberculosis	Several weeks	*Mycobacterium tuberculosis*	Fever, weight loss, cough, fatigue
Tularemia	1–14 days	*Francisella tularensis*	Fever, pleuro-pneumonitis, respiratory failure, shock

[a]These diseases are treatable with antibiotic therapy; drug-resistant strains pose a problem in some cases.

the health care system, resulting in decreased childhood vaccination. It takes a lot of money to maintain an immunization program.

You received your first immunization at about the age of 2 months. It was actually a triple vaccine, referred to as **DPT;** D stands for diphtheria, P stands for pertussis, and T stands for tetanus.

Pertussis (Whooping Cough)

Whooping cough is a highly infectious and potentially lethal disease caused by *Bordetella pertussis* and is a particular threat to babies and young children; the majority of cases occur in children under 4 years of age. Children with whooping cough are the primary source of the disease. The bacilli bind to ciliated epithelial cells that line the upper respiratory tract and secrete toxins that cause damage to the cells whose function is to clear mucus from the air passages. The net effect is a buildup of mucus and obstruction of these passages. Initially, the disease presents with the symptoms of a common cold followed by paroxysms (spasms) of violent, hacking, persistent, and recurrent coughing, usually about 15 to 20 coughs, in an attempt to cough up the mucus (Fig. 8.6). These episodes result in a need for oxygen, triggering deep and rapid inspirations throughout the partially

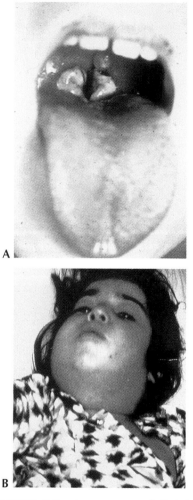

A

B

FIGURE 8.5 Appearance of diphtheria. **(A)** Inflammation of tonsils; **(B)** "bull neck" characteristic. (Source: Immunization Action Coalition.)

FIGURE 8.6 A child with pertussis coughing violently. (Source: Immunization Action Coalition.)

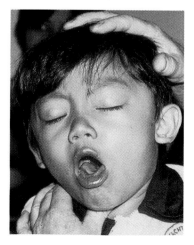

obstructed passages that are responsible for the characteristic **"whoop."** The coughing may be so violent as to cause vomiting, hemorrhage, and even brain damage. The bacteria are spread by talking, coughing, sneezing, and laughing. Whooping cough may last for several weeks or months.

Since 1940, a pertussis vaccine has been administered, in combination with vaccines against diphtheria and tetanus, dramatically decreasing the case rate in children in the United States from 200,000 to fewer than 1,000 per year. In the 1980s an upswing of the case rate from that seen in approximately 1968 began—an example of reemergence of an old disease. The pertussis vaccine has been controversial and accused of causing brain damage, but this has never been conclusively proved. Nevertheless, the "bad press" accounts for fewer pertussis immunizations and a resultant increase in the disease. Recently, a new vaccine utilizing only a part of the bacterial cell has been developed. This new pertussis vaccine produces fewer side effects and, like the old vaccine, is administered along with the diphtheria and tetanus vaccines.

Pertussis vaccine does not produce lifelong immunization, nor does recovery from whooping cough, contrary to popular belief. Persistent coughs in older children and adults may well be whooping cough, usually present in a milder form. As an example, in December 1999, six cases of whooping cough erupted in a high school in a small town in Massachusetts (In the News 8.2). Whooping cough is a devastating disease in nonimmunized populations. For example, in 1996 in a village in Guatemala, 42 people, mostly children, died of this disease. The population of this village was never immunized; they lived in an atmosphere of political unrest for 36 years because of armed civil strife. The disease was worsened by extreme poverty, leading to poor nutrition and overcrowding; a one-room hut in a Guatemalan village may house as many as 10 people. Obviously, the link between poverty and disease is not limited to whooping cough. The World Health Organization estimates that as many as 350,000 people worldwide are not vaccinated against pertussis.

Streptococcal Infections

Streptococci are a large group of microbes, the most significant one of which is *Streptococcus pyogenes*. The particular manifestation of disease is related largely to the specific toxin produced. Infections range from generally mild, the most common of which is **"strep throat,"** to a variety of life-threatening conditions, including **necrotizing fasciitis,** more dramatically called **"flesh-eating strep,"** and streptococcal **toxic shock syndrome.** The organisms reside in the nose and throat and are transmitted by respiratory droplets from infected persons or by contact with wounds and sores on the skin.

Strep throat, more correctly termed streptococcal pharyngitis or tonsillitis, is a common and usually mild infection predominantly found in children 5 to 15 years of age and is characterized by a red and sore throat, fever, and headache. Definitive diagnosis is by a throat swab culture. In some cases, **scarlet fever** results; it is caused by a streptococcal strain producing an **erythrogenic toxin.** The disease is characterized by a red rash and "strawberry tongue" (Fig. 8.7). Some streptococci are invasive and cause serious and life-threatening infections. Streptococcal toxic shock syndrome is an example and is characterized by multiorgan failure. Puppeteer Jim Henson, creator of the Muppets, died of streptococcal pneumonia in 1990 at the age of 53, only 20 hours after entering a hospital with flulike symptoms. Necrotizing fasciitis is another form of invasive streptococcal disease; in 1994 several outbreaks of this rare disease, dubbed by the media

Six Cases of Whooping Cough Reported at Local School

State Govt. Urges College Freshmen To Get Meningitis Shots

The Rhode Island Depar[t] Health recommended yes[terday] that all first-year college stude[nts] in the state get a vaccination ag[ainst] meningitis unless they have al[ready] been vaccinated. Receiving a s[econd] dose of the vaccine may increa[se] one's susceptibility to the dise[ase]

Infant meningitis vaccine approved

Fourth Patient Dies in Harford Legionnaires' Outbreak

"It's in the Air!"

"flesh-eating bacteria" and "killer bacteria," occurred in England and the United States and aroused unwarranted fears multiplied by news accounts. One story was headlined "The bacteria that ate my face"! The disease is a rarity; it is horrendous and is the result of infection of a minor wound with an invasive strain of streptococci that elaborates flesh-destroying enzymes and toxins. The disease progresses rapidly and can lead to extensive damage, disfigurement, amputation, and death.

Glomerulonephritis, a kidney disease, and **rheumatic fever,** a condition involving the heart and joints, are possible long-term complications of repeated early childhood streptococcal infections. The incidence of these diseases has decreased drastically in the past 25 years as a result of the advent of antibiotics, but unfortunately, drug-resistant strains are a concern.

Bacterial Meningitis

Meningitis is an inflammation of the meninges, the three membranes covering the spinal cord and the brain. This condition is potentially serious, since microbial invasion of the nervous system is involved. The key to survival is early diagnosis; several hours can make the difference between survival and death. Meningitis can be caused by a variety of bacteria, viruses, protozoans, and fungi. This section deals only with bacterial meningitis. Individuals with meningitis initially suffer coldlike symptoms which then progress quickly to more definitive complaints, including fever, possibly delirium, and stiffness in the neck and back.

A number of bacteria can infect the meninges, but the two most

FIGURE 8.7 Characteristic "strawberry tongue" of a patient with scarlet fever. Children with this disease frequently have an inflamed tongue along with a characteristic red rash. (Source: Armed Forces Institute of Pathology.)

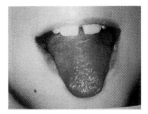

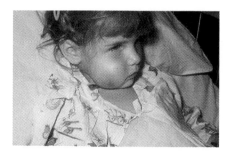

FIGURE 8.8 *H. influenzae* type b infection. Cellulitis is present in the child's cheek. (Source: Immunization Action Coalition.)

common species are *Neisseria meningitidis* and *Haemophilus influenzae* type b, the latter frequently referred to as Hib. The causative organisms live in the back of the nose and throat in healthy carriers and in infected individuals and are spread by respiratory droplets between people. Coughing, sneezing, kissing, and sharing eating utensils allow for passage. The organisms are very fragile outside the body, which limits the disease. The bacteria initially grow on the nasal pharynx and may invade the blood, from which they enter the meninges (Fig. 8.8).

Hib meningitis occurs primarily in children under 5 years of age. Although it is a less serious form of meningitis, it should not be regarded lightly. Once the leading cause of bacterial meningitis in the United States, its incidence has dramatically decreased since the introduction of the Hib vaccine in the early 1990s.

Meningococcemia, caused by *N. meningitidis,* is very serious. Fortunately, the infection responds to antibiotic therapy, although some individuals are left with severe, irreversible damage. There is no vaccination in general use, but a vaccine is available in threatened epidemics. In 1998, three fatal cases of meningitis occurred in Rhode Island and sparked a massive inoculation program of thousands of children. College students living in dormitories appear to be particularly susceptible to bacterial meningitis; in the last few years, episodes have occurred on several college campuses. A 17-year-old student at Johnson & Wales University in Providence, R.I., died of meningitis (In the News 8.2) in November 1999. The jury is still out as to whether college-bound students should be immunized against this potentially deadly disease in view of their future barracks-like living conditions. Box 8.3 describes the horrors of meningitis.

Lower Respiratory Tract Infections

Figure 8.4 indicates that the lower respiratory tract consists of the larynx (voice box), trachea (windpipe), bronchial tubes, and lungs. The tissue of the lungs terminates in millions of air sacs termed alveoli. The lungs are covered by a membrane called the pleura. The terms **laryngitis, tracheitis,** and **bronchitis** refer to the anatomical location; "itis" means inflammation. **Pleurisy** is an inflammation of the pleura, and **pneumonia** is an inflammation of the lungs. Lower respiratory tract infections are caused by a variety of bacteria, viruses, fungi, and protozoans.

Legionnaires' Disease (Legionellosis)

In 1976, the American Legion held its 58th state convention at the historic Bellevue-Stratford hotel in Philadelphia; this convention was a particularly significant one, since it marked the 200th anniversary celebration of the founding of our nation. The occasion was marred by a mysterious pneumonia in 182 of the Legionnaires, 29 of whom died. After an exhaustive investigation, it was discovered that the disease was bacterial and was disseminated through the air-conditioning ducts of the hotel. The organism was later identified and termed *Legionella pneumophila.* The publicity and the panic that ensued over the next few years were dramatic and inspired musician and composer Bob Dylan in 1981 to write a song about **Legionnaires' disease** (Box 8.4).

Isolated and unreported cases of legionellosis occur in the general population; symptoms of the disease are fever, muscle pain, cough, and pneumonia. These are symptoms shared with other respiratory infections, making definitive diagnosis difficult. The infections can be acquired by persons

BOX 8.3   Meningitis, a Big Dread on Campus

The following article is adapted from the October 12, 1999, *Washington Post* (Health section, p. 7), with permission of the author, Kathleen Phalen. The article text was obtained from the Immunization Action Coalition.

Life on Virginia Military Institute's (VMI) freshman "rat line" is rough, regimented and leaves little room for gim-riders, the tag given to those who spend too much time on the sick list. So when Cadet John Fuller started feeling sick and couldn't hold his head in a rat's traditional chin-down pose, he started to worry.

"It was a Wednesday night; in 15 minutes I went from feeling fine to feeling horrible," said the 19-year-old student from Anderson, Ind., now in his second year at the military college in Lexington, Va. "I was unbelievably cold. It felt like it was 30 degrees below—I thought I was getting the flu."

Fuller went to the school's infirmary, he learned he didn't have a fever, he got some Tylenol and went to bed. The next morning, while dressing, he couldn't push his chin to his neck. "I think that was the most noticeable thing, because I knew we were required to hold that position," he said. "By the time I got to class, I had red dots on my arms."

Barely 24 hours later, Fuller was in the intensive care unit at the University of Virginia hospital in Charlottesville. He had meningococcemia, a form of meningitis caused by the bacterium *Neisseria meningitidis.*

"It moved super-fast," said Fuller about this brush last February with this sometimes fatal disease. "I was sweating and the headache, the pain, was so unbelievable—I was conscious the whole time, but it felt like I was drunk, I was talking, but I wasn't coherent."

Fuller joins the rising ranks of college students infected with meningo-coccal disease. It often strikes those living in close conditions, such as college dorms and military barracks.

There have been recent cases in the Washington area. Two weeks ago, a 20-year-old student at the University of Maryland was hospitalized with bacterial meningitis but is now recovering. The same day the student was hospitalized, a Manassas Park Middle School teacher, Jane Dimitrious, 47, died of bacterial meningitis.

The disease has two common forms: meningococcal meningitis, an inflammation of the membranes surrounding the brain and the spinal cord, or meningococcemia, when the bacteria move in to the bloodstream. There is also a type of meningitis caused by a virus, but it is usually less severe than the bacterial form.

The bacterium causing meningitis is common and people often harbor it in their nasal passages and throats without effect. But when it gets into the bloodstream and moves to the brain, it causes problems. The bacteria can be spread by simply sneezing or coughing. Kissing and sharing utensils, drinking glasses or cigarettes can also contribute to its onset. Yet despite that, the bacterium is not especially contagious, and very close contact is usually necessary to spread the disease.

Particularly alarming to health officials is an increase in cases among college freshman and sophomores living in dormitories, a group the CDC reports are more than five times more likely to develop meningitis than college students in general. More than 600 cases of meningitis were reported in 1996 among people age 15 to 24, twice as many as a decade ago.

"Freshmen come into crowded conditions and are exposed to bacteria they have not previously seen. And then their lifestyles—binge drinking, smoking, spending time in bars, lack of sleep—compromise their immune systems," said Turner. Three years ago, Turner began recommending that U-Va. students be vaccinated after seeing five cases in a 20-month period at the university. The university hasn't had any cases since then.

Fuller was lucky. A rapid infusion of potent antibiotics stopped the progression of the bacteria. Nonetheless there are those receiving similar treatment who don't make it or are severely impaired. Others mistake the symptoms—high fever, headache, stiff neck, nausea, vomiting, confusion, and sleepiness—for the flu and never seek treatment. A cadet at VMI died in 1996 from meningococcemia. Reporting to the campus hospital with flu-like symptoms, Cadet Scott Hickey of Staunton, Va., died 12 hours later. The onset was similar to Fuller's.

There are no words to describe how Lynn Bozof felt when she discovered there was a vaccine that could have saved her son, Evan. "We were in disbelief," said Bozof, of Marietta, Ga. Evan, a 20-year-old honor student at Georgia Southwestern University, called his mom on a Wednesday afternoon to say he had a horrible headache.

"Migraines run in our family, so we weren't that alarmed," she recalls. "I checked on him around 5 and he had a blinding headache and couldn't keep anything down. I told him to have a friend take him to the emergency room."

Evan was hospitalized and slipped into a coma and because the disease ravages the circulatory system, his legs and arms became gangrenous. He, too, had meningococcemia and died 26 days later. His family is working with Georgia officials to educate students and parents about the disease and the vaccine.

BOX 8.4   "Legionnaire's Disease," by Bob Dylan

Some say it was radiation, some say
there was acid on the microphone,
Some say a combination that turned
their hearts to stone,
But whatever it was, it drove them to
their knees.
Oh, Legionnaire's disease.

I wish I had a dollar for everyone that

died within that year,
Got 'em hot by the collar, plenty an old
maid shed a tear,
Now within my heart, it sure put on a
squeeze.
Oh, that Legionnaire's disease.

Granddad fought in a revolutionary
war, father in the War of 1812,

Uncle fought in Vietnam and then he
fought a war all by himself,
But whatever it was, it came out of the
trees.
Oh, that Legionnaire's disease.

of any age, but middle-aged and older persons are the most susceptible, as are cigarette smokers and those with chronic lung diseases. When death occurs, it is usually attributable to shock and kidney failure.

Transmission does not involve person-to-person contact. Inhalation of aerosols that come from a water source contaminated with *L. pneumophila* is the mode of transmission. Air-conditioning cooling towers, spas, decorative fountains (e.g., in malls and hotel lobbies), vegetable mists in markets, showers, and whirlpools (Fig. 8.9) have all been implicated (Table 8.6). The organisms grow well in the warm and stagnant waters afforded by these environments and are then aerosolized. Ironically, hospital outbreaks have occurred on frequent occasions because the temperature in the hot water lines is kept relatively low (around 120°F) for safety reasons, allowing microbial multiplication. Furthermore, this microbe is more resistant than most to disinfection by chlorine. Fortunately, home and automobile air-conditioning units have never been implicated.

Legionellosis has been associated with gardening and the use of potting soil in Australia and Japan. In 2000, the CDC reported on three cases (one each in California, Oregon, and Washington) transmitted by potting soil. Health care providers need to be aware of this association.

Pontiac fever, first reported in Pontiac, Mich., is also caused by *L.*

FIGURE 8.9 Whirlpools are relaxing, but the aerosols they generate may be a source of the bacterium that causes Legionnaires' disease. (Author's photo.)

TABLE 8.6   A sampling of Legionnaires' disease outbreaks

Year	Location	Source	Outcome
1976	American Legion convention, Philadelphia, Pa.	Air-conditioning ducts in a hotel	182 taken ill, 29 deaths
1989	Stafford Hospital, England	Unknown	98 taken ill, 10 deaths
1992	Orlando, Fla.	Unknown	
1998	University of Washington Medical Center	Water-cooling tower	3 deaths in 5 months
1999	The Netherlands	Decorative water display at a bulb flower show	79 taken ill, 7 deaths
1999	Havre de Grace Hospital and Hartford Memorial Hospital, Connecticut	Water supply	4 deaths
1999	U.S. Naval Academy	Air-conditioning system	*L. pneumophila* found during a routine check; no illness

pneumophila. The symptoms resemble those of a variety of other respiratory illnesses and last a few days. No deaths from Pontiac fever have been reported.

Prevention of Legionnaires' disease can be accomplished by improvements in the design and maintenance (better housekeeping) of cooling towers, whirlpools, spas, and other sources of warm, stagnant waters in which there is the potential for aerosolization.

Tuberculosis

Mycobacterium tuberculosis (the tubercle bacillus) is the causative agent of **tuberculosis,** a disease of antiquity with its origins in the Middle East about 8,000 years ago. Evidence suggests that human tuberculosis was acquired from cattle and has affected civilization through the centuries. Tuberculosis is the leading cause of deaths resulting from microbes worldwide. Estimates are that about one-third of the world's population is infected, that over 2 million people die of tuberculosis annually, and that about 8 million new cases develop annually. The World Health Organization estimates that almost 1 billion people will become infected by 2020 if control measures are not improved. Tuberculosis is a current epidemic and is discussed further in chapter 15.

Tularemia

Francisella tularensis is the causative agent of tularemia. It is one of the most virulent of all bacteria; inhalation of as few as 10 organisms is sufficient to establish disease. Its extreme infectivity and ease of dissemination make it a potential biological weapon. Tularemia is a zoonotic disease; its natural reservoir consists of small mammals, including mice, voles, rabbits, and squirrels. Humans acquire tularemia through inhalation, contact with infected animal tissues, ingestion of contaminated food or water, or arthropod bites. There is no evidence of person-to-person transmission.

SEXUALLY TRANSMITTED DISEASES

Unlike most other microbial diseases, the **sexually transmitted diseases (STDs)** carry with them the stigma of human wickedness because they are associated with personal sexual behavior; the attitude of some people is that those who acquire them "deserve what they got." This idea of retribution, however, is unfair and unacceptable, particularly since gonorrhea and syphilis (like AIDS, which is a viral infection) may be passed from mother to child in utero. STDs were formerly known as **venereal diseases, or VD.** Ironically, these miserable diseases, recognized in the 1600s, were named after Venus, the Roman goddess of love. Before the germ theory of disease, it was believed that some females passed onto their male partners a "noxious substance" that resulted in VD (a flagrant example of sexism).

Microbes that cause STDs are directly transmitted from the warm, moist mucous membranes of one individual to the membranes of another individual during sexual practices, including vaginal and anal intercourse and oral sex. Semen and vaginal discharges are the major sources. The male and female genital urinary tracts are diagrammed in Fig. 8.10. Promiscuous individuals are particularly prone to STDs because of their sexual behavior. Unfortunately, one STD is not exclusive of the others, so individuals may have more than one disease simultaneously. Table 8.7 summarizes the STDs.

The advent of public education programs and the appearance of penicillin in 1943 led to a dramatic decrease in the incidence of syphilis and gonorrhea; predictions were that they would largely be diseases of the past. Unfortunately, this declaration of victory was premature. In recent times, the STDs have become a serious public health problem, particularly because of the emergence of antibiotic-resistant strains. The 1960s were a time of sexual revolution, sexual liberation, and an "anything goes" attitude. Views about sex that had persisted since the Victorian era were on the decline, and birth control pills separated procreation from lovemaking. "Free love" was the name of the game, promiscuity was heralded, the use of condoms and spermicides decreased, and STDs were on the rise. The current concern with AIDS, a viral infection (chapter 9), overshadows the continuing problem of other STDs. Because vaccines are not available, prevention efforts focus on abstinence, monogamy, and safe sex.

FIGURE 8.10   Anatomy of the human urinary system **(A)** and male **(B)** and female **(C)** reproductive systems.

A Urinary system

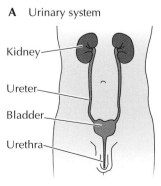

Kidney

Ureter

Bladder

Urethra

B Male reproductive system

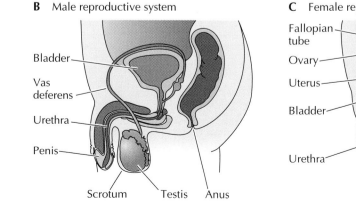
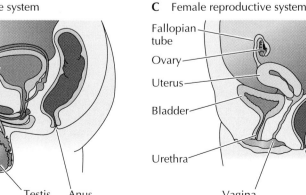

Bladder

Vas deferens

Urethra

Penis

Scrotum   Testis   Anus

C Female reproductive system

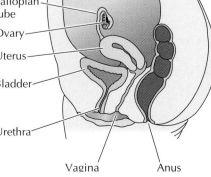

Fallopian tube

Ovary

Uterus

Bladder

Urethra

Vagina   Anus

TABLE 8.7   Sexually transmitted bacterial diseases[a]

Disease	Incubation period	Causative microbe	Symptoms
Syphilis	16–28 days (possibly as long as 10 weeks)	*Treponema pallidum*	Painless sores on the penis or cervix, rashes on the palms and soles, and eventual paralysis and insanity
Gonorrhea	2–7 days	*Neisseria gonorrhoeae*	Burning urination, cervical and urethral infection, abdominal pain, sterility
Chlamydia	2–6 weeks	*Chlamydia trachomatis*	Abnormal pregnancy, infertility, inflammation of the testicles
Lymphogranuloma venereum	1–4 weeks	*Chlamydia trachomatis*	Sores on the penis and vagina that develop into painful buboes
Chancroid	3–5 days	*Haemophilus ducreyi*	Painful ulcers on the penis, labia, or clitoris

[a]These diseases are treatable with antibiotic therapy; drug-resistant strains pose a problem in some cases.

Syphilis

Girolamo Fracastoro, a poet and physician, wrote a poem in 1530 about a shepherd boy named Syphylis who unknowingly offended Apollo, the sun god; Apollo bestowed his wrath onto the boy with a "loathsome" disease. Ultimately, this new disease became known as *syphilis sive morbus gallicus,* or the French (some said the Italian) disease. The causative agent of **syphilis** is the spirochete-shaped bacterium *Treponema pallidum.* It frequently coinfects with pathogens that cause other STDs, including AIDS. Humans are the only known reservoirs. In the United States in the last 25 years, the disease has been most common in the 15- to 34-year-old population, and males are more than three times more susceptible than females. The disease is sometimes referred to as "the great imitator," since, in its later stages, it mimics other diseases. Syphilis progresses through a series of three stages (Fig. 8.11 and 8.12): **primary, secondary,** and **tertiary.** The primary stage is manifested by the appearance of painless **chancres** (sores) on the penis, or on the cervix, which may be undetected. These chancres shed **treponemes** continuously. The chancres disappear in about 4 to 6 weeks and give false hope of recovery to those infected. In untreated individuals, over the next approximately 5 years, symptoms of a rash, particularly on the palms and soles, appear and disappear. This secondary stage is systemic, meaning that the bacteria multiply and spread throughout the body. About one-third of those untreated progress to tertiary syphilis, an

FIGURE 8.11   Stages and time line of syphilis. Times are approximate and subject to considerable individual variation.

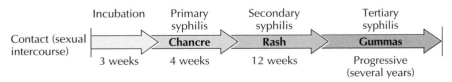

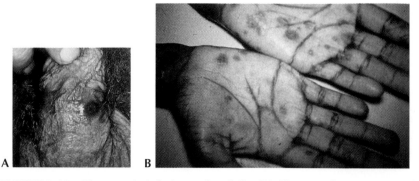

FIGURE 8.12 Characteristic lesions of syphilis. **(A)** Chancre of primary syphilis on penis. **(B)** Lesions of secondary syphilis on palms; these lesions may occur on other body surfaces. (Source: CDC.)

advanced stage which develops over the next 40 years, during which numerous organs and tissues, particularly those of the cardiovascular and nervous systems, show degenerative changes. **Neurosyphilis** is particularly disabling and can result in a shuffling walk (tabes), paralysis (paresis), and insanity. Prior to the antibiotic era, individuals with tertiary syphilis constituted a significant portion of the population of mental institutions. A poem by an anonymous author describes the progression of syphilis (Box 8.5). In the poem, the term **gummas** refers to tumorlike lesions that develop during the tertiary stage; these gummas destroy nerve or skin tissue.

Although sexual contact is the usual means of transmission of syphilis, the organism can be passed in saliva and poses a threat to dentists, dental hygienists, and "big kissers." Of particular concern is that infants can acquire **congenital syphilis** as the result of spirochetes that pass across the placenta from mother to baby. Such infants develop serious problems, including **saber shins,** a condition in which the shin bone develops abnormally.

On July 20, 1998, a group of scientists at the Institute for Genome Research in Maryland and collaborators at the University of Texas reported that they had completed a genetic map of *T. pallidum,* opening the possibility of new efforts to develop a vaccine.

Gonorrhea

Gonorrhea, sometimes referred to as "clap" or "drip," is caused by *Neisseria gonorrhoeae;* like syphilis, the only reservoir is the human. It is estimated that 800,000 cases of this disease occur annually in the United States and involve an expenditure of approximately $1.1 billion. *N. gonorrhoeae* may be transmitted via vaginal, oral, or anal sex. Pili (chapter 4) enable the organisms to attach to cells lining the urethra, allowing them to hang on during the passage of urine.

A major problem of gonorrhea is that carriers often remain without symptoms after infection but continue to transmit the disease for over 10 years. As many as 10% of males and 30% of females may not show symptoms. In males, the disease is characterized by a pus-containing drip from the penis and burning during urination. In females, the disease may remain hidden; during campaigns to halt this disease, huge billboards, particularly in urban areas, were prominently displayed with the slogan "Gonorrhea Hides in Women." In females, the cervix and urethra are the most common sites of infection. **Pelvic inflammatory disease (PID)** occurs in about 50% of untreated females and is characterized by abdominal pain and, possibly, sterility.

An Anonymous Poem, *Syphilis*

There was a young man from Black Bay
Who thought syphilis just went away
He believed that a chancre
Was only a canker
That healed in a week and a day.

But now he has "acne vulgaris"—
(Or whatever they call it in Paris);
On his skin it has spread
From his feet to his head.
And his friends want to know where his
  hair is.

There's more to his terrible plight;
His pupils won't close in the light
His heart is cavorting,
His wife is aborting,
And he squints through his gun-barrel
  sight.

Arthralgia cuts into his slumber;
His aorta is in need of a plumber;
But now he has tabes,
And saber-shinned babies,
While of gummas he has quite a number.

He's been treated in every known way,
But his spirochetes grow day by day;
He's developed paresis,
Has long talks with Jesus,
And thinks he's the Queen of the May.

Source: L. M. Prescott, J. P. Harley, and D. A. Klein, *Microbiology*, 4th ed. (McGraw-Hill, New York, N.Y., 2000).

N. gonorrhoeae can be transmitted into the eyes of newborns during delivery, causing corneal damage and possibly blindness, a condition referred to as **ophthalmia neonatorum.** As a preventive measure, silver nitrate eyedrops or erythromycin, tetracycline, or antibiotic ointment is placed into the eyes of newborns at birth. The American Academy of Pediatrics and the CDC recommend prophylactic treatment of the eyes of all infants shortly after birth, a treatment which is required by law in most states.

Chlamydia

Chlamydia, caused by *Chlamydia trachomatis,* is the most common STD, but most people are not even aware of its existence. It is sometimes referred to as "the silent epidemic." This disease is particularly prevalent in young adults and teenagers. Even those with the disease may be unaware that they are infected, since approximately 70% of infected females and 30% of infected males lack symptoms. They are, therefore, particularly vulnerable to complications, including PID (as in gonorrhea), abnormal pregnancy, and infertility. Males are subject to inflammation of the testicles and infertility. As in gonorrhea, newborns whose mothers are infected have a 50-50 chance of developing an eye-threatening condition; therefore, antibiotics are administered at birth. (Silver nitrate, although effective against *N. gonorrhoeae*, is not effective against *C. trachomatis* and hence has been largely replaced by antibiotics.) Chlamydia is the easiest STD to treat, and a single dose of antibiotic usually cures the disease in 1 week.

Lymphogranuloma Venereum and *Haemophilus ducreyi* Disease

Lymphogranuloma venereum and *H. ducreyi* disease are primarily associated with tropical and semitropical areas and are rarely found in the United States (Fig. 8.13).

CONTACT DISEASES (OTHER THAN STDs)

There are several contact diseases, only three of which are discussed here (Table 8.8).

Peptic Ulcers

Students sometimes complain that the stress of exams, papers, (boring) lectures, and all of the other facts of academic life are "giving them **ulcers.**"

AUTHOR'S NOTE *From 1956 to 1958, I was a young army officer stationed in a medical laboratory outside Tokyo. Gonorrhea was rampant among the military personnel, most of whom were 18 to 20 years old. The diagnosis was simple: look for the presence of a drip from the penis, and place a drop of the discharge on a glass slide. The slide was then Gram stained and examined under a microscope for the presence of* Neisseria *organisms. The examination lasted less than 5 minutes, and, fortunately, penicillin was the cure. I took my turn in the VD clinic, behind a curtain, checking one young man after another.*

FIGURE 8.13 Characteristic groin bubo of a patient with lymphogranuloma venereum. (Source: Armed Forces Institute of Pathology.)

TABLE 8.8   Contact diseases (other than STDs)[a]

Disease	Incubation period	Causative microbe	Transmission	Symptoms
Leprosy	3–6 years	*Mycobacterium leprae*	Skin contact; air droplets	Development of lepromas; neurological damage
Staphylococcal skin infections	3–4 days	*Staphylococcus aureus*	Skin contact	Pimples, abscesses, and peeling on skin; systemic symptoms in toxic shock syndrome
Peptic ulcers		*Helicobacter pylori*	Unknown, but probably person to person, possibly food and contaminated water	Burning or gnawing in epigastrium, particularly when stomach is empty; possibly bleeding; nausea, vomiting

[a]These diseases are treatable with antibiotic therapy; drug-resistant strains pose a problem in some cases.

It is true that student life is stressful (so is a professor's!), but it won't give you ulcers. The popular myth is that ulcers are directly associated with stress. Take a look at Table 8.9, and you will probably be surprised to learn that those engaged in stressful occupations are no more at risk of getting ulcers than those in occupations that are considered nonstressful. Ulcers were also once thought to be the result of eating spicy foods. Treatment focused on hospitalization, bed rest, bland diet, and consuming large quantities of milk, cream, and antacids to coat the lining of the stomach and the

TABLE 8.9   Stress and ulcers: a myth[a,b]

10 most stressful jobs	10 least stressful jobs
1. Inner city high school teacher	1. Forester
2. Police officer	2. Bookbinder
3. Miner	3. Telephone line worker
4. Air traffic controller	4. Toolmaker
5. Medical intern	5. Millwright
6. Stockbroker	6. Repairperson
7. Journalist	7. Civil engineer
8. Customer service/complaint worker	8. Therapist
9. Secretary	9. Natural scientist
10. Waiter	10. Sales representative

[a]Do you think that stressful jobs lead to ulcers? Which of these people listed above are most likely to have an ulcer? The answer may surprise you. All the workers in this list are just as likely to get an ulcer as any others you can imagine. An inner city high school teacher's stomach is no more likely to be riddled with ulcers from the stress of dealing with troubled teens (not to mention the lunchroom's tacos) than a forester's stomach. While stress and diet can irritate an ulcer, they do not cause it. Ulcers are caused by the bacterium *H. pylori* and can be cured with a 1- or 2-week course of antibiotics, even in people who have had ulcers for years.
[b]According to *Health Magazine.* Source: CDC.

duodenum (the first part of the small intestine). In some cases, surgery was performed to remove the diseased area. But relief was only temporary.

In 1982, two Australian physicians, Robert Warren and Barry Marshall, claimed that *Helicobacter pylori* (a comma-shaped bacterium), not stress or diet, was the cause of peptic ulcers. This finding knocked the medical establishment off its feet. Marshall went to the extreme of drinking cultures of *H. pylori* to prove his claim to a skeptical medical community, and, sure enough, he developed ulcers. Finally, in 1994, a National Institutes of Health Consensus Development Conference affirmed a strong association between *H. pylori* and peptic ulcers and recommended antibiotics for treatment. The role of *H. pylori* in causing ulcers stimulated research, now under way, to look for causal relationships between microbes and a variety of chronic diseases (see chapter 16).

How does *H. pylori* survive in the harsh acidic environment of the stomach and duodenum that is hostile to most microbes? The answer is a wonderful example of biological adaptation. The bacterium produces the enzyme **urease,** which breaks down the compound **urea,** a metabolic product of the body, into carbon dioxide and ammonia, an alkaline compound which neutralizes the acidity. This clever survival strategy protects *H. pylori.*

According to the CDC, the mechanism of transmission of *H. pylori* is not clear, nor is there an explanation why some persons infected with the bacterium develop ulcers, while others do not. Direct person-to-person contact appears to be the most plausible route; humans are the primary reservoir. There is also the possibility that food and water are involved in transmission.

The diagnosis of peptic ulcers is based on several methods, including the presence of antibodies specific for *H. pylori* and the culturing of tissue, obtained by biopsy, to demonstrate the presence of the bacterium. A test called the **breath test** is about 94 to 98% accurate and is based on the organism's production of urease. For this test, the patient drinks a preparation containing isotopically labeled urea. If present, *H. pylori* breaks down the urea, resulting in the formation of carbon dioxide, which is exhaled. Measurement of the labeled carbon dioxide in the breath determines the presence or absence of *H. pylori.*

H. pylori infection can be cured in 2 to 3 weeks with appropriate antibiotics and other medications. The recurrence rate after antibiotic treatment is about 6%, compared to 80% when only antacids are used. The bottom line is that ulcers are caused by a curable infection.

Studies have indicated that long-term *H. pylori* infection is associated with the development of gastric cancer, the second most common cancer worldwide.

Leprosy

Among all those diseases carrying a social stigma, **leprosy** is at the top of the list (Boxes 8.6 and 8.7). Biblical accounts of "affliction" point to leprosy; through the ages, the disease conjured up dreadful images of rejection and exclusion from society, disfiguring skin lesions, and fingers and toes falling off. Lepers were required to carry and ring bells and to shout "unclean" as they approached others. Fear in the minds of healthy individuals led to terrible tales of cruelty. Today, the term "leper," because of its stigma, is not socially acceptable, and the disease is now known as **Hansen's disease.** The story of a Belgian priest, Father Damien, is one of great dedication and compassion. In 1870, Damien established a leper colony on the island of Molokai in Hawaii as a refuge for those with the disease. He spent his life

BOX 8.6   The Politics of the Leprosy Bacillus

In August 2001, I visited the Leprosy Museum at St. Jørgens Hospital in Bergen, Norway. This might appear to be an unusual activity; most tourists are attracted to the other museums and cultural sites that Norway has to offer. (In fact, not to my surprise, I was the only tourist there!) A pamphlet describing the museum reads as follows:

Founded around 1411 and run continuously until 1946, St. Jørgens Hospital is one of the oldest hospital institutions in Scandinavia. Leprosy was a common health problem in the western parts of Norway until the late 19th century, and thousands of Norwegian lepers have lived out their lives at St. Jørgens. The museum presents the hospital and its daily life and also the Norwegian contributions to international leprosy research. The famous Norwegian physician Armauer Hansen discovered the leprosy bacillus in Bergen in 1873.

While visiting the museum, I met Sigurd Sandmo, its curator, and spent several hours with him over a 2-day period. Sigurd is a medical historian with a vast amount of knowledge. I was so impressed with this young man that I invited him to write the following piece about leprosy.

Today Norway is a fully developed country and among the wealthiest nations in the world. But 150 years ago one would consider Norway to be a poor outpost in Europe, with many of the problems to be found only in developing countries today. One of the most conspicuous problems was leprosy, a disease which was looked upon with great interest and concern by Norway's physicians and authorities. In the second half of the 19th century, the city of Bergen had the highest concentration of leprosy patients in Europe, and the city was an international center for leprosy research.

For most other European countries, where the disease had ceased to be a problem around 1500, leprosy had during the 19th century more or less become a subject for missionaries. In the reports from work among African and Asian lepers, the Levitical meaning of the disease was now reproduced. An almost forgotten medieval disease had renewed its religious and racial actuality. In Europe only Norway and Iceland experienced a second bloom of leprosy in the 19th century. The Norwegian distribution of leprosy was rather local, and the disease was mainly found among poor fishermen and peasants in the western parts of the country. With an incidence of more than 3% of the population in some districts, this was a shame for the young Norwegian state. In the 1870s a Swedish physician referred to the problem as "the waves of shame on the shores of Norway."

The national authorities offered the medical circles in Bergen both economic freedom and political influence to have the problem solved. In return they expected visible results. From our modern point of view, we may see several scientific breakthroughs. In 1847 Daniel C. Danielssen published the first symptomatology of leprosy, which represents the birth of modern leprosy research. In 1856 Ove G. Hoegh founded the Norwegian Leprosy Registry, probably the first national disease registry in the world. But the value of these events was hard to recognize at the time. Much more spectacular was Armauer Hansen's discovery of the leprosy bacillus in 1873. Hansen's publication from 1874, *Preliminary Contributions to the Characteristics of Leprosy,* is the earliest description of a microorganism as the cause of a chronic disease.

Hansen's discovery represented the kind of breakthrough the public health authorities were waiting for. It was the microscope against the Old Testament and the scientist's rationalism against the Levitical myth. And ever since, health workers have used the story of the discovery to fight the myth where it has survived. Hansen has become a symbol of dignity, humanism, and a scientific approach toward patients who are still suffering from Hansen's disease. When we come across Hansen stamps from Thailand or a Hansen monument in the Vietnamese jungle, we easily recognize the impact Hansen's discovery once had. The authorities, the mob, the church, the peasants, and the patients could all see how Hansen's scholarly approach challenged the 2,000-year-old stigma.

However, Hansen's discovery had immediate political implications too. At the time, Norway did not have any leprosy legislation. Admission to all hospitals was voluntary, as the common opinion was that the disease was either inheritable or a nonspecific condition caused by bad living conditions. During the first years after the discovery, Hansen's theory of a *contagium vivum* was met with a certain amount of skepticism from his colleagues, both in Norway and abroad. When Hansen presented his proposal for a new act requiring isolation of leprosy patients in Norway, many of his colleagues felt that this was an inhumane law, branding the patients as criminals. Hansen defended himself as a pragmatic scientist: How can we act humanely toward persons who have a contagious disease? His opponents read Hansen's contributions to the discussion with regret. However, the law was passed in 1885 and served as a model and inspiration for leprosy legislation and public action in disease colonies in several other countries, such as Greece, Hawaii, and Japan. Today the remains of colonies in Spinalonga, Greece, and in Molokai are sad sights for visitors. In Japan surviving patients now receive compensation for their sufferings caused by harsh legislation.

In the discussions concerning the public health work and legislation in Norway in the 1870s and 1880s, we can see how Hansen's work gradually became politicized. The discovery itself was internationally accepted during the 1880s, but many of Hansen's opponents and colleagues saw that the political answer to the leper question was complicated and that the new legislation in a way revitalized the old stigma. In one of Norway's medical journals, the discussion went on for several years. After years of heated debate with irritated colleagues, Hansen might have considered whether his fundamentalistic positivism and rationalism were, after all, suitable for discussing public health work.

The Politics of the Leprosy Bacillus *(continued)*

Several elements from the discussion on leprosy legislation in the 1880s were repeated with regard to another disease 100 years later in Norway, as in most other countries; the AIDS discussions of the 1980s were surprisingly similar. Again, the possibility of establishing disease colonies was discussed and society's rights and responsibilities with respect to infected individuals were presented as a question of yes or no. And, again, the evidence of contagion was used by different groups with different interests to justify their political and religious opinions. Once again, scientific discoveries of microorganisms turned out to be powerful but unpredictable political weapons.

We often think of biological and medical discoveries as non-political statements. And when we look at the single participant, we are usually right. But like the discovery of the leprosy bacillus, scientific breakthroughs are usually caused by priority given by political actors, and scientific results will often take on a political dimension once they leave the laboratory. Both the discovery of the leprosy bacillus and the discovery of human immunodeficiency virus remind us how vulnerable and risky this dimension can be.

Source: Sigurd Sandmo, curator, Leprosy Museum, St. Jørgens Hospital, Bergen, Norway.

---

there and ultimately died of leprosy. In more recent times, Mother Teresa stated, "When I am washing the wounds of a leper, I feel I am doing great service to the Lord."

Mycobacterium leprae is the causative agent of Hansen's disease; note that it belongs to the same genus, *Mycobacterium,* as the tubercle bacillus. The disease is transmitted by skin contact and respiratory droplets over a long period of time; it is not considered to be highly infectious. The incubation time is long, and it takes about 3 to 6 years of close contact with an individual with the disease for it to be acquired. Children appear to be more susceptible. The disease is characterized by a variety of physical manifestations (Fig. 8.14), including disfiguring tumorlike skin lesions **(lepromas)** and neurological damage to the cooler peripheral areas of the body, such as the hands, feet, face, and earlobes. This may lead to curling of the fingers

---

BOX 8.7 *En Klagesang (A Lament)*

The following are excerpts of the poem *En Klagesang (A Lament)* by Peder Olsen Feidie, a patient in St. Jørgens Hospital for lepers, Bergen, Norway, from 1832 to 1849.

I dream of when I was a lad,
Of all the happy times I had—
A joy it was to live.
But fortune quickly changed her face
And sorrow then did joy replace.
For me and many more
This fate has lain in store.

I was not yet fifteen years old,
My mind was full of joys untold,
Then were they all cut short.

Pain overcame me and did start
Quickly to pierce marrow bone and heart.
Oh! It was hard to bear
This burden laid on me.

We lepers can no doctors get:
Here must we stay and wait and fret,
Until our time is up.
Peter from prison did escape
Because on God's grace he did wait.
O God, break now the chains
Which bind our limbs with pains.

One is covered with sore on sore,
Another is dumb—speaks no more,
A third hobbles on crutches.
A fourth no daylight now can see,

A fifth has lost all his fingers.
Surely now it is clear
What we must suffer here.

In St. George's Hospital here,
Suffering over a hundred bear,
And wait to be set free.
O holy Ghost our Helmsman true,
Steer us all our sufferings through,
And to heaven lead us,
For there are we set free.

Source: P. Richards, *The Medieval Leper and His Northern Heirs* (D. S. Brewer, Rowman and Littlefield, Cambridge, N.J., 1977).

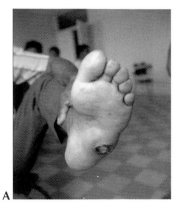
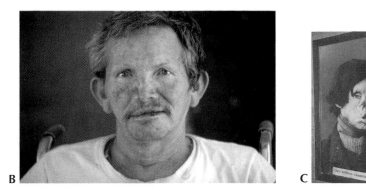

A   B   C

FIGURE 8.14   Leprosy, now usually called Hansen's disease. Disfiguring tumorlike lesions appear on the skin, and neurological damage occurs in the cooler areas of the body. (Sources: World Health Organization [**A** and **B**]; author [**C**].)

(described as "claw hand"), thickening of the earlobes, collapse of the nose, and possibly blindness. The individual may lose the ability to perceive pain in the fingers and toes, resulting in serious deformities; accidental burns may occur. Fortunately, drugs are now available to arrest the disease, and patients with Hansen's disease can look forward to a normal life.

Staten Island University Hospital in New York has a center dedicated to treating patients with Hansen's disease. It is estimated that there are over 500 cases in New York City and in excess of 10,000 in the United States. According to a staff member at the hospital, "It [leprosy] gets to New York through immigration. We've had a number of people from Central and South America and Asia. It is now a very treatable disease, and no longer a death sentence." There is a Hansen's disease facility in the continental United States under the auspices of the Public Health Service, located in Carville, La. There were once more than 400 residents at the facility, but now there are fewer than 75. A recent decision to close the hospital has caused great concern among some of the older resident patients who have spent their lives there. These residents have been referred to specialized outpatient facilities throughout the country and have been offered financial subsidies by the government. The leprosy center in Molokai has also been closed, except for a few of the older residents who choose to remain there.

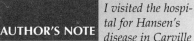

AUTHOR'S NOTE   *I visited the hospital for Hansen's disease in Carville in 1962. The atmosphere in this modern hospital, located in a parklike setting, was remarkably upbeat, as was the attitude of the people with the disease. I photographed several of the patients (Fig. 8.14) but chose not to include some of the more disfiguring cases.*

Staphylococcal Infections

Bacteria of the genus *Staphylococcus* are normal inhabitants of the skin, mouth, nose, and throat; the skin is the largest organ in the human body. Usually, these organisms present no problem. However, when the skin is broken, as can result from a wound, burn, or other circumstance, its normal integrity as a barrier is lowered. Certain strains of *S. aureus* are the most frequent staphylococci involved in infections. Nosocomial infections (chapter 7) are a major problem, and staphylococci account for approximately 13% of the 2 million hospital infections reported each year.

Staphylococci are frequent causes of localized skin infections, occurring as pimples, abscesses, and inflamed lesions filled with a core of pus. Abscesses can progress to produce **boils,** which can develop into **carbuncles.** Carbuncles are larger and deeper lesions and may, in fact, reach baseball size. These lesions are extremely painful and are dangerous, since they can progress into **systemic** (blood-borne and widespread) **infections** throughout the body. A relatively common manifestation of staphylococcal infection is **impetigo.** This is a superficial infection of the skin and usually occurs around the mouth in the form of blisters that ooze a yellowish liquid. More frequently, impetigo occurs in very young children, particularly following a runny nose, which sets up irritation in the surrounding tissue. Impetigo is annoying but not particularly serious, other than the fact that it is spread easily from child to child. Impetigo can also be caused by streptococci. **Scalded skin syndrome** is also caused by staphylococci and tends to be more prevalent in children with infection of the stem of the umbilical cord. In these cases, the skin can become blistery as a result of the production of an **exfoliative toxin** that peels away the skin to expose a red layer.

Toxic shock syndrome came to light in the late 1970s and caused considerable fear, particularly among menstruating females who used a particular brand of tampon made of a superabsorbent material. A major outbreak occurred in 1980 and led to a recall of the tampons in question. Some strains of *S. aureus* secrete an exotoxin which produces high fever, nausea, vomiting, peeling of the skin (particularly on the palms and the soles), and a dangerous drop in blood pressure that leads to life-threatening shock. Toxic shock syndrome can also occur in men and in nonmenstruating females and is sometimes associated with surgical wound infections. In 1981, almost 1,000 cases of toxic shock syndrome were reported; in 1997, approximately 100 cases were reported. Toxic shock syndrome is also caused by streptococci.

SOILBORNE DISEASES

A few of the most common soilborne diseases are presented in Table 8.10. Anthrax and tetanus are spore-forming bacilli and survive in soil for many years.

TABLE 8.10   Soilborne bacterial diseases[a]

Disease	Incubation period	Causative microbe	Transmission	Symptoms
Anthrax	1–15 days	*Bacillus anthracis*	Contact with endospores in soil and in occupations involving handling of wool, hides, meats	Cutaneous form produces skin lesion, headache, nausea, fever
Tetanus	4 days to several weeks	*Clostridium tetani*	Contact with spores from soil, animal bites, gunshot wounds	Lockjaw, muscle stiffness, and spasms due to production of tetanospasmin
Leptospirosis	7–13 days	*Leptospira interrogans* (spirochete)	Urine, soil, or water contaminated with urine from dogs, cats, sheep, rats, mice	Flulike with possible complications in liver and kidney

[a]These diseases are treatable with antibiotic therapy; drug-resistant strains pose a problem in some cases.

FIGURE 8.15 A cutaneous anthrax lesion. (Source: Armed Forces Institute of Pathology.)

Anthrax

Anthrax, a potentially deadly disease caused by *Bacillus anthracis,* has captured considerable media attention as the most probable weapon for biological warfare (see chapter 14). (The organism is a sporeformer; recall from chapter 4 the extreme resistance of bacterial spores.) Spores can be readily spread by missiles and bombs and can cause anthrax in humans and in other organisms, including those used as a food source. The Department of Defense is now calling for anthrax vaccination of all U.S. military personnel.

Anthrax is primarily found in large, warm-blooded animals, mostly sheep, cattle, and goats, and is acquired by the ingestion of spores during grazing. Infection can also occur in humans and is most prevalent in agricultural regions of the world. There are three varieties of anthrax, which vary by the route of transmission and subsequent symptoms. **Inhalation anthrax** is an occupational hazard for humans exposed to contaminated dead animals and animal parts, e.g., tanners and sheep shearers, who, in the course of their work, are most likely to inhale spores; this condition is also called woolsorter's disease. Inhalation anthrax is the most severe form of the disease and is the greatest threat with the deadliest consequences in terms of biological warfare (see chapter 14). The incubation period is 1 to 6 days. Early symptoms are coldlike and progress to severe breathing problems within several days, followed by death 1 to 2 days later. A second form of anthrax is **cutaneous anthrax** (Fig. 8.15), also acquired by contact with bacilli or spores via wool, hides, leather, or hair products. About 20% of untreated cases result in death; death is rare with appropriate antibiotic therapy. A third form, **gastrointestinal anthrax,** is the result of the ingestion of inadequately cooked meat contaminated with *B. anthracis,* leading to acute inflammation of the gastrointestinal tract characterized by abdominal pain, vomiting of blood, and severe diarrhea. Approximately 25 to 60% of untreated cases result in death. Antibiotics are effective against all three forms of anthrax, but early intervention is necessary. A vaccine is available for animal workers because their occupation classifies them as a high-risk group. The use of anthrax vaccine in military personnel is a controversial issue that is addressed in chapter 14. Person-to-person transmission does not occur.

Tetanus

Tetanus, also called **lockjaw,** conjures up terrible images and is usually associated with stepping on a rusty nail. Tetanus is the only disease that is preventable by a vaccine and is not communicable. It is acquired by exposure to spores of *Clostridium tetani.* The tetanus organism produces **tetanospasmin,** a neurotoxin considered to be the second most deadly bacterial toxin; estimates are that as little as 100 micrograms can kill an adult. (The botulinum toxin [see above] is considered the most potent bacterial toxin.)

The tetanus bacilli are present worldwide and are abundant in soil, manure, and dust. They are normal inhabitants of the intestinal tracts in horses and cattle, and, more rarely, in humans. The fact that these organisms produce spores and the hardiness of the spores explain their longevity and abundance in the soil. Tetanus develops when spores gain access into the body through wounds. These spores then germinate into bacilli, which multiply in deep wounds that tend not to bleed a lot (puncture wounds). Hence, in the "stepping on a rusty nail" scenario, the nail produces a relatively deep puncture wound, which in itself is of little concern; "rusty" implies that the nail has been in the soil for a long time. The crevices in the rust provide for the adherence of minute amounts of soil, which potentially contain large numbers of spores. Gunshot wounds, animal bites, knife stabbings, and injury on the prongs of a pitchfork all serve as grisly mechanisms of puncture wounds, possibly allowing the entrance of tetanus spores. The incubation period is anywhere from 3 days to 3 weeks, but usually about 8 days.

C. tetani, unlike the majority of bacteria dealt with in this chapter, has no invasive ability and is not transmitted from person to person. During growth, tetanus toxin is produced which interferes with the relaxation phase of muscle contraction, resulting in uncontrollable muscular contraction (Fig. 6.11). Stiffness in the jaw (lockjaw) is an early symptom due to the contraction of facial muscles; as more toxin is produced, the neurotoxin continues to spread, leading to contraction in other muscles, particularly in the limbs, stomach, and neck. Tetanus survivors report excruciating pain resulting from the spasms of muscle contractions, which can be strong enough to break bones. The extreme contractions in the back and rib muscles cause the body to arch severely, so that only the victim's head and heel are in contact with the surface, a position referred to as **opisthotonos**.

Death occurs by suffocation because of the victim's inability to breathe; imagine taking a very deep breath and not being able to blow out in order to take another breath.

In the United States, 47 of the 50 states require schoolchildren to be immunized against tetanus, and all of the states require children in day care facilities to be immunized. Nevertheless, 201 cases of tetanus were reported in the United States from 1991 to 1994. A study released in 1998 reported that two out of three senior citizens may lack immunity to tetanus because of failure to comply with immunization guidelines.

Tragically, **neonatal (newborn) tetanus** is a common manifestation of tetanus in newborns in developing countries (Fig. 8.16C), and the course of this disease follows the same pattern as tetanus. Symptoms occur within 2 weeks after birth; the infant fails to suck properly on the mother's breast,

AUTHOR'S NOTE *The photograph in Fig. 8.16A was taken in Central America. The boy was approximately 10 years old and was brought into the hospital near death; there was no possible intervention. He died several hours later, a horrible outcome from, presumably, a minor wound. Immunization would have prevented this.*

FIGURE 8.16 Tetanus. The tetanus toxin causes muscle contraction and rigidity. **(A)** Opisthotonos (author's photo); **(B)** opisthotonos (source: CDC); **(C)** neonatal tetanus (source: National Library of Medicine); **(D)** risus sardonicus (sardonic grin) (source: National Library of Medicine).

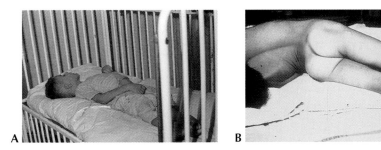

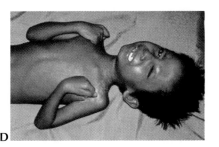

becomes irritable, and has convulsions. The disease is prevalent in the poorest nations of the world and occurs when a baby is delivered under unsanitary conditions to a non-tetanus-immunized mother.

Imagine a scenario in which a nonimmunized woman delivers a baby on the dirt floor of a small hut in a remote village in some impoverished area. An untrained birth attendant, unaware of the need for sanitary conditions, cuts the umbilical cord with an unclean razor blade, thereby introducing tetanus. In some traditions, the baby's umbilical stump is treated with cow dung, ash, mustard oil, or other unsterilized "folk" products.

Neonatal tetanus is a real killer in the developing world. In 1998, about 215,000 newborns died of the disease, but only one of the deaths was in the United States. Box 8.8 describes a case of neonatal tetanus in Montana in which the infected baby was treated in a hospital and survived. Prevention is centered on vaccination of all women of childbearing age (since their protection is passed to their newborns), improved delivery and postdelivery practices, and education.

Leptospirosis

Jane Seymour, a well-known actress and star of the TV show *Dr. Quinn, Medicine Woman,* contracted **leptospirosis** in 1997 while filming in Puerto Rico. In Costa Rica in 1996, a group of 26 adventuresome whitewater rafters were shocked when five members of their party became ill with leptospirosis.

What is leptospirosis? This disease, also known as **swamp fever,** is caused by the spirochete *Leptospira interrogans*. These spirochetes are harbored in nonhuman hosts, including dogs, rodents, and a variety of wild animals, and is spread through their urine. Rats appear to be the most significant source of disease, probably because there is greater opportunity for contact with rat urine than with that of other animals. In fact, there is concern that inner-city residents may be at particular risk for leptospirosis, along with farmers, sewer workers, and workers in other occupations who may be exposed to infected rat urine. Additionally, bodies of water, river banks, and vegetation may become contaminated as a result of runoff from surrounding soil. The spirochetes penetrate the human skin, enter the bloodstream, and rapidly invade virtually all organs. The infection can also be acquired by swallowing water from sources where rats may have urinated.

Leptospirosis can be a serious and potentially fatal disease, with symptoms including jaundice, fever, headache, nausea, and chills, and may require hospitalization. Diagnosis is based on blood tests.

Leptospirosis is most common in tropical countries. Approximately 100 cases per year are reported in the United States, but the actual number is considered to be much higher since many cases are not reported.

ARTHROPOD-BORNE DISEASES

Recall from chapter 7 the major role that arthropods play in the transmission of microbial diseases. Vector control is vitally important in breaking the complicated transmission cycle of these diseases. Numerous species of fleas, mosquitoes, flies, ticks, and lice are responsible for a variety of devastating diseases that have plagued humankind for centuries and have influenced the course of civilization. Remember that spirochetes and rickettsiae are bacteria and were briefly described in chapter 4. Table 8.11 summarizes the arthropod-borne bacterial diseases presented in this chapter.

BOX 8.8 Neonatal Tetanus, Montana, 1998

Montana Newborn of an Unvaccinated Mother Contracts Neonatal Tetanus after Application of Nonsterile Clay to the Umbilical Cord

Neonatal tetanus (NT) is a severe, often fatal disease caused by a toxin of *Clostridium tetani*, a ubiquitous spore-forming bacterium found in high concentrations in soil and animal excrements. NT is associated with nonsterile delivery and umbilical cord-care practices for newborns of mothers with antitoxin levels insufficient to protect the newborn by transplacental transfer of maternal antibody. In 1997, NT accounted for an estimated 277,400 deaths worldwide but is rare in the U.S. During 1995–1997, of 214 tetanus cases reported in the United States, only one occurred in a neonate. This report summarizes the investigation in March 1998 of an NT case by Missoula City-County Health Department (MCCHD) and the Montana Department of Health and Human Services (MDHHS). The findings indicated that tetanus in a newborn of an unvaccinated mother occurred after the application of nonsterile clay to the umbilical cord.

On March 21, 1998, a 9-day-old newborn, who had no previous medical problems, was taken to a hospital by her parents who reported a 10-hour history of an inability to nurse and difficulty in opening her jaw. Her parents also had noticed a foul-smelling discharge from her umbilical cord during the preceding 1–2 days. No other symptoms were noted by the parents. On admission, the newborn had trismus, increased general muscle tone, and hyperresponsiveness to external stimuli. The umbilical cord was covered with dried clay, which when retracted revealed a foul-smelling yellow-green discharge. Culture from the umbilical cord grew several anaerobic (*C. perfringens*, *C. sporogenes*) and aerobic (*Staphylococcus*, *Streptococcus*, and *Bacillus* sp.) bacterial species. NT was diagnosed based on the clinical characteristics.

The newborn was treated with tetanus immune globulin (500 units intramuscularly) and penicillin G (300,000 units per kilogram per day intravenously) for 10 days. On March 24, she required mechanical ventilation and remained ventilated for 12 days. She was discharged on April 10, with no apparent neurological sequelae and was developing normally on follow-up at age 7 months.

The mother, a 32-year-old non-Hispanic white woman born in the United States, had never been vaccinated because of her family's philosophic beliefs. She had no complications during her pregnancy and was attended throughout her pregnancy by a licensed "direct-entry" midwife from her community. The newborn was delivered in a local hospital by cesarean section. While in the hospital, she received standard umbilical care with isopropyl alcohol. The newborn was discharged at 3 days of age. For home umbilical cord care, the parents applied a "Health and Beauty Clay" powder provided by the midwife. This clay powder was applied to the umbilical cord up to three times daily with a clean cotton-tipped swab. The family lived in a rural area in a house adjacent to a horse pasture. Although the newborn and her mother stayed primarily indoors, the family's dog often ran between the house and the pasture.

The "Health and Beauty Clay" was a bentonite clay from Death Valley, California. According to the manufacturer, it had been sold for 21 years as a cosmetic product without reported adverse health outcomes. The manufacturing process of the clay did not include sterilization. The clay was shipped in 2-pound containers, sold by weight in a local store, and dispensed to local midwives in smaller containers. The midwives would further aliquot the clay into 2-ounce, presumably clean vials for distribution to their patients. The use of the clay for umbilical cord care was common among local direct-entry midwives because they believed it accelerated drying of the umbilical cord.

On April 9, MCCHD distributed a health-care advisory to more than 60 health-care providers in the area that emphasized the importance of tetanus toxoid vaccination, particularly for pregnant women, and cautioned against using nonsterile products for umbilical cord care. Following this case, use of clay for umbilical cord care was discontinued by midwives in the community. The mother of the case patient has since been vaccinated with tetanus and diphtheria toxoids (Td), but as of October 1998 has not initiated vaccination for her infant because of concern about potential adverse effects.

Source: Immunization Action Coalition. Reprinted from B. Goode et al., *Morbidity and Mortality Weekly Report* **47:**928–929, 1998.

Plague

Plague is also known as the **Black Death** and conjures up terrible images of disease throughout the course of history. The term "plague" is frequently used in a general sense to describe an explosive outbreak with a high death rate. More strictly, however, plague refers to a specific bacterial disease, caused by a bacillus known as *Yersinia pestis,* one of the most virulent bacteria known. All it takes is a single bacillus to establish infection.

TABLE 8.11   Arthropod-borne bacterial diseases[a]

Disease	Incubation period	Causative microbe	Transmission	Symptoms
Plague		*Yersinia pestis* (bacillus)	Fleas	Hemorrhages under skin (Black Death), fever, buboes
Bubonic plague	2–6 days			
Pneumonic plague	2–3 days			Pneumonia
Lyme disease	3 days–1 month	*Borrelia burgdorferi* (spirochete)	Ticks	Possibly expanding bull's-eye rash, flulike symptoms (headache, fatigue, fever, muscle pain); in later stages, arthritis-like neurological impairment, heart inflammation
Relapsing fever		*Borrelia recurrentis* (spirochete)		High fever, headaches, muscle pain, weakness
Endemic	5–10 days		Ticks	
Epidemic	5–10 days		Body lice	
Rocky Mountain spotted fever	2–6 days	*Rickettsia ricketsii*	Ticks	Rash which begins on days 2–4, appears noticeably on wrists and ankles, and then spreads upward
Typhus fever				Rash characterized by dark spots on days 5–6 on upper trunk which spreads over most of body, fevers, chills, muscle pain, fatigue, cough, headache
Epidemic typhus	1–2 weeks	*Rickettsia prowazekii*	Body lice	
Endemic typhus	1–2 weeks	*Rickettsia typhi*	Fleas	
Ehrlichiosis	7–10 days	*Ehrlichia chaffeensis* and *E. canis*	Ticks	Like a "bad flu" with fever, chills, headache, muscle pain, nausea; rash only rarely present

[a]These diseases are treatable with antibiotic therapy; drug-resistant strains pose a problem in some cases.

Medieval Europe was plagued from 1347 to 1351 with *Y. pestis* when homes were inhabited by fleas carrying the plague bacillus. In the short span of only 4 years, one-third of the population, estimated at 25 million people, were ravaged and died of the Black Death. The dead were piled up and fed upon by rats, resulting in even more infected rats whose fleas spread the disease. The plague swept through populations; in the 1300s, at the peak of the epidemic in Paris, as many as 800 people died each day, cutting the population in half. It was once believed that carrying sweet-smelling herbs and flowers and holding them to the nose protected against the poisonous vapor of plague. Figure 8.17 depicts a person in protective clothing; the birdlike mask contains the plant material. *Y. pestis* was used as a weapon of biological warfare by the Japanese army during World War II; biological warfare is the subject of chapter 14. A major plague epidemic that killed over 12 million people began in Asia in 1890 and was carried to San Francisco by rat-infested ships at the turn of the century.

Plague is considered a reemerging infection because of its increase throughout the world. Plague resurfaced in 1994 in Surat, India, about 400 miles from Bombay, and resulted in the infection of several hundred people and about 50 deaths. In the 14th century, the Black Death required 17 years to cross the trade routes from China to Iceland. Today it takes only about 10 hours to fly from Bombay to London's Heathrow Airport. This is less than the incubation time, allowing an infected person to make the trip without any symptoms of illness. In the United States, sporadic cases occur, particularly in the Southwest, because of contact with infected rodents. For more than 3 decades before 1965, only a few cases per year occurred in the United States, but this rate of incidence gradually increased to 30 to 40 cases in the mid-1980s before decreasing again (though not to pre-1965 levels).

Plague is a zoonosis; over 200 species of mammals, primarily rodents (including gophers, ground squirrels, mice, and wild rats), serve as reservoirs. Rat fleas are the usual vectors, and transmission occurs primarily from infected rats to other animals, including humans, as a result of contact with a dead plague-infected animal or, more usually, being bitten by an infected flea. In **bubonic plague,** the best-known form of the disease, the bacilli localize in the lymph nodes, particularly in the nodes of the groin, armpits, and neck, and cause the nodes to swell to the size of eggs. These enlarged nodes, known as **buboes,** are hard, red, and painful. The bacilli invade the bloodstream, liver, lungs, and other sites. Hemorrhages occur under the skin, and the dried blood turns black, hence the term Black Death. The mortality rate in untreated cases is over 50%. Box 8.9 is a descriptive account of bubonic plague. Disease control, as in all zoonotic diseases, centers on interrupting the transmission between reservoirs and humans. Bubonic plague, although it may be severe enough to kill its victims, is not normally infectious.

A second form of the plague, known as **pneumonic plague,** occurs when individuals with the bubonic form develop pneumonia and transmit the bacteria to others by coughing and through saliva. The bacteria invade the victim's lungs, which become filled with a frothy, bloody fluid. Pneumonic plague approaches 100% fatality without early detection and treatment. A vaccine is available but offers protection for only a few months and is only given to persons at high risk. **Septicemic plague** is a third form of the disease; it results from the spread of infection from the lungs to other parts of the body, but it can also be acquired by direct contact of contaminated hands, food, or objects with the mucous membranes of the nose or throat. This form of plague is considered to be 100% fatal.

FIGURE 8.17 Costume that was formerly thought to be protective against plague. (Source: National Library of Medicine.)

Lyme Disease

The residents of the small New England town of Old Lyme, Conn., are probably not thrilled that an emerging zoonotic disease bears its name because it was first described in their community in 1975. This tick-borne disease is now present throughout the United States, particularly in the Northeast, the upper Midwest, and the Pacific Coast, and its incidence has dramatically increased (Fig. 8.18). It is the most common vector-borne disease in the United States and is also found in Europe, Australia, the former Soviet states, China, and Japan.

The biology of Lyme disease is particularly complex, since five closely interrelated organisms must be present at appropriate times. They are (i) the spirochete *Borrelia burgdorferi,* (ii) the tick that serves as the vector (the particular species is a function of the geographic area), (iii) deer that serve as hosts for the maturation of tick eggs to the adult stage, (iv) mice that allow the further development of the tick, and (v) humans or other final hosts. The cycle requires 2 years to complete. It is important to understand the life cycle so as to target mechanisms that will break the transmission cycle (Fig. 8.19).

Adult ticks feed and mate on large mammals, particularly deer (Bambi

BOX 8.9 The Plague: Death Defined

What was it like for a victim of the plague?

It started with a headache. Then chills and fever, which left him exhausted and prostrate. Maybe he experienced nausea, vomiting, back pain, soreness in his arms and legs. Perhaps bright light was too bright to stand.

Within a day or two, the swellings appeared. They were hard, painful, burning lumps on his neck, under his arms, on his inner thighs. Soon they turned black, split open, and began to ooze pus and blood. They may have grown to the size of an orange.

Maybe he recovered. It was possible to recover. But more than likely, death would come quickly. Yet...perhaps not quickly enough. Because after the lumps appeared he would start to bleed internally. There would be blood in his urine, blood in his stool, and blood puddling under his skin, resulting in black boils and spots all over his body. Everything that came out of his body smelled utterly revolting. He would suffer great pain before he breathed his last. And he would die barely a week after he first contracted the disease.

The swellings, called buboes, were the victim's lymph nodes, and they gave the bubonic plague its name. But the bubonic form of the disease was only one manifestation of the horrible pandemic that swept Europe in the 1340s. Another form was pneumonic plague. The victims of pneumonic plague had no buboes, but they suffered severe chest pains, sweated heavily, and coughed up blood. Virtually no one survived the pneumonic form.

The third manifestation was septicemic plague. This sickness would befall when the contagion poisoned the victim's bloodstream. Victims of septicemic plague died the most swiftly, often before any notable symptoms had a chance to develop. Another form, enteric plague, attacked the victim's digestive system, but it too killed the patient too swiftly for diagnosis of any kind.

Medieval Europeans had no way of knowing any of this. The causes of plague were not discovered until the late 19th century.

Plague is carried by rodents like rats and squirrels, but it is transmitted to humans by the fleas who live on them. A flea, having ingested plague-infected blood from its host, can live for as much as a month away from that host before he needs to find another warm body to live on. When a blood-engorged flea attempts to find another victim, it invariably injects into that victim some of the blood already within it. If the injected blood contains the bacterium *Yersinia pestis*, the result is bubonic plague. Fleas were, alas, such a part of everyday life that no one noticed them much. In this invisible manner the plague spread from rat to human and to cat and dog, as well.

Pneumonic plague is airborne. It is contracted by breathing the infected water droplets breathed (or coughed) out by a victim of the disease. The pneumonic form was much more virulent and spread much more quickly and just as invisibly.

Plague is occasionally transmitted by direct contact with a carrier through open sores or cuts, directly into the bloodstream. This could result in any form of the plague except pneumonic, although it is likely that such incidents most often resulted in the septicemic variety. The septicemic and enteric forms of the plague killed most quickly of all and probably accounted for the stories of individuals going to bed apparently healthy and never waking up.

People died so swiftly and in such high numbers that burial pits were dug, filled to overflowing and abandoned; bodies (sometimes still living) were shut up in houses which were then burned to the ground; and corpses were left where they died in the streets.

Source: © Melissa Snell ("Death Defined"), licensed to About.com, Inc. Used by permission of About.com, Inc., which can be found on the Web at www.about.com. All rights reserved.

---

may not be so cute!) in the fall and early spring. Female ticks become engorged with blood, fall off the deer, and lay their eggs on the ground; by summer, the eggs hatch into **larvae.** Larvae feed on small mammals and birds in the summer and early fall and are inactive until the next spring, when they molt into **nymphs.** The nymphs feed on small rodents and other small mammals and birds in the late spring and summer and mature into adults in the fall, thereby completing the 2-year cycle. The larval and nymph stages become infected with the spirochetes as they take a blood meal from their infected intermediate hosts, particularly white-footed mice; the spirochetes remain in the tick during the tick's developmental stages. The nymph is the major source of transmission to humans and other animals. Two or more days of feeding are required by the nymphs, during which the ticks transmit the spirochetes. The poppy seed size of the nymphs allows them to escape being removed since they are virtually undetectable. Adults (Fig. 8.20) are about the size of sesame seeds.

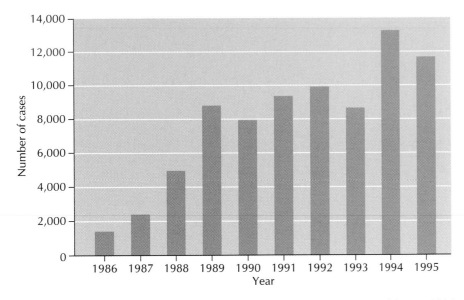

FIGURE 8.18 Number of reported cases of Lyme disease in the United States, 1986 to 1995. Lyme disease has been on the increase since it was first discovered in Lyme, Conn. (Redrawn from a graph provided by CDC.)

FIGURE 8.19 The cycle of Lyme disease is complicated and involves several stages. (Redrawn from a diagram provided by the American Lyme Disease Foundation.)

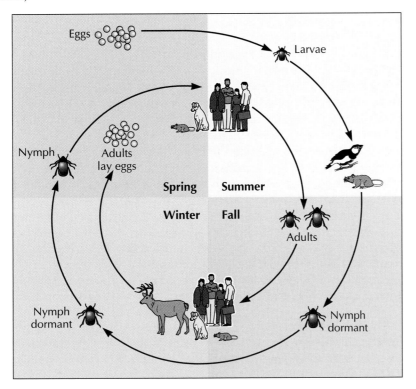

FIGURE 8.20 A deer tick, the vector of Lyme disease. (Source: Bernard Furnival, American Lyme Disease Foundation.)

In its early stages, the disease causes flulike symptoms, including fever, headache, and muscle and joint pain. Some individuals have a spreading "bull's-eye" skin rash (Fig. 8.21). These are the "lucky" people, since the vague flulike symptoms apply to many diseases, making the early diagnosis of Lyme disease difficult. Unfortunately, the rash may not occur until 1 month or more after the tick bite, or it may never occur; other symptoms, including arthritis and numbness, may occur years later. Diagnosis includes history of tick exposure, symptoms, and blood tests to look for the presence of the antibodies to the spirochetes. A not unusual story illustrating the problems of Lyme disease diagnosis is that of a man who lived with a diagnosis of multiple sclerosis for 9 years before being correctly diagnosed with Lyme disease. Infected individuals treated early with appropriate antibiotics usually recover completely and quickly. Those diagnosed with late and chronic Lyme disease may experience arthritis-like symptoms and neurological difficulties.

Intervention is based on avoiding tick-infested areas, particularly in May through July, when ticks are particularly active and looking for a blood meal. Measures of protection include use of the insect repellent deet (*N*,*N*-diethyl-*m*-toluamide), appropriate clothing, personal inspection for and removal of ticks after outdoor activity, removal of leaves, and clearing of brush and tall grasses near houses (particularly if there are young children in the area), since ticks live on the tips of grasses and shrubs. A 1998 issue of the *New England Journal of Medicine* reported on two Lyme disease vaccines with at least 76% protection rates; both require three shots and 1 year to achieve full effectiveness.

Relapsing Fever

The two varieties of relapsing fever, caused by the spirochete *Borrelia recurrentis*, are **endemic relapsing fever** and **epidemic relapsing fever.** The endemic form is transmitted by ticks and occurs sporadically in most areas of the world. The mortality rate is less than 5%. The epidemic form is transmitted directly from human to human by body lice, which bite and cause itching; infection results from scratching the skin and, thereby, crushing the spirochete-infected lice into the skin. This disease currently exists in eastern and central Africa, China, and the Peruvian Andes and has a mortality rate of approximately 30% in untreated cases. Relapsing fever is usually associated with overcrowding (as in refugee camps), war, poverty, and other conditions which lead to social breakdown. The best forms of prevention are the delousing of people and their clothing and improved hygienic conditions; the implementation of these measures is dehumanizing and difficult and presents serious or even overwhelming difficulties in catastrophic conditions. The spirochetes undergo antigenic switching, an important mechanism of virulence for avoiding host antibody defenses, to escape immunological detection (chapter 6). This phenomenon accounts for the recurrence of symptoms and is the basis for the name relapsing fever. Relapses occur approximately every 2 to 4 days; in the endemic form there may be three or four recurrences, whereas in the epidemic form there are typically only two or three relapses. There are reports, however, of over 10 relapses!

Rickettsial Diseases

As mentioned in chapter 4, rickettsiae are atypical bacteria in the sense that, like viruses, they are obligate intracellular parasites and require eukaryotic host cells. They are carried by arthropods (with a single exception),

FIGURE 8.21 Bull's-eye rash of Lyme disease. About 1 month or more after the tick bite, this rash may appear, but some patients never develop a rash. (Reprinted from P. H. Gilligan, M. L. Smiley, and D. S. Shapiro, *Cases in Medical Microbiology and Infectious Diseases*, 2nd ed., ASM Press, Washington, D.C., 1997.)

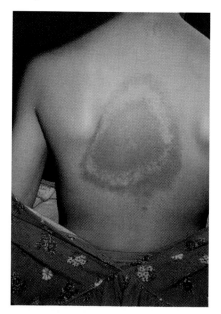

including ticks, body lice, and fleas. The symptoms generally include a rash, which varies depending upon the particular rickettsial infection, and flulike symptoms. Antibiotic therapy is effective, and further methods of control are directed at minimizing the contact between the vector and humans.

Rocky Mountain Spotted Fever

The tick-borne disease **Rocky Mountain spotted fever** is misnamed, because it is more common in the southeastern part of the United States than in the Rocky Mountains, where it was first reported. The causative organism, *Rickettsia rickettsii*, is transferred by a tick feeding on a human or other mammalian host. As in all tick-borne diseases, early removal of the tick is vital, since transmission occurs during the feeding cycle. Estimates are that there is a higher than 6% death rate when treatment is delayed for more than 3 days after the symptoms first appear; if treatment is started earlier, the death rate is 1.3%. Up to 20% of those infected but not treated die; the elderly population is particularly vulnerable. Approximately 1,500 cases of this disease occur annually in the United States.

Typhus Fever

There are two varieties of typhus fever. **Endemic typhus,** also known as **murine typhus,** is caused by *Rickettsia typhi* and is transmitted by fleas; **epidemic typhus** is caused by *Rickettsia prowazekii* and is transmitted by body lice.

In the endemic form, rickettsiae are found on mice and rats and are transmitted to humans by the bites of rat fleas. Those whose jobs or residences are in rat-infested areas are particularly at risk for endemic typhus. Rodent control is primary in breaking the disease cycle.

Epidemic typhus is caused by *R. prowazekii*. This organism was named after Howard Ricketts and Stanislaus von Prowazek, who in the early 1900s discovered the rickettsial organism and the vector. Both men died of epidemic typhus. The disease is transmitted by the human body louse, and the human is the only reservoir of the disease. As the louse feeds, it has the nasty habit of defecating; the bite causes an itch, and as the individual scratches, the rickettsiae and the louse's feces are inoculated into the bite. (This is the same process as in the epidemic form of relapsing fever). Typhus is truly a "lousy" disease. The infected louse dies usually within a few weeks, and its entire life cycle (from egg to adult) can take place on the clothing of an infected person. No epidemics of this disease have occurred since 1922, but there have been sporadic cases, with flying squirrels serving as possible reservoirs for the disease. Like relapsing fever, also a louse-borne disease, conditions of war, crowding, unsanitary conditions, and little opportunity to bathe or to wash clothing are conducive to this disease. As with all vector-borne diseases, environmental control of the vector is important.

Ehrlichiosis

Ehrlichiosis (Fig. 8.22 and Table 8.12) is an emerging tick-borne infection similar to Lyme disease that was first reported in humans in 1986, although the disease had been known for years to exist in horses, dogs, cattle, and other mammals. Some biologists have suggested that the ticks may be moving from dogs to humans. More than 500 cases of this flulike illness have been reported in the United States since 1986, with approximately 10 deaths,

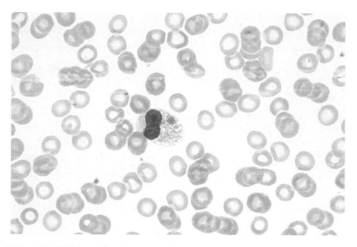

FIGURE 8.22   Ehrlichiae, the bacteria that cause ehrlichiosis, hide in white blood cells. The ehrlichia in the figure is stained purple. (Source: Gregory Storch, Washington University School of Medicine.)

primarily along the Atlantic coast and in south-central areas of the United States. In the summer of 1995, ehrlichiosis struck a dozen residents in the New York City suburbs. There are two forms of the disease, both of which respond well to antibiotic therapy, but about 5% of untreated cases result in death. **Human monocytic ehrlichiosis (HME)** and **human granulocytic ehrlichiosis (HGE)** appear to be the same disease but are caused by different species. HME is caused by *Ehrlichia chaffeensis;* the particular species that causes HGE is not definitely known, but it may be *Ehrlichia canis.*

OVERVIEW   A variety of diseases caused by bacteria are presented and organized on the basis of their mode of transmission, namely, food borne and waterborne, airborne, sexually transmitted, contact, soil, and arthropod borne. Some bacterial diseases are diseases of antiquity, while others are new, emerging, or reemerging. Some bacteria

TABLE 8.12   Clinical presentation of human ehrlichioses[a]

Clinical signs and symptoms	Severe clinical spectrum
Fever	Disseminated intravascular coagulation
Malaise	Pancytopenia (blood cell deficiency)
Headache	Encephalitis
Myalgia, arthralgia	Meningitis
Anorexia	Pulmonary infiltrates
Chills, sweating	Gastrointestinal bleeding
Nausea, vomiting	Respiratory failure
Rash	Renal failure
Cough	Death
Diarrhea	
Abdominal pain	

[a]Source: CDC.

have caused major epidemics and pandemics throughout the centuries, and their death tolls have influenced the course of history. These diseases are described with attention to the microbial cycle of disease, pathogenicity, and other related factors.

The information presented in this chapter makes it clear that, despite advances in hygiene, immunization, and drug therapy in the 20th century, humans remain threatened by pathogenic bacteria. Food-borne and water-borne intoxications and infections are characterized by gastroenteritis; despite the watchdog efforts of the U.S. Department of Agriculture, they continue to have a significant effect on the health of Americans. Airborne bacteria cause disease categorized into upper and lower respiratory tract infections. Immunization and antibiotic therapy have been instrumental in reducing the morbidity and mortality in those countries that are able to implement these practices. STDs caused by bacteria have been overshadowed by the pandemic of AIDS, a virally caused disease. STDs were on the decline after the appearance of penicillin in the 1940s, but the sexual revolution in the 1960s resulted in an increase in their prevalence. Contact diseases (other than those transmitted sexually) remain a problem; the necessity for direct contact limits their spread. The relatively recent discovery that most gastrointestinal ulcers are caused by bacteria, and not by stress as previously believed, is a major breakthrough in the treatment of ulcers. Direct person-to-person contact is thought to be the most plausible route of transmission. Pathogenic bacteria may be present in the soil; those that are spore forming survive for very long periods of time. Circumstances leading to their transmission into the body result in infection. Arthropods are vectors in the transmission of several bacterial diseases, including those caused by spirochetes and rickettsiae, that are responsible for major epidemics. Mosquitoes may well qualify as "public enemy number 1"! ■

SELF-EVALUATION

PART I Choose the *single* best answer.

1. Families with pet iguanas or turtles are particularly at risk for
   a. *Campylobacter* infection   c. salmonellosis
   b. listeriosis   d. shigellosis

2. One of the most potent known toxins is associated with
   a. *Salmonella*   c. staphylococci
   b. botulism   d. cholera

3. Oral rehydration therapy is associated with
   a. *Salmonella*   c. cholera
   b. meningitis   d. Legionnaires' disease

4. The beautiful fountain in a shopping mall may be a source of
   a. listeriosis   c. Legionnaires' disease
   b. leptospirosis   d. leprosy

5. Which of the following toxins is a neurotoxin?
   a. botulinum toxin   c. cholera toxin
   b. diphtheria toxin   d. more than one of the above

6. Which one of the following is an upper respiratory tract infection?
   a. diphtheria   c. legionellosis
   b. tetanus   d. tuberculosis

PART II Match the statement with a disease from the list below. A letter may be used once, more than once, or not at all.

These items are based on soilborne and arthropod-borne diseases.
1. a rickettsial disease transmitted by lice
2. possibly transmitted by dogs and cats
3. strong candidate for biological warfare
4. named after town in Connecticut
   a. leptospirosis
   b. anthrax
   c. Lyme disease
   d. epidemic typhus
   e. endemic typhus

(continues)

SELF-EVALUATION (continued)

These items are based on airborne infections.
5. can cause obstructive membrane in pharynx
6. characterized by severe cough
7. associated with Philadelphia
    a. Legionnaires' disease
    b. diphtheria
    c. meningitis
    d. pertussis

These items are based on sexually transmitted and other contact diseases.
8. syphilis
9. primarily associated with tampons
10. "unclean, unclean"
11. gonorrhea
    a. "great imitator"
    b. toxic shock syndrome
    c. virus associated
    d. clap or drip
    e. Hansen's disease

PART III Answer the following.

1. Distinguish between food poisoning (intoxication) and food infection. Name two diseases in each category.

2. Identify the following terms:
    a. oral rehydration therapy
    b. opisthotonos
    c. epidemic relapsing fever
    d. PID

3. You are about to travel to a foreign country where sanitation is at a lower level than you are accustomed to. What precautions will you take during your stay?

9

VIRAL DISEASES

The first day or so we all pointed to our countries. The third or fourth day we were pointing to our continents. By the fifth day we were aware of only one Earth.
SULTAN BIN SALMAN AL-SAUD
DISCOVERY SPACE SHUTTLE MISSION

PREVIEW This chapter presents the major viral diseases of humans. As pointed out in chapter 1, emerging viruses, including Ebola virus, human immunodeficiency virus, Lassa virus, and hantavirus, to mention only a few, seem to have appeared suddenly from nowhere and threaten our existence. Viruses are assigned to a particular family based on their shape, nucleic acid content, size, and other properties; those that cause human disease fall into 13 RNA and 6 DNA virus families.

For each viral disease presented, particular attention is paid to its mechanism of transmission and other factors involved in the microbial cycle of disease (chapter 7): pathogenicity, incubation time, treatment, and (in some cases) immunization. With many bacterial diseases (chapter 8), the affected area of the body frequently correlates with the mechanism of transmission and the route of entry of the bacteria into the body. For example, food-borne bacteria primarily infect the digestive system, and airborne bacteria primarily infect the respiratory tract. In the case of viruses, there may be little connection between the route of entry and the particular organs and tissues of the body involved. The measles and chicken pox viruses, for example, are airborne and enter the body via the respiratory route, but the skin is their major target. Rabies is acquired as a break in the skin from the bite of an infected animal, but the virus attacks the nervous system. In many cases, there may be more than one route of entry and/or transmission, making many viral diseases difficult to pigeonhole.

The chapter is divided into five sections based, as in chapter 8, on the mechanisms of transmission: (i) airborne, (ii) arthropod borne, (iii) food borne and waterborne, (iv) sexually transmitted, and (v) contact (other than sexually transmitted). As you read through this chapter, you will note that the naming of viruses does not follow the genus and species system estab-

187

lished for bacteria (e.g., *Escherichia coli*). Viruses are named after the location in which they were first noted, the primary body site involved, symptoms, or the arthropod vector. Some examples are Ebola virus, named after the Ebola River in the Democratic Republic of the Congo (formerly Zaire); hepatitis (liver inflammation) viruses, named after the major body sites involved; yellow fever virus, named for the jaundice (yellowish tinge) characteristic of the disease; and Colorado tick fever virus, named after the virus's vector. Further, you will note that immunization is available against some viral diseases but not against others. Keep in mind that antibiotics are not effective against viruses, but there are some antiviral agents (not antibiotics) that are used in specific cases. Antibiotics and antiviral agents are discussed in more detail in chapter 12. As you study each of these viral diseases, recall the five stages of viral replication (chapter 5): adsorption, penetration, replication, assembly, and release. Remember that these stages occur in infected cells, causing damage to the host, which may be anywhere from subclinical to mild to severe to fatal. Frequently, the diagnosis of a particular viral disease is based on the clinical symptoms, which may be readily apparent, such as swollen jaws in mumps and the characteristic rash of measles or chicken pox. However, laboratory diagnosis, based on demonstration of viral antigens or antibodies and culture of the specific virus (chapter 5), may be the only way to definitively diagnose many viral diseases. Because of the reporting time necessary and the expense involved, these procedures are not always routine. Frequently, it is assumed that the condition will simply run its course.

AIRBORNE VIRAL DISEASES

The primary source of airborne diseases is respiratory droplets from an infected person to a susceptible person. Hence, having an "infectious laugh" may not be a desirable trait. A summary of the major airborne viral diseases is presented in Table 9.1.

TABLE 9.1   Airborne viral diseases[a]

Disease	Incubation period	Virus category	Symptoms	Immunization and comments
Influenza	1–2 days	ssRNA	Coldlike: chills, fever, muscle aches and pains, coughing, sneezing, sore throat	Annual flu shot for senior citizens and other at-risk persons
Mumps (rubeola)	10–21 days	ssRNA	Coldlike, Koplik's spots; rash	MMR vaccine
Mumps	10–20 days	ssRNA	Swollen jaw	MMR vaccine
German measles (rubella)	12–32 days	ssRNA	Fever, rash	MMR vaccine; prenatal transmission
Respiratory syncytial virus	4–5 days	ssRNA	Coldlike	No vaccine; life-threatening; most prevalent cause of respiratory infection under 6 months of age
Common cold	1–3 days	ssRNA	Coldlike	No vaccine
Chicken pox and shingles	10–23 days (chicken pox), few to many years (shingles)	dsDNA	Chicken pox: rash; shingles: intense pain along nerve	Vaccine now available; lifelong immunity after infection; prenatal transmission
Hantavirus	1–3 days	ssRNA	Influenzalike	No vaccine; transmitted via aerosolized fecal material, saliva, or urine from rodents; high fatality rate

[a] MMR, measles-mumps-rubella; ss, single stranded; ds, double stranded.

Influenza

The year was 1918, and the world was engaged in the horrors of World War I. As if that were not tragic enough, along came the **influenza** virus, an enemy to all sides engaged in the conflict—a biological agent knowing no political or national loyalties. The enemy could not be defeated by the conventional warfare of tanks, infantry, and artillery. Influenza took a toll in human life greater than the ravages of the war itself; the pandemic that it created is regarded as one of the most devastating of all time (In the News 9.1 and Fig. 9.1). Little was known about viruses; it was not until about 20 years later that viruses "came into their own." Over 1 million people died in the short span of only 120 days, including an estimated 500,000 in the United States (almost as many people as live in Alaska today). Estimates are that the virus killed more than 20 million people worldwide, including 12.5 million in India. The virus killed rapidly; a person could be fine in the morning, feel sick in the afternoon, and be dead by nightfall! Bodies piled up, and burying of the dead became a major public health problem. The overworked, underpaid, and at-risk grave diggers, perhaps in an effort to keep their sanity during those terrible times, developed their own jargon, stories, and poems. One poem that surfaced was:

> I lost a little bird
> And its name was Enza
> I opened the window
> And in-flu-enza!

Some influenza viruses (type A) infect not only humans but other species, including seals, dogs, pigs, and birds (particularly ducks).

The term "flu" is certainly familiar to you and is used somewhat loosely to describe coldlike respiratory symptoms accompanied by muscle

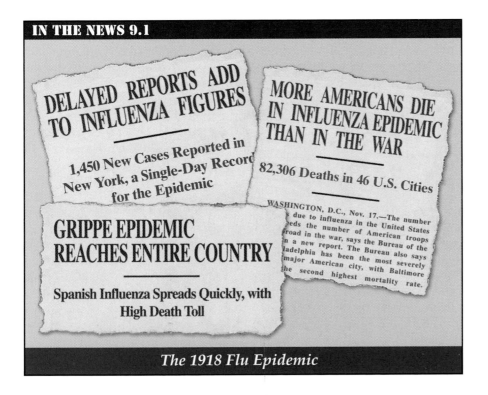

IN THE NEWS 9.1

DELAYED REPORTS ADD TO INFLUENZA FIGURES

1,450 New Cases Reported in New York, a Single-Day Record for the Epidemic

GRIPPE EPIDEMIC REACHES ENTIRE COUNTRY

Spanish Influenza Spreads Quickly, with High Death Toll

MORE AMERICANS DIE IN INFLUENZA EPIDEMIC THAN IN THE WAR

82,306 Deaths in 46 U.S. Cities

WASHINGTON, D.C., Nov. 17.—The number ... due to influenza in the United States ...eds the number of American troops ...road in the war, says the Bureau of the ...n a new report. The Bureau also says ...adelphia has been the most severely ...major American city, with Baltimore ...he second highest mortality rate.

The 1918 Flu Epidemic

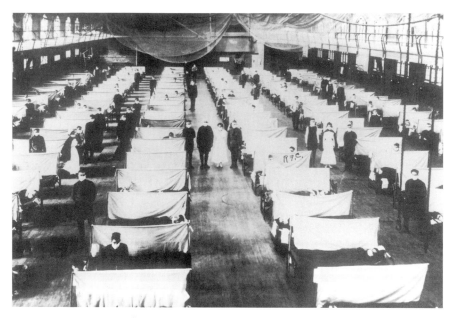

FIGURE 9.1 A ward of patients with influenza. The 1918 influenza was catastrophic and killed millions of people all over the world. (Source: Public Health Service.)

pain, but in most cases, this is probably not the "true" flu. "Stomach flu" is another vague term used to describe symptoms associated with gastroenteritis; the flu virus does not cause gastroenteritis. Influenza is caused by an RNA virus which is as distinct from the cold virus and the gastroenteritis-causing virus as dogs are from cats. So the term "flu" is frequently used inappropriately (particularly by students on exam days!). The "grippe" and the "24-hour" viruses are common but meaningless expressions.

There are three categories of flu viruses, all RNA viruses, designated A, B, and C. **Influenza A virus** causes epidemics, and occasionally pandemics, and is associated with animal reservoirs, particularly birds. **Influenza B virus** is less severe, causing only epidemics, and there is no animal reservoir, while **influenza C virus** does not cause epidemics and produces only mild respiratory illness. Influenza control is directed against types A and B.

Influenza is primarily an airborne infection and is acquired from droplets and aerosols; fomites play a secondary role. The incubation time is 24 to 48 hours. Young children are particularly significant sources of the virus because of their relatively nonhygienic practices. Conditions of crowding and close intermingling favor transmission, as in theaters, classrooms, nursing homes, and barracks. The disease usually peaks from about the middle of December through early March, the months in temperate climates when "fresh air" indoors is not possible. There is no seasonal variation in tropical areas. The real flu is characterized by exaggerated coldlike symptoms, including headache, high fever, muscle pain (particularly in the back and legs), severe cough, and congestion. The disease usually disappears in a week or two, and treatment, including bed rest and fluids, is aimed at relieving the symptoms.

Each year, in early October, preparation for the flu season begins; those over 65 years of age and other high-risk groups, including people with immunodeficiency diseases and, in some cases, pregnant women, are urged to get a flu shot. Why is this necessary each year? Lifetime, or at least long-term, prevention is usually the case in immunization—but not so with influenza! The explanation lies in the biology of the influenza virus. Protruding through the viral membrane are **hemagglutinin (H)** spikes, of which there can be as many as 500 per virus, and **neuraminidase (N)** spikes, of which there are about 100 per viral particle (Fig. 9.2). Both of these serve as virulence factors (chapter 6). The H spikes are essential in attaching the virus to the epithelial (lining) cells of the respiratory mucosa and in aiding their penetration into these cells. The N spikes play a role in the last stage of the viral replication cycle, namely, release of new virions, allowing for spread to other cells. Both of these spikes are antigens, which means that they provoke the production of specific antibodies as part of the body's defense mechanism (chapter 11). Now, here is the reason why influenza vaccination must be given on a yearly basis. These spikes undergo variation in their antigenic structure. The different forms are assigned numbers, e.g., H1, H2, H3, N1, N2, and N3. There are 13 types of H and 9 types of N, allowing for 117 possible combinations! Hence, antibodies produced against H1 will react against H1 spikes but not against H2 spikes. Think of it as a lock-and-key arrangement—a key made against lock H1 will not fit into the modified H2 lock but only into the original H1 lock (Fig. 9.3). Influenza viruses continually undergo two types of changes, known as **antigenic shift** and **antigenic drift.** Antigenic shift is a major, abrupt antigenic change in the H or N spikes that results from the recombination of genetic material from different viral strains, creating a new influenza virus strain. Consider that if a pig were to be coincidentally infected by both a human strain and a bird strain, a new hybrid strain could develop that could then infect a human; the pig served as a **blender.** Flu pandemics commonly originate in China, where millions of pigs, birds, and people live in close quarters, allowing for new combinations of strains (Fig. 9.4). Possibly, the deadly 1918 influenza arose as a result of antigenic shift. Antigenic drift, on the other hand, is a minor change in the H and N spikes occurring over a period of years. The upshot of both antigenic shift and antigenic drift is that they enable the virus to evade the antibody defense mechanisms of the host. Since these changes continually occur, the viral strains that make up the vaccine from year to year need to be adjusted. Antibody against last year's viral strains will be effective only to the extent that some of these strains may

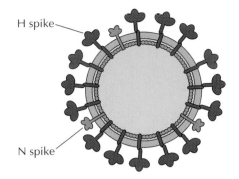

FIGURE 9.2   Influenza virus H and N spikes.

FIGURE 9.3   Lock-and-key arrangement of antigen and antibody.

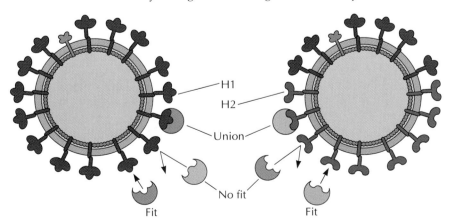

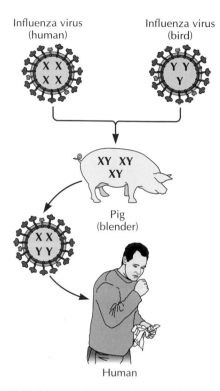

Influenza virus (human)   Influenza virus (bird)

Pig (blender)

Human

FIGURE 9.4 Origin of a new influenza virus strain.

coincidentally be the same. If these changes in the virus did not occur, as in polio, measles, and mumps, a flu shot would provide long-term, maybe lifetime, immunity.

Antigenic shift occurs periodically and only with influenza type A. Historically, type A was responsible for the occurrence of pandemics, because people the world over had no antibody protection against a new strain, leaving them vulnerable. During the 20th century, major pandemics occurred in 1918, 1957, and 1968. The full nomenclature of a virus reflects the virus group (A, B, or C), where and when the virus was isolated, and the H and N types. An example is the A/Philippines/82/H3N2 influenza virus.

A study of influenza epidemics and pandemics that have occurred over the centuries indicates a cyclical pattern of influenza outbreaks. In fact, public health experts fear that we are overdue for an epidemic, possibly a pandemic, of influenza at this time. This accounts for the near panic that occurred in December 1997, when headlines and television broadcasts throughout the world warned us of the so-called "bird flu," the "Hong Kong flu," or the "chicken flu" that was occurring in Hong Kong and caused six deaths. As a public health measure to break the transmission chain, the Hong Kong government slaughtered all of the territory's 1.4 million chickens. Tourists were scared away from Hong Kong, a major tourist site and one of Asia's most modern and crowded cities, with over 7 million people living in a small area. After a 6-week ban, the health authorities allowed chickens to be brought back into the city, but, as a precaution, health workers conducted random tests among the 35,000 chickens before allowing them to be transported into the city to a wholesale poultry market. All of the chickens tested negative for the bird flu virus (now that is something to cluck about!). The threatened epidemic could not have come at a worse time of the year; it deprived many families of the traditional chicken dishes during the Chinese New Year festivities. Some celebrated the return of the chickens with dinner parties featuring (you guessed it) chicken! The fear on the part of the public health authorities the world over was warranted.

Although the time is ripe for a major flu outbreak, today's health infrastructure is better equipped—in terms of surveillance, diagnostics, and treatment—to handle an onslaught of influenza and to avoid the horrors of the 1918 influenza pandemic. For a massive outbreak to occur, a large segment of the population would have to be nonimmunized. How can this happen? As pointed out, if a new strain of the influenza virus were to arise by antigenic shift, the world population would be nonimmunized and, hence, susceptible.

How do health officials determine almost a year in advance of the flu season which viral strains are to be included in the vaccine cocktail for the coming year? The World Health Organization, in cooperation with other agencies, has over 100 surveillance sites around the world that gather information regarding the circulating flu types. These data are assembled in the fall of each year and used to design next year's vaccine. Vaccine-manufacturing pharmaceutical companies begin production in February for the following flu season. The predominant strain during the 1999–2000 flu season turned out to be type A (H2N2), as had been the case for the previous 2 years. Thus, the vaccine was well matched, indicating the reliability of early surveillance.

Influenza outbreaks are costly. The Centers for Disease Control and Prevention (CDC) estimated that 90 million Americans (about 35% of the population) came down with the flu in 1994. Twenty thousand deaths and an inestimable number of people becoming ill led to a staggering number of lost workdays. Each year, the cost of vaccine preparation, vaccine administration, and loss of productivity runs into the billions of dollars in the United States alone. These figures do not take into account what happens on a worldwide basis. The numbers would be shocking.

It seems as though the threat of a flu outbreak is always present, requiring that the CDC and other health agencies be constantly on the alert (In the News 9.2). In 1999 U.S. health officials, for the second year in a row, warned tourists, particularly those over age 65 or with chronic respiratory or heart problems, to be wary of traveling to Alaska and the Yukon because of an outbreak involving hundreds of tourists. In a span of only about 10 weeks, 428 cases were reported among tourists on two cruise ships. The flu (and other diseases) is easily transmissible on cruise ships because of the relatively crowded quarters.

The 1918 flu pandemic remains a mystery in terms of understanding why this virus was so virulent. Many victims of the disease died because their lungs filled with fluid, and they literally drowned. Mostly older people die from influenza, but in the 1918 pandemic the disease struck young adults. Because flu pandemics appear to be cyclical, for the sake of future readiness it is important to understand why the 1918 flu pandemic was so devastating. This was the reason that Johan Hultin, a 73-year-old pathologist from San Francisco, traveled to the Alaskan tundra to exhume the bodies of Eskimos who died during the 1918 global influenza outbreak and brought back tissue specimens for study. The Armed Forces Institute of Pathology is deciphering the virus's genetic blueprint from these tissue specimens in an attempt to discover why this virus was so lethal. Hopefully, this information will be of value in designing a vaccine should this virus, or a similar one, emerge again.

IN THE NEWS 9.2

DNA research sheds light on 1918 flu epidemic

Pig and human flu virus may have combined

Officials Prepare for Killer Flu Pandemic

Hospitals Inundated with Patients in Early Flu Outbreak

Flu shots may be a good idea for children

FDA Approves New Flu Drug

Tamiflu Can Help Recovery but Has Side Effects, Must Be Taken Promptly

Influenza

There is reluctance on the part of many individuals, including those in the 65-year-or-older age group, who are the most susceptible, to be vaccinated against influenza. Their refusal of the vaccine is risky, since **secondary bacterial infections,** which may lead to a fatal pneumonia, are more likely to occur in the older population. This is why many physicians prescribe antibiotics for their older patients who have been diagnosed with the flu. Historically, it was thought that influenza was caused by a bacterium, subsequently named *Haemophilus influenzae,* but it turned out not to be the case. In fact, *H. influenzae* is a less frequent cause of secondary bacterial pneumonia in influenza patients than are staphylococci and streptococci.

Relief is on the way. There are relatively few antiviral drugs, but in 1999, the Food and Drug Administration (FDA) approved two new antiflu drugs, **zanamivir** (Relenza) and **oseltamivir** (Tamiflu), that are effective against both types A and B influenza viruses; other agents are in the pipeline. These drugs do not prevent infection, nor are they cure-alls, but if taken within about 48 hours after the appearance of symptoms, they can shave off a few days of illness and reduce the severity of a bout with the flu. These drugs are not intended as substitutes for the vaccines, and, as with all medicines, they have side effects. A big problem is that individuals would need to recognize very early flu signs (which mimic many other microbial diseases) and visit their physician almost immediately to get a prescription for these drugs. It is expected that, in the near future, home do-it-yourself flu detection kits will be available (a great idea for a Christmas stocking-stuffer!).

On a second front, **"plug drugs,"** which will act to both treat and prevent the flu, are on the horizon. They act to plug the part of the N spike that is common (invariant) to all N spikes and, therefore, not influenced by antigenic shift or drift. This action halts the spread of the virus by preventing it from traveling from one cell to another.

Respiratory Syncytial Virus

The RNA virus **respiratory syncytial virus** causes a disease which is present worldwide. It has an incubation period of 4 to 5 days and is particularly life-threatening to infants under 6 months of age; in fact, it is the most prevalent cause of respiratory infection in this age group. The symptoms of the disease are nonspecific, which makes diagnosis difficult; they include fever, runny nose, ear infection, and pharyngitis. The disease can progress to serious lower respiratory tract infection, including obstructed airways. It is estimated that approximately 91,000 children are hospitalized each year with approximately 4,500 deaths. In older children and in adults, the infection is manifested as a common cold. The virus is so named because in the respiratory tissues, a syncytium—a network of large, abnormal cells with multiple nuclei—may be present. Outbreaks are a threat in pediatric wards and in nurseries, and the results can be devastating. Transmission of disease is decreased, as is so often the case, by frequent and careful hand washing. Treatment is largely supportive; oxygen therapy may be necessary in some cases. The antiviral drug **ribavirin** may be administered as an aerosol. Research is under way to develop a vaccine that can be administered intranasally, but so far, attempts have been unsuccessful.

Common Cold

The **common cold** is the most frequent infection, with the highest loss of workdays and school days, establishing an economic burden. This seems odd, considering the complex technological advances that have taken place in recent years, including the Human Genome Project, gene therapy, and the projected eradication of polio and measles. Hence, the common cold, although not considered life-threatening, is "nothing to sneeze at" (pun intended!). Next time you are in a drugstore or a supermarket, take a look at the dozens of remedies marketed to relieve the symptoms of the common cold (Fig. 9.5). Manufacturers of these remedies are ever wary to avoid use of the term "cure" for fear of legal liability, but they come very close to the edge in their claims.

The symptoms of sneezing, coughing, sore throat, stiffness, and that general "blah" feeling are all too familiar. Colds are popularly described as "summer colds," "head colds," "chest colds," and "winter colds." These expressions have no medical basis, but they do indicate a seasonality and hint at the variety of symptoms. This reflects the fact that colds are caused by a large variety of distinctive groups of viruses; within each group, there may be over 100 different strains. This is the reason that immunization is a problem despite the fact that vaccine technology exists. Which strain, or strains, should be chosen for development of a vaccine? There needs to be a common denominator, but, to date, one has not been identified. Recent research reports suggest that hope is on the way in this regard.

Cold viruses are transmitted not only by respiratory droplets but also, to a large extent, by fomites, including hands, doorknobs, and faucets. The incubation period is about 1 to 3 days. Notice in your classrooms the fantastic number of coughs and sneezes (let alone the saliva-disseminating talk by the professor which never seems to end) that takes place during a 50-minute lecture. Some individuals carry a handkerchief or tissues into which nasal and cough secretions and saliva are discharged (sorry if I am grossing you out!). Others sneeze or cough directly into their hands and have no reluctance in offering their hand for a friendly and infectious handshake. Three-ply tissues were introduced several years ago; the middle ply was impregnated with a compound designed to kill the cold virus. However, for a variety of reasons, including cost, efficiency, and the abrasiveness of the tissue, these tissues were not popular with the public.

As mentioned, there is a large variety of common cold viruses responsible for colds, but about half of colds are caused by two groups, the **rhinoviruses** and the **coronaviruses**. Most of the cold viruses possess mechanisms of adhesion which allow them to attach to the nasal pharynx and avoid being entirely eliminated during vigorous coughing and sneezing. This permits replication and spread to neighboring cells. Those viruses that escape become airborne.

Diagnosis of a cold is symptom based. The illness is generally mild and self-limiting, although a nuisance. Laboratory tests are not necessary or worthwhile, except for epidemiological purposes.

Measles (Rubeola), Mumps, and German Measles (Rubella)

Measles, mumps, and German measles are distinct "childhood diseases," and each is caused by a specific RNA virus. In 1968, a **measles-mumps-rubella** vaccine called **MMR** was introduced and drastically reduced the incidence of these diseases. If vaccination were halted, many new cases would result; the public needs to be aware of this, because a false sense of security can lead to failure to immunize, resulting in a nonimmunized and vulnerable population. The use of MMR in those countries fortunate enough to implement immunization has dramatically reduced the

FIGURE 9.5   Cold relief medicines. The common cold is a source of misery for millions of persons each year. Shelves are stocked with medicines to relieve the symptoms. (Author's photo.)

incidence of these diseases. Although the triple vaccine is composed of live, attenuated viruses, complications are rare, and the risks of acquiring one or more of these diseases is much greater than are the risks of complications. The vaccine is given at 12 to 15 months of age and again at 4 to 6 years of age. There is no specific treatment for these three diseases, and prevention by immunization is the key.

Measles

It should be emphasized that measles is distinct from what is commonly known as German measles, which is discussed below. The measles virus causes one of the most infectious viral diseases; estimates are that there is a >98% chance of becoming infected if exposed directly to someone with measles. The mechanism of transmission is by respiratory droplets, and the disease is fostered by overcrowding, low levels of immunity in the population, malnutrition, and inadequate medical care. Humans are the only reservoir for this disease (as is true of smallpox, chicken pox, and mumps), making it a target for eradication. The symptoms of measles are coldlike, with the development of characteristic **Koplik's spots** in the mouth early in the disease, followed by a red rash (Fig. 9.6) that starts on the face and spreads to the extremities and over most of the body. The disease is usually mild and self-limiting, but 1 in 500 children with measles develops potentially serious and even fatal complications, including pneumonia, ear infections, brain damage, and seizures. Worldwide, measles is significant and is ranked the seventh most frequent cause of death.

In developing countries, more than 15% of children who contract measles die of the disease or its complications. The disease is a killer of malnourished children and in populations lacking immunization. Immunization resulting from having had the disease is considered to be lifelong for survivors, whereas the MMR vaccination affords approximately a 20-year protection. There are "susceptibles" who slip through the regimen of immunization because of a false sense of security that these diseases are gone or because of the burden of health care costs, particularly in Third World countries. Therefore, outbreaks of measles will continue to occur.

Mumps

The hallmark of mumps is a large swelling on one or both sides of the face, resulting from infection of the **parotid gland,** one of three pairs of salivary glands, which is located at the junction of the upper and lower jaw (Fig. 9.7). Humans are the only natural hosts for this disease, which has an incubation period of 10 to 20 days. In temperate climates, the disease usually occurs in late winter and early spring, most commonly in children under the age of 15 years. Many children are asymptomatic, that is, they show no symptoms, but nevertheless are rendered immune.

Symptoms other than swelling of the parotid glands (which may not occur) include fever, nasal discharge, and muscle pain. The virus can spread to other structures, including the testes, ovaries, meninges, heart, and kidneys. Complications are unlikely, but there has been some concern that in young males, inflammation of the testes **(orchitis)** could lead to sterility. This has not proved to be the case, although this temporary complication is painful. Rarely, permanent deafness occurs and is usually confined to one ear.

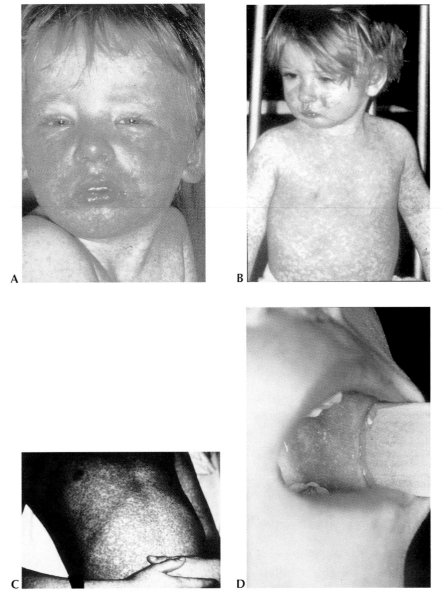

FIGURE 9.6 Measles (rubeola). **(A)** Characteristic rash with red and runny eyes and nose. (Source: CDC.) **(B)** Diffuse red rash over body. (Source: Armed Forces Institute of Pathology.) **(C)** Rash on approximately day 4 covering arms and stomach (Source: CDC.) **(D)** Koplik's spots inside the mouth. (Source: CDC.)

Rubella (German Measles)

Of the several human viral diseases that cause a rash, rubella is the mildest disease and has an incubation period of 12 to 32 days. In fact, for generations, **rubella,** caused by an RNA virus, was thought to be a mild form of measles (rubeola). In 1829, a German physician recognized that these were two different diseases; hence, the term "German measles" came into use. The disease is endemic worldwide and is highly infectious; it is spread largely through respiratory secretions, but urine from infected individuals can also transmit rubella. Some individuals are asymptomatic and transmit the virus without knowing it. The characteristic rash starts on the face and progresses down the trunk and to the extremities; it resolves in about 3 days (Fig. 9.8).

FIGURE 9.7 Mumps. This child has swollen parotid glands. (Source: CDC.)

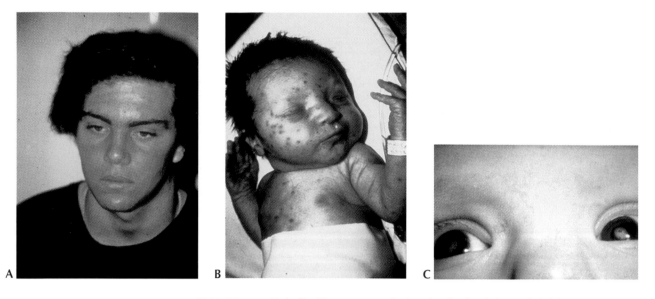

FIGURE 9.8 Rubella (German measles) strikes both adults and children and is very dangerous to babies. **(A)** Teenager with rash on face; **(B)** infant born with rubella; **(C)** newborn with thickening of the lens of the eye that causes blindness. (Source: CDC.)

In itself, rubella is not regarded as a serious disease. However, prenatal transmission can occur, particularly if the mother is infected during the first trimester of pregnancy, even if she is asymptomatic. Major consequences include cardiac lesions, deafness, ocular lesions, mental and physical retardation, and glaucoma. Approximately 15% of those exposed may escape infection during their childhood years, posing a risk that females may enter their childbearing years without having had measles. Therefore, it is particularly important that females be immunized against rubella.

Chicken Pox (Varicella) and Shingles (Herpes Zoster)

The expression "brothers under the skin" aptly describes **chicken pox** and **shingles,** two seemingly different diseases that cause skin eruptions. Historically, they were thought to be caused by different viruses, but it has long been established that they are manifestations of the same DNA virus. The virus, therefore, is called **varicella-zoster virus.** Chicken pox is a disease usually associated with childhood; shingles typically occurs after about the age of 45 to 50 in some individuals who had chicken pox in their earlier years, but it has been reported in children as young as 8 years old.

Humans are the only hosts for chicken pox and shingles. Chicken pox has an incubation period of 10 to 20 days and is transmitted by airborne droplets or by contact with the fluid in the blisterlike skin lesions **(vesicles)** which develop. The disease is highly infectious, especially before the emergence of the rash. Early signs include fever, headache, and generalized aches and pains, followed in a few days by an itchy rash with fluid-filled vesicles appearing on the scalp, face, trunk, and extremities (Fig. 9.9). More than 100 vesicles can appear on the body at any one time; usually, adults with chicken pox develop more vesicles than do children. The vesicles tend to occur in a succession of crops over a 2- to 4-day period, and an individual may have a combination of newly emerging and old, "crusty" vesicles.

In most cases, the disease is self-limiting, but in some cases antiviral agents and antibody therapy may be indicated.

Chicken pox, with rare exceptions, confers lifelong immunity. In 1995, the FDA approved a chicken pox vaccine with the recommendation that it be administered to all children after 1 year of age; its implementation appears to be catching on as it becomes mandatory in an increasing number of states. The potential seriousness of chicken pox can be underestimated. Box 9.1 provides dramatic evidence of the value of chicken pox immunization. It is estimated that from 1920 to 1990, 3.3 million American children (under the age of 15 years) per year developed chicken pox, resulting in about 7,400 hospitalizations annually and about 50 deaths per year. The disease exists worldwide, and nearly all children are infected during their early years. Like measles, chicken pox can cross the placenta and cause serious fetal damage.

Shingles is not life threatening, but painful complications occur in some individuals (Fig. 9.9) with a history of chicken pox. The condition is described as causing one of the most intense pains of any disease, which is not surprising since the virus infects nerve fibers. During the initial chicken pox infection, viruses take up residence in a latent state in collections of nerve cells **(ganglia)** located along the spinal column or in nerves supplying the face. The spinal nerve which girdles the trunk at about the belt line is frequently involved, resulting in this area's being a common site for shingles. These viruses "lie in state" for years but may be triggered into an active replicating state by a variety of agents, including X rays, certain drugs, and immunodeficiency, causing a painful eruption on the skin along the path of the nerve.

An individual with shingles can transmit chicken pox to a person who has not had the disease. More rarely, shingles can be triggered in an

AUTHOR'S NOTE *My son arrived home from college during his freshman year for the Christmas recess. On his first night, he awakened hot, feverish, and covered with an itchy rash. It was obvious that he had chicken pox and, in thinking back, his mother and I realized that he did not have this disease as a child. A few days later, we received a letter from his college advising us that an outbreak of chicken pox had started just before the semester break. Recovery was complete in about 10 days.*

FIGURE 9.9 Chicken pox and shingles. These two diseases are caused by varicella-zoster virus and are manifested by characteristic rashes. **(A and B)** Chicken pox. (Sources: CDC **[A]**; American Academy of Pediatrics **[B]**.) **(C)** Shingles. This painful condition is caused by the presence of the virus along sensory nerves, where they replicate and produce painful skin eruptions along the path of the nerve. (Source: VZV Research Foundation.)

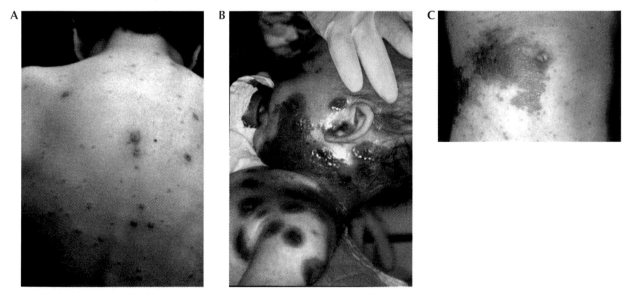

BOX 9.1　Chicken Pox: More Than Just a Childhood Disease

The Immunization Action Coalition's publication IAC EXPRESS received the following case report via e-mail from a Canadian physician describing the death of a 3½-year-old boy from varicella encephalitis. At the time of his death, a vaccine against varicella was not yet available in Canada. The physician's e-mail is reprinted as follows:

A 3½-year-old boy developed chickenpox April 5, 1998. His 7-year-old brother had it at the same time. The younger child had a mild case with relatively few lesions. Four days before admission the 3½-

year old became sleepy and developed a headache. Two days later he developed increasing lethargy, vomiting, drowsiness and disorientation. He was taken to our community hospital on April 11. He had a lowered level of consciousness, responding slightly to pain. The next morning he had shaking movements, probably due to acute herniation of the brain due to swelling. He became comatose, was transferred to a major medical center, and pronounced brain dead on April 13. Life support was discontinued, and he died. The autopsy confirms diagnosis of varicella encephalitis.

At the time of his illness, varicella vaccine was not available in British Columbia.

A footnote: The mother of the child was devastated by his death. She has refused to set foot in our hospital again because of the unbearable memories, and plans to deliver the child she is now carrying in another city.

Dr. Kirsten Emmott
Comox, British Columbia, Canada

Source: Immunization Action Coalition.

individual with a history of chicken pox upon exposure to a person with chicken pox.

Hantavirus Pulmonary Syndrome

The first known U.S. outbreak of a **hantavirus**, an RNA virus that causes **hantavirus pulmonary syndrome (HPS),** occurred in 1993 in the Four Corners area (where New Mexico, Colorado, Utah, and Arizona meet), with 24 cases of a severe influenzalike respiratory illness complicated by respiratory failure and, in some cases, death. Hantaviruses were known to be present in Africa since the 1930s but were associated with hemorrhagic (bleeding) symptoms. In the United States, the disease has been confirmed in at least 14 states and has infected over 500 people (Fig. 9.10). It is "hot" on the list of emerging viral diseases.

A case of HPS occurred in a 61-year-old, previously healthy Vermont

FIGURE 9.10　Distribution of cases of HPS (January 1997) and of the deer mouse (*Peromyscus maniculatus*). The presence of HPS is linked to the presence of deer mice. (Source: CDC; deer mouse range information from W. H. Burt and R. P. Grossenheider, *A Field Guide to the Mammals: North America North of Mexico*, 3rd ed. [Houghton Mifflin, New York, N.Y., 1980].)

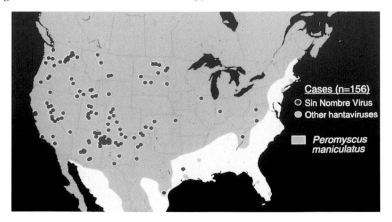

resident in February 2000. This was somewhat unusual, since only 15 (5%) of the 284 cases confirmed by the CDC have occurred east of the Mississippi River. According to the report, the Vermont Department of Health conducted an on-site investigation and observed mouse droppings under the kitchen counter and in the cellar. During the 2 months preceding hospitalization, the patient had cleaned a mouse nest from a woodpile, observed a mouse in the basement, and trapped two mice under the kitchen cabinets. As a follow-up, the wildlife services program of the U.S. Department of Agriculture captured 46 rodents of several species. Forty-three rodents were tested, and two of five deer mice were positive for hantaviral antibodies; all other rodents were negative. This report underscores the necessity of being alert to avoid exposure to mice and mouse droppings.

The fatality rate for HPS is at least 60%. The disease primarily strikes young, healthy adults, and death occurs in only several days after infection because of accumulation of fluid in the lungs, which interferes with oxygen diffusion. Oxygen therapy is frequently required in an attempt to avert death.

The HPS virus is carried by rodents, especially the deer mouse (Fig. 9.11) and the cotton rat, both inhabitants of the Southwest. Because of the explosive nature of HPS, the mouse carrier has been referred to as "the mouse that roared." Transmission of the virus to humans is the result of exposure to dried and aerosolized fecal material, saliva, or urine from infected rodents. It is also possible to contract the disease when fresh or contaminated rodent droppings get into the skin, eyes, food, or water or as the result of a rodent bite; the human is a dead-end host.

Prevention is based on avoiding contact with rodents and their excreta. Since the virus is primarily spread by airborne contaminated droplets, entering barns or other buildings that are frequently closed for extended periods may be hazardous to your health, as is disturbing rodent-infested structures. Diagnosis of hantavirus disease is difficult, since the early flu-like signs are characteristic of a variety of viral and bacterial diseases. Definitive diagnosis requires laboratory procedures.

Smallpox

Smallpox was declared eradicated by the World Health Organization in 1980, as discussed in chapter 13. Over the centuries, it has been an agent of biological warfare, and it continues to be a threat (chapter 14).

ARTHROPOD-BORNE DISEASES

Arthropods as vectors were discussed in chapter 7, and arthropod-borne bacterial diseases were described in chapter 8. A variety of viral diseases are also transmitted by arthropods (Table 9.2). (Arthropod vectors are also involved in some diseases caused by protozoans and worms and will be discussed in chapter 10.) Arthropod-borne viruses are called arboviruses. They are maintained in nature by biological transmission to susceptible vertebrate hosts when blood-sucking infected arthropods take a blood meal.

Dengue Fever

Dengue fever, also called **breakbone fever,** can result from four dengue viruses. The disease is worldwide in tropical areas, and over 100 million cases are reported per year. There is an ongoing epidemic of dengue fever in Latin America that has increased to about 200,000 cases, with over 5,000

FIGURE 9.11 A deer mouse. This mouse species is a vector of HPS; it sheds the virus in urine, feces, and saliva. (Source: CDC.)

TABLE 9.2 Arthropod-borne viral diseases[a]

Disease	Virus category	Symptoms	Comments
Arboviral encephalitides	ssRNA	Meningitis, coma, tremors, convulsions, paralysis, brain damage	EEE, WEE, St. Louis encephalitis, West Nile virus (all mosquito borne), tick-borne Colorado fever
Yellow fever	ssRNA	Fever, bloody nose, black vomiting, jaundice	Immunization by live attenuated virus; relatively long-lasting immunity; used for those at high risk
Dengue fever	ssRNA	Flulike symptoms	DHF can develop; *Aedes aegypti* and *Aedes albopictus*
DHF	ssRNA	Hemorrhages in skin, gums, and other areas	Potentially fatal

[a]DHF, dengue hemorrhagic fever; EEE, eastern equine encephalitis; ss, single stranded; WEE, western equine encephalitis.

deaths in recent years. The vectors are the *Aedes aegypti* and *Aedes albopictus* mosquitoes. *A. albopictus*, also referred to as the **Asian tiger mosquito,** was introduced from Asia in 1985 in used tire casings. The first outbreak of dengue fever in the United States in 40 years occurred in 1980. The recent appearance of *A. albopictus* in the United States is of great concern because this species is aggressive, and its biting habits could bring dengue fever to a new area. The CDC is on the alert for the possibility of dengue fever and monitors states along the southern border. Eighteen cases were reported in Laredo, Tex., in 1999, resulting in implementation of mosquito control activities; no cases were reported in 2000.

Dengue fever is usually self-limiting, and although the disease can be debilitating, recovery occurs in about 10 days. **Dengue hemorrhagic fever** is a serious and potentially fatal infection that can be caused by a dengue virus strain different from the one causing the initial infection. The condition is manifested by hemorrhages occurring in the skin, gums, and other areas within the body. Shock may develop, requiring immediate treatment to prevent death, and even with intensive supportive measures, as many as 40% of those infected may die. There is no immunization against any of the four dengue virus strains. Since the early characteristics of the disease are flulike, a definitive diagnosis is difficult and requires laboratory confirmation, either through virus isolation or through detection of virus-specific antibodies in the blood.

Yellow Fever

Yellow fever, also called yellow jack, is caused by an RNA virus. The disease once had a widespread distribution, but mosquito control measures have resulted in elimination in many countries, including the United States. In 1793, a yellow fever epidemic devastated Philadelphia, America's early capital city. The disease is still present in areas of South America, Central America, and Africa. In jungle areas, monkeys serve as reservoirs, and the incidence of disease is highest in these areas because mosquitoes bite both monkeys and humans. The mosquito vector is *A. aegypti,* a mosquito which has adapted to humans; it bites by day and breeds in stored water. Old tire casings are particularly significant in this regard. The disease is manifested by fever, bloody nose, headache, nausea, muscle pain, (black) vomiting, and jaundice. In about a week, the infected individual is either dead or in the process of recovery.

Immunization against yellow fever is available with live, attenuated viruses. Protection is relatively long-lasting, and immunization is required for travelers to certain destinations.

The history of the Panama Canal is linked with yellow fever and malaria (Box 9.2 and Fig. 9.12). The construction of the canal was an engineering triumph thwarted for years by a virus and by a protozoan (the malaria parasite). Hence, its construction was also a triumph over microbes. It was a case of David defeating Goliath until biologists turned the tide, and Goliath was the winner.

Arboviruses

The four most common **arboviruses** in the United States are **eastern equine encephalitis, western equine encephalitis, St. Louis encephalitis,** and **Colorado tick fever** viruses. Note that Colorado tick fever is, as the name states, tick borne and, in fact, is the most common tick-borne viral fever in the United States. The other arboviruses are all carried by mosquitoes. Humans do not serve as reservoir hosts; these viruses cycle between wild animals, primarily wild birds and mosquitoes, with humans and horses serving as dead-end hosts (Fig. 9.13). When the virus invades the spinal cord, the **meninges** (wrappings around the spinal cord and brain) become inflamed and may result in coma, convulsions, tremors, paralysis, memory deficits, and possibly permanent brain damage.

In areas where the disease may be a problem, sentinel animals (caged rabbits or chickens) are left in mosquito-infested areas; blood is drawn from these animals on a periodic basis and tested for the presence of antibodies, which indicate the presence of viruses. Also, mosquitoes are collected and tested to determine whether they are carrying encephalitis viruses.

A scare occurred in June 1998 in Westerly, R.I., when tests on a group of bird-biting mosquitoes showed the presence of eastern equine encephalitis, which can be fatal to humans. Public health workers urged residents to take appropriate control measures, including using mosquito repellent, wearing protective clothing, and eliminating breeding grounds such as buckets, old tires, wet leaves in roof gutters, and other wet areas that allow for the breeding of the mosquitoes.

New York City has claimed success in reducing the rate of murder and in improving the quality of life, but in the early fall of 1999, a new menace—St. Louis encephalitis—occurred, resulting in three deaths (In the News 9.3). Insecticide spraying and other mosquito abatement measures were initiated. Mayor Rudolph Giuliani was quoted as saying "the more dead mosquitoes, the better."

It appears that mosquitoes have a vengeance against New York. In late August 1999, on the heels of the St. Louis encephalitis outbreak, New Yorkers were plagued by yet another arbovirus, **West Nile virus,** first described in the 1930s in Africa. This virus is carried primarily by the same mosquito that transmits St. Louis encephalitis. In a period of only several weeks, 56 cases, including 7 deaths, had been reported in the New York City area. Mosquito control measures, including aerial spraying, were quickly implemented.

It is not certain how the virus arrived in New York, but experts believe that it probably came on a bird or mosquito imported on a jet aircraft. This is another example of how disease can be spread from one continent to another.

Initially, the disease was somewhat of a mystery, since outbreaks of illness and deaths occurred in horses and also caused the deaths of hundreds

AUTHOR'S NOTE *In the spring of 1999, during a field trip in the Amazon area of Brazil, in one particularly remote area where our boat docked, there was a young native girl, obviously there to greet the tourists, with a monkey perched on one shoulder and a second monkey cradled in her arms. Within only several feet from her was a tree with, I would guess, about 15 monkeys. This girl was feeding the monkeys and playing with them as if they were her pets. I could not help but think that this could be the scenario in which yellow fever virus, human immunodeficiency virus, Ebola virus, or other viruses made the leap from monkey to human.*

BOX 9.2   Construction of the Panama Canal: a Triumph in Engineering and Biology

The construction of the Panama Canal linking the Atlantic and Pacific Oceans was a monumental undertaking over a 35-year span. In 1914, the canal's waterways were finally opened to traffic. The obstacles encountered in completing this massive project were tremendous and seemingly impossible to overcome. There were political, financial, engineering, and health problems of gigantic proportions, but one by one they were resolved. As it turned out, the toughest challenge was waging and winning the war against mosquitoes, carriers of yellow fever and malaria. Two excellent books tell the story of the agonies and triumphs involved in the creation of the canal. Their titles, *And the Mountains Will Move* and *The Path between the Seas,* suggest the huge scale of the project. The task of moving thousands of tons of earth was overshadowed by the threat of microbial enemies that could not be seen, namely, the virus of yellow fever and the protozoan of malaria. The canal is a winding path stretching 51 miles through a series of locks that raise and lower ships 64 feet. One set of locks is said to have used enough concrete to build a wall 8 feet thick and 12 feet high stretching from Cleveland to Pittsburgh.

Interest in building a canal between the Atlantic and the Pacific dates back to the 16th century. But it was not until 1879 that France's Ferdinand de Lesseps, having successfully completed the Suez Canal in Egypt 10 years earlier, embarked on building the Panama Canal. He planned to construct a sea-level canal, based on the construction of the Suez, but that idea was abandoned in favor of a lock canal. de Lesseps never finished the job. He had underestimated the power of mosquitoes, malaria, and yellow fever. The workers were plagued by these natural enemies to the extent that the project was abandoned after a 10-year battle. In the 1880s, a French resident of Panama warned de Lesseps, "If you try to build the Canal, there will not be trees enough on the Isthmus to make crosses for the graves of your laborers." His prediction, unfortunately, proved to be correct.

In 1903, the United States acquired the Panama Canal during the presidency of Theodore Roosevelt. In 1904, William C. Gorgas, an American military physician, went to Panama. Epidemics were in full force at the time, but in less than a year, Gorgas cleared the area of disease by draining the swamps in order to deprive mosquitoes of breeding places. Gorgas implemented other sanitary measures that he had successfully enforced in Havana, Cuba. Although numerous other U.S. officials were involved in the project over the next 10 years, Gorgas was the only one who stuck with the project from beginning to end. It was his success in controlling the mosquito population that led to the completion of the canal.

Dramatic passages from *The Path between the Seas* vividly describe the hardships.

Yet in the tropics, malaria was taken as an inevitable fact of life, part of the landscape. Yellow fever, by contrast, came and went in vicious waves, suddenly, mysteriously. In those places where it was most common—Panama, Havana, Veracruz—it was the stranger, the newcomer, who suffered worst, while the native was often untouched. Wherever or whenever it struck, it spread panic of a kind that could all but paralyze a community. It was a far more violent and hideous thing to see; a more gruesome way to die.

The mortality rate among those who contracted the disease could vary enormously, from 12 to 15 percent to as much as 70 percent. Generally speaking, however, a yellow-fever patient in Panama in the 1880's had a less than fifty-five percent chance of survival. As with malaria, the patient was seized first by fits of shivering, high fever, and insatiable thirst. But there were savage headaches as well, and severe pains in the back and legs. The patient would become desperately restless. Then, in another day or so, the trouble would appear to subside and the patient would begin to turn yellow, noticeably in the face and in the eyes.

In the terminal stages the patient would spit up mouthfuls of dark blood, the infamous, terrifying *vómito negro,* black vomit. The end usually came swiftly after that. The body temperature would drop, the pulse fade. The flesh would become cold to the touch, "almost as cold as stone and [the patient] continues in that state with a composed sedate mind." Then, as a rule, in about eight to ten hours, the patient would die. "And so great was the terror the disease generated that its victims were buried with all possible speed."

You are going to have the fever,
  Yellow eyes!
In about ten days from now
Iron bands will clamp your brow;
Your tongue resemble curdled cream,
A rusty streak the center seam;
Your mouth will taste of untold
  things,
With claws and horns and fins and
  wings;
Your head will weigh a ton or more,
And forty gales within it roar!

Source: D. McCullough, *The Path between the Seas* (Simon & Schuster, New York, N.Y., 1977).

FIGURE 9.12 The Panama Canal. Construction of the Panama Canal was possible only when yellow fever and malaria were conquered. (Source: Edwin P. Ewing, Jr., CDC.)

FIGURE 9.13 Cycle of transmission of arthropod-borne viral encephalitides. EEE, eastern equine encephalitis; WEE, western equine encephalitis.

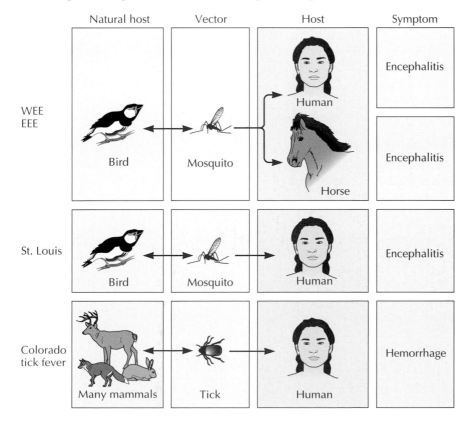

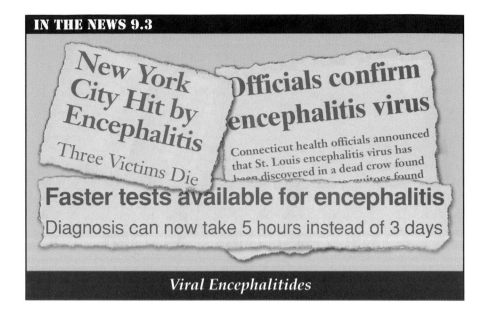

IN THE NEWS 9.3

New York City Hit by Encephalitis
Three Victims Die

Officials confirm encephalitis virus

Connecticut health officials announced that St. Louis encephalitis virus has been discovered in a dead crow found

Faster tests available for encephalitis
Diagnosis can now take 5 hours instead of 3 days

Viral Encephalitides

of crows in the area. At the Bronx Zoo, several rare birds died and others were ill, including "a trumpeter swan doing the backstroke." Mosquito abatement measures, along with the onset of cooler temperatures, were effective in breaking the transmission cycle.

FOOD-BORNE AND WATERBORNE VIRAL DISEASES

A number of food-borne and waterborne bacterial infections were described in chapter 8. The major food-borne and waterborne viral diseases are summarized in Table 9.3. They contribute to the public health problem of food-borne and waterborne diseases and cause significant morbidity and mortality. The CDC estimates an annual incidence of food-borne and waterborne diseases (both bacterial and viral) in the United States at 6.5 million to 76 million cases, with as many as 5,000 to 9,000 deaths. Chapter 12 describes public health measures in effect to minimize this problem.

Gastroenteritis

Gastroenteritis is characterized by a stomach ache, diarrhea, vomiting, and abdominal cramps as a result of inflammation of the stomach and intestinal

TABLE 9.3   Food-borne and waterborne viral diseases

Disease	Incubation period	Virus category[a]	Symptoms	Immunization and comments
Gastroenteritis	2–10 days	ssDNA	Stomach ache, diarrhea, dehydration, vomiting	RotaShield oral vaccine withdrawn
Hepatitis A	30 days	ssRNA	Jaundice, abnormal liver function in tests	Vaccine available for those at high risk
Poliomyelitis	7–14 days	ssRNA	Usually asymptomatic but can cause lifelong paralysis	Vaccine available since 1950s; disease about to be eradicated

[a]ss, single stranded.

tract. The term **stomach flu** is popular but meaningless. An exact diagnosis of gastroenteritis is difficult, since these symptoms are caused by a variety of viruses, bacteria, and protozoans. The symptoms usually last a few days, and recovery is uncomplicated. Dehydration is a potential problem, particularly in infants and in the elderly, because of fluid loss resulting from excessive diarrhea and vomiting. Transmission is by the fecal-oral route, resulting from contamination of food and beverages by careless food handlers. Raw or undercooked shellfish from contaminated waters are also a potential threat. Outbreaks, following a 1- to 2-day incubation period, are most likely to occur in schools, child care centers, cruise ships, dormitories, and nursing homes.

A group of viruses known as **rotaviruses** is the most common cause of seasonal viral gastroenteritis in children under 5 years of age; it is sometimes called "winter diarrhea." CDC estimates that in the United States, rotavirus infections account for 20 to 40 deaths, about 55,000 hospitalizations, and as many as 500,000 visits to physicians annually, with a cost exceeding $1 billion. Globally, rotaviruses cause about 600,000 deaths per year. In August 1998, the FDA licensed **RotaShield**, a live oral vaccine against rotavirus, culminating 25 years of basic research. Unfortunately, in October 1999, the vaccine was withdrawn after only a little more than a year on the market because of reports of bowel blockages in a small number of children who had received the vaccine. **Norwalk** and **Norwalk-like** viruses are frequent causes of gastroenteritis in older children and adults.

Hepatitis A and E

Hepatitis is an inflammation of the liver and is commonly caused by one of five viruses designated hepatitis A, B, C, D, and E viruses. **Hepatitis A virus,** like the other hepatitis viruses (with the exception of hepatitis B virus), is an RNA virus. Jaundice frequently occurs in hepatitis virus infections, along with abnormal liver function test results. (Hepatitis B, C, and D are presented below based on their route of transmission.)

Hepatitis A, formerly called infectious hepatitis, is usually a mild and self-limiting disease with an abrupt onset. Its average incubation time is approximately 30 days, but the incubation period may be as long as 2 months. Unlike other hepatitis infections, recovery is usually complete without chronic infection. The virus is found in feces, and transmission is by the fecal-oral route, most frequently resulting from contamination of food and drinking water. Close personal contact, including oral sex, contributes to transmission. Approximately 50% of cases are subclinical. The CDC estimates that approximately 150,000 persons per year are infected in the United States, with approximately 100 deaths per year, and annual costs exceed $200 million based upon lost workdays and medical costs. Good personal hygiene, emphasizing hand washing and good sanitation, is the most effective means of control.

Diagnosis is based on laboratory detection of antibodies in the blood serum; there is no specific treatment. An effective vaccine is available and recommended for use in high-risk individuals, including those traveling to foreign countries where the disease is endemic. Injections of blood products rich in hepatitis A virus antibodies are also used pre- and postexposure to hepatitis A virus.

Hepatitis E virus, like hepatitis A virus, is transmitted by the fecal-oral route. It is uncommon in the United States but is endemic in Africa, Central America, India, and Asia.

Poliomyelitis

A generation back, the word **poliomyelitis** (usually shortened to **polio**), also known as **infantile paralysis,** struck fear into the hearts of parents the world over. Now polio sits on death row and is expected soon to join small-pox, the first eradicated disease, in extinction. In 1988, the World Health Assembly adopted a resolution to eradicate poliomyelitis globally by 2000, and although the year has passed, victory is around the corner. The fact is that, in most parts of the world, polio has been drastically reduced; the last case in the United States was in 1979. All of this has taken place since the introduction of the first polio virus vaccine, the **Salk vaccine,** in the 1950s (In the News 9.4), followed a few years later by the **Sabin vaccine.**

Poliomyelitis is a highly transmissible infectious disease. Most cases of polio are asymptomatic, and only a small number result in paralysis. Transmission is from person to person by direct fecal-oral contact, by indirect contact with infectious saliva or fecal material, or by contaminated sewage or water. The virus replicates in the pharynx and the intestinal tract, and it may invade the blood and be disseminated to the nervous system. Replication in nerve cells causes **paralytic poliomyelitis,** resulting in severely deformed limbs (Fig. 9.14). Paralysis occurs in about 7 to 21 days from the time of initial infection; in about 2 to 10% of the cases, death may result. **Bulbar poliomyelitis** is an extremely dangerous form of polio because patients have difficulty in swallowing and breathing because of muscle paralysis; they must spend long periods, perhaps their entire lives, in **iron lungs** (Fig. 9.15). Electrical failures were a nightmare; attendants had to mechanically pump the bellows of an iron lung to keep a patient breathing. An iron lung has a diameter slightly larger than that of an oil drum and a length approximately equal to that of two drums. What a terrible fate to lie in such a contraption year after year! After exposure to poliovirus, more than 90% of susceptible contacts become infected and have lifetime immunity but only to the particular viral type (of which there are three) that caused the infection. Humans are the only reservoir for poliovirus, as is the case with

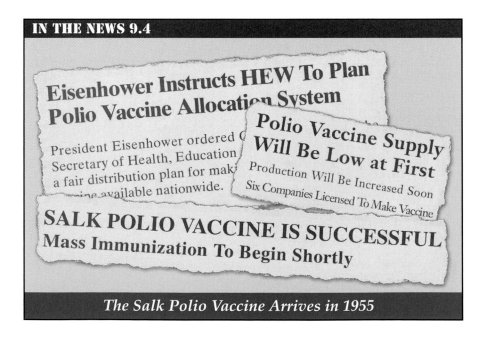

IN THE NEWS 9.4

Eisenhower Instructs HEW To Plan Polio Vaccine Allocation System

President Eisenhower ordered (Secretary of Health, Education a fair distribution plan for mak ...ine available nationwide.

Polio Vaccine Supply Will Be Low at First

Production Will Be Increased Soon Six Companies Licensed To Make Vaccine

SALK POLIO VACCINE IS SUCCESSFUL Mass Immunization To Begin Shortly

The Salk Polio Vaccine Arrives in 1955

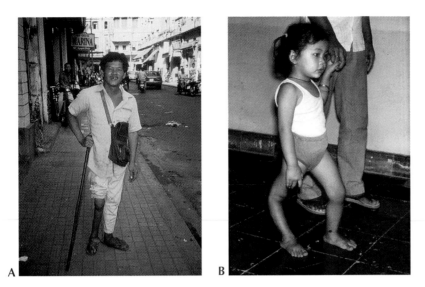

FIGURE 9.14   The crippling effects of polio. **(A)** A man with a weak, withered leg; **(B)** a girl with a major deformity in one leg. (Source: CDC.)

smallpox and measles, making possible the eradication of these diseases. (Measles is targeted as the third disease to be eradicated.)

Franklin Delano Roosevelt, president of the United States from 1933 to 1945, fell victim to polio in 1921 at age 39. The disease left this future president a paraplegic. He had been "an athlete, a man who had loved to swim and sail, to play tennis and golf, to run in the woods, and to ride horseback in the fields." For the remainder of his life, he never walked under his own

FIGURE 9.15   Life in an iron lung. Many victims of polio had to spend their lives in an iron lung because they could not breathe on their own. (Source: Public Health Service.)

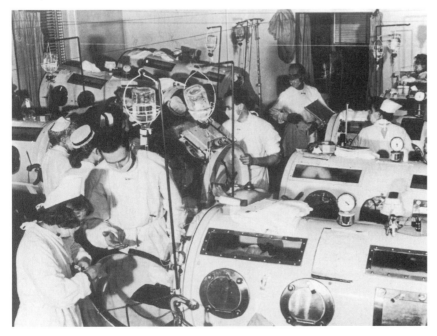

AUTHOR'S NOTE   *I remember the panic that ensued during my childhood when polio, or even suspected polio, threatened a neighborhood. Theaters, swimming pools, playgrounds, and all other places where young children might congregate were closed. Children were kept indoors. Throughout much of the year, the March of Dimes campaign to support the National Foundation for Infantile Paralysis was relentless in fundraising to support research toward combating the crippling disease of polio. In movie theaters, containers bearing the label "March of Dimes" were passed around, and theater patrons were encouraged to drop in coins to support research aimed at developing a vaccine against polio. In 1952, Jonas Salk, a University of Michigan virologist, announced the development of a vaccine, employing dead viruses of all three polio strains, that could safely be used in humans; implementation effectively reduced polio in the next several years. In 1960, the first U.S. test of a second polio vaccine, developed by Albert Sabin of the University of Cincinnati and composed of live but attenuated (weakened) viruses, was held. In the United States and in some other countries, the Salk vaccine was replaced by the Sabin vaccine; in some countries, the Salk vaccine continued to be used. The advantages and disadvantages of a live attenuated vaccine versus a vaccine made with dead microbes are discussed in chapter 11. The American Academy of Pediatrics' 2002 immunization schedule recommends four doses of inactivated (Salk) polio vaccine at the ages of 2 months, 4 months, 6 to 18 months, and 4 to 6 years.*

FIGURE 9.16   Franklin D. Roosevelt in a wheelchair. Roosevelt was a victim of polio during his presidential years and was instrumental in the March of Dimes campaign leading to development of the polio vaccine. (Source: Franklin D. Roosevelt Library.)

power. Throughout his 4 years as governor of New York and his 12 years as president of the United States, his legs were encumbered by heavy metal braces. This courageous and highly visible individual lent his name and influence to the National Foundation for Infantile Paralysis and to the March of Dimes fund-raising campaigns to combat polio. On May 2, 1997, in Washington, D.C., a 10-foot-tall memorial showing FDR (Roosevelt's nickname) seated in a wheelchair was dedicated to his memory. The statue is located on Cherry Tree Walk beside the Tidal Basin. Make sure to visit if you have the opportunity.

SEXUALLY TRANSMITTED VIRAL DISEASES

There are three major sexually transmitted viral diseases: genital herpes, genital warts, and acquired immunodeficiency syndrome (AIDS) (Table 9.4). AIDS is discussed in chapter 15.

Genital Herpes

By this time, you should realize that there are a variety of herpesviruses and that they are all "bad actors." Herpes is a highly infectious disease caused by **herpes simplex virus (HSV),** a DNA virus, of which there are two closely related types. HSV type 1 (HSV-1) is usually associated with painful sores around the mouth and lips, called cold sores or fever blisters, and,

TABLE 9.4 Sexually transmitted diseases

Disease	Incubation period	Virus category[a]	Symptoms	Comments
Genital herpes	4–10 days	dsDNA	Painful and itchy sores on the penis or on the labia, vagina, or cervix; can be asymptomatic	Lifetime infection with recurrent lesions
AIDS	Up to 12 years, maybe more	ssRNA	Weight loss, tiredness, loss of appetite, repeated infections	Always fatal, but longevity has been substantially improved; causes severe immunosuppression
Human papillomavirus and genital warts	1–6 months	dsDNA	Presence of warts; can be asymptomatic	Genital warts; cervical cancer and other genital cancers

[a]ds, double stranded; ss, single stranded.

occasionally, on the throat and the tongue. HSV type 2 (HSV-2) characteristically causes genital herpes, manifested as painful and itchy sores on the penis in males or on the labia, vagina, and cervix in females. HSV-1 and HSV-2 have an incubation period of 4 to 10 days. Oral sex can result in herpes in both the oral and genital regions. Both viruses are passed by direct contact with herpes sores from one person to another, or from one part of the body to another. **Latency,** a period of "taking up residence and hiding," is characteristic of herpesviruses in general, including varicella-zoster virus (discussed above). The viruses hide out in collections of nerve cells situated just outside the spinal column and are triggered into an active replicating state by a variety of factors, including stress, fever, sunlight, colds, and menstruation. The herpes lesions are highly infectious until the sores heal within about 2 to 4 weeks. Once acquired, infection is lifelong, and episodes of recurrent, painful genital ulcers occur. Generally, within the first year of infection, four or five symptomatic recurrences take place.

Transmission occurs through direct contact, including kissing and vaginal, anal, and oral sex. Frequently, the virus is transmitted by asymptomatic individuals who have no knowledge of their disease; it is important to realize that these individuals, as well as those who are aware that they are harboring the virus in a latent state, can transmit the disease. The disease is usually mild, but initial infection with HSV-1 and HSV-2 can be accompanied by high fever and large numbers of painful sores. Prompt treatment with antiviral drugs is somewhat effective. The disease can be severe in individuals with AIDS or other immunosuppressive conditions.

Pregnant HSV-infected females are at risk for miscarriage during the course of their pregnancy. In those carrying full-term infants, a Caesarean section is performed to minimize the risk of infection to the newborn during passage through an HSV-infected birth canal. Finally, HSV-infected females are at a greater risk of developing cervical cancer than uninfected females.

In 1997, the CDC estimated that 45 million people in the United States were infected with genital herpes. The disease is more common in females (1 out of 4) than in males (1 out of 5) and is more common in blacks (45.9%) than in whites (17.6%). According to a 1998 report in the *New England Journal of Medicine,* 90% of American adults infected with HSV-2 do not know that they are infected. The disease may be diagnosed by visual inspection or by culturing tissue from herpes sores to determine the presence of the virus. There is no cure, but antiviral medications may help reduce the duration and severity of occurrences. There is no vaccine, and prevention is based on safe sex practices.

Genital Warts

Human papillomaviruses (HPVs) are responsible for **common warts,** which are benign, painless elevated growths that occur most frequently on the fingers; **plantar warts,** which are deep and painful warts on the soles of the feet; and **genital warts.**

Genital warts (also called venereal warts) are now one of the most common sexually transmitted diseases (STDs) in the world; in the United States, an estimated 24 million people are infected with HPV. The disease is on the increase, with as many as 1 million new cases in the United States each year. There are more than 60 types of HPV, of which more than one-third are sexually transmitted and invade genital tissue. Genital warts are highly contagious, and about two-thirds of people whose sexual partners have genital warts develop warts within 3 months.

In females, warts can occur on the outside and inside of the vagina, on the **cervix** (the opening to the uterus), and around the anus. In males, warts may occur on the shaft or head of the penis, on the scrotum, and around the anus. Oral sex with a person with genital warts can result in warts in the mouth or throat. The warts may disappear without treatment, but they can develop into large fleshy growths that resemble pieces of cauliflower.

A major concern is the link between some strains of HPV and cancer, including cervical, vulvar, and anal cancer and, more rarely, cancer of the penis. The best preventative is to avoid direct contact with the virus. Clinical trials are now in progress for a vaccine to prevent disease and for a vaccine to treat cervical cancer.

CONTACT DISEASES (OTHER THAN STDs) AND BLOOD-BORNE DISEASES

Several viral contact diseases (Table 9.5) are currently of considerable significance in the United States and around the world.

Infectious Mononucleosis

Mononucleosis, often colloquially called "mono" or "kissing disease," is a frequent and unwelcome guest on college campuses. As many as 50 out of every 100,000 Americans have symptoms of **infectious mononucleosis,** and about 80% of all cases occur between the ages of 15 and 30. Most cases

TABLE 9.5   Contact diseases other than STDs

Disease	Incubation period	Virus category[a]	Symptoms	Comments
Infectious mononucleosis	4–7 weeks	DNA (EBV)	Headache, loss of appetite, fever, sore throat, swollen glands	"Kissing disease"; ages 15–30 years most affected
Hepatitis B,C, and D	45–180 days (HBV), 2–22 weeks (HCV)	dsDNA (HBV), ssRNA (HCV), ssRNA (HDV)		Blood, blood products, sexual practices; frequently associated with AIDS
Rabies	1–2 months	ssRNA	Encephalitis, anxiety, drooling, agitation, "hydrophobia"	Almost always fatal if not treated; immunization available; zoonotic with many wild animal reservoirs
Ebola hemorrhagic fever	4–16 days	ssRNA	Massive hemorrhage	

[a]EBV, Epstein-Barr virus; ds, double stranded; ss, single stranded.

of mononucleosis are caused by **Epstein-Barr virus** (EBV), a DNA virus. The virus infects and replicates in salivary glands and is transmitted by saliva and mucus during kissing, coughing, and sneezing. The incubation period is approximately 4 to 7 weeks after exposure. An infected person poses little risk to household members or college roommates, assuming there is no direct contact with the infected person's saliva.

Symptoms are vague, making diagnosis difficult. In fact, in many people, the disease is asymptomatic, or so mild that they are not even aware that they are infected. The early symptoms include "not feeling great," headache, fatigue, and loss of appetite, followed by the later triad of fever, sore throat, and swollen glands, particularly in the neck. Enlargement of the lymph nodes and spleen is a more serious complication.

Diagnosis can be definitively established only by laboratory tests based on the detection of antibodies and abnormal white blood cells. A test for antibodies can be done within a matter of minutes in a physician's office. There is no specific treatment, and symptoms generally disappear in 4 to 6 weeks. EBV, like herpesviruses discussed earlier in this chapter, remains latent in the body for life and is held in check by the immune system, although recurrences can occur. It is now known that a condition called **chronic fatigue syndrome** is not caused by EBV, as was formerly suspected.

EBV is the cause of **Burkitt's lymphoma,** a malignant cancer of the jaw and abdomen that occurs in children in central and western Africa. Why this lymphoma is found predominantly in Africa remains a mystery, although there is speculation that having malaria predisposes a person to this condition.

Hepatitis B, C, and D Viruses

As previously mentioned, besides hepatitis A and E viruses, there are other hepatitis viruses, which are transmitted primarily by blood and blood products and by sexual practices. Hepatitis B and hepatitis C, along with other diseases, are frequently found in persons with AIDS. In fact, hepatitis B is sometimes referred to as an "AIDS twin" because of their similarities of transmission.

Hepatitis B is more transmissible than AIDS; an individual pricked with a needle from an individual who has both hepatitis B and AIDS has a 40% chance of acquiring hepatitis B but only a 0.5% chance of acquiring AIDS. Both hepatitis B and hepatitis C are transmissible by blood and blood products; can be asymptomatic; can be manifested as chronic diseases; result in serious lifelong liver problems (Fig. 9.17), possibly necessitating a liver transplant; and are transmissible from infected pregnant women to their infants. Although the clinical symptoms caused by all the hepatitis viruses are somewhat similar, their biology, circumstances of transmission and infection, and consequences of infection are different. Laboratory tests are required for a definitive diagnosis of hepatitis.

Hepatitis B virus (HBV) has a long incubation time, ranging from 45 to 180 days, with an average of about 75 days. HBV is a common cause of cancer, second only to tobacco. In the Far East and in sub-Saharan Africa, where HBV is widespread, liver cancer is a common cause of death. HBV is also an occupational hazard for health professionals, including physicians, medical and dental hygienists, dentists, and others who have contact with blood. These individuals may serve as sources of HBV to the population at large. Infection can also be acquired by sharing needles, tattooing or body art (Fig. 1.19), acupuncture, and ear piercing.

FIGURE 9.17   Hepatitis B. This woman has a hepatoma (liver cancer) resulting from chronic hepatitis B virus infection. (Source: Patricia Walker, Ramsey Clinic Associates, St. Paul, Minn.)

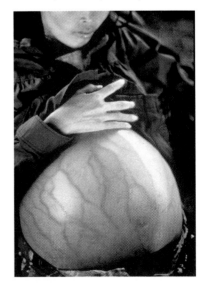

Estimates are that each year in the United States, more than 200,000 people of all ages get hepatitis B and approximately 5,000 die of the disease. Although a vaccine was introduced in the early 1980s, the incidence of disease reduction was disappointing. More recently, a more effective and safer vaccine was developed, resulting in the 1992 recommendation by the American Academy of Pediatrics of universal immunization for newborns. Older children, particularly adolescents, should also be vaccinated, particularly if they are sexually active. Assuming that immunization is practiced as recommended, the threat of HBV infection should be greatly reduced. More than 80 countries have begun inoculating children against HBV. In Saudi Arabia the infection rate dropped from 7 to 0.5% in only 8 years as the result of an HBV immunization program. One of the problems is that the vaccine is expensive, particularly considering that three shots are required, and, for now, beyond the reach of poor countries. One feasible approach is to vaccinate infants of infected mothers.

In the United States the cost per dose of vaccine is $7.60. It would take approximately $120 million per year to eliminate hepatitis B from the United States and about 15 years for reduction of the disease to be evident. It is a high figure, but not when you consider that the annual cost for treating persons who contract hepatitis B is $700 million.

Representative John Joseph Moakley of Massachusetts died of leukemia in May 2001, after a long and distinguished career in public service. In the 1980s he was diagnosed with hepatitis B, and he was at death's door by 1995, but a liver transplant saved his life.

Hepatitis C virus (HCV) was first identified in 1988 and was responsible for almost all cases of blood transfusion-transmitted (non-A, non-B) hepatitis. Hepatitis C is a chronic blood-borne infection with an incubation period of 2 to 22 weeks; it may be subclinical or mild, but approximately 50% of the cases progress to chronic hepatitis. The first reliable screening tests for HCV were not implemented until 1992; therefore, many people may be infected with HCV but unaware that they have the disease, particularly considering that it can take 20 years for the appearance of symptoms. An estimated 2.7 million persons in the United States are infected, and 8,000 to 10,000 deaths occur each year from cirrhosis (scarring) and liver cancer. Intravenous drug users constitute the vast majority of hepatitis C victims, and as many as 300,000 may have contracted it from blood transfusions before screening tests were available. Hepatitis C is considered to be the major reason for liver transplants in the United States. As a result, in 1998 the Department of Health and Human Services announced that people who received transfusions before 1992 from donors who later proved positive for hepatitis C would be notified. Obviously, this will not reveal individuals who were infected by dirty needles or by sexual contact. Keep in mind that there is no immunization against hepatitis C. Drugs, including **alpha interferon** and ribavirin, are available for the treatment of HCV, but, unfortunately, the results are disappointing. Further clinical studies are under way, and it now appears that the combination of interferon and ribavirin clears the virus from about 40% of patients, whereas only 20 to 30% are helped by interferon alone. Researchers are actively looking for new drugs against HCV; a major problem is that the virus does not grow in human cell cultures or in laboratory animals. In 1996 the molecular map of HCV was completed, opening up new possibilities of designing effective drugs.

Hepatitis D virus, also called hepatitis delta virus, causes severe liver disease. It has a close relationship with HBV, and both may be transmitted together, establishing a coinfection. In fact, without HBV, hepatitis D virus cannot replicate and is therefore called a "defective virus."

To conclude this discussion of hepatitis, there is good news and bad news. The good news is that this discussion of hepatitis is now finished; the bad news is that there are additional viruses causing hepatitislike symptoms.

Rabies, the "Kiss of Death"

Death is almost a certainty in untreated cases of **rabies;** few survive this virus-caused encephalitis, because there is no treatment once symptoms appear. Thanks to the brilliance, courage, and pioneering work of Louis Pasteur, a French chemist, a rabies vaccine is available for both prevention and early treatment of this disease.

Rabies virus is an RNA virus; the disease is zoonotic, with a worldwide distribution in diverse wild animal reservoirs, including coyotes, skunks, cats, bats, foxes, and raccoons, as well as domestic animals. Although control measures of immunization have substantially reduced the incidence of domestic animals as vectors in the United States and in other countries, dogs remain the major source of rabies in Asia, Africa, and Latin America, causing thousands of deaths each year. In the United States there has been a dramatic shift over the past 40 years in the principal wildlife carrier. In 1966, foxes were the major reservoirs, followed several years later by skunks; since about 1993, raccoons have emerged as the principal reservoir. They have been dubbed "garbage can gourmets" because they frequent neighborhoods in search of food. Further, rabies-infected animals may bite dogs and cats, who then become a menace to the human population. In the United States, the number of cases of rabies in wild animals has significantly increased, but there has been a dramatic decrease in the incidence of rabies-related human deaths over the last 50 years. Only about 5 to 15% of individuals bitten by rabid animals even develop the disease.

The bat is a rabies vector and has acquired a bad but unearned reputation. The novel *Dracula* by Bram Stoker, published in 1897, was responsible for the reputation of bats as evil, blood-sucking creatures capable of turning their victims into horrible life-forms and transmitting rabies. A Broadway show titled *Dracula* is a big hit. Bats have been further dramatized by the popularity of Batman as he wheels around in his Batmobile. Expressions such as "bats in the belfry," "batty," "blind as a bat," and "crazy as a bat" are part of our jargon and indicate our fear of bats. But let's clear up a few misconceptions: bats are not blind, most species do not suck blood (an exception being vampire bats of South America), and they do not particularly like to "get in your hair." Actually, they are remarkable creatures (some say "beautiful," but that is too much) and play an important role in ecosystems around the world by eating insects, including agricultural pests. The last two cases of rabies in America were transmitted by bats. The first case was that of a 71-year-old man in Texas who encountered a bat in his motel room. He managed to get rid of the animal and had no recollection of having been bitten, but about 2 months later, he died of rabies. The second case occurred in a 32-year-old New Jersey man into whose living room a bat had flown. He captured the animal by hand and flung it out the window; he died of rabies 2 months later. These cases indicate that individuals exposed to bats should be immunized, even if there has been no obvious bite.

The most common mode of rabies transmission is through the bite of an infected animal whose saliva contains the virus. The virus can also be transmitted into the eyes, nose, and respiratory tract. The early symptoms of rabies are nonspecific and flulike. There may be pain and tingling at the site of the bite, followed by symptoms of anxiety, confusion, agitation, delirium, abnormal behavior, hallucination, and drooling. An old term for rabies is **hydrophobia** (fear of water). Rabies has been described as follows. "The patient is tortured at the same time by thirst and a revulsion toward water." The mere sight of water results in uncontrollable spasms in the muscles of the mouth and pharynx, leading to spitting and choking.

Before the work of Pasteur, the public's fear of hydrophobia and the prevailing attitude that the disease could be transmitted through the saliva or breath of rabies victims were so overwhelming that those unfortunate victims who were bitten by rabid dogs lived in terror of what their fate might be. Records indicate that violent modes of death, including suffocation, were inflicted upon those who had suffered bites of a rabid animal. Such events must have been quite frequent, since, in 1810, a bill in France was written in the following terms: "It is forbidden under pain of death, to strangle, suffocate, bleed to death, or in any other way, murder individuals suffering from rabies, hydrophobia, or any disease causing fits, convulsions, furious and dangerous madness." Early treatments were terrible. **Cauterization** was the most frequently employed method; if the wounds were somewhat deep, it was recommended to use long, sharp, and pointed needles and to push them well in, even if the wound was on the face. Another practice involved the sprinkling of gunpowder over the wound and setting a match to it.

The incubation time for the onset of symptoms is related to the extent of the wound and its closeness to the brain. The incubation time for wounds on the hands is about 8 weeks, whereas for wounds on the face, it is about 5 weeks. The average incubation period is 1 to 2 months, with extremes of 1 week and more than 1 year. Initially, the virus multiplies in the muscle and connective tissue at the site of the bite; then it migrates along nerves to the central nervous system.

A scare occurred at a petting zoo in Iowa when an infected 5-month-old black bear cub named Chief died of rabies. The bear previously had not shown any symptoms of rabies; it was docile enough so that visitors routinely petted, fed, and played with it. According to a newspaper article, "Some adults put potato chips in their mouth to feed the bear mouth-to-snout; others put their hands in the bear's mouth. One little girl even shared a piece of bubble gum with the bear." Public health officials advised people who had visited the bear cub and might have had contact, depending upon the circumstances, to take the vaccine. According to the state epidemiologist, "Once you come down with symptoms of rabies you progress and die. We've got to get these people the vaccine as fast as possible."

The treatment of rabies has come a long way since Pasteur's first treatment, in 1885, of a boy who had been bitten by a rabid dog; his second success (Box 9.3 and Fig. 9.18) was even more dramatic, because about 6 days had elapsed prior to immunization. Since that time, improved vaccines have been developed. The vaccine currently in use consists of inactivated virus cultured in tissue culture. Treatment consists of one dose of immunoglobulin (blood serum from individuals containing high levels of antibodies against rabies) and a first dose of the vaccine given as soon as possible after exposure, with the remaining four doses given on days 3, 7, 14, and 28, respectively. The injections are relatively painless and are given in the arm in the same way that other immunizations are given. Earlier versions of the vaccine were administered into the abdomen and were painful.

BOX 9.3 Pasteur and the Development of the Rabies Vaccine

When one thinks of the famous names associated with the history of science, the name Louis Pasteur ranks among the greatest; his biography is an account of his distinguished scientific career. Pasteur's phenomenal success in diverse fields of research is without parallel, but his development of the rabies vaccine was his crowning glory. Nobel laureate Selman A. Waksman, discoverer of streptomycin, the first antibiotic effective against tuberculosis, wrote in the *New York Times* on February 5, 1950, "Pasteur was not only the great scientist who was largely responsible for the creation of the science of microbiology, he was its high priest, preaching and fighting for the recognition of its importance in health and in human welfare."

In the early 1870s, at about the age of 50 and having already suffered a cerebral hemorrhage which endangered his life and caused a permanent paralysis of his left arm and leg, Pasteur began to conduct research related to human infection. By this time he was well renowned and respected in France and beyond by the scientific community for his work on crystallography, fermentation, diseases of wine and beer, spontaneous generation, silkworm disease, cholera, and anthrax. It is not possible to cite with certainty the event or events that resulted in Pasteur's entry into rabies research. One story describes how, as a young boy, Pasteur had witnessed and heard screams of a victim bitten by a rabid wolf; the victim was undergoing cauterization of the wounds with a red-hot iron. Rabies, which has its origin in antiquity, is considered one of the most dreaded diseases of humans.

Before the mid-1880s, Pasteur succeeded in cultivating the virus of rabies in the brain stem of rabbits. He then attenuated the virus (weakened its virulence) by suspending a fragment of the rabbit virus-infected brain stem from a thread in a sterilized vial. As the brain stem preparation dried, virus virulence gradually decreased, and by 14 days these crude preparations were totally without virulence. Dogs protected by injection (immunized) with this crude vaccine and experimentally bitten by rabid dogs, or even submitted to the application of active rabies virus directly onto their brains, did not develop rabies. Pressure had already been brought to bear on Pasteur in France and in other countries to use the vaccine on humans, but Pasteur resisted the temptation, realizing that the time was not yet right for a trial on human subjects.

Monday, July 6, 1885, was a momentous day for Louis Pasteur and for the history of medicine. It was on that day that Joseph Meister, a 9-year-old boy, was brought to Pasteur's laboratory by his mother. On his way to school, 2 days earlier, Joseph Meister had been attacked by a mad dog and suffered 14 wounds on his hands, legs, and thighs. Some of the wounds were so deep that walking was difficult. The child's wounds had been treated with carbolic acid 12 hours after the incident by a local physician, who advised that Joseph be brought to Paris to be seen by Pasteur. The boy's mother pleaded to Pasteur for help for her doomed son; Joseph Meister was about to become the first human to receive the rabies vaccination. Pasteur's reputation was on the spot. Here was this mother pleading with him to save her child, but he worried that his vaccine, which had been tried only on dogs, would not be effective. The upshot was that young Meister received 13 inoculations of Pasteur's rabies vaccine over the next several days and survived his ordeal. The Joseph Meister story has a tragic ending unrelated to rabies. Pasteur and Meister had kept in close touch throughout the years, and, according to one account, in 1940—55 years after the incident that gave him a lasting place in medical history—he committed suicide rather than open Pasteur's burial crypt, of which he was the keeper, to the German invaders.

Jean-Baptiste Jupille, age 15, was the second patient treated by Pasteur, under less favorable circumstances. Jupille was in a group of six shepherd boys who were attacked by a rabid dog. One of the many biographers of Pasteur describes the event:

> The children ran away shrieking, but the eldest of them, Jupille, bravely turned back in order to protect the flight of his comrades. Armed with his whip, he confronted the infuriated animal which flew at him and seized his left arm. Jupille wrestled the dog to the ground and succeeded in kneeling on him, forcing his jaws open in order that he might disengage his left hand, and in so doing, his right hand was seriously bitten in its turn; finally, having been able to get hold of the animal by the neck, Jupille called his little brother to pick up his whip which had fallen in the struggle and secured the animal's jaws with the whip. He then took his wooden shoe, with which he battered the dog's head.

Source: R. Vallery-Radot, *La vie de Pasteur* (Hachette et Cie, Paris, France, 1900).

Jupille was brought to Pasteur for treatment 6 days after being bitten, in contrast to the 2-day interval in the Meister case. Pasteur was particularly reluctant to treat Jupille because of the extended time that had elapsed since the attack. Nevertheless, Pasteur again put his reputation on the line and injected Jupille with the vaccine. Jupille did not develop rabies, and his act of bravery was commemorated with an impressive statue depicting the struggle between a boy and a rabid dog, which stands today on the grounds of the Pasteur Institute (Fig. 9.18). (Note the terror in the eyes of the boy and the dog as well as the wooden shoe at the base of the statue.)

FIGURE 9.18 Statue of Jean-Baptiste Jupille at the Pasteur Institute in Paris. Jupille was attacked by a rabid dog while attempting to protect younger children from the animal. He was the second person to receive the vaccine. (Author's photo.)

Whether the individual needs to be vaccinated is carefully evaluated and depends on the circumstances. Laboratory tests are available to determine whether the animal does, in fact, have rabies; the big problem is that if a wild animal is involved, then that animal is usually not available for study. If the bite is from a domestic animal, the animal can be quarantined and observed for symptoms. It may be necessary to kill the animal and examine the brain for the presence of Negri bodies (chapter 5). It may have occurred to you that, for most diseases, immunization is usually given as a preventive measure and not as a treatment; if it has, you are correct (and very astute). However, the situation with rabies is somewhat unusual in that postexposure vaccination, as a treatment, is effective because of the long incubation time generally associated with rabies. On the other hand, one case of human rabies developed in only 10 days after the person had been bitten. The presumption is that if an individual is bitten by a wild animal, therapy is recommended. After an animal bite, the wound should be thoroughly cleansed. Preexposure vaccination is recommended for those in high-risk groups, including veterinarians, animal handlers, and laboratory personnel engaged in work with rabies.

Ebola Virus

Following are vivid excerpts about **Ebola hemorrhagic fever** from *The Coming Plague*, by Laurie Garrett:

> And he was bleeding. His nose bled, his gums bled, and there was blood in his diarrhea and vomit. . . . At her side, vomiting blood and bleeding from her eyes, was her husband, Ekombe Mongawat. . . . They pumped Antoine full of antibiotics, chloroquine, vitamins, and intravenous fluid to offset his dehydration. Nothing worked. . . . The horror was magnified by the behavior of many patients whose mind seemed to snap. Some tore off their clothing and ran out of the hospital, screaming incoherently. . . . Some, the huts of the infected, were burned by hysterical neighbors.

Here is a quote from an article in *Newsweek* published on May 22, 1995:

> Then, as the virus starts replicating in earnest, the victim's capillaries clog with dead blood cells, causing the skin to bruise, blister, and eventually dissolve like wet paper. By the sixth day, blood flows freely through the eyes, ears, and nose, and the sufferer starts vomiting the black sludge of his disintegrating internal tissues. Death usually follows by day nine.

These descriptions are enough to make you break out in a sweat. It is no wonder that near panic resulted in the United States in 1989, 1990, and again in 1996 when Ebola virus was introduced into primate quarantine facilities in Pennsylvania, Texas, and Virginia, from monkeys imported from the Philippines. Several individuals developed antibodies, but no human cases of disease were identified. One episode occurred in an animal caretaker at the Reston, Va., facility who accidentally cut himself with a virus-contaminated scalpel. Ebola virus was isolated from his blood, and antibodies were later detected. The media hype was exaggerated, and the public was both fascinated and terrorized. The movie *Outbreak,* based on the Reston episode, was a hit, as was the bestseller *The Hot Zone.*

Ebola virus, the cause of Ebola hemorrhagic fever, is named after the Ebola River in the African nation of Zaire (now the Democratic Republic of the Congo), where it was first detected in 1976; it causes sporadic outbreaks of severe infection with a high fatality rate in humans and nonhuman primates (monkeys and chimpanzees). There are four identified subtypes, each named after the geographic region where the disease was first identified: Ebola-Zaire, Ebola-Sudan, Ebola-Ivory Coast, and Ebola-Reston. The first three subtypes produce disease in humans, whereas Ebola-Reston causes illness only in nonhuman primates. The disease is considered to be zoonotic, although the reservoir of the virus is not yet known.

As noted in the foregoing accounts, Ebola hemorrhagic fever is a horrendous disease. Symptoms, including fever, chills, muscle aches, headache, stomach pain, sore throat, and abdominal pain, appear after an incubation period of 4 to 16 days. The blood fails to clot, resulting in massive hemorrhage throughout the body, both externally and internally. Transmission requires close contact with an individual who is ill with the disease; there are no documented cases of the disease being acquired from nonhuman primates. Aerosol transmission has been documented in nonhuman primates but has not been reported in humans. Since the virus may persist in genital secretions for a brief time, sexual contact may also convey the disease. Humans are "incidental" hosts—they do not "carry" the virus. The natural reservoir, i.e., where the virus "hides," is not known, but based on the epidemiology, an animal reservoir native to the African continent is suspected.

An outbreak of Ebola hemorrhagic fever occurred in Kikwit, Democratic Republic of the Congo, in 1995. Kikwit is 240 miles east of Kinshasa, the capital of that nation. The factors regarding transmission and control that are now known were not realized at the time. The outbreak originated in a hospital. As is the case in many developing countries, syringe needles are reused because of underfinanced health care systems, and this, along with other factors, promulgated the outbreak. The hospital personnel walked freely among the patients and unknowingly spread the disease from one to another; isolation procedures were not carried out. In many countries, especially poor countries, families play an intimate and crucial role in caring for hospitalized relatives and, in so doing, may become victims of the disease. Upon returning to their own villages, they further spread the virus.

Definitive diagnosis of Ebola virus infection is based on the detection of antibodies or of Ebola virus genetic material. There is no standard treatment available against Ebola virus; supportive measures are used.

FIGURE 9.19 Biosafety cabinets. Highly virulent microbes must be handled in biosafety cabinets in order to protect the laboratory personnel and the environment. (Source: CDC.)

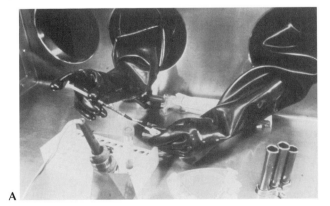

A

B

Control measures in a community or hospital environment center on avoidance of direct contact with infected persons and their blood or secretions and with the bodies of deceased patients. Barrier nursing precautions are called for, including the use of gowns, goggles, masks, and gloves by physicians, nurses, and other health care workers, the restriction of visitors, and proper disposal of wastes and corpses. Researchers and other laboratory workers handling Ebola virus and other viruses that cause hemorrhagic fevers must work in biosafety level 4 cabinets, the most heavily protected type of laboratory (Fig. 9.19), in order to be protected from the threat of infected tissues.

OVERVIEW Humans serve as hosts for a variety of viral diseases. West Nile and Ebola virus infections are regarded as "new" diseases, whereas rabies and plague are diseases of antiquity. The viral diseases discussed in this chapter are categorized on the basis of their primary mode of transmission: airborne, arthropod borne, food borne and waterborne, sexually transmitted, and contact. Each of these diseases is described with attention to factors involved in the microbial cycle of disease, pathogenicity, and other considerations of interest.

On the basis of complementary receptor molecules on the surface of viruses and host cells, viruses target a specific type of host cell or, in some cases, a variety of cell types. Some viruses cause asymptomatic or mild infection, while others are almost always fatal; some are zoonoses, and some exhibit latency. Fever, muscle aches, and respiratory distress are common symptoms and are sometimes too vague to make a definitive diagnosis. Skin rashes accompany some viral infections and are significant in establishing a clinical diagnosis. Several viruses either cross the placenta or are acquired by newborns during delivery and can cause serious damage. ■

SELF-EVALUATION

PART I Choose the *single* best answer.

1. What is the distribution of families of RNA and DNA viruses that cause disease?
   a. 13 RNA and 16 DNA families
   b. 16 RNA and 13 DNA families
   c. mostly DNA families
   d. none of the above

2. Which virus does not belong in the following group?
   a. RSV   c. EBV
   b. influenza virus   d. hantavirus

3. Which one of the following is zoonotic?
   a. rubella   c. rubeola
   b. RSV   d. none of the above

4. HSV-1 causes _____, and HSV-2 causes _____.
   a. cold sores; genital herpes
   b. fever blisters; cold sores
   c. fever blisters; shingles
   d. chicken pox; genital herpes

5. Congenital herpes is usually acquired through
   a. contaminated hands   c. placental transfer
   b. other babies   d. an infected birth canal

6. Chicken pox and shingles are caused by
   a. varicella-zoster virus
   b. different strains of varicella-zoster virus
   c. herpes simplex and herpes zoster
   d. contaminated foods

7. Which of the following is a defective virus?
   a. hepatitis E virus   c. EBV
   b. hepatitis D virus   d. HPS

(continues)

SELF-EVALUATION (continued)

PART II Match the statement on the left with the disease (or microbe) on the right by placing the correct letter in the blanks.

_____ 1. kissing disease

_____ 2. most frequent encephalitis in United States

_____ 3. tick-borne encephalitis

_____ 4. blood-borne

_____ 5. common under 6 months of age

a. St. Louis encephalitis
b. hepatitis C
c. dengue fever
d. RSV
e. Colorado tick fever
f. hepatitis A
g. infectious mononucleosis

PART III Answer the following.

1. Individuals who were vaccinated against influenza last year need to be revaccinated each year. Why is this the case?

2. (a) A virus is designated A/Singapore/1982/H3N1. What does this all mean? (b) With regard to the "A," are there any substitutes, and, if so, what are they?

3. The pig has been described as a "blender" regarding influenza. What does this mean?

10

PROTOZOANS AND HELMINTHS AND THE DISEASES THEY CAUSE

If a man take no thought about what is distant, he will find sorrow at home.

CONFUCIUS

PREVIEW Malaria, leishmaniasis, sleeping sickness, and giardiasis are examples of protozoan diseases. Ascariasis, river blindness, and tapeworm infection are examples of diseases caused by helminths (worms). Why are protozoans and worms lumped together in this chapter? What do protozoans and worms and their diseases have in common? Protozoans are microscopic and unicellular, while helminths are macroscopic and multicellular. The overriding characteristic that links them is that they are all eucaryotic cells or are composed of eucaryotic cells. The protozoans are unicellular eucaryotes, and the helminths are multicellular eucaryotes. The helminths are really not microbes but are included because some cause infection (sometimes referred to as infestation) and challenge the immune system. The term "parasite" is reserved by some biologists for the multicellular parasites—the worms—while others use it to include both protozoans and worms. In this book, the term parasite is used in an inclusive sense to indicate all the categories of pathogenic microbes and worms.

Many parasitic protozoans exhibit complicated life cycles involving more than one host and, in some cases, strategies for survival outside the host. There are at least 20,000 species of protozoans, but relatively few are pathogens for humans. Protozoan diseases are more common in the tropics but are increasingly significant in the United States because of immigration and tourism. Pathogenic protozoans are transmitted by food and water, vectors, and sexual contact.

Helminths are classified into two morphological categories, roundworms and flatworms. Although worms are at the systems level of biological organization in many cases, it is primarily their reproductive systems that are well developed, a feature ensuring large quantities of eggs. Some helminths exhibit complex life cycles involving two (or more) intermediate hosts prior to their becoming infective for humans. The life cycle of many helminths limits them to tropical climates, but some are prevalent worldwide. Helminths are transmitted to human hosts by food and water, vectors, or direct contact.

BIOLOGY OF PROTOZOANS

Protozoans are eucaryotic microorganisms. The term "protozoan" is derived from the Greek *protos,* meaning first, and *zoon,* meaning animal. Their distribution is worldwide, and they are virtually everywhere. They are found in fresh water and in marine habitats, in mud, drainage ditches, water-filled tires, in the guts of termites, and in soil.

Protozoans are unique, as are bacteria, in that their unicellular characteristic requires that each cell bear the total burden of "staying alive" on its own; there is no cellular specialization and differentiation allowing for sharing of functions, as is the case for multicellular organisms. Like all eucaryotes, protozoans have a cell membrane, a membrane-bound nucleus, and other membrane-bound organelles within the cytoplasm. Some species have a protective rigid cover, the **pellicle,** outside the cell membrane; freshwater species continually take in water and eliminate it by contractile vacuoles which push water out. Most protozoans are heterotrophic and aerobic. Many ingest food particles by phagocytosis; the particles are then enclosed within food vacuoles in which digestion takes place.

Many parasitic protozoans have complicated life cycles involving more than one host and, in some cases, survival in nature outside of a host. **Encystation** allows for outside survival and is a process resulting in a dormant resting stage **cyst** surrounded by a thick capsule. In some helminthic life cycles, transmission between hosts takes place by ingestion of cysts. **Excystation** to the **trophozoite** form (the reproductive and feeding stage) occurs after ingestion of the cyst by the host.

Asexual reproduction by binary fission is the usual mode of reproduction in protozoans. A primitive form of sexual reproduction, called **conjugation,** occurs in some protozoans and allows for the direct exchange of genetic material during physical contact of mating pairs. (Yes! Some have a sex life.)

At least 20,000 species of protozoans exist, but relatively few are pathogens for humans. Some are pathogens for animals and plants. The majority are free living and constitute a sizeable portion of **plankton,** a primary food source for many aquatic organisms and the base of food chains. Although protozoan (and helminthic) diseases are more common in the tropics, they are significant in temperate zones, particularly with increased immigration and tourism. Some very significant diseases of humans are caused by protozoans and are included in the World Health Organization's (WHO's) list of the most prevalent tropical diseases. For example, malaria, a protozoan mosquito-borne disease, kills more than 1 million people in the world each year. Not all of the protozoan diseases are unique to the tropics; in 1993 in Milwaukee, **cryptosporidiosis,** a disease transmitted by fecally contaminated drinking water, caused the largest waterborne epidemic in the United States in history. In 1996, in the United States, an epidemic of **cyclosporiasis,** a protozoan disease, occurred as a result of contaminated raspberries from Guatemala.

Classification of the protozoans is complicated and controversial. There is no agreement between protozoologists on a universally acceptable classification, but from a practical and medical point of view, four groups are significant and are classified on the basis of their mechanism of locomotion (Table 10.1 and Fig. 10.1): **sarcodina, mastigophora, ciliata,** and **sporozoa.**

PROTOZOAN DISEASES

Protozoan diseases are organized by their mode of transmission (Table 10.2), following the pattern for bacterial diseases (chapter 8) and viral diseases (chapter 9).

Food-Borne and Waterborne Diseases

Giardiasis

Consider this hypothetical scenario. A group of Boy Scouts returns from a hiking trip in the Rocky Mountains tired, hungry, and dirty, but in good health and in good humor. The next day, disaster strikes: several of the scouts fall ill with cramps, nausea, and diarrhea. The diagnosis is likely to be **giardiasis,** a common protozoan intestinal infection in the United States acquired by drinking contaminated water. If you are a hiker or a camper, beware of drinking from "pristine" mountain streams and other bodies of water. The days are long gone when you could drink directly from clear, cool, refreshing lakes and streams without running the risk of acquiring giardiasis, an infection caused by the flagellated protozoan *Giardia lamblia.* The intestinal disease caused by this parasite has been dubbed **"hiker's diarrhea"** and **"beaver fever."** Over the past 15 years, giardiasis has been recognized as one of the most common waterborne human diseases in the United States (Fig. 10.2), and it is the most commonly diagnosed intestinal parasite in public health laboratories. The parasite is found throughout the United States and throughout the world and has been isolated from humans and a variety of animals, including beavers, coyotes, cats, and cattle; it is acquired by drinking water contaminated with *Giardia* cysts. After excystation of the cysts to the vegetative and multiplying trophozoite forms in the intestinal tract, attachment to the lining of the intestine occurs by means of **sucking disks.** The parasites feed on mucous secretions and undergo reproduction, resulting in large populations. Typical symptoms include diarrhea, abdominal pain, and large amounts of gas. Although the diarrhea may be extensive, it is not bloody because the protozoans do not usually invade deeply into the lining of the intestine. Symptoms usually appear about 2 weeks after infection and may last 6 weeks or longer. Transmission also occurs through fecally contaminated food, contaminated environmental surfaces, and person-to-person contact by the fecal-oral route, as might occur in day care nurseries. Swallowing water from contaminated swimming pools, lakes, rivers, or springs or eating foods contaminated with sewage or feces from humans or animals is another source. It is

TABLE 10.1 Classification of the protozoans

Group	Locomotion	Disease(s)
Sarcodina	Ameboid-like	Amebic dysentery
Mastigophora	Flagella	Chagas' disease, African sleeping sickness, giardiasis, trichomoniasis, leishmaniasis
Ciliata	Cilia	Balantidiasis
Sporozoa	No locomotion	Malaria, babesiosis

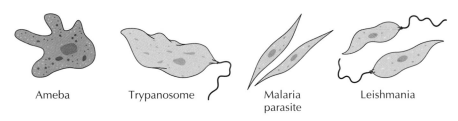

Ameba   Trypanosome   Malaria parasite   Leishmania

FIGURE 10.1 Some protozoan parasites.

imperative that uncooked vegetables and fruits be thoroughly washed in uncontaminated water. Further, the parasites may be accidentally acquired from contaminated toys and diaper-changing tables. Good hand-washing practices are essential.

Giardia epidemics have broken out in child care centers, with as many as 70% of the children infected with *Giardia* parasites. The finding of cysts or trophozoites is diagnostic for giardiasis, but this is problematic because the organisms may be shed intermittently, necessitating the examination of stool specimens over several days. New tests utilizing probes for the detection of cysts in stool samples are under development. A number of drugs are effective in treating this disease.

A major problem in the control of giardiasis is the resistance of the parasite to the chlorine concentration used in municipal water supply systems. Adding to the problem is that *Giardia* has a very low infectious dose: as few as 10 cysts are enough to establish giardiasis. Persons at the greatest risk of exposure are children in day care centers, men who have sex with men,

TABLE 10.2 Major protozoan diseases

Organism	Classification	Disease	Transmission	Principal site(s)
Food borne and waterborne				
Giardia	Mastigophora	Giardiasis	Water, direct contact	Intestinal tract
Entamoeba	Sarcodina	Amebiasis	Water, direct contact, food	Intestinal tract
Cryptosporidium	Sporozoa	Cryptosporidiosis	Water	Intestinal tract
Toxoplasma	Sporozoa	Toxoplasmosis	Food; contact with cat feces resulting in fecal-oral transmission	Brain, heart, lungs; possible transfer to fetus
Arthropod borne				
Trypanosoma brucei	Mastigophora	African sleeping sickness	Tsetse fly	Blood
Trypanosoma cruzi	Mastigophora	Chagas' disease (South American trypanosomiasis)	Kissing bug	Heart
Leishmania	Mastigophora	Leishmaniasis	Sand fly	Liver, spleen, mucocutaneous membranes, skin
Plasmodium	Sporozoa	Malaria	Mosquito	Blood
Babesia	Sporozoa	Babesiosis	Tick	Blood
Sexually transmitted				
Trichomonas	Mastigophora	Trichomoniasis	Sexual contact	Urogenital tract

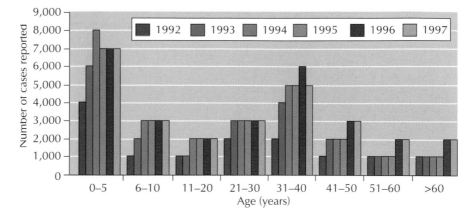

FIGURE 10.2 Giardiasis case reports, United States, 1992 to 1997. The incidence of giardiasis increased during this period. Giardiasis is one of the most common waterborne diseases in the United States. [Redrawn from B. W. Furness, M. J. Beach, and J. M. Roberts, *Morbidity and Mortality Weekly Report* **49**(SS-07):1–13, 2000.]

backpackers and campers, travelers to areas where the disease is endemic, and persons drinking water from shallow wells.

In August 2000, the Centers for Disease Control and Prevention (CDC) issued a report on giardiasis surveillance in the United States from 1992 to 1997 and documented the first nationwide look at giardiasis. A summary of the report indicates that transmission occurs in all major geographic areas of the country and that the seasonal peak coincides with the summer recreational water season, reflecting the heavy use by young children of communal swimming sites such as lakes, rivers, swimming pools, and water theme parks. Estimates based on state surveillance data indicate that as many as 2.5 million cases of giardiasis occur annually in the United States. When infection is suspected but multiple stool examinations are negative, an alternative procedure is the **Enterotest;** the patient swallows a lead sinker in a gelatin capsule attached to a string, and after 4 hours the string is withdrawn and the capsule is examined for the presence of trophozoites (Fig. 10.3).

Cryptosporidiosis

In September 1993, the *Milwaukee Journal* published an article entitled "Fatal Neglect" 5 months after the cryptosporidiosis outbreak in Milwaukee. The problem was due to the contamination of the water supply system by the sporozoan *Cryptosporidium parvum,* which resulted in about 100 deaths and 400,000 illnesses.

FIGURE 10.3 Enterotest for giardiasis.

C. parvum is found in a variety of mammals, birds, and reptiles. Infectious **oocysts** are discharged into the water in fecal material, and their ingestion initiates infection. The oocysts undergo excystation to **sporozoites,** which penetrate the intestinal cells where multiplication results in a new batch of oocysts. The release of these oocysts into the environment completes the life cycle. In patients with acquired immune deficiency syndrome (AIDS) and in other immunocompromised hosts, potentially life-threatening diarrhea often occurs. Up to 25 bowel movements occur per day, resulting in a loss of over 10 liters of fluid and severe dehydration. Other symptoms include weakness, fever, nausea, and abdominal pain. The incubation period ranges from several days to a few weeks. Outbreaks of cryptosporidiosis are difficult to prevent because oocysts are not inactivated by the usual doses of chlorination, their small size (about 5 micrometers in length) allows passage through sand filtration, and as few as 10 oocysts are sufficient to cause disease. In field tests only about 10% of water samples seeded with oocysts were detected, indicating the difficulty of identification and surveillance. In the 1993 Milwaukee cryptosporidiosis outbreak, cattle were initially suspected to be the prime source of contamination, but later studies indicated that human waste contaminated the city water supply. As a result, Milwaukee has spent over $74 million in improvements to their two water plants, including new filters and a new disinfection system. Also, the water intake pipe was moved nearly a mile further into Lake Michigan and away from polluted water flowing from the Milwaukee River. Two other examples emphasize the frequency of cryptosporidiosis: an outbreak resulted from a contaminated pool at a summer party in Oregon, and in Minnesota an outbreak was traced to contaminated food (Fig. 10.4). Outbreaks of cryptosporidiosis continue to occur, and constant vigilance is required. The appearance of clean water does not guarantee its safety for drinking.

Amebiasis

You might be fascinated by the movements of amebas in a drop of pond water examined under the microscope. Unfortunately, these "cute" amebas have disease-producing cousins. **Amebiasis** is primarily caused by the protozoan *Entamoeba histolytica* and occurs worldwide, especially in regions with poor sanitation. In the United States, amebiasis is most prevalent in immigrants from developing countries, in people who have

FIGURE 10.4 Cryptosporidiosis. **(A)** History of cryptosporidiosis outbreaks caused by contamination of water supplies; **(B)** *Cryptosporidium* oocysts (left) and a warning sign on a fountain (right). (Source: CDC.)

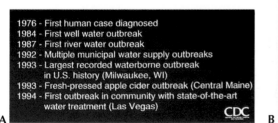
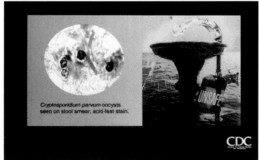

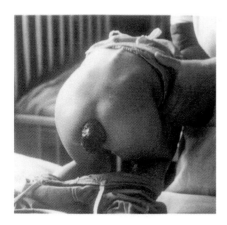

FIGURE 10.5 Prolapsed rectum. The dysentery resulting from amebic infection is so severe that the frequent and intense straining has caused the child's rectum to protrude from the anus. (Author's photo.)

traveled to developing countries, in people who live in institutions where it is difficult to maintain good sanitary conditions, and in those who practice oral-anal sex.

The life cycle is initiated by the ingestion of cysts, which undergo excystation in the intestinal tract. Four trophozoites emerge from each cyst, move into the large intestine, and attach to the wall, where they mature, multiply, and feed. About 90% of infected individuals remain asymptomatic or suffer from only mild disease characterized by diarrhea and stomach pain. However, severe amebiasis can result, accompanied by fever, bloody stools, and stomach pain due to destruction of the lining of the large intestine. Symptoms usually occur about 1 to 4 weeks after infection and include dysentery (bloody, mucus-filled stools), weight loss, fever, and fatigue. The dysentery may be severe enough to cause the rectum to extrude through the anus (prolapsed rectum) (Fig. 10.5). In some cases, amebas invade the kidneys, skin (Fig. 10.6), brain, spleen, and liver. Invasion of the liver can result in amebic hepatitis and liver abscesses (Fig. 10.7). The fatality rate is about 10% in severe cases, and estimates are that over 100,000 people die each year.

Fecal-oral amebic infections are more common in countries where sanitation is poor; 10 to 50% of the population in tropical countries may be infected with *E. histolytica*. In the United States, amebiasis is relatively rare and frequently remains undiagnosed, indicating that there may be more cases of amebiasis than are reported.

Diagnosis of amebiasis is based upon finding trophozoites or cysts in an examination of fecal smears and can be difficult, since nonpathogenic intestinal amebas need to be differentiated. Other, more definitive diagnostic laboratory tests are available to help in diagnosis.

Several drugs are available for the treatment of amebiasis. As in the case of giardiasis and cryptosporidiosis, chlorination of water supplies may fail to kill amebic cysts. Practicing good personal hygiene, including thorough hand washing after using the toilet and before handling food, minimizes the risk. Individuals traveling to countries in which poor sanitary conditions prevail need to be particularly careful about their source of water and about fruits and uncooked vegetables, which may have been washed with contaminated water.

FIGURE 10.6 Perianal cutaneous amebiasis. The amebas have caused erosion of the tissues around the anus. (Author's photo.)

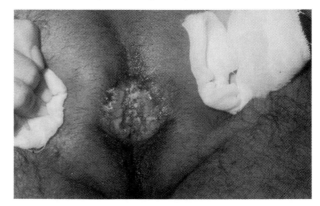

Toxoplasmosis

Toxoplasmosis is caused by the parasite *Toxoplasma gondii* and has a worldwide distribution. The disease is a zoonosis and occurs in over 200 species of birds and mammals; members of the feline (cat) family, both domestic and wild, serve as the primary reservoir and host. Toxoplasmosis can be acquired by humans after the accidental ingestion of oocysts present in cat feces or by eating meat that contains cysts. Exposure to infective oocysts from cat feces usually occurs when people are careless in their hand-washing habits and ingest minute amounts of cat feces containing oocysts in cat litter or in sandboxes or when people work in garden soil in which cats have defecated. In moist surroundings, the oocysts are capable of remaining viable and infective for several months.

Additionally, humans may acquire toxoplasmosis through the ingestion of cysts in pigs, sheep, cattle, and poultry that have picked up oocysts in the soil. The oocysts develop into cysts in the tissue of these animals and may then be consumed by humans. Studies have indicated that pork and lamb are frequently contaminated with cysts. Eating raw or undercooked meats is a common source of toxoplasmosis; cysts are killed by heating above 60°C. Most cases of toxoplasmosis are asymptomatic or produce only mild symptoms, including sore throat, low-grade fever, and lymph node enlargement. However, in those individuals with AIDS or with immune systems weakened from other causes, toxoplasmosis is frequently a rapidly fatal disease resulting from massive invasion of the parasites into the brain.

There is about a 30% chance that the disease can be transmitted during pregnancy to the fetus, resulting in stillbirth or in serious fetal defects, including brain damage, convulsions, retinal damage leading to blindness, and death. Pregnant women should not change a cat's litter box and should minimize the touching of cats, since the possibility of picking up oocysts on the hands and transferring them into the mouth exists. The complicated life cycle of *Toxoplasma* is illustrated in Fig. 10.8.

Diagnosis of toxoplasmosis is difficult because the disease resembles infectious mononucleosis. Definitive diagnosis can be accomplished by isolating the parasites, by identifying them in samples of infected tissues, or by detecting the presence of antibodies. Drugs are available to treat the disease and may need to be taken for as long as 1 year to prevent recurrent infection.

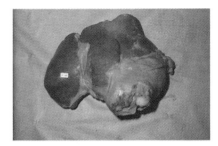

FIGURE 10.7 Amebic abscesses in the liver. Amebas may invade from the intestine and produce lesions outside the intestine. Hepatic (liver) amebiasis, a serious disease, results from invasion of the liver by trophozoites. (Author's photo.)

Arthropod-Borne Diseases

Trypanosomiasis

African trypanosomiasis, also known as **sleeping sickness,** is an ancient disease and one of many that have plagued Africa for millennia. For those affected, this potentially fatal disease is hardly a joking matter. There are two types of African trypanosomiasis named for the area of Africa in which they exist. West African trypanosomiasis is caused by *Trypanosoma brucei gambiense,* found in the rain forests of western and central Africa; the East African variety, caused by *Trypanosoma brucei rhodesiense,* is found in the upland savanna of East Africa. Both species are transmitted by the bite of infected tsetse flies, which are large and aggressive flies that inflict painful bites. The geographic distribution of the two varieties of African trypanosomiasis is a reflection of the different ecological niches occupied by the tsetse fly vectors.

AUTHOR'S NOTE *The expression "sleeping sickness" is sometimes used in a jocular manner; students are sometimes accused of having "sleeping sickness" during lectures. On some days, the condition appears to be epidemic!*

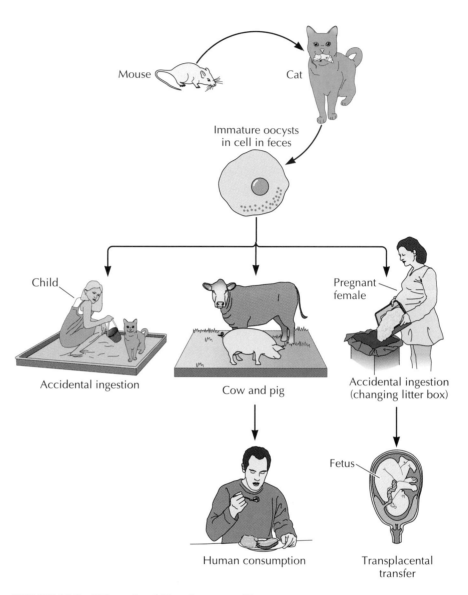

FIGURE 10.8   Life cycle of *Toxoplasma gondii*.

The cycle of trypanosomiasis starts when a tsetse fly takes a blood meal from an infected reservoir host; wild animals (e.g., hyenas, lions, and wild pigs), domestic animals (e.g., goats, cows, and dogs), and humans are reservoirs. The trypanosomes multiply and migrate from the gut of the fly to the salivary glands, where further development takes place. As many as 50,000 parasites can be injected by a single fly bite, far in excess of the approximately 500 that are required to establish infection. After the bite, a chancre (red sore) develops at the site, and from there the parasites move into the blood, the spinal fluid, lymph nodes, and the brain. Early symptoms include fever, fatigue, swollen lymph nodes, and aching muscles and joints. As the disease progresses over a few months to a few years, the trypanosomes invade the brain and cause personality changes, progressive confusion, difficulty in walking, and altered sleep patterns which become worse with time. Sleeping for long periods of the day (hence, the name "sleeping sickness"), accompanied by insomnia at night, is common. If the disease is left untreated, death occurs within a few months to several years after infection.

Trypanosomes exhibit antigenic variation—they "change their coats" as a strategy of invasion of the host's immune defense mechanisms. This is a property that interferes with development of a vaccine.

West African trypanosomiasis is on the rise in areas of Sudan despite some success in capturing the flies in traps baited with water buffalo urine. In 1997 a team of epidemiologists examined almost 1,400 persons in 16 villages and reported the presence of the disease in every village surveyed. Further, they reported an overall increase in disease prevalence of 16% in the last 10 years or so.

The life cycle of *Trypanosoma cruzi,* the causative agent of American trypanosomiasis, or **Chagas' disease,** is similar to that of the trypanosomes that cause sleeping sickness. Wild animals, including rodents, opossums, and armadillos, serve as reservoirs for *T. cruzi.* The vector is a **triatomid insect,** commonly referred to as a **"kissing bug"** (Fig. 10.9), which lives in the thatched roofs (Fig. 10.10), cracks in the mud walls, and other dark places characteristic of adobe huts, close to humans. The vector harbors the trypanosomes in its hindgut and discharges them in its feces. It has the nasty habit of defecating near the site of its bite, resulting in inoculation of the trypanosomes into the skin when the area is scratched; the insect bites at night.

Chagas' disease occurs in Central and South America. The parasite causes widespread tissue damage, particularly to the heart, causing it to enlarge and impairing its function. Death occurs within 30 years of infection if the disease is untreated. In addition to the usual diagnostic methods, **xenodiagnosis** is sometimes used; in this technique, kissing bugs known to be free of the trypanosomes are allowed to feed on the individual suspected of having Chagas' disease, and a few weeks later the insects are examined for the presence of the parasites (Fig. 10.11). Estimates are that approximately 12 million individuals in South and Central America suffer from Chagas' disease.

Chagas' disease was endemic in Uruguay as a result of a lack of screening of blood before transfusion and the presence of kissing bugs in 80% of the houses. Through an intensive program of vector (kissing bug) reduction by both indoor and outdoor spraying, and replacing thatched roofs with metal, infection rates fell to below 0.1%, except in a single area. This was a decline of

FIGURE 10.9 A triatomid "kissing bug," the vector of *Trypanosoma cruzi.* (Author's photo.)

AUTHOR'S NOTE *Figure 10.10 was taken on a field trip to a village on the outskirts of San Jose, Costa Rica, with a high incidence of Chagas' disease and kissing bugs. In one hut, over 50 bugs were isolated from a wooden bed frame, the only piece of "furniture" in the one-room hut. The bugs were examined for the presence of trypanosomes, and most of the insects were positive.*

FIGURE 10.11 Xenodiagnosis. This technique is useful in detecting low levels of parasites in individuals thought to be infected. It has been most commonly used in the diagnosis of Chagas' disease. Laboratory-raised noninfected triatomids feed on a person. The triatomids are then dissected and examined for the presence of the parasite. (Source unknown.)

FIGURE 10.10 Thatched-roof huts. This type of structure is an ideal habitat for triatomids. (Author's photo.)

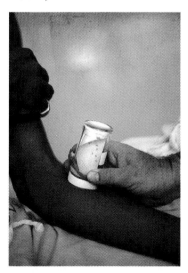

FIGURE 10.12 *Phlebotomus dubosci*, a sand fly vector of *Leishmania* parasites, taking a blood meal through human skin. (Source: Special Programme for Research and Training in Tropical Diseases, WHO. Photographer: Sinclair Stammers.)

80% in only 16 years. Uruguay was declared free of Chagas' disease in 1997, a great triumph of public health measures. Trypanosomes are highly infectious blood-borne pathogens. Laboratory workers and health care providers need to be aware of this and need to exercise appropriate precautions.

Leishmaniasis

Leishmaniasis, a globally widespread disease, is caused by several species of leishmania and is endemic in most tropical and subtropical areas. Transmission occurs by the bite of female **sand flies** that become infected while taking a blood meal from wild and domestic animal reservoir hosts. In the Mediterranean region, including North Africa, jackals and dogs are important reservoirs, and dogs are significant reservoirs in Brazil. In India, there are no animal reservoirs, and the disease is transmitted by sand flies from human to human. Donor blood is not universally screened for the presence of leishmanias, thereby contributing to the problem.

When an infected sand fly bites a human (Fig. 10.12), the parasites enter the skin and are engulfed by cells called **macrophages.** These cells are "professional" phagocytes and are of prime importance to the host immune system (chapter 11). The parasites multiply and are released when a macrophage bursts. The life cycle is illustrated in Fig. 10.13.

There are three human manifestations of leishmaniasis, determined by the particular parasite species, the geographic location, and the host immune response. **Visceral leishmaniasis,** also called **kala-azar,** is the most severe form of the disease, with close to a 100% mortality rate within 2 to 3

FIGURE 10.13 Life cycle of *Leishmania*. In most areas, wild and domestic animals serve as reservoirs. In India, the parasite is transmitted from human to human.

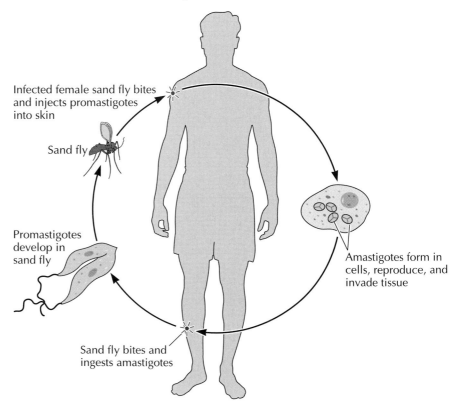

Infected female sand fly bites and injects promastigotes into skin

Sand fly

Promastigotes develop in sand fly

Sand fly bites and ingests amastigotes

Amastigotes form in cells, reproduce, and invade tissue

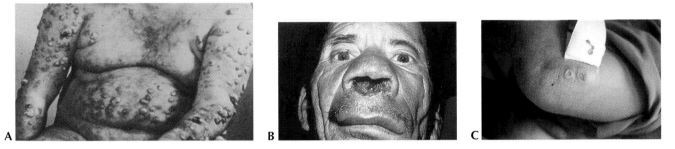

FIGURE 10.14 Cutaneous leishmaniasis. A variety of skin lesions are caused by infection with certain species of *Leishmania*. They are described as dry or moist and diffuse, raised, or ulcerated. (Sources: Armed Forces Institute of Pathology [**A** and **B**]; author [**C**].)

years if not treated. The parasites invade the liver and spleen and cause characteristic symptoms, including irregular bouts of fever, weakness, weight loss, anemia, and protrusion of the abdomen due to the swelling of the spleen and liver. The other two varieties of leishmaniasis are more localized and are seldom fatal. In **mucocutaneous leishmaniasis,** also called **espundia,** the parasites invade the skin and mucous membranes, causing destruction of the nose, mouth, and throat. **Cutaneous leishmaniasis** (Fig. 10.14), also called **oriental sore,** results in mild to disfiguring skin lesions, primarily on exposed parts of the body, particularly the face, arms, and legs. Typically, only a few *Leishmania* lesions appear on an individual, but there may well be over 100.

People with oriental sores rarely get kala-azar; this has led to the practice of self-vaccination on inconspicuous areas of the body. Some parents purposely infect their children with leishmania strains that cause oriental sores as a prevention against getting kala-azar, a potentially fatal form of the disease. Although this type of vaccination may prevent cutaneous lesions, it is not clear that this procedure is effective in preventing kala-azar.

Estimates are that approximately 12 million people in the tropics and subtropics suffer from leishmaniasis. This disease was of concern in military personnel serving in the Gulf War in the early 1990s, since the disease is endemic in the Persian Gulf region. Approximately 50 servicemen contracted the disease.

Malaria

Ask almost anyone to name a tropical disease, and the chances are that they will name **malaria.** Malaria has been around for over 2,000 years, and despite many years and many dollars spent on trying to control this disease, it remains a scourge on humankind. It causes, without a doubt, the world's biggest burden of tropical disease and, with the exception of tuberculosis, kills more people than any other microbial disease (though it will soon be overtaken by AIDS). The term "malaria" is derived from the Italian words *male* (bad) and *aria* (air). In the 1940s, it appeared that malaria was almost eradicated by the spraying of the insecticide dichlorodiphenyltrichloroethane (DDT), but malaria has since returned with a vengeance as a result of the emergence of DDT-resistant, other insecticide-resistant, and antimalarial drug-resistant strains. The disease is not limited to Africa, although the burden of malaria falls heavily on that continent. Gro Harlem Brundtland, director general of WHO, stated, "Malaria is hurting the living

> **AUTHOR'S NOTE** *Figure 10.14 shows different kinds of* Leishmania *lesions. Figure 10.15A was taken on a field trip into the Sinai Desert to search for rodent reservoirs.*

FIGURE 10.15 A search for *Leishmania* rodent reservoirs in the Middle East. **(A)** A typical desert scene. (Author's photo.) **(B)** *Psammomys obesus*, a rodent that can be a *Leishmania* reservoir. (Source: CDC.)

A

B

standards of Africans today and is also preventing the improvement of living standards for future generations. This is an unnecessary and preventable handicap on the continent's economic development." Studies have indicated that Africa's gross domestic product would now be substantially higher if malaria had been eliminated years ago. Here are some facts about malaria that should shock you:

- Annually, more than 1 million people die, and another 300 million to 500 million new cases occur (about two-thirds of them in Africa); 700,000 of these deaths occur in children under 5 years of age.
- Pregnant females have a high mortality rate.
- Forty percent of the world's population lives in countries with malaria.
- About 1,200 cases of malaria are diagnosed in the United States each year, mostly in immigrants and in travelers returning from areas where malaria is endemic, particularly sub-Saharan Africa and the Indian subcontinent.
- Many strains of the malaria parasite are resistant to antimalarial drugs.

Malaria is transmitted by the bite of female *Anopheles* mosquitoes infected with the *Plasmodium* protozoan parasite. The four most common species are *Plasmodium falciparum, P. malariae, P. ovale,* and *P. vivax. P. falciparum* is the most deadly and kills 1 to 2% of those infected. Malaria can also be acquired by shared needles, blood transfusions, and mother-to-fetus transmission.

Many protozoans exhibit complicated life cycles, but the malaria life cycle is one of the most complex, consisting of an **asexual** and a **sexual** stage (Fig. 10.16). The sexual stage occurs in female *Anopheles* mosquitoes, and the asexual stage occurs in the liver and red blood cells of the infected individual. When a plasmodium-infected mosquito bites an individual, the infective forms of the parasite, called **sporozoites,** pass from the salivary glands of the mosquito into the person's bloodstream. Within about an hour, the sporozoites travel to the person's liver, where asexual multiplication takes place, resulting in forms known as **merozoites** which eventually leave the liver. A single sporozoite can produce thousands of merozoites. The merozoites enter **erythrocytes** (red blood cells), where further asexual multiplication occurs. This releases more merozoites into the bloodstream, where they may infect other erythrocytes. If the mosquito bites while the plasmodium parasites are in the blood, the parasites are ingested and develop into male and female **gametes** (sex cells), initiating the sexual cycle. This results in sporozoites that migrate into the mosquitoes' salivary glands. The cycle is now complete.

While the parasites are in the liver, the infected individual usually experiences no symptoms. Bouts of shaking chills and burning fever are associated with malaria and result from the destruction of large numbers of red blood cells accompanying the release of merozoites into the bloodstream. This event is frequently synchronized during an attack, and many new merozoites are released at one time. It is as though the parasites can tell time! In the case of *P. malariae,* chills and fever, resulting from erythrocyte lysis, occur about every 72 hours, while in the other three species, these symptoms occur about every 48 hours. The release of toxins accompanying

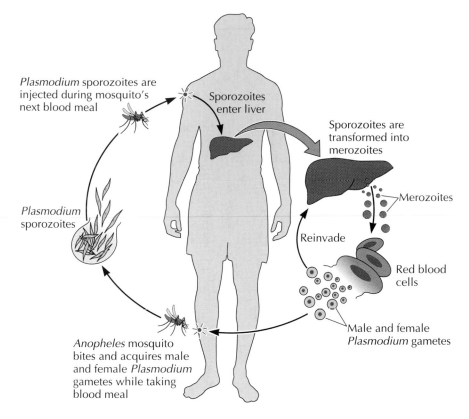

Plasmodium sporozoites are
injected during mosquito's
next blood meal

Sporozoites
enter liver

Sporozoites are
transformed into
merozoites

Plasmodium
sporozoites

Merozoites

Reinvade

Red blood
cells

Male and female
Plasmodium gametes

Anopheles mosquito
bites and acquires male
and female Plasmodium
gametes while taking
blood meal

FIGURE 10.16   Life cycle of the malaria parasite.

the bursting of red blood cells results in the characteristic chills and fevers associated with malaria.

After a person is bitten by a malaria-infected mosquito, symptoms usually appear in 2 to 4 weeks but may not appear for 1 year. Malaria can result in anemia if large numbers of red blood cells are lysed. Muscle aches, fatigue, diarrhea, and nausea may accompany the chills and fever. Several episodes of chills and fever constitute an attack (paroxysm); between attacks the individual feels normal. *P. falciparum* is the most dangerous of the four species, since it can cause **cerebral** (brain) malaria, which may be fatal. Malaria is a major factor in areas where it is endemic, contributing to maternal deaths because it causes severe anemia. Further, pregnant females who are infected with malaria and are positive for human immunodeficiency virus are more likely to infect their unborn child with the virus.

In the early 1900s, malaria was present in the United States, primarily in the Southeast, but massive mosquito abatement programs and a concomitant reduction in the number of human carriers have been effective; the disease is no longer endemic in the United States. Because of an increase in immigration and travel to malarious areas, there are now about 1,000 new cases a year. **Airport malaria** has been described; plasmodium parasites and infected mosquitoes have been known to survive flights from countries where malaria is present and can transmit the disease in the vicinity of an airport.

The diagnosis of malaria is based on the detection of malaria parasites in red blood cells during microscopic examination of a drop of blood smeared

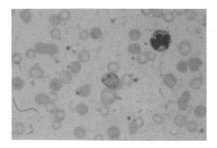

FIGURE 10.17 Blood smear showing a malaria parasite. *Plasmodium* parasites infect red blood cells. The cell in the center of the photo is infected with the parasite; the cell with a red nucleus is a normal white blood cell. (Author's photo.)

onto a slide (Fig. 10.17). Several drugs are available for the treatment and prevention of malaria; the particular drug prescribed is dependent upon the circumstances, including the malaria species identified, severity of disease, and age of the individual. Malaria is a curable disease if promptly diagnosed and treated with appropriate drugs. Drug resistance is a serious problem, and international travelers and travel clinics need to be aware of this and appropriately advised. During January to March 2000, two U.S. citizens died from malaria as the result of receiving malaria-preventive drugs to which the parasite was resistant in the area of their travels (In the News 10.1). Protection of children in several areas of Africa by their sleeping under bed nets regularly treated with recommended insecticides reduced the incidence of child deaths by about 30%. Active research is in progress to develop a malaria vaccine, and researchers are optimistic. The techniques of recombinant DNA technology are particularly promising.

Over 2,000 athletes from Nigeria, Ethiopia, Senegal, South Africa, and Tanzania joined WHO's Roll Back Malaria program in an effort to raise awareness about the disease. Their dedication is based on the realization that rolling back malaria and marathon running share the common challenge of dedication, effort, and endurance.

Babesiosis

Babesiosis, sometimes referred to as "the Babe" by biologists, is caused by the protozoan parasites *Babesia microti* and *Babesia divergens*. **Babesiosis** is transmitted by ticks that become infected by feeding on infected vertebrate hosts (rodents, cattle, and wild animals). The parasites multiply and develop in the ticks and are transmitted to the next vertebrate host, where they invade the red blood cells. As in malaria, the parasite undergoes cycles of multiplication and reinvasion of red blood cells. Based on its similarity to malaria, babesiosis can be easily misdiagnosed as malaria in areas where malaria is endemic. Cases may be asymptomatic, but when symptoms occur, they include severe headache, high fever, and muscle pain. Since red blood cells are destroyed, anemia and jaundice

IN THE NEWS 10.1

Student Who Spent Semester in Africa Dies of Malaria

Chlorine in Swimming Pools Is Not Strong Enough To Kill Parasites

Babesiosis kills two in Massachusetts

Protozoa in the News

can also appear. The symptoms may last for several weeks, and, in some cases, a prolonged carrier state may develop. Almost all untreated cases of *B. divergens* babesiosis are fatal, whereas cases of *B. microti* babesiosis are seldom fatal in healthy persons. Avoiding tick bites is the best means of protection. Babesiosis is distributed worldwide and, in the United States, has been most frequently identified in the Northeast and Midwest.

Several arthropod-borne protozoan diseases have been presented. Although they are most common in tropical and semitropical areas, they remain a constant threat to other areas of the world, including the United States (In the News 10.1). Measures of control center around programs designed to control populations of the arthropod vectors involved and include spraying (Fig. 10.18) and avoidance of standing water in which mosquitoes breed.

Sexually Transmitted Diseases

Trichomonas vaginalis causes **trichomoniasis.** It is estimated that as many as 7 million cases occur annually in the United States and close to 200 million cases occur annually worldwide, making this disease one of the most common sexually transmitted diseases. Rarely, it is also transmitted by contaminated towels and articles of clothing. The human urogenital tract is the reservoir, and many infected persons serve as asymptomatic carriers. The organism has no cyst stage and, therefore, cannot survive for long periods outside the host. It frequently accompanies infection with chlamydia and gonorrhea, both bacterial diseases.

In females, symptoms include intense itching, urinary frequency, pain during urination, and vaginal discharge. In males, the disease may be characterized by pain during urination, inflammation of the urethra, and a thin milky discharge, but most infections are asymptomatic. Diagnosis is accomplished by direct microscopic examination of the discharge and searching for the presence of actively swimming trichomonads. The infection is successfully treated by the oral administration of appropriate drugs, and it is recommended that both sex partners be treated simultaneously to prevent "ping-pong" reinfection.

FIGURE 10.18 Spraying as part of a mosquito abatement program. Mosquitoes are vectors of malaria and a number of other microbial diseases. Their control is an important function of various health organizations. (Source: Public Health Service.)

Diagnosis and Treatment

In the diagnosis of bacterial diseases, the patient's history and symptoms, culturing of bacteria, and serological examination are diagnostic tools. The fact that most bacteria can be easily cultured in laboratories greatly facilitates a definitive diagnosis (chapter 4). Viral diseases are diagnosed by the patient's history and symptoms and by serologic tests; viral isolation (chapter 5) is not routinely carried out. The diagnosis of protozoan (and helminthic) diseases is also based on the patient's history and symptoms, particularly the travel history. Examination of fecal material for the presence of parasites and their eggs or larvae is the major diagnostic tool for protozoans that are transmitted by food and water, while examination of blood smears is diagnostic for the vector-borne blood parasites. In some cases, bone marrow and skin biopsies and examination of spinal fluid may be diagnostic.

Preventive immunization is not available for protozoans, but active research to develop vaccines is under way, particularly for malaria and leishmaniasis. In the case of malaria, some protection can be accomplished by taking certain drugs beginning about 2 weeks prior to travel to countries in which malaria is present. All of the protozoan diseases are, to some extent, treatable by drugs; the success of treatment is based on a number of factors, including the particular parasite, drug resistance, time after exposure, age of the individual, and overall health of the individual. Patients receiving these drugs frequently complain that the treatment may be worse than the disease. This is not surprising, since the cells of both the parasites and their human inhabitants are eucaryotic, and, therefore, drugs toxic for the parasites are also toxic for the host's cells.

BIOLOGY OF HELMINTHS

The worms crawl in,
the worms crawl out,
in your belly,
and out your snout.

ANONYMOUS

This may sound like a cute little nursery rhyme, but in reality it describes a gruesome picture. Live worms of several species have been known to crawl out the mouth, nose, umbilicus (belly button), and anus of infected individuals and have been discharged into toilet bowls and onto the ground, particularly in impoverished countries where poor sanitation is frequent. As is the case with bacteria and viruses, it should be emphasized that the majority of helminths are not parasites, but there are a number of worms which cause disease in humans, in other animals, and in plants.

The presence of harmful bacteria, viruses, and protozoans in the body is a chilling thought, but the presence of worms evokes an even stronger response. The idea of a worm, a eucaryotic, multicellular animal, possibly slithering around within your body is repugnant. All of the microbes discussed thus far are subcellular, procaryotic, or eucaryotic single-celled microscopic biological agents. They may, in fact, be more serious but are not necessarily viewed as loathsome because they are not visible. In other words, what you cannot see may not be so bad, but looking at worms is another story!

Helminthic diseases are major problems in tropical countries and are of lesser concern in temperate climates. Their presence in the United States and in other temperate areas is becoming of increasing concern because of immigration, travel, and AIDS, which increases one's susceptibility to protozoans and *Strongyloides* worms (discussed below). Infection with worms may result in death, but more frequently, worm diseases are chronic and debilitating and may result in entire communities' leading a lower quality of life. In some areas, as many as 90% of the population have at least one type of worm; they are almost like normal fauna. In some cases, their life cycles are very complex and include both larval and adult forms as well as multiple intermediate hosts. The mode of transmission varies with the particular species and may include food and water, arthropods, and direct contact.

The worms are divided into two morphological categories, **roundworms** and **flatworms.** Worms are at the systems level of biological organization (see chapter 2) and possess circulatory, nervous, reproductive, excretory, and digestive systems. Depending on the species, some systems may be lacking or may be rudimentary. The parasitic mode of existence of worms eliminates, or at least greatly minimizes, the need for a sophisticated digestive system, nervous system, and method of locomotion. They are surrounded by absorbable nutrients already digested, have little to react to, have no place to go in the quest for food, and are passively transferred from host to host. On the other hand, their reproductive system is quite developed, ensuring the production of very large quantities of fertilized eggs. In some species, males and females are **dioecious** (the two sexes have different reproductive organs), whereas in others, single animals are **hermaphroditic** and produce both sperm and eggs; cross-fertilization or self-fertilization may occur, depending upon the species.

HELMINTHIC DISEASES

Helminthic diseases are organized by their mode of transmission (Table 10.3), as are bacterial diseases (chapter 8), viral diseases (chapter 9), and protozoan diseases (Table 10.2).

Food-Borne and Waterborne Diseases

Ascariasis

Ascariasis is caused by *Ascaris lumbricoides,* large roundworms that resembles earthworms. These worms set up housekeeping in the small intestine. Females can grow to over 1 foot in length; adult males are 8 to 10 inches long.

Infective eggs are passed onto the soil in the feces of infected individuals. People become infected when they ingest food or water contaminated with the eggs. The eggs are very hardy; they resist drying and thrive in warm, moist soils. Once in the intestine, the eggs hatch into larvae (immature worms), which first penetrate the intestinal wall and begin their journey through the body. They enter blood vessels and are carried along by the blood flow into the heart and then into the capillaries of the lungs, migrate up the respiratory tree, and enter the back of the throat. They are swallowed and returned to the small intestine, where they mature into male and female worms over a period of 2 to 3 months. Female worms produce 200,000 or more eggs per day, and adults live approximately 2 years.

Many people experience no symptoms when infected with *Ascaris*, but, in some, damage to the lungs may occur as a result of the worms migrating

TABLE 10.3 Major helminthic diseases

Organism	Classification	Disease	Transmission[a]	Principal site(s)
Food borne and waterborne				
Ascaris	Roundworm	Ascariasis	Food, water	Intestinal tract, lungs
Trichuris	Roundworm	Whipworm	Food, water	Intestinal tract
Trichinella	Roundworm	Trichinellosis	Pork consumption	Intestinal tract
Dracunculus	Roundworm	Guinea worm	Water	Skin
Taenia saginata	Flatworm	Beef tapeworm	Beef consumption	Intestinal tract
Taenia solium	Flatworm	Pork tapeworm	Pork consumption	Intestinal tract
Vector borne				
Wuchereria	Roundworm	Filariasis	Mosquitoes	Lymph vessels
Onchocerca	Roundworm	River blindness	Blackflies	Skin, eyes
Contact				
Ancylostoma, Necator	Roundworm	Hookworm	Contact	Intestinal tract, lungs, lymph
Enterobius	Roundworm	Pinworm	Contact	Intestinal tract
Schistosoma	Flatworm	Schistosomiasis	Water contact	Skin, bladder, blood vessels
Strongyloides	Roundworm	Strongyloidiasis	Contact	Small intestine, skin, bladder, small blood vessels

[a]Transmission may occur by more than one mechanism.

through the lung tissue, causing a cough and the threat of secondary bacterial infection. Rarely, adult worms may be coughed up or block the pharynx, causing suffocation. Further, malnutrition may result from a large number of worms feeding on intestinal contents (Fig. 10.19). When the worm burden is high, intestinal obstruction and perforation of the intestinal tract may occur, leading to death.

Estimates are that worldwide as much as 25% of the population is infected with *Ascaris,* particularly in tropical and subtropical areas, as a result of poor sanitation and hygiene. In some areas of the United States, up to 50% of the children are infected and are burdened with 5 to 10 worms. Individuals practicing poor hygiene may suffer repeated bouts of infection.

The presence of *Ascaris* eggs or worms in the feces is diagnostic; an adult worm may be coughed up or passed in fecal material. A number of drugs are effective to rid the body of adult worms; they work by temporarily relaxing the worms, resulting in their passage through the anus.

Whipworm Infection

Trichuris trichiura resembles a whip, hence its English name "whipworm." Whipworm is morphologically distinct from *Ascaris* but has a similar life cycle. Eggs passed in fecal material onto moist, warm soil become infective about 3 to 6 weeks after ingestion. The eggs hatch in the small intestine and release larvae that mature, migrate to the large intestine, and take up residence. Following sexual maturation and fertilization, the females begin to

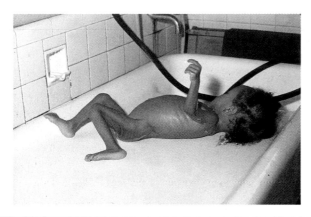

FIGURE 10.19 Malnutrition associated with ascariasis. Lack of food or a poor diet results in malnutrition, a problem particularly present in Third World countries. Most people with malnutrition are infected with one or more species of worms. The swollen belly of the child in the photo is characteristic of malnutrition accompanied by worm infection. (Author's photo.)

lay eggs 60 to 70 days after infection and produce as many as 5,000 eggs, which can be discharged onto the soil, per day. The life span of the adults is about 4 to 7 years.

In most cases, the infection produces no symptoms, but it can cause abdominal pain, diarrhea, and rectal prolapse, particularly in small children. Identification of whipworm eggs in feces is proof of infection. Drugs are available for treatment. Trichuriasis is the third most common roundworm disease of humans and is more frequent in the tropics and subtropics. It is estimated that nearly 300 million people, including some in the southeastern United States, carry these worms.

Trichinellosis

Beware of eating raw or undercooked pork, because it may be infected with *Trichinella spiralis* and cause **trichinellosis.** For that matter, if you like to eat wild game, including bear, boar, walrus, or seal, make sure that the meat is well cooked. (Remember that, the next time you order a walrus sandwich!) When pork or other meat that contains trichinella cysts is ingested, the hard covering of the cysts is dissolved by digestive juices, and worms emerge in the small intestine. Maturation occurs in 1 to 2 days, and, after mating, adult females lay eggs which develop into larvae. Larvae migrate through the blood and lymph vessels and are transported to muscles, including the

AUTHOR'S NOTE *The photos in Fig. 10.20 were taken in a hospital in San Jose, Costa Rica. The cause of death was due to intestinal blockage caused by a massive ball of entangled live and writhing worms. During my professional career, I have witnessed many gross events, but that scene may have been the worst. The autopsy further revealed that a number of worms had migrated into the bile duct and into the liver.*

FIGURE 10.20 Intestinal obstruction due to ascarid worms. **(A)** A tangle of ascarid worms caused intestinal blockage, resulting in death. **(B)** Worms invaded the gall bladder. **(C)** Worms invaded the liver. (Author's photos.)

eye, tongue, diaphragm, and chewing muscles, within which they form cysts. The human host is a dead end for trichinellas, since the person will not be eaten. Infected rats and rodents are fed upon by meat-eating animals, including pigs, which therefore play a major role as reservoirs in maintaining trichinellosis.

Early symptoms of trichinellosis occur within only 1 to 2 days after infection and include nausea, diarrhea, vomiting, fatigue, and fever, followed by later symptoms of headaches, chills, aching muscles, and itchy skin. If the burden of worms is heavy, the individual may experience cardiac and respiratory problems; in severe cases, death may occur. Mild cases may never be diagnosed and are limited to flulike symptoms.

The incidence of trichinellosis has markedly declined through the years. According to the CDC, in the late 1940s an average of 400 cases and 10 to 15 deaths occurred in the United States each year; from 1982 through 1996, the number declined to an average of 57 per year and a total of 3 deaths. The decline reflects a combination of factors, the primary one of which is the passage of laws prohibiting the feeding of **offal** (the organs and trimmings of butchered animals), raw meat, and uncooked garbage to pigs. Therefore, it is highly unlikely that you will get trichinellosis from pigs raised in the United States. Awareness on the part of the public regarding the danger of eating raw or undercooked pork or wild game has also contributed to the decline of trichinellosis. Infection occurs worldwide but is mostly found in areas where raw or undercooked pork and pork products (ham and sausage) are consumed. A muscle biopsy and a blood test can reveal the presence of trichinellosis. Several drugs with limited effectiveness are available to treat this infection.

The admonition about eating undercooked wild game may be more serious than you realize. In October 1999, an outbreak of trichinellosis traced to infected wild boar meat occurred in four adults living in a small town in southeastern France. They had all dined on undercooked barbecued wild boar meat that had been hunted in the Camargue, a swampy region in the Rhone River Delta. (I hope this does not *bore* you!)

Dracunculiasis (Guinea Worm Disease)

The thought of watching a 3-foot worm, the diameter of a paper clip or a piece of thin spaghetti, emerging from a blister on your leg over a 3-month period and causing excruciating and debilitating pain during its emergence is horrifying. You cannot even pull the worm out and endure the agony just to get it over with, because the chances are that a part of it will be left inside your leg, leading to a potentially life-threatening bacterial infection. You cannot "grin and bear it" but can only bear it, and the most you can do is to coil the worm around a small stick as it emerges from the blister and hope that the agony will end in 2 to 3 weeks and not drag on for as long as 3 months. Surgical removal prior to the formation of a blister is possible, but local conditions usually make medical intervention impossible.

What is **dracunculiasis?** The disease is caused by the parasitic roundworm *Dracunculus medinensis* and is present in poor communities in Africa lacking access to safe drinking water. The life cycle (Fig. 10.21) is initiated by drinking water contaminated with dracunculus larvae; small **copepods** ("water fleas") feed on these larvae. Copepods are small crustaceans in the phylum *Arthropoda* (Table 7.5). When water containing infected water fleas is ingested, the copepods are digested but not the worm larvae, which penetrate the stomach and intestinal wall and enter the abdominal cavity. Over

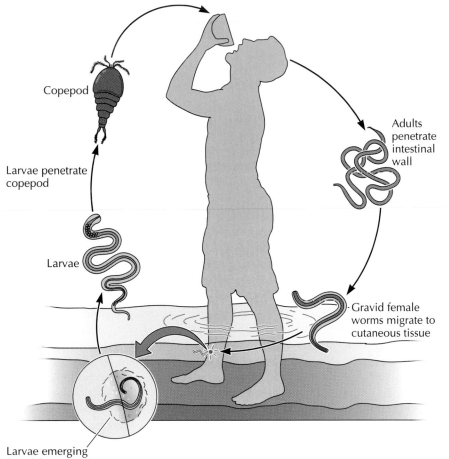

Copepod

Larvae penetrate
copepod

Larvae

Adults
penetrate
intestinal
wall

Gravid female
worms migrate to
cutaneous tissue

Larvae emerging
from blister

FIGURE 10.21   Life cycle of guinea worms.

the next year, maturation into adult worms occurs, copulation takes place, the male worms die, and the females migrate toward the skin surface, most commonly the foot or the leg. The females reach adult size of up to 3 feet in length in about a year. A painful blister occurs at the site, which eventually ruptures, causing a painful, burning sensation; in an attempt to gain some relief, persons stand in water or immerse the blistered area in water. As a result of the temperature change, the blister breaks open, exposing the adult worm, which begins to bore its way out through the skin, cutting the flesh (Fig. 10.22) during its migration and causing terrible pain. The female releases a milky white fluid that contains millions of larvae into the water, a process that continues each time the worm comes in contact with water. The larvae are ingested by copepods and mature in 2 weeks (and after two molts) into infective larvae, completing the cycle.

Symptoms of infection occur about 1 year after the ingestion of water contaminated with infected water fleas. A few days to several hours before rupture of the blister and emergence of the worm, fever and pain and swelling in the area occur. Diagnosis is made clinically and does not require laboratory confirmation. No anthelmintic medication is available to prevent or end infection. Healing at the blister site takes place in about 8 weeks, but secondary bacterial infection and permanent crippling and disability may result.

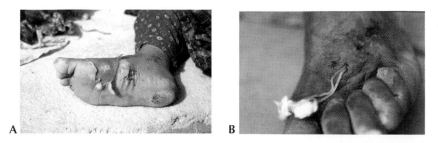

FIGURE 10.22 Guinea worm, the "fiery serpent." **(A)** After a female worm matures, it induces formation of a blister at the site from which it will emerge. **(B)** A worm emerging from a foot, a painful process which can take several weeks. (Source: WHO.)

In addition to the human misery associated with dracunculiasis, the disease poses an economic and financial burden for families. While the worm is emerging, an infected person is often unable to work and to resume daily activities for a few months. Difficulty in walking, loss of appetite, and exhaustion are common symptoms. The inability to farm and to harvest results in heavy crop losses, and children are often kept out of school in order to work in the fields.

Once the larvae are ingested, guinea worm disease cannot be cured. Prevention is based on education of villagers in affected communities, many of whom are illiterate, to practice the simple act of filtering their water through cloth or nylon filters to rid the drinking water of copepods. Additionally, people need to avoid seeking relief from the burning sensation of emerging worms by immersing their feet or other body areas in reservoirs of drinking water.

Guinea worm disease is on a path toward eradication as a result of the collaboration of the public and private sectors that serve as a model for eradication of other diseases. A coalition of organizations, initiated by the Global 2000 Program of the Carter Center and the CDC, in concert with WHO and UNICEF, the World Bank, numerous bilateral and multilateral agencies, private corporations, and foundations, is working toward the eradication of guinea worm disease. In May 2000, at the 53rd World Health Assembly (the highest governing body of the WHO), the Bill and Melinda Gates Foundation awarded a grant of $28.5 million to accelerate the eradication of dracunculiasis. Former President Jimmy Carter took the initiative in 1986 in marshaling global resources against dracunculiasis and was knighted in Mali, Africa, in April 1998 in recognition of his efforts (Box 10.1). The eradication campaign reduced the number of reported cases by 97%, from an estimated 3.2 million in 1986 to fewer than 100,000 in 1999. The number of countries with guinea worm disease has been reduced from 20 to 13, and Asia is now reported to be free of the parasite. Education about filtering drinking water, avoidance of entering water when worms are emerging, and treating water sources with compounds to kill copepods is important in the eradication of guinea worm. Further, the provision of clean drinking water is essential.

Tapeworms

People who complain that they cannot gain weight sometimes say, "I must have a tapeworm." There is no scientific basis for that statement. (Most people have the opposite problem of gaining too much weight!) There are

BOX 10.1 Former President Jimmy Carter and Guinea Worm Disease

It is now former president *Sir* Jimmy Carter. On April 3, 1998, former U.S. president (1977–1981) Jimmy Carter was knighted in Mali, western Africa, for recognition of his work toward the eradication of guinea worm disease. In 1986, the Carter Center launched the Global 2000 Program with an effort against the crippling malady of guinea worm disease in Pakistan and in Ghana. The center continues to lead the fight against this parasite. During its 16-year battle against guinea worm, reported cases have been reduced from 3.5 million worldwide to about 60,000. The African country of Sudan is a remaining stronghold of the disease because of a long-standing civil war, which has interfered with efforts to combat the disease. In other parts of Africa, eradication is at hand; WHO has certified India and Pakistan as having eradicated guinea worm.

In March 1995, Carter and Sudan's President Omar al-Bashir announced a cease-fire, with the Sudanese People's Liberation Movement/Army (SPLM/A) and the South Sudan Independent Movement/Army (SSIM/A) signing on a few days later. The announcement came during a trip to Africa by former president Carter and his wife, Rosalynn, to assess progress toward guinea worm eradication. In late May, the government, the SPLM/A, and the SSIM/A extended the cease-fire for another 2 months. Carter said, "The primary purpose of the cease-fire is to permit the leaders and citizens of Sudan, working with others, to carry out a major effort to eradicate guinea worm disease, prevent river blindness, and immunize children against polio and other diseases." This historic cease-fire negotiated by former president Carter allowed health care workers access to previously inaccessible remote areas. The agreement is the longest cease-fire ever negotiated to fight disease and to implement public health programs.

Within a few months after the cease-fire began, the Carter Center announced that the following had been accomplished:

- visits to 2,253 villages where guinea worm disease was present
- distribution of 115,425 cloth filters to households
- vaccination of 34,481 children for polio and 40,000 children for measles
- administration of vitamin A supplements to 35,000 children
- treatment of 9,031 children with oral rehydration therapy for diarrhea
- delivery of 200,000 ivermectin tablets, donated by the American pharmaceutical company Merck & Co. for distribution in the field to treat river blindness

The staff of the center focuses on educating the villagers not to drink unfiltered water and to filter their water and ensures the availability of filters to do so. Folklore prevails, and some villages believe "the fiery serpent" to be the result of sorcery. An important part of the eradication program is to apply frequent and regular larvicide treatments to stagnant water. The three major ingredients for successful eradication are political will, financial support, and technical expertise, all provided by the Carter Center and its global partners. The center estimates supplies needed to treat guinea worm in 5,000 villages as follows:

- 6,000 forceps
- 6,000 scissors
- 30,000 tablets of painkiller
- 6,000 plastic bags
- 6,000 gauze pads

Additionally, the center's partners have provided filter cloth valued at more than $14 million and larvicide valued at more than $2 million. The Carter Center's guinea worm eradication program has received a huge boost from Johnson & Johnson and Home Depot. Johnson & Johnson is donating enough medical supplies to assemble 6,000 health kits to be used in the treatment of guinea worm disease. Home Depot, a longtime Carter Center partner, is contributing storage facilities, shipping supplies, and volunteers to assemble the kits before they are shipped to Africa.

President Carter said, "This generous support from Johnson & Johnson and Home Depot will enable us to supply healthy kits to all remaining villages while arming the Carter Center and village health workers for the final assault on guinea worm disease....With less than 3% of the original caseload left to tackle, this significant new corporate commitment from Johnson & Johnson provides critical momentum for total eradication."

In 1986, more than 3.2 million people in Africa and Asia were afflicted with guinea worm disease, and another 1.2 million were at risk of infection. Guinea worm is a parasitic disease, recorded as far back as 3,000 years ago, that rarely makes headlines but is so painful and debilitating that victims are unable to work, attend school, care for children, or harvest their crops. There is no cure or vaccination for guinea worm; only health education combined with straining water through a special nylon filter, providing clean water from a borehole well, or chemically treating ponds can prevent the disease. The gauze and other supplies provided in the medical kits will ease the pain and prevent contamination of water sources. The Carter Center continues to spearhead eradication efforts in the 13 African countries still afflicted.

several species of tapeworms, but the two most common are *Taenia saginata* (the beef tapeworm) and *Taenia solium* (the pork tapeworm). The life cycles of these two parasites are similar and are initiated by the ingestion of undercooked beef or pork containing encysted larvae. The beef tapeworm is the larger of the two and may reach 20 to 25 feet in length, while the pork tapeworm reaches approximately 15 to 20 feet in length. It is awesome to think that individuals with tapeworms might have a 20-foot animal dangling into their intestinal tract! People become infected by eating meat (muscle) containing larvae. The larvae have been encysted into forms known as **cysticerci.** Upon ingestion, the cysticerci are digested except for the **scolex** (the head), which attaches to the intestinal wall by suckers. The scolex produces compartmentlike segments known as **proglottids,** each of which is a reproductive bag containing both testes and ovaries. The tapeworm body continues to lengthen as new proglottids are produced, and the worms reach adult size. Each worm has anywhere from 1,000 to 2,000 proglottids, which mature and contain 80,000 to 100,000 eggs per proglottid. Proglottids detach from the worm and migrate to the anus, and approximately six per day are passed in the stool. The eggs can survive for long periods in the environment; cattle and other **herbivorous** (plant-eating) animals become infected by grazing on contaminated vegetation. Larvae hatch from the eggs, penetrate through the intestinal wall, and become encysted as cysticerci in the muscle of the animal. When humans (or other animals) ingest these cysticerci, the cycle is completed.

Both *T. saginata* and *T. solium* produce only mild abdominal symptoms. Laboratory diagnosis is made on the identification of eggs and proglottids in the feces. It takes about 3 months following infection for adult tapeworms to develop; therefore, laboratory confirmation is not initially possible. Eggs of *T. saginata* and *T. solium* cannot be differentiated microscopically, but examination of proglottids and, rarely, the scolex allows species identification.

Both species are worldwide in distribution. *T. solium* is more common in poor communities where people may live in close contact with pigs and eat undercooked pork. Medication is available for the treatment of tapeworms.

A serious disease called **cysticercosis** can result from the ingestion of pork tapeworm eggs (as opposed to ingesting cysticerci). These eggs are passed in the feces of an infected person and may contaminate food, water, or surfaces. Ingesting contaminated water and foods or putting contaminated fingers into the mouth may result in cysticercosis. Those infected with pork tapeworm can reinfect themselves by poor sanitary habits. The tapeworm eggs hatch inside the stomach, penetrate the intestine, and travel through the bloodstream; ultimately, they may cause cysticerci in the muscles, brain, or eyes. Lumps may be present under the skin, and infection in the eyes may cause swelling or detachment of the retina. **Neurocysticercosis** (cysticerci in the brain or spinal cord) is serious and may cause death. Seizures and headaches are the most common symptoms and can occur months, or even years, after infection. Antiparasitic drugs are available to treat cysticercosis, and their use is dependent upon the number of lesions judged to be present and the individual's symptoms.

Figure 10.23 illustrates the common practice of hanging sides of beef in open, unrefrigerated butcher shop stalls in some countries, and Fig. 10.24A shows a tapeworm-contaminated piece of beef from a Costa Rican butcher's stall. In the United States, the Department of Agriculture is responsible for the inspection of beef that is intended for human consumption.

FIGURE 10.23 A butcher shop stall in a Third World country. The unrefrigerated pieces of meat are open to the environment and contamination. (Source: WHO.)

You may be dismayed to learn that there are other species of tapeworms that can be acquired by humans. Fish tapeworm infections are common in Scandinavia, Russia, and other areas of the world, including the Great Lakes area of the United States. The disease is acquired by humans through the ingestion of raw or undercooked contaminated fish. Sushi eaters beware (Box 10.2 and Fig. 10.25)!

Arthropod-Borne Diseases

Lymphatic Filariasis (Elephantiasis)

Figure 10.26 is a dramatic example of **elephantiasis.** (The term "elephantitis" is not correct, but as illustrated, the grotesque swelling in the legs is elephantlike, and the texture and the appearance of the skin resemble the hide of an elephant.) In the figure, note the swelling in the legs and hands.

What causes elephantiasis and brings about this deformity? The disease is caused by tiny adult threadlike worms, called **filarial worms,** which block the **lymphatic vessels** and cause the accumulation of large amounts of lymphatic fluid in the limbs, particularly in the legs. In males, the

FIGURE 10.24 Tapeworm cysticerci. **(A)** A tapeworm larva-contaminated piece of beef in a butcher shop in Costa Rica. **(B)** Tapeworm cysticerci resemble small bladders. (Author's photos.)

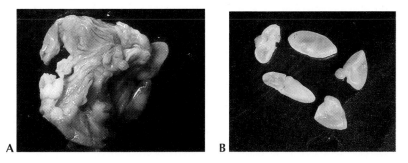

A   B

BOX 10.2 Sushi Eaters Beware

Sushi: either you like it or you hate it! Sushi lovers the world over enthusiastically devour these tasty and exquisitely presented little cakes of rice topped with a filet of raw fish and can hardly wait to return another day to brandish their chopsticks. On the other hand, non-sushi eaters grimace and shudder at the thought of putting a piece of raw fish in their mouths; they find the appearance and texture disgusting. Like it or not, we are living in the midst of a sushi craze. Sushi bars can be found in just about all large cities and small ones, too. There is even a sushi bar in Las Cruces, N.Mex.! Platters of these delicacies, creatively arranged and presented, are found on buffet tables at weddings, anniversary parties, and other festive occasions. You can even buy freshly made sushi at many local markets.

The making of sushi dates back to at least the 1600s; it was introduced in Southeast Asia as a method of preservation of fish. The fish were packed with rice, which fermented and produced lactic acid, resulting in their being pickled. The modern version of sushi, sliced raw fish served on a cake of rice, is credited to Hanaya Yohei in 1824. Why is sushi so popular? By no means is it inexpensive; a single little tidbit will cost you anywhere from $2.50 to $6.00, or maybe even more if you crave the exotic. Sushi fits in well with the current emphasis on reducing meats

and carbohydrates in our diets. It is rich in protein and in a variety of minerals, and the vinegar in sushi rice is claimed to have antibacterial properties, reduce fatigue, and reduce blood pressure. Nutritionists recommend two to three servings of fish per week, since fish is an excellent source of omega-3 fatty acids, substances that are limited in the typical American diet. These fatty acids are thought to be effective in the control of heart disease, rheumatoid arthritis, cancer prevention, and maybe even depression. In short, sushi, according to sushi lovers, is a health food.

What about the dangers of acquiring worms from eating raw fish? It is true that there are risks, not so much in the fish itself, but in the process of handling the fish, as is the case for all food items. Sushi chefs appear to be meticulous and frequently wash their hands, counters, and utensils. Statistics indicate that about 2,000 Japanese acquire worms each year from raw fish; statistics are not kept in the United States. *Diphyllobothrium latum* is a fish tapeworm that can reach 20 feet in length; it grows at a fantastic rate in a person's intestinal tract. More commonly, *Anisakis simplex*, a slender half-inch-long worm, may be found in sushi. You can even buy a key chain with this worm embedded in plastic at the museum store in the Parasitological Museum in Tokyo! The presence of these worms can cause abdominal pain that

is frequently misdiagnosed as appendicitis, resulting in numerous unnecessary appendectomies. Adding to its miserable reputation, the worm excretes waste from its face and has a borer-like tooth that enables it to drill holes into the intestinal wall. Fortunately, the human is a dead-end host, resulting in the death of the worm after several days. On the other hand, if the *Anisakis* worm is lucky enough to be swallowed by a seal, it is able to complete its life cycle. The fish used for sushi are usually frozen, frequently on the deep-sea fishing boats, to keep them from spoiling on the long trip home; it is a good practice because freezing kills the parasites, affording a measure of protection to potential sushi eaters. For this reason, many sushi restaurants routinely freeze their fish before its appearance on cakes of rice.

So the bottom line is this: choose your sushi restaurant carefully (as you should any restaurant), and then go ahead and order whatever strikes your fish fancy. You might choose the shiromi (whitefish), hamachi (yellowtail), uni (sea urchin), or ikura (salmon roe). The biggest danger is that you might ingest too much wasabi and feel like you are choking to death! For the uninitiated, wasabi is made from horseradish, and a minute amount is used on the sushi. It is even more potent than curry and will cause gasping, bring tears to your eyes, and burn your lips. Enjoy!

FIGURE 10.25 Sushi (cakes of rice adorned with raw fish). It is highly unlikely that the fish is contaminated with worms. (Author's photo.)

scrotum and the penis may be involved, and scrotal swelling can be so severe that a wheelbarrow is used to support the scrotum to allow limited walking. In females, the breasts may become considerably enlarged. Whatever the areas affected, they may swell to several times their normal size. The lymphatic vessels function in the body's fluid balance and return **lymph** back into the circulation. Hence, blockage of these vessels will result in a "backup" of lymph, causing swelling. Individuals with filariasis are prone to bacterial infections.

Wuchereria bancrofti and *Brugia malayi* are the two major species responsible for lymphatic filariasis. They are transmitted by mosquitoes. The cycle of infection is initiated by the bite of mosquitoes carrying filarial larvae. The larvae migrate into the lymphatic vessels, where, within about a year, they grow into adult worms with a life span of up to several years. The adult worms mate, and the females release into the blood millions of **microfilariae** which may be ingested by mosquitoes taking a blood meal. The mosquitoes transmit the microfilariae to other humans, completing the cycle.

The early symptoms are chills, fever, and inflammation of lymph nodes and lymphatic vessels; some individuals remain asymptomatic and may never develop elephantiasis. Despite the elephantlike deformities that may develop, the disease is not usually fatal. WHO ranks filariasis as the second leading cause of permanent and long-term disability worldwide. The disease carries a social stigma because of its extreme disfigurement, and individuals are frequently shunned to the point that they are neglected by their families. Their disability may cause them to be unable to work. Filariasis is most common in tropical and subtropical areas and is fostered by poor sanitation and population growth, which create increased breeding areas for mosquitoes. (Population growth as a factor in emerging and reemerging infections is discussed in chapter 1.)

Examination of blood samples allows the identification of microfilariae. In patients with elephantiasis, swollen limbs are wrapped in compressive bandages in an effort to force lymph from them to reduce discomfort and swelling.

Filariasis affects 120 million people in 73 countries where it is endemic, including numerous countries in Africa, Asia, and the western Pacific and at least seven countries in the Americas. More than 41 million of these cases occur in Africa. Further estimates are that debilitating genital swelling occurs in 25 million men and that elephantiasis of the leg occurs in 15 million persons. Further, an additional 900 million people are estimated to be at risk for filariasis; approximately one-third live in India. About half of the people with lymphatic filariasis have overt clinical disease; the remainder harbor infections with millions of microfilariae and dozens of adult worms in their bodies but with internal damage undetected and untreated. But there is hope. In 1993 the International Task Force for Disease Eradication evaluated 94 infectious diseases for the feasibility of eradication, and lymphatic filariasis is one of only six diseases considered potentially eradicable. At the 50th World Health Assembly in May 1997, a resolution proposing the elimination of lymphatic filariasis was passed. New advances in diagnosis and the availability of simple, safe, inexpensive, and conveniently delivered drugs that kill the microfilariae and adult worms are reasons for optimism. As a result of an effective single-dose treatment, global elimination of lymphatic filariasis is now possible. In January 1998, the pharmaceutical company SmithKline Beecham announced a massive donation program of several billion doses of the drug **albendazole** in support of this effort. The Carter Center's Global 2000 Program is working to help eliminate lymphatic filariasis in Nigeria, the African country with the greatest number of persons infected. This collaboration, like that described for guinea worm, is a wonderful example of the cooperation of the private and public sectors. (River blindness, to be discussed next, is still another example.)

Onchocerciasis (River Blindness)

Figure 10.27, a dramatic statue entitled *A Little Child Shall Lead*, depicts a young child leading his blind elder by the end of a long stick. The blind adult is a victim of **onchocerciasis,** also called **river blindness,** the world's second leading cause of blindness; it exists in many parts of Africa and Central America and is caused by *Onchocerca volvulus,* a parasitic worm that can

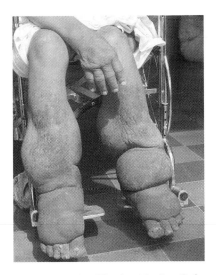

FIGURE 10.26 Elephantiasis. Infection with filarial worms can block lymphatic vessels, causing severe edema (swelling). (Author's photo.)

FIGURE 10.27 *A Little Child Shall Lead.* Infection with the parasitic worm that causes onchocerciasis can result in blindness. In some African villages, a child leading a blind person is a common sight. Onchocerciasis is also called "river blindness" because the blackfly vector lives in rivers and the disease is most common among people who live near rivers. This photo was taken at WHO headquarters in Geneva, Switzerland. Replicas of the statue stand at the Carter Center in Atlanta, Ga., and at the headquarters of the World Bank in Washington, D.C. (Author's photo.)

live up to 14 years in the body. Onchocerciasis can devastate entire communities. In many small villages, nearly all of the villagers past the age of 40 are blind, presenting a serious obstacle to socioeconomic development.

The disease vector is a blackfly which abounds in fertile river banks and breeds in fast-flowing rivers and streams. When the fly bites, larval worms are deposited under the skin. The larvae mature into adult worms about a year later and live for as long as 14 years. They produce millions of microfilariae (microscopic larvae) that migrate throughout the body, causing rashes, intense itching, depigmentation of the skin, and nodules which form around the adult worms. Some migrate into the eyes and invade the cornea, causing blindness. These manifestations begin to occur about 1 to 3 years after injection of the larvae by the flies. The cycle (Fig. 10.28) is perpetuated when a biting blackfly ingests microfilariae from the blood of an infected person while taking a blood meal and then injects larvae into a new host.

Diagnosis is difficult, and there is no blood test to detect the presence of microfilariae. The presence of nodules under the skin is diagnostic, and their removal by minor surgery reduces the number of microfilariae produced. Several anthelmintic drugs are available; ivermectin is particularly effective as it kills migrating microfilariae. In practice, ivermectin is the only drug suitable for treatment, since other drugs cause severe reactions.

WHO estimates that 123 million people worldwide, of whom 96% are in Africa, are at risk for onchocerciasis. Over 6.5 million suffer from severe itching and skin disease, 270,000 are blind, and 500,000 are severely visually impaired. The disease is endemic in 37 countries, of which 30 are in sub-Saharan Africa and 6 are in the Americas.

FIGURE 10.28   Life cycle of *Onchocerca volvulus*.

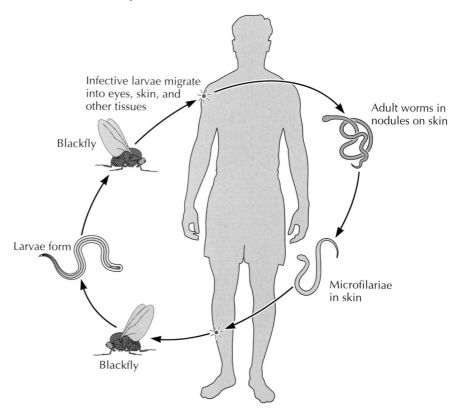

Infective larvae migrate into eyes, skin, and other tissues

Blackfly

Larvae form

Blackfly

Adult worms in nodules on skin

Microfilariae in skin

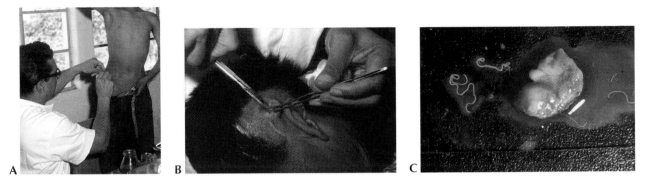

FIGURE 10.29 Onchocerciasis intervention. A bite by an infected blackfly causes the formation of a nodule containing the larval form of the worm. Migration of worms into the eyes can result in blindness. **(A)** A health care worker examining nodules; **(B)** surgical removal of worms from nodules; **(C)** appearance of worms. (Author's photos.)

Until the 1980s, control focused on the use of chemicals to kill immature blackflies in rivers. A renewed and major effort to eliminate river blindness was initiated in 1987 when the drug **ivermectin** was licensed by Merck & Co. and provided free of charge for treating the disease. The Carter Center, the Lions Club Sight First Project, the CDC, and Merck are collaborating in an effort to curb river blindness.

AUTHOR'S NOTE *The photos in Fig. 10.29 were taken on a field trip in Costa Rica. Figure 10.29B demonstrates nodules under the skin which were easily removed by minor surgery. I took my turn with the scalpel.*

Direct-Contact Diseases

Hookworm Disease

In the United States, rural areas of the South were, at one time, burdened with **hookworm disease,** a disease that insidiously saps the strength of those parasitized, leading to the stereotype of southerners as being slow, lazy, ignorant, and lacking in ambition. Thanks to the pioneering work of the Rockefeller Sanitary Commission, the incidence of this major parasitic disease markedly declined as the result of improved sanitation, particularly the installation of indoor plumbing, and a better way of life, including the wearing of shoes. Hookworm disease is caused by two roundworm species, *Ancylostoma duodenale* (the Old World hookworm) and *Necator americanus* (the New World hookworm). *N. americanus* was widespread in the southeastern United States, and *A. duodenale* is found in southern Europe, northern Africa, northern Asia, and areas of South America.

Infection occurs by direct contact with soil contaminated with hookworm larvae while walking barefoot, touching soil with bare hands, or accidentally swallowing bits of contaminated soil (Fig. 7.7). The mature adult worms live and mate attached to the wall of the small intestine (Fig. 10.30) and produce thousands of eggs, which are passed in the feces. Eggs require warm, moist, shaded soil and hatch in 1 to 2 days into barely visible larvae. After 5 to 10 days, the larvae reach the infective stage and penetrate the skin; they are carried to the lungs, where they are coughed up and swallowed and reach the intestinal tract, in about a week. In the small intestine, the larvae develop into half-inch-long adult worms which produce eggs that ultimately reach the soil, completing the life cycle (Fig. 10.31). The cycle takes about 5 to 6 weeks from larval invasion to egg deposition by female worms. Person-to-person contact is not possible, since development of the larvae in soil is part of the life cycle.

FIGURE 10.30 Hookworm in wall of intestine. (Author's photo.)

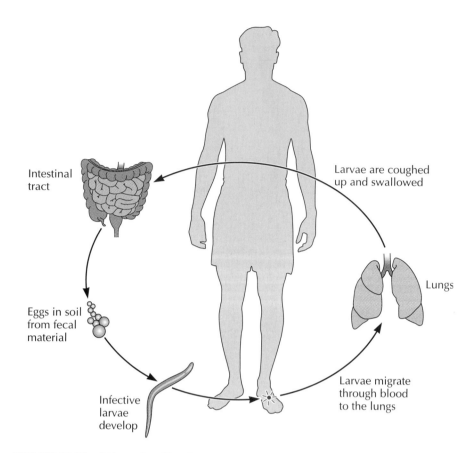

Intestinal tract

Larvae are coughed up and swallowed

Lungs

Eggs in soil from fecal material

Infective larvae develop

Larvae migrate through blood to the lungs

FIGURE 10.31   Life cycle of hookworms.

The adult worms have mouth parts, called **biting plates,** by which they attach and suck blood, causing anemia, protein deficiency, and fatigue; in children, continual infection can cause retarded growth and mental development as the direct result of blood loss. Serious infection can be fatal in infants and causes children to be malnourished. The severity of symptoms appears to be directly related to the hookworm burden. Fewer than 25 worms do not usually cause disease, while 500 to 1,000 worms result in severe and often fatal damage.

The CDC estimates that about 20% of the world's population is burdened by hookworms, particularly in tropical and subtropical climates where about 1 billion individuals carry hookworms. Diagnosis of hookworm is confirmed by the presence of eggs in a fecal sample. Medication is available for treatment.

Strongyloidiasis

Strongyloidiasis is caused by the nematode *Strongyloides stercoralis*, sometimes called the threadworm because of its minute size. Around 800 million people worldwide are burdened with this parasite. *S. stercoralis* is unique in that the females produce eggs by **parthenogenesis,** a process which does not require fertilization by males (tough luck, boys). Other animals, including some insects, amphibians, and reptiles, share this unusual biological property. *S. stercoralis* is also unique in that it exhibits two alternative life cycles. The **parasitic cycle** is completed within the host. Infected larvae in the soil penetrate the skin and are sequentially transported by blood to the heart, lungs, and bronchial tree and into the pharynx, where they are coughed up and swallowed, finally reaching the small intestine. There, they mature into adult female worms and produce eggs parthenogenetically. Some eggs develop into larvae which penetrate the small intestine, or the skin around the anus, and repeat the cycle through the heart, lungs, bronchial tree, and pharynx and to the small intestine where they mature into adult worms. This repeated cycle results in internal **autoinfection,** a process unique among worms in humans. Autoinfection accounts for persistent infections for over 20 years in infected persons who have moved to areas where the parasite is not present. Further, autoinfection can result in hyperinfection, characterized by the dissemination of the parasite within the body.

The free-living cycle is an alternative strategy. It results from the passage of eggs in the feces that can develop into free-living worms that produce large numbers of infected larvae. The free-living cycle is an important reservoir of human infection.

Strongyloides infection may be mild and go unnoticed, but it can be severe, resulting in nausea, vomiting, anemia, weight loss, and chronic bloody diarrhea. In 1958, a Japanese biologist infected himself by the intradermal injection of larvae in order to follow the course of the disease. The severity of infection appears to be related to the number of worms, the site of involvement, and the immune status of the host. In individuals who are immunocompromised, large numbers of invasive larvae can cause extensive damage leading to death.

As with many worm infections, definitive diagnosis is established by the presence of larvae in the feces and duodenal contents. A reliable blood test to detect antibodies to *Strongyloides* has been developed. Drugs are available for the effective treatment of strongyloidiasis.

Enterobiasis (Pinworm Disease)

Pinworm disease is the most common helminthic disease in the United States, with an estimated 40 million people infected according to the CDC. *Enterobius vermicularis,* about the length of a staple, is the culprit.

The life cycle is a simple one. The adult male and female worms live and mate in the cecum. At night, while the infected person sleeps, egg-bearing females exit through the anus, deposit eggs around the skin in the anal area, and then die. As many as 15,000 eggs may be deposited. Egg laying is triggered by the lower temperature outside the body and by other factors. Infective larvae develop within the eggs in about 4 hours. The eggs are swallowed and hatch in the intestine, releasing larvae that mature into adults, and become established in the intestine. The interval from ingestion of larvae to egg-producing females is about 1 month; the life span of the adults is about 2 months.

Self-infection is a common mode of transmission, particularly in children. Pinworms cause an intolerable itch, resulting in intense scratching. Eggs cling to the fingers, are lodged under the fingernails, and are then swallowed, reinitiating infection. Further, an infected person can transfer eggs through direct contact, on bedclothes, and even by inhalation of airborne eggs.

Anal itching, restlessness, and loss of appetite are common and relatively mild symptoms; the disease is rarely debilitating. School-age children have the highest rate of infection, and in some child care centers and other institutional settings, over 50% of children are infected. If one member of a family has pinworm, there is a good chance that other family members are infected.

Diagnosis is based on the "Scotch tape test"; a piece of tape is applied to the anal region during the night or early morning hours. If eggs are present, they stick to the tape and are identified under a microscope. Treatment involves a two-dose course of an anthelmintic drug; the second dose should be given 2 weeks after the first. It is usually wise to treat all members of a family to break the cycle. According to the CDC, cleaning and vacuuming the entire house and washing sheets every day are probably not necessary or effective.

Schistosomiasis

Two hundred million people are infected with **schistosomiasis** in 74 countries of Africa, South America, the Caribbean, the Middle East, and the Far East. It is caused by *Schistosoma japonicum, Schistosoma haematobium,* or *Schistosoma mansoni;* these worms are frequently called **blood flukes.** Certain species of snails serve as intermediate hosts in these geographical regions; they release immature, free-swimming microscopic forms, called **cercariae,** resembling miniature tadpoles, that can survive in the water for about 48 hours. Exposure to water (Fig. 10.32) containing cercariae allows their penetration into the skin. This penetration is followed by migration into blood vessels and, in about a week, entrance into the liver, where the organisms achieve sexual maturity. The adults have evolved a clever strategy of coating themselves with antigens of the host, thereby escaping detection by the host's immune system. Female worms reside in a groove in the body of the larger male, and the two remain entwined as a (happy) mating pair for up to 10 years. The pair migrate into small blood venules (veins) of the intestinal tract or of the bladder, depending on the particular schistosome species. Ancient Egyptian writings describe males with blood in

FIGURE 10.32 Schistosomiasis. Larval forms called cercariae are present in water and penetrate the skin, initiating an infection. **(A)** Two boys wading in water despite a warning sign; **(B)** a boy with a swollen belly resulting from schistosomiasis. (Source: CDC.)

A

B

their urine as "menstruating"; possibly, this was the result of schistosomes in their urinary tract. The worms feed upon blood, and the females produce eggs within several weeks; the eggs are discharged by infected persons defecating or urinating into the water. The eggs hatch into actively motile larvae that penetrate the snails and develop into cercariae, completing the cycle.

Itchiness develops at the penetration site within several days and may be followed by fever, chills, cough, and muscle aches within the next several weeks, but some people experience no symptoms during this early stage. The major consequence of schistosomiasis is a result of the presence of eggs in the liver, intestine, or bladder. The eggs cause allergic reactions, contributing to the damage caused by these blood parasites. More rarely, eggs may move into the brain or spinal cord, causing seizures, paralysis, or inflammation of the spinal cord. The adult worms can damage the spleen, liver, lungs, intestine, and bladder, causing the abdomen to become distended (Fig. 10.32B). The adult worms escape detection by the immune system and thus can survive for years.

Diagnosis is confirmed by the presence of eggs in the urine or stool. Safe and effective drugs are available for treatment, but they are relatively expensive. Prevention through improved sanitation and control of the snail vectors is a more efficient and cost-effective approach. In some areas of Africa, biological control to reduce the cercarial population is being practiced by the introduction of a species of small fish that feeds on snails.

Chapter 1 cited technological development as a causative factor of new

BOX 10.3   An Outbreak of Cercarial Dermatitis (Swimmer's Itch)

On October 28, 1991, an employee of the Division of Public Health (DPH), Delaware Department of Health and Social Services, reported that her son and at least 10 other persons who had recently participated in a high school biology class field trip to Cape Henlopen State Park had contracted pruritic (itchy) dermatitis. The Delaware Health Monitoring and Program Consultations Section conducted an investigation to confirm the reported presumptive diagnosis of cercarial dermatitis and to assess the extent of this outbreak.

On October 19, 1991, 37 students aged 13 to 16 years and their teacher spent about 3 hours wading at low tide in a shellfishing area. The weather was sunny and unseasonably warm; the air temperature was 80°F. Within the next 9 days, 29 students developed pruritic dermatitis; 11 visited their physicians for treatment.

On November 7, all of the students and their teacher were examined and interviewed by staff from the DPH. A case was defined as pruritic dermatitis that developed in a person after ≥12 hours of seawater exposure. Of 37 persons who had had contact with seawater, 30 (81%) met the case definition of cercarial dermatitis.

On November 7 and 8, mudflat snails were collected from the site of the field trip. Snails were isolated in culture dishes and examined after 24 hours for the presence of released cercariae. Two specimens were found in a dish containing two snails. Their morphology was consistent with that of an avian schistosome implicated as a causative agent of cercarial dermatitis.

The DPH recommended to park officials that they advise persons wading in the shellfishing area to wear protective clothing (e.g., long pants, hip boots, and waders) and that this advisory should remain in effect until December 31, when cold-water temperatures would reduce unprotected water contacts.

Note
Outbreaks of cercarial dermatitis occur unpredictably, and routine surveillance for cases and outbreaks is impracticable. A timely public health response depends on early recognition of cases by local physicians who diagnose or suspect cercarial dermatitis. Appropriate preventive measures include posting signs that warn of the risk of contracting cercarial dermatitis in the affected area and recommending protective clothing and curtailment of water-related activities.

Source: Adapted from R. Wiley, D. Wolfe, R. Flahart, C. Konigsberg, and P. R. Silverman, *Morbidity and Mortality Weekly Report* **41**:225–228, 1992.

FIGURE 10.33 "Man's best friend." Dogs and other pets can be sources of infection with worms, bacteria, and viruses. Good sanitation and hygiene and proper care of pets are keys to minimizing the risk. (Author's photo.)

and emerging infections and referred to the building of the Aswan Dam in Egypt in the 1960s. An unfortunate and unpredictable consequence of this project was the significant increase in schistosomiasis because of the favorable conditions created for the intermediate snail hosts. Schistosomiasis does not occur in the United States because the required intermediate snail host is not present.

A condition known as **swimmer's itch,** or **cercarial dermatitis,** occurs sporadically and unpredictably in certain localities in the United States (Box 10.3), particularly in the northern lakes, and in Mexico and Canada. Swimmer's itch is due to penetration of the skin by cercariae from eggs of water birds infected with the avian species of schistosomes; humans are an accidental host, and development does not proceed beyond the skin. The body's immune system destroys the cercariae, and, in the process, allergenic products are released which are responsible for the itching and rash.

Travelers to countries in which schistosomiasis is endemic need to be aware of this disease and to avoid swimming or wading in fresh water in which schistosomes are present. Acute schistosomiasis has developed in travelers returning from Africa.

TRANSMISSION OF WORMS FROM PETS

A variety of worms have been described here, including several exotic species that you will probably never encounter. However, our pet cats, dogs (Fig. 10.33), and birds (but not goldfish), with whom we share our homes

---

BOX 10.4   Man's Best Friend

Few would argue that dogs are not man's best friend. Dogs have been domesticated for thousands of years and have been used as sheep and cattle herders and as guard dogs. They conjure up visions of St. Bernards trudging through deep snow in an effort to rescue stranded mountaineers; Dalmatians accompanying fire engines; Disney's *101 Dalmatians,* Lassie, and Rex; scenes of children playing with dogs; and dogs alerting household members of a fire. They are known for their gentleness, loyalty, and affection. No matter how stressful a dog owner's day may be, upon returning home the owner will be greeted by an excited, happy animal anxious to bestow kisses. A pat on the head, a soothing "good dog," or a treat is all that is asked for.

Dogs come in many sizes, shapes, colors, and temperaments. I have two dogs, a 55-pound, cream-colored standard poodle named Clementine and a 20-pound miniature black poodle called Oscar (Fig. 10.33).

Unfortunately, there can be another side to having a dog as a pet. They can be pests, such as when they demand to "go out" after having just "gone out." It seems that at times your schedule revolves around your pet's bowels and urinary tract. Dog owners dread the day that their beloved companion will have to be "put to sleep" as an act of kindness and in respect for their dignity. Moreover, dogs, and other pets, can be a source of disease. Fortunately, most microbial diseases, including worms, that dogs acquire cannot be transmitted to humans, and, fortunately for dogs, most human microbial diseases are not suffered by dogs. Nevertheless, there are exceptions.

Most puppies are born with *Ascaris* worms that can cause visceral and ocular larva migrans. Children are particularly susceptible because of their play in outdoor sandboxes and in the soil, increasing the possibility of the accidental ingestion of ascarid eggs. The eggs develop into larval forms which migrate throughout the body, particularly into the brain and optic nerve, causing disturbances in vision. The incidence of this complication is low.

A variety of bacteria, including *Salmonella, Campylobacter,* and *Escherichia coli* O157:H7, all of which are described in chapter 8, have been documented to be transmitted from pets to humans.

Recommendations to minimize the risk of pet-to-human transmission include the practice of good personal hygiene, particularly hand washing; deworming of the dogs as necessary; frequent cleaning up of the area where the dogs relieve themselves; and, most important, not feeding pets raw or undercooked fish or meats.

and sometimes beds, can be sources of worm infections and other infections that may be worse than their bite! Simple measures of personal hygiene and eliminating worms from your pets greatly minimize transmission from pets to people (Box 10.4).

OVERVIEW Protozoans are microscopic, eucaryotic, unicellular organisms; helminths are macroscopic, eucaryotic, multicellular organisms. Both groups contain species that are pathogens of humans. Depending upon the specific organism, these parasites are transmitted by food and water, vectors, sexual contact, and direct (other than by sexual) contact.

Many of the protozoan and helminthic diseases are found in the rural sections of tropical and subtropical areas that suffer from great poverty and poor sanitation. Strategies of prevention and control are targeted at vector abatement, minimizing contact between humans and parasites, and implementation of sanitary measures. Sanitation is extremely important and includes the use of latrines, provision of safe drinking water, avoidance of nightsoil (human feces as fertilizer), proper cooking and preparation of foods, and the wearing of shoes.

Drugs are available for the treatment or prevention of these diseases, but the treatment may be worse than the disease. Protozoans and worms are eucaryotic organisms, as are humans, meaning that drugs toxic to them are also toxic to humans. Hence, it is a juggling act to choose drugs that are more toxic to the parasite than to the human host. ■

SELF-EVALUATION

PART I Choose the *single* best answer.

1. Mosquitoes are vectors for
   - **a.** leishmaniasis
   - **b.** filariasis
   - **c.** malaria
   - **d.** more than one of the above

2. Intestinal blockage is most likely to occur after infection with
   - **a.** *Ascaris*
   - **b.** tapeworms
   - **c.** earthworms
   - **d.** filarial worms

3. Which of the following was a problem during the Gulf War?
   - **a.** malaria
   - **b.** amebiasis
   - **c.** leishmaniasis
   - **d.** trypanosomiasis

4. Drinking what appears to be good, clean water from a mountain stream may put you at risk for infection with
   - **a.** guinea worms
   - **b.** *Giardia*
   - **c.** *Cryptosporidium*
   - **d.** tapeworms

5. Which parasite can be acquired from ingestion of eggs?
   - **a.** tapeworm
   - **b.** guinea worm
   - **c.** *Ascaris*
   - **d.** more than one of the above

6. Which disease can be acquired from stepping on larvae in the soil?
   - **a.** hookworm
   - **b.** tapeworm
   - **c.** guinea worm
   - **d.** elephantiasis

PART II Match the statement on the left with the disease (or microbe) on the right by placing the correct letter in the blanks.

_____ 1. worm migrates out through skin
_____ 2. elephantiasis
_____ 3. dysentery
_____ 4. ingested cysts
_____ 5. neurological
_____ 6. river blindness
_____ 7. heart condition
_____ 8. outbreak in Milwaukee (contaminated water)
_____ 9. ingestion of eggs
_____ 10. chills and fever
_____ 11. larval form enters body through skin
_____ 12. several cases in Gulf War

- **a.** onchocerciasis
- **b.** malaria
- **c.** *Giardia*
- **d.** tapeworm
- **e.** Chagas' disease symptoms
- **f.** cryptosporidiosis
- **g.** leishmaniasis
- **h.** amebiasis
- **i.** hookworm
- **j.** dracunculiasis

(continues)

SELF-EVALUATION (continued)

Fill in the vector for each disease.

Disease	Vector (Transmission)
1. leishmaniasis	_____
2. Chagas' disease	_____
3. sleeping sickness	_____
4. malaria	_____
5. onchocerciasis	_____
6. dracunculiasis	_____
7. filariasis	_____

PART III Answer the following.

1. Choose a protozoan disease that is transmitted by food or water, and describe its cycle. A diagram would be helpful.

2. Pretend you are a health worker involved in the eradication of guinea worm disease. Prepare an article to be handed out to the inhabitants of a village telling them about the disease and how to avoid becoming infected.

3. You are the director of public health services in an African country that is plagued with a variety of protozoan and helminthic diseases. Prepare a document outlining your plans to reduce the number of people who will become infected over the next years. (A budget will be developed based on your agenda.)

PART **II**

MEETING THE CHALLENGE

THE IMMUNE RESPONSE

PREVIEW The word immunity, in its broadest sense, means freedom from a burden, be it legal action, taxes, or, in the present context, disease. For example, Monica Lewinsky was granted immunity in the 1998 Clinton sex probe in exchange for her testimony at a grand jury hearing. In the hit television show *Survivor*, Richard Hatch wore the "immunity necklace," which meant that his fellow castaways could not vote him off the island. Boris Yeltsin dramatically announced his resignation as president of Russia on the last day of 1999; one of the incoming president's first presidential decrees provided bodyguards, a pension, and total immunity for Yeltsin. In the early years of the colonization of the United States, the battle cry of the colonists was "immunity from taxes."

Chapter 6 characterizes the ultimate outcome of microbial disease as the result of the dynamic interplay between two opposing forces, a seesaw effect summarized by the expression $D = nV/R$, and its explanation serves as the basis for this chapter. In the equation, D refers to microbial disease, the numerator (n = number of microbes; V = microbial virulence) refers to the microbe and its virulence factors, and the denominator (R = resistance = immunity) is a function of the immune system of the host, the subject of this chapter. Truly, a cell war is taking place with all the violence of *Star Wars*; the outcome of the battle will determine whether no infection, mild infection, or severe and possibly fatal infection will result. The early hours of encounter between the microbe and the immune system are particularly crucial in determining the eventual outcome.

The immune system functions to recognize and destroy foreignness as embodied in invading microbes and their products. This is accomplished by nonspecific (inherent) immunity and specific (acquired) immunity, functions which are accomplished by a variety of tissues and organs. Nonspecific immunity is characterized by physiological defenses which act to prevent microbes from gaining access into the body and facilitating the elimination of those that have penetrated into the body. Specific immunity targets microbes for elimination, based on the recognition of their foreign antigens, by antibody-mediated and cell-mediated immunity.

AUTHOR'S NOTE *In writing this chapter, I was initially in a dilemma. Should the microbes and their virulence mechanisms precede the body's immune defense, or the other way around? It is the "what came first, the chicken or the egg?" puzzle. Since a strong focus in this text is on disease prevention, it seemed more logical to first present what it is that the immune system is combating. I hope you agree!*

Impairment of the immune system by acquired immune deficiency syndrome (AIDS), leukemia, immunosuppressive drugs, or congenital disease renders the immunocompromised individual subject to repeated life-threatening infections.

BASIC CONCEPTS

The immune system's function is to kill, destroy, or eliminate the microbial invaders. (Were it not for the viruses, "kill" would be a perfectly appropriate term. Recall that viruses are subcellular and, therefore, not alive.) There is no one good way to measure the effectiveness of the immune system because of its many complex interactions. In fact, ability to cope with the constant barrage of microorganisms is not constant, nor is it measurable. Age, sex, nutrition, and "good health," along with physical and mental stress, are factors known to play a significant role in immune status. Think about final exam time. Lights are on late in dormitories and in study halls around the campus, and fast foods are consumed at an even more rapid rate than usual as you review (or, as too frequently happens, peruse your study materials for the first time!) a semester's work for four or five courses in about a week's time. There is no doubt that your immune system is compromised, making you more vulnerable to infection.

Immune status varies not only within the individual but between individuals. Consider, for example, that a few hundred people may be crowded into a lecture hall, movie theater, or restaurant and that all are exposed to circulating cold viruses. Some will "catch the cold" and others will not. Obviously, the numerator in the equation $D = nV/R$ is the same for all who are exposed, and, therefore, whether you end up with an infection or not is largely a function of R, your immune status at that time. A recent news article cited two deaths that occurred from babesiosis, a tick-borne protozoan disease (chapter 10), which is rarely fatal unless the individual has an underlying health problem. Both of the patients who died had such health problems. Genetics, too, plays a key role; it is well established that there are ethnic and racial differences in susceptibility to microbial infection.

While the primary role of the immune system is as a mechanism of defense against microbial infection, it also acts as a system of surveillance leading to elimination of tumor cells and destruction of old, worn-out, and damaged red blood cells. These cells, like microbes, are recognized as **foreign** and trigger an immune response. This "recognition and response" strategy to foreign molecules is the hallmark of the immune system and is the basis for defense against microbial intruders and internal surveillance.

The immune system is regarded as beneficial; for the most part, this is true, but there is another side to the coin. Many people suffer from **allergies,** which are adverse immune responses to molecules associated with pollens, dust, foods, mites, antibiotics, and bee stings. These allergies run the gamut from being a nuisance (sneezing, watery eyes, and runny nose) to life-threatening episodes. A dramatic example is the case of a student at Johnson & Wales University in Providence, R.I., who died of an allergic reaction after eating shrimp in the dining hall at his dormitory. Your physician, before prescribing an antibiotic, should ask you whether you are allergic to any antibiotics and will not prescribe one to which you are allergic. Otherwise, the consequences could be worse than the infection for which you are being treated.

Autoimmune disease is another example of a clearly nonbeneficial aspect of the immune system. The immune system functions to recognize and destroy that which is foreign, including invading microbes and their secretions (e.g., toxins and enzymes). However, failure to distinguish between foreign (nonself) and nonforeign (self) is the basis for autoimmune diseases. In other words, the immune system is attacking the body's own cells and tissues as though they were foreign. This leads to a host of debilitating autoimmune diseases, including rheumatoid arthritis, lupus erythematosus, and a variety of anemias. Paul Ehrlich, an early immunologist, referred to this dysfunction as "horror autotoxicus," a fear of self-poisoning.

The major obstacle to transplantation of organs is rejection of transplanted organs by the immune systems of their new hosts. Should this be considered a nonbeneficial aspect of the immune system? Most immunologists would not think so, since, after all, the recipient is responding to foreign molecules of the donor cells in a manner similar to that which nature evolved to thwart microbial invasion. Transplantation technology is not a natural phenomenon but is the result of medical science.

ANATOMY AND PHYSIOLOGY

Having now established a few basic concepts involving the immune system, attention needs to be focused on anatomical and physiological considerations. Perhaps in a precollege course in biology you dissected a frog; recall that the digestive system was a series of defined anatomical structures starting with the oral cavity, which led to the esophagus, which opened into the stomach, and so forth. The same organization can be said for all the systems of the body. There is a hierarchy of cells, tissues, and organs in biological systems. The immune system is, however, unique in that the anatomical structures are, for the most part, shared with more defined systems. The spleen, in most anatomy manuals, would be described with the circulatory system, and the lymph nodes would be considered components of the lymphatic system yet these structures are also parts of the immune system. The major structures of the immune system are shown in Fig. 11.1.

Functionally, the immune system consists of two components: (i) **nonspecific** or innate (inherent) immunity and (ii) **specific** or acquired (after birth) immunity (Fig. 11.2). Both of these mechanisms are ways in which the body eliminates "foreignness," including microbes.

Blood

Several components of blood are vital to the defense of the body and contribute to either nonspecific immunity or specific immunity or both. A summary of the components of blood is presented in Table 11.1. The globulin fraction is particularly important to the function of the immune system; a subfraction, **gamma globulin,** contains most of the **antibodies.** Antibodies are molecules produced in response to **antigens** and act to target them for removal. Conversely, antigens are molecules which bring about antibody production. Antigens and antibodies are described more fully in a later section.

There are three categories of blood cells, the **red blood cells** (erythrocytes), the **white blood cells** (leukocytes), and the **platelets** (thrombocytes); there are five categories of white blood cells. The red blood cells and platelets play no direct role in immunity and, therefore, are not further described. Table 11.2 presents the five types of white blood cells and their normal distribution. A **differential count** reflects the ratio of the white blood cell components and is an important tool in the diagnosis of suspected bacterial infection. A drop of blood, procured by jabbing a finger or from a sample obtained by inserting a needle into a vein, is smeared onto a microscope slide, covered with a special stain, rinsed, and examined under the microscope to perform

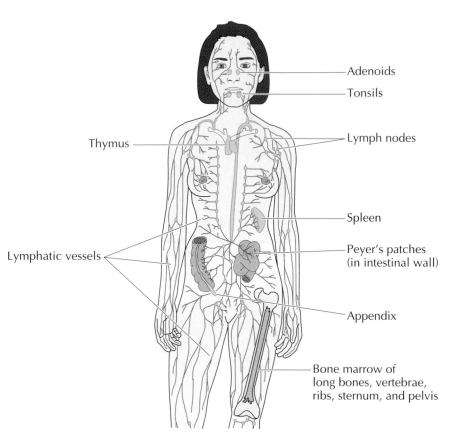

Adenoids

Tonsils

Thymus

Lymph nodes

Spleen

Peyer's patches
(in intestinal wall)

Lymphatic vessels

Appendix

Bone marrow of
long bones, vertebrae,
ribs, sternum, and pelvis

FIGURE 11.1   Anatomy of the immune system.

a differential count. Frequently, in bacterial (versus viral) infections, an elevated neutrophil count is found, which might justify antibiotic therapy while a more definitive laboratory diagnosis is awaited. By comparison, in infectious mononucleosis, a viral disease commonly called "mono" (chapter 9) that is characterized by fatigue, fever, and sore throat, there is an increase in lymphocytes. Fig. 11.3 illustrates a blood smear composite.

Lymphocytes, neutrophils, and **monocytes** have distinct immune functions; the last two are **phagocytic cells.** Phagocytosis, or "cell eating," is described in more detail in a later section. The lymphocytes, of which there are two categories, the **B lymphocytes** and the **T lymphocytes,** are the key cells of immunity. The B lymphocytes are the only producers of antibodies but are dependent, in most cases, on **T-helper** (T_H) cells for this function. Some T cells are capable of becoming sensitized to ("angry at") the foreign molecules (antigens) carried by the invading microbes and striking back against these invaders, bringing about their destruction. The activity of sensitized T cells is also dependent on T_H cells. Collectively, the antibodies produced by B lymphocytes and the activity of sensitized T lymphocytes bring about destruction of microbes and that, in the final analysis, is the primary function of the immune system. Human immunodeficiency virus (HIV), which is responsible for AIDS, destroys T_H cells, resulting in the failure of an infected person to mount an appropriate immune response either by antibody formation or by the development of sensitized lymphocytes (because of their dependency on T_H cells). This fact explains the immune devastation associated with HIV; the individual is said to be

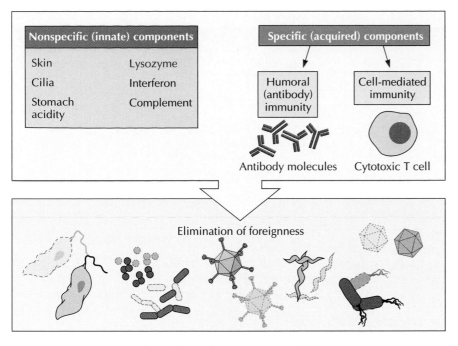

FIGURE 11.2 Nonspecific and specific components of the immune system.

immunocompromised and suffers one infection after another or may even have several infections at one time.

The role of **basophils** and **eosinophils,** the last two categories of white blood cells, is not entirely clear. Eosinophilia (an elevated eosinophil count) is associated with the presence of helminth infections and with allergies.

TABLE 11.1 Composition of blood

Component	Function
Plasma	
Water (80–90%)	Solvent
Proteins	
Albumin	Blood volume
Globulins	Antibody molecules
Fibrinogen	Blood clotting
Complement	Immune amplification
Interferon	Antiviral properties
Sugar, lipids, amino acids, vitamins, oxygen, carbon dioxide	Metabolism, cellular respiration, homeostasis
Blood cells[a]	
Erythrocytes (red blood cells)	Oxygen and carbon dioxide transport
Leukocytes (white blood cells)	Antibody formation, cell-mediated immunity, phagocytosis
Thrombocytes (platelets)	Blood clotting

[a]In each microliter of blood, there are about 4.5 million to 5.5 million erythrocytes, 5,000 to 10,000 leukocytes, and 250,000 to 400,000 thrombocytes.

TABLE 11.2   Categories of leukocytes

Cell type	% of total leukocytes	Function(s)
Agranulocytes		
Lymphocytes	25–35	Antibody formation, cell-mediated immunity
Monocytes	3–8	Phagocytosis
Granulocytes		
Neutrophils	60–70	Phagocytosis
Eosinophils	2–4	Inflammatory response, limited phagocytosis
Basophils	0.5–1	Not clear; contain histamine

Basophils are rich in granules of histamine, which are released during allergic reactions and are responsible for watery eyes, itchy throat, sneezing, and runny nose. The symptomatic treatment is antihistamine medication, of which there are many over-the-counter choices available. You may be one of the unlucky sufferers. Don't feel too bad: approximately 40% of the population has allergies.

In addition to blood cells, there are blood proteins (Table 11.1) which contribute to immune function. **Complement,** a series of blood proteins, is a significant defense mechanism against potential disease-causing microbes; it is a system of **"biological amplification."** Consider the role of the amplifier in your music system as an analogy. It does not actually produce the music; if you were to unplug the amplifier, the music would continue, but the quality would suffer. Complement does not cause but, rather, enhances phagocytosis. Further, it can bring about **lysis** (destruction by chemical attack) of the bacterial cell wall, causing it to "spill its guts" through the now-leaky membrane. (What a way to go!)

Interferon, another component of blood, was discovered in 1957 and was so named because of its ability to interfere with viral replication. It is not a single component but a group of related compounds. The product is released from virus-infected cells and acts to warn other cells of impending danger by triggering them to release virus-blocking enzymes. Unfortunately, interferon has not lived up to the lofty expectation that it would become a powerful antiviral drug. It was hoped that interferon would act

FIGURE 11.3   Several types of blood cells.

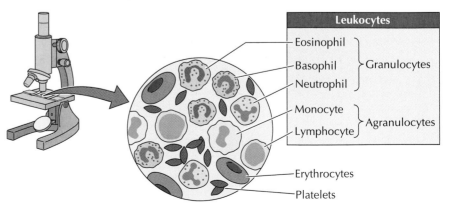

on viruses to the degree that antibiotics are effective against bacteria. It has been only partially successful in the treatment of certain forms of hepatitis. On a more positive note, interferon enhances phagocytosis and has been found to be somewhat effective in cancer treatment.

The Lymphatic System

The **lymphatic system** is anatomically intertwined with the blood circulatory system, and it contributes to the body's defense mechanisms. **Lymph,** a tissue fluid derived from blood but without blood cells, is circulated throughout the lymphatic capillaries and veins, which open into the blood circulatory system. Hence, the two systems are interdependent: lymph is derived from blood and drains back into blood (Fig. 11.4). Situated along the path of the lymphatic veins are the **lymph nodes.** You may have experienced swollen lymph nodes in your armpits, groin, or neck, indicating that these nodes have filtered out invading microorganisms; a war is in progress. These nodes are highly significant in that they contain both B and T cells, enabling antibody production. Lymph nodes also contain phagocytic cells, which are significant in the removal of microbes and damaged or worn-out cells of the body. Thus, these structures contribute to both the nonspecific and specific immune systems.

Primary Immune Structures

Bone Marrow

The **bone marrow,** a **primary immune structure,** plays a highly significant role in immune defense mechanisms; it is the source of all the blood cells (red blood cells, white blood cells, and platelets) and produces them from **stem cells.** These stem cells are located in the red bone marrow of the sternum (breastbone), vertebrae, and the upper ends of the long bones of the body. There are two pathways of blood cell maturation (Fig. 11.5), the **myeloid**

FIGURE 11.4   Interrelationship of the lymphatic and circulatory systems.

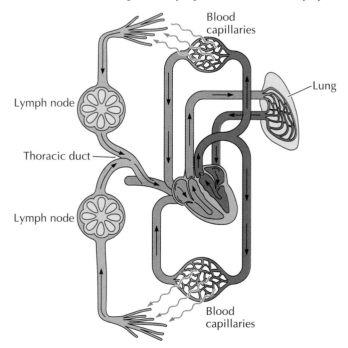

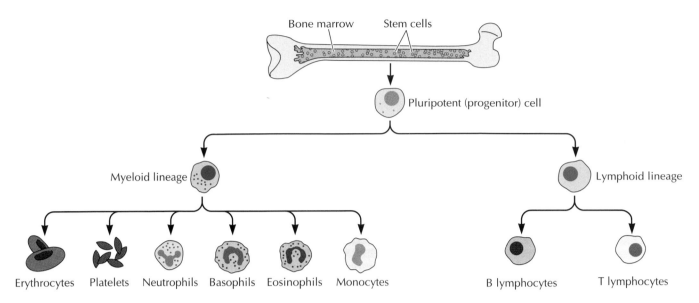

FIGURE 11.5   Pathways of blood cell maturation.

path and the **lymphoid path.** The myeloid lineage leads to platelets, red blood cells, and four of the five categories of leukocytes (monocytes, neutrophils, eosinophils, and basophils), while the lymphoid lineage leads to lymphocytes. Ultimately, all the categories of blood cells are released into the general circulation as mature cells. Lymphocytes further mature into two distinct functional categories, B cells and T cells. B lymphocytes complete their maturation in the marrow and are released into the circulation, from which they seed other (secondary) structures of the immune system.

The Thymus Gland

Some lymphocytes are released from the marrow prior to their maturation and end up in the **thymus gland** (not thyroid [Fig. 11.1]), an organ located behind the sternum (breastbone) and just above the heart. In this gland these cells complete their maturation into T cells, which seed other (secondary) structures of the immune system.

At birth, the thymus is proportionately a large organ, and it attains its maximum size by puberty. These early years are the ones in which numerous childhood diseases and immunizations are experienced; the immune system is maturing, and the thymus plays a significant role. Some children are born without a thymus gland, a condition called **DiGeorge syndrome,** and they suffer the consequences of deficient immunity throughout their lifetimes, as manifested by one potentially life-threatening infection after another. After adolescence, the thymus gradually degenerates and assumes a lesser significance; its function is taken over by the bone marrow.

To recapitulate, the thymus is the site of maturation of T cells; the bone marrow is the site of maturation of B cells. These are the primary structures of the immune system and serve as the beds for the maturation of T and B lymphocytes; these mature cells are then "seeded" into the secondary structures of the immune system (Fig. 11.6).

Secondary Immune Structures

The **spleen, tonsils, adenoids,** lymph nodes, and patches of tissue associated with the intestinal tract are included in the **secondary immune**

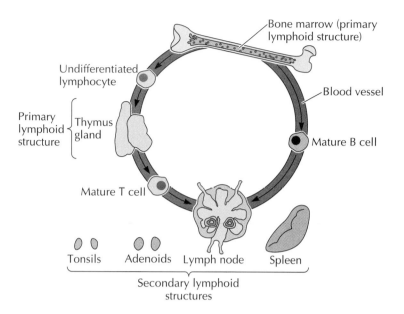

FIGURE 11.6   Maturation and "seeding" of B and T lymphocytes.

structures. The spleen contains phagocytic cells and both mature T and B cells, endowing it with immunologic functions. It is a spongy, fist-sized organ located in the upper left portion of the abdominal cavity. Because of its location and its relatively thin connective tissue covering, it is subject to injury, as might be sustained in an automobile accident. Severe trauma to this organ may result in its rupture, with severe hemorrhaging necessitating its removal (splenectomy).

The tonsils and adenoids are located at the back of the throat, just in front of the pharynx. Their lymphocytes play a role in protection against microbes entering through the nose and throat. A generation or so back, it was routine for young children to undergo removal of their tonsils (tonsillectomy) because of the frequency of throat and ear infections resulting from infected tonsils; this is no longer as common because of antibiotics. Doctors often performed these operations in the home "on the kitchen table." Occasionally, tonsillectomy is performed on individuals with repeated infections that do not respond well to antibiotics. (As an additional benefit, you are indulged with a lot of ice cream and popsicles to help alleviate a severe sore throat!) Patches of tissue, similar to the tonsils and adenoids, are distributed in the lining of the gastrointestinal, respiratory, and urinary tracts and contribute to immunity. Almost all antigens entering the body end up in lymph or blood and are then carried to lymph nodes or to the spleen, where they encounter cells of the immune system.

DUALITY OF IMMUNE FUNCTION: NONSPECIFIC AND SPECIFIC IMMUNITY

Nonspecific Immunity

On the basis of the foregoing presentation, it is clear that blood, bone marrow, the thymus, and the secondary immune structures are the key elements in the body's protection against microbial invasion. As mentioned earlier in this chapter, the immune system functions in (i) nonspecific (innate, or inherent at birth) immunity and (ii) specific (acquired after birth) immunity. Specific immunity, in turn, has two arms: (i) humoral, or antibody-mediated, immunity and (ii) cell-mediated immunity.

Nonspecific immunity is characterized by physiological defenses which operate either to prevent microbes in the external environment from gaining access into the body or to eliminate those that have penetrated into the body. Table 11.3 summarizes nonspecific immunity. These defense mechanisms are present at the time of birth and do not involve specific recognition of the microbe: they act against all microbes in the same fashion.

The skin is the first line of defense against microbial infection. It is a physical barrier which, when intact, blocks the entry of microorganisms into the body based on its structural composition. Obviously, there are circumstances in which the skin is broken, allowing possible penetration. Cuts and abrasions, insect bites, and injections with hypodermic needles are familiar examples. Staphylococci, normal residents on the surface of the skin, are particularly prone to enter the body via these breaks and establish infection. Many diseases, including malaria, yellow fever, babesiosis, and sleeping sickness (other than in the classroom) are transmitted by insects whose bites penetrate the skin. Mucous membranes line the internal body cavities and act as internal barriers. For example, the digestive tract running from the mouth to the anus is lined by a mucous membrane, which protects the body from invasion by microbes that reside within the digestive tube. Penetration of this mucous membrane by intestinal microbes, even though they normally reside within the digestive tract, is extremely serious and leads to a condition known as **peritonitis.** Surgeons are very much aware of this when performing procedures on the digestive system; nicking the lining of the intestinal tract may allow microorganisms to spill out into body cavities where they do not belong, creating a potentially life-threatening situation. As another example, the airways in certain areas of the respiratory system have cells with hairlike projections, called **cilia,** that propel the microbes into the pharynx, where they are swallowed and, after proceeding through the digestive tract, removed with fecal material. This system is called the mucosal-ciliary escalator system. Smokers are more prone to respiratory infections than nonsmokers, since smoke damages this system. In the eye, the enzyme **lysozyme** is present in tears and is an important defense mechanism; it destroys some bacterial cell walls. The acidity of the stomach is detrimental to most microbes. Certainly, food and eating utensils are not sterile, and hence large numbers of microbes may gain access to the stomach.

A number of nonspecific internal mechanisms, including the previously described complement and interferon, work against those microbes

TABLE 11.3   Nonspecific immunity

Mechanism	Function
Skin and mucous membranes	Mechanical barriers
Cilia	Found along respiratory tract and have "upward" motion pushing microbes up to pharynx, where they are swallowed
Phagocytosis	A system of phagocytic cells in the blood and scattered throughout the body
Lysozyme	Found in tears; breaks down bacterial cell walls
Interferon	Found in blood; has antiviral properties
Acid pH of stomach	Many microbes are killed by strong acid environment

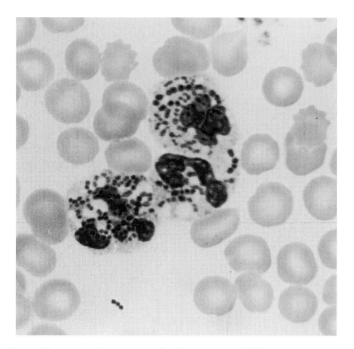

FIGURE 11.7   Phagocytosis. Bacterial cells are engulfed by certain categories of cells of the host's immune system and destroyed; this is an important nonspecific defense mechanism. (Author's photo.)

that breach the nonspecific external defense mechanisms. **Inflammation,** characterized by redness (rubor), heat (calor), swelling (edema), and pain (dolor), mobilizes phagocytes to the inflammatory site. Consider the following familiar scenario: you get a splinter in your finger, and within several hours the area is red, hot, swollen, and painful. Microbes from the skin, most likely staphylococci, invade through the site of injury. The injury causes these small blood vessels to become engorged with blood (vasodilation), accounting for the redness and the heat; some of the blood leaks from the vessels into the surrounding tissue spaces, causing swelling and pain as a result of increased pressure and the effect of products released by the injured tissue on nerve endings. What is the significance of the inflammatory reaction? It hardly seems like a defense mechanism! The answer lies in the fact that in these blood-engorged capillary beds, monocytes and neutrophils, both phagocytic cells, pile up and stick to the vessel walls surrounding the site of the injury. These cells are attracted to the injured tissue and migrate out of the capillaries; the upshot of the inflammatory reaction is, therefore, that phagocytes in large numbers are attracted to the inflamed site and "mop up" the bacteria, thus preventing or minimizing infection.

Phagocytosis is a highly significant nonspecific defense mechanism by which monocytes and neutrophils engulf and destroy foreign substances, including microbes (Fig. 11.7). It is an extremely efficient host defense mechanism in its own right but is rendered even more efficient in the presence of antibodies and complement. The discovery of phagocytosis by Elie Metchnikoff in 1884 was a landmark in immunology for which Metchnikoff received a Nobel Prize in 1908 (Fig. 11.8). Phagocytosis is not limited to cells of the bloodstream; a variety of other cells have phagocytic capabilities and are strategically located throughout the body, establishing an efficient system of surveillance. **Macrophages** are monocytes which have migrated out of the blood but retain their phagocytic capability. Some

FIGURE 11.8   Elie Metchnikoff (1845–1916). The concept of phagocytosis is credited to Metchnikoff. This statue is on the grounds of the Pasteur Institute in Paris. (Author's photo.)

macrophages become fixed at particular sites; other macrophages are called **wandering macrophages** because they move freely about the tissues. Collectively, monocytes, neutrophils, and macrophages are referred to as "professional phagocytes"; they are located strategically throughout the body and are extremely important in the destruction of microbes.

Specific Immunity

The specific immune system responds to the presence of foreign and potentially harmful microbes that breached the external and internal nonspecific defense mechanisms (Fig. 11.9). At this point, an all-out war between microbes and host is in effect, as represented by $D = nV/R$.

FIGURE 11.9 Mechanisms of specific immunity. **(A)** Protective effects of antibodies binding to antigens (humoral immunity); **(B)** protective effects of cell-mediated cytotoxicity (CMI).

A Protective effects of antibodies binding to antigens

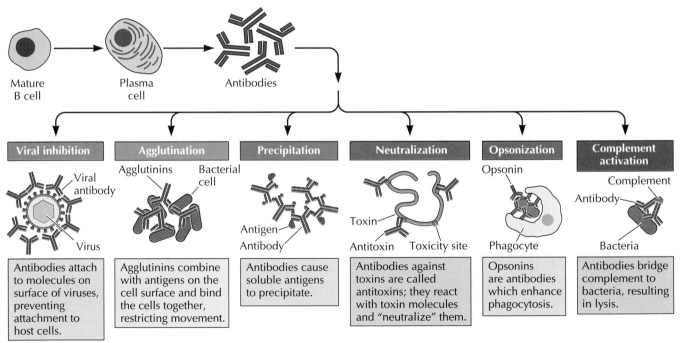

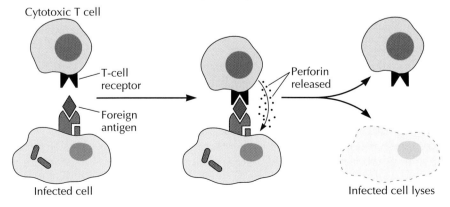

Antibody-mediated (humoral) immunity and cell-mediated immunity (CMI) are the two categories of the specific immune system, and they can be considered the "big guns." While these systems operate in very different ways, the projected goal is a common one, and that is to eliminate foreign invaders or antigens. Specific immunity is based on the recognition of foreign antigens. Most antigens are large protein molecules associated with microbes, tumor cells, damaged cells, pollens, dust, and foods. They (i) cause the production of antibodies specific for that antigen (humoral immunity) or (ii) bring about the production of sensitized (angry) lymphocytes directed against that antigen (CMI).

A key to understanding immunity is the concept of specificity. Think of antigen-antibody specificity as the complementarity that exists between a lock and a key or between two pieces of a puzzle that fit together (Fig. 9.3). Humoral immunity is antibody mediated; the antibodies are produced by B cells (with the aid of T_H cells) in response to antigens. CMI is mediated by T cells that have become sensitized (angry) to a specific antigen and react only against that antigen. (Think of CMI in this way: when you are angry at an individual, you vent your anger against only that person.)

Antibody-Mediated (Humoral) Immunity

The role of antibody immunity in the prevention and recovery from certain diseases has been realized for centuries. Before the 1700s, vaccination against what later was termed smallpox was practiced. In this procedure, scabs were picked from pox-infected individuals, dried, pulverized, and introduced into the nose of those susceptible. In 1796, Edward Jenner introduced a vaccination against smallpox by inoculation of the cowpox virus, which causes a benign infection in humans. The inhalation of scabs and inoculation with the cowpox virus resulted in antibodies which were protective against the smallpox virus. This disease, truly a scourge of humankind, is the first and, currently, the only disease to be eradicated, as declared by the World Health Organization in 1979.

B lymphocytes that are "revved up" by contact with antigen and T_H lymphocytes and release antibodies are called **plasma cells.** Estimates are that a single plasma cell can produce thousands of specific antibodies per second. There are actually five categories of antibodies, or **immunoglobulins (Igs),** termed **IgG, IgA, IgD, IgE,** and **IgM,** each with particular properties. IgG accounts for approximately 80% of the antibody molecules and is the best characterized. The antibody molecule is a four-chained structure (Fig. 11.10); it binds with antigens on bacterial cells, viral particles, or toxins, resulting in a "kill, destroy, or eliminate" outcome (Fig. 11.11).

As previously stated, while it is true that antigen-driven B cells are the only antibody producers, T_H cells are required to help these B cells produce antibodies; they do so by releasing molecules called **cytokines.** How do B cells know which antibody specificity to produce? The **clonal selection theory** (Fig. 11.12) is the explanation. According to this theory, a population of B cells for every possible antigen exists at the time of birth. Each B cell displays many copies of a (single) specific antibody molecule on its surface. By random contact, antigens "dock" with surface antibodies of corresponding specificity (lock and key), thus stimulating those B cells to undergo cellular reproduction. Hence, the population of B cells that was selected by the antigen is expanded. Antibodies that are copies of the surface antibody are produced and released. After a lag time of approximately 10 to 12 days, antibodies in amounts large enough to be detected in the blood are present.

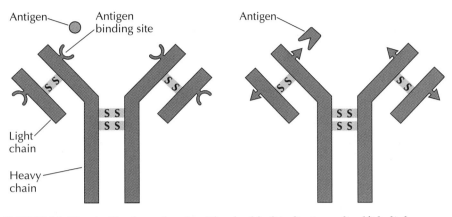

FIGURE 11.10 Antibody molecules. The double S indicates a disulfide linkage.

FIGURE 11.11 Antibody- and cell-mediated immunity and the final resolution.

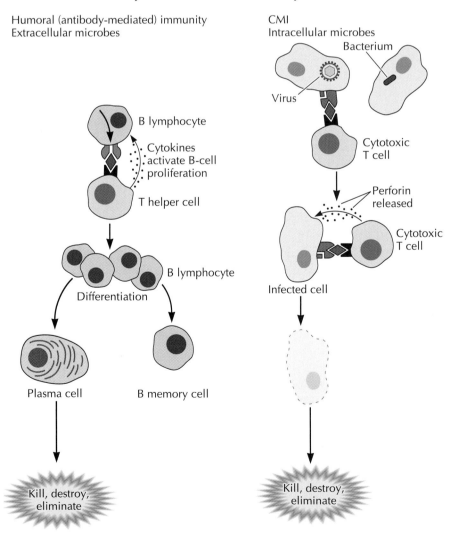

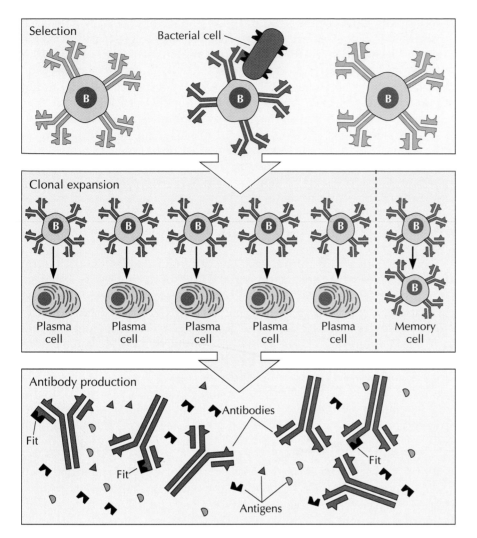

FIGURE 11.12   Clonal selection theory. B indicates a B cell.

Some of these B cells may not, in fact, progress to antibody-producing plasma cells but may remain for years as **memory cells.** This accounts for the fact that, except in rare circumstances, you can get certain so-called childhood diseases (measles, German measles, mumps, and chicken pox) only once in a lifetime despite repeated exposure. Assume, for example, that you had measles as a child. In later years your child has measles, and, despite the extremely close contact, you do not acquire the disease from your child. The explanation is based on the fact that when you had measles your B cells made antibodies against this virus, but a part of the B-cell population was reserved as memory cells. Now every time, usually throughout your life, that you come in contact with measles virus, these preprogrammed memory cells respond in a matter of hours, not days, and pour out measles antibodies at a rate and quantity sufficient to target the virus for destruction.

CMI

Antibody-mediated immunity is the body's defense strategy for coping with **extracellular microbes,** but what about those **intracellular bacteria** and viruses (all of which are intracellular), once they penetrate cells?

Bacterial pathogens, for the most part, are extracellular; that is, they take up residence on the surface of the cells. For example, streptococci, the causative agents of strep throat, colonize the surface of the throat and pharynx, and the organisms do not penetrate the cells. Antibodies play little role in protection against microbes which are intracellular. It should be emphasized, however, that viruses, during their early, extracellular stage prior to cell invasion, are targeted by antibodies. (It would be pointless to immunize against viral diseases if this were not the case.) So what is the defense? CMI is the major defense strategy against intracellular bacteria, protozoans, viruses, and tumor cells (Fig. 11.11), all of which possess antigens to which the body responds by delivering a "knockout" punch. To further complicate an already complex system, there is more than one category of T cells, collectively referred to as the **T-cell subset;** each has a specific role and bears distinctive **cluster of differentiation (CD)** molecules (Table 11.4).

The categories of T lymphocytes are distinguished by distinctive receptor molecules which are acquired in the thymus during the T-cell maturation process. The T_H cells, identified by **CD4 receptor molecules,** are the regulatory or control cells. The cytotoxic T (T_C) cells, identified by **CD8 receptor molecules,** are the effectors of CMI. The T_H cells are crucial in that the B and T_C lymphocytes are functionally crippled in the absence of or in the reduction of T_H cells. During antigenic stimulation, T_H cells secrete molecules, known as cytokines, which "turn on" B and T_C cells. As previously stated, HIV attacks T_H lymphocytes, resulting in impairment of both antibody-mediated immunity and CMI. The infected individual is severely **immunocompromised,** i.e., has a weakened immune system and suffers from one infection after another, eventually dying from the condition. Blood from AIDS-infected individuals is routinely monitored to determine the level of T_H lymphocytes. The prognosis is very grave when the number of T_H cells drops significantly, signaling that the individual does not have the capacity to combat invading microbes via the specific immune system. When you hear that an individual with AIDS has a low CD4 level, this refers to a low level of T_H cells. AIDS is discussed in chapter 15.

Unlike the direct presentation of extracellular antigen to receptor molecules on B lymphocytes (clonal selection theory), antigens from intracellular bacteria, viruses, and tumor cells are presented in a more complex manner. First, the intracellular microbes bearing the foreign antigens are

TABLE 11.4   The T-cell subsets

Cell type	Abbreviation	Clusters of differentiation (receptors)	Function(s)
T-helper cell	T_H	CD4	Activates B cells to produce antibodies; activates cytotoxic T cells
Cytotoxic T cell	T_C	CD8	Killer cell that works directly against cells with "foreign" intracellular antigens, including viruses and bacterially infected cells
T-suppressor cell	T_S	CD8	Regulatory cell that probably works in concert with T_H cells
T-delayed-type hypersensitivity cell	T_{DTH}	CD4	Plays role in allergic responses; activates macrophages

phagocytized by **antigen-presenting cells (APCs),** including macrophages, within which they are "chewed up." Pieces of these processed antigens are presented on the surface of molecules known as **major histocompatibility molecules** to receptor molecules on T_C cells. The now-activated T_C cells produce a membrane-penetrating protein, **perforin,** which directly leads to the lysis of the host cell harboring the viruses or intracellular bacteria, thereby halting microbial multiplication and spread. If the cell is a tumor cell, then its destruction is similarly accomplished; this is an important defense mechanism against cancer. Hence, as with antibody-mediated immunity, the end point, or final resolution, of CMI to kill, eliminate, or destroy (Fig. 11.13) is accomplished by T_C lymphocytes.

CLINICAL CORRELATES

Leukemia, unfortunately, may be a familiar term to you. It is a form of cancer characterized by uncontrolled reproduction of white blood cells. There are two major categories of leukemia reflecting the two pathways of blood cell maturation (Fig. 11.5). **Myeloid leukemia** is the result of overproduction of monocytes and granular leukocytes. Immature neutrophils, for example, are produced in excess and do not have the capacity to carry out phagocytosis, resulting in the individual's being immunocompromised.

FIGURE 11.13   The immune response and the final resolution.

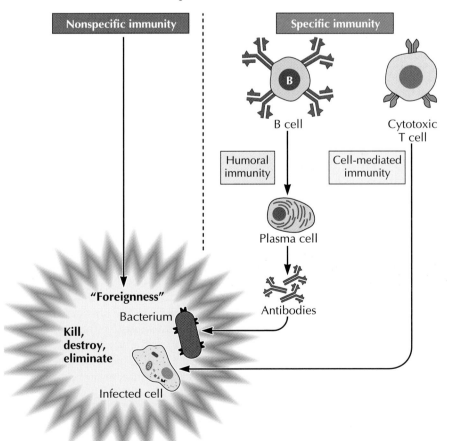

Lymphocytic leukemia is the result of overproduction of lymphocytes, leading to abnormally large numbers of immature and nonfunctional cells. The cells responsible for the overproduction of leukocytes tend to spread by **metastasis** from the bone marrow and reproduce in tissues throughout the body, crowding out normally functioning cells. Leukemia can be classified as acute or chronic. In **acute leukemia,** the symptoms appear suddenly and progress rapidly; death occurs in a few months unless the condition is treated. **Chronic leukemia** can remain undetected for many months, and life expectancy without treatment is somewhere around 3 years. Acute lymphocytic leukemia is the most common form of leukemia in children, but, fortunately, treatment of this condition has a high success rate. The Dana-Farber Cancer Institute in Boston is world renowned for its success in treating children with leukemia. This facility was originally the Jimmy Fund Building, and fans of baseball history will be delighted to know that baseball legend Ted Williams of the Boston Red Sox was a frequent visitor and contributor to this institution.

Despite these elaborate mechanisms of defense against potential pathogens, people do become ill with infectious diseases and sometimes die of them. It may be that the immune system is defeated in the dynamic interplay between microbe and immunity in a virtual tug-of-war, by a deadly infectious agent such as Ebola virus or rabies virus or the bacterium that causes meningococcemia. An individual may be immunocompromised as the result of AIDS, leukemia, or the use of immunosuppressive drugs.

A variety of immunodeficiencies may be present at birth as a result of genetic defects. As described in this chapter, a properly functioning immune system requires the interplay of both nonspecific and specific mechanisms; an impairment results in an immunocompromised individual subject to repeated life-threatening infections. Table 11.5 gives a sampling of immunodeficiency disorders, two of which are described here.

Severe combined immunodeficiency (SCID) is a particularly devastating congenital disease: individuals lack functional T and B cells and therefore cannot mount either an antibody- or cell-mediated immune response. Death is a certainty, resulting from repeated infections. This disorder is commonly known as "bubble boy disease," because infants born with this condition must be kept in a germ-free environment, a plastic bubble. One of John Travolta's early movies, *The Boy in the Plastic Bubble*, which was made for television in 1976, dramatized the plight of a boy who spent his first 12 years in a germ-free bubble because he was born with SCID. Those afflicted cannot even have contact with their parents, and all of the items introduced into the bubble, including air, food, and water, must be sterilized. SCID is the result of a genetically caused enzyme deficiency. In recent years, gene therapy and early bone marrow transplants (within 3 months of birth) have proved to be effective in some cases.

DiGeorge syndrome is a disorder resulting from the absence or incomplete development of the thymus. T-cell maturation is abnormal, with resulting impairment of both humoral immunity and CMI.

Chronic granulomatous disease is an inherited disorder of phagocytes. Because of the inability of phagocytes to kill bacteria and fungi, serious, life-threatening, and persistent infections result.

Allergic reactions and autoimmune diseases are cited earlier in this chapter as adverse reactions of the immune system. Allergic reactions are reactions of hypersensitivity (exaggerated responses). Autoimmune diseases are characterized by the misrecognition of self as nonself (foreign), resulting in the immune system's mounting an attack against the self.

TABLE 11.5 A sampling of immune deficiency disorders

Deficiencies in complement
C3 (complement factor 3) deficiency
C5 (complement factor 5) deficiency
Deficiencies in phagocytosis
Chronic granulomatous disease
Chediak-Higashi syndrome
Deficiencies in B lymphocytes
X-linked agammaglobulinemia
Bruton's syndrome
Deficiencies in T lymphocytes
DiGeorge syndrome
Wiskott-Aldrich syndrome
Ataxia telangiectasia
Deficiencies in both B and T lymphocytes
Severe combined immunodeficiency disorder (SCID)

OVERVIEW The immune system is a defense mechanism against invading microbes which minimizes the burden of microbial diseases. The ultimate outcome of exposure to pathogenic microbes is the result of the dynamic interplay between the disease-producing (virulence) mechanisms of the parasite and the immune system of the host. The recognition of foreignness is the key element in triggering mechanisms of immune defense. The immune system may inappropriately target self as foreign and mount an immune response against self, resulting in a variety of autoimmune diseases.

Anatomically, the immune system is unique in that it shares cells, tissues, and organs with other functional systems of the body. Blood and blood cells, thymus gland, bone marrow, tonsils, lymph nodes, and spleen are key components of the immune system. The nonspecific immune system includes the skin, phagocytic cells, components of blood, and ciliated cells that line part of the respiratory tract. Specific immunity responds to the presence of foreign and potentially harmful microbes by antibody-mediated (B) cells and by cell-mediated (T_C) cells. Both specific immune mechanisms target microbial invaders for destruction.

Disorders of the immune system may be the result of infection, cancer, or immunosuppressive drugs or may be congenital. ■

SELF-EVALUATION

PART I Choose the *single* best answer.

1. Which type of white cell is primarily involved in phagocytosis?
   a. leukocytes
   b. eosinophils
   c. basophils
   d. lymphocytes

2. Antibodies are associated primarily with which blood fraction?
   a. albumin
   b. globulins
   c. fibrinogen
   d. complement

3. DiGeorge syndrome is associated with
   a. humoral immunity
   b. phagocytic dysfunction
   c. complement dysfunction
   d. more than one of the above

4. Which represents an innate, nonspecific immunity?
   a. Ig molecules
   b. acidity of stomach
   c. phagocytosis
   d. more than one of the above

5. The AIDS virus specifically attacks
   a. T_H cells
   b. red blood cells
   c. T_C cells
   d. all of the above

6. Myeloid stem cells give rise to all of the following except
   a. mast cells
   b. neutrophils
   c. platelets
   d. basophils

7. Which of these statements is true regarding the antibody molecule?
   a. Ig = antibody
   b. a four-chain antigen
   c. is specific for an antigen
   d. all of the above

8. Which one of the following is not considered a secondary immune structure?
   a. spleen
   b. tonsils
   c. bone marrow
   d. lymph nodes

PART II Fill in the following.

1. B cells mature in the _____.

2. Which lymphocyte category is necessary for both CMI and antibody-mediated immunity? _____

3. Name a secondary lymphoid structure.

   _____

4. Chronic granulomatous disease is manifested by

   _____.

PART III Answer the following.

1. What are autoimmune diseases? Discuss the immunological basis of these diseases.

2. Describe the clonal selection theory. What does it explain?

3. An individual with AIDS has a low CD4 count. What does this mean? What is the significance of this low count?

CONTROL OF MICROBIAL DISEASES

Prevention is much better than healing because it saves the labor of being sick.
THOMAS, A 17TH-CENTURY PREACHER

PREVIEW "Are you better off today than you were 8 years ago?" If you followed the political rhetoric for the 2000 presidential campaign, you recognize this as a question frequently posed by Vice President Albert Gore in his bid to become president. But what about matters of health, particularly as related to diseases caused by microbes? The question to be asked is, Is your generation better off than your parents' and grandparents' generations? Most decidedly, your response would be "yes," despite the current problem of new and reemerging infections, a problem which has existed for only the past few decades. Society, particularly in developed countries, is no longer "plagued by plagues," except for the current pandemic of acquired immune deficiency syndrome (AIDS). Even in developing countries, the burden of disease has been reduced, although certainly not to the degree enjoyed by the richer nations of the world. You need to be aware, however, that thousands of citizens in the United States live in pockets of poverty and disease that are no better, and may be even worse, than some areas in Africa.

Consider that a person born in the United States in the early 1900s could look forward to an average life span of 45 years. At that time, the death rate at birth was slightly higher than 10%; among those surviving, tuberculosis was, ultimately, the leading cause of death. Today, life expectancy in the United States is about 75 years. The World Health Organization (WHO) recently ranked life expectancy in countries around the world (Table 12.1) based not on the usual number of years spent alive, but, rather, on healthy life expectancy. These projections are based on **disability-adjusted life years (DALYs),** a new statistic which takes into account years of life lost to death and disability. The emphasis is on a dual standard—living longer and living in good health. Japan is at the top of the list. Sierra Leone is at the bottom; its people have a dismal 26 years of good health. Surprisingly, the United States, which spends the most on health,

TABLE 12.1 WHO rankings of healthy life expectancy[a]

Country and ranking	Disability-adjusted life years
1. Japan	74.5
2. Australia	73.2
3. France	73.1
4. Sweden	73.0
5. Spain	72.8
6. Italy	72.7
7. Greece and Switzerland (tie)	72.5
9. Monaco	72.4
10. Andorra and San Marino (tie)	72.3
24. United States	70.0
171. Sierra Leone	26.0

[a]Source: WHO.

ranks only 24th overall (70 years), although wealthy Americans are the world's healthiest people. However, the bottom 2.5% of Americans have healthy life expectancies characteristic of sub-Saharan Africa in the 1950s, according to WHO. Christopher Murray of WHO stated, "Native Americans, poor rural black populations, and some of the inner city populations have terrible health status." According to the Centers for Disease Control and Prevention (CDC), there are unacceptable marked disparities in health outcomes by race, ethnicity, and socioeconomic status. A dramatic example is evident in four counties of South Dakota where the life expectancy is 56.5 years—equal to the life expectancy in Bangladesh and Sudan.

We are, without a doubt, better off than the previous generation and, even more so, than the generation before that. Over the past century, the average life span in the United States has increased by about 30 years, 25 of which are the result of advances in public health. Attesting to this, the CDC quoted an eminent historian as stating, "The retreat of the great lethal diseases was due more to urban improvements, superior nutrition, and public health than to curative medicine." In 1999, the CDC published a report entitled *Ten Great Public Health Achievements in the United States, 1900–1999* (Table 12.2); three of those achievements focused directly on a reduction of infectious diseases (Fig. 12.1). This chapter considers these achievements under the following sections: "Sanitation and Hygiene," "Food Safety," "Clean Water," "Immunization," and "Antibiotics."

SANITATION AND HYGIENE

Part of your daily routine each morning when you awake and prepare for your day is attending to matters of sanitation and personal hygiene, including showering, brushing your teeth, and using a clean flush toilet. You do not even need to consider where the water goes when you flush the toilet. It is good that you are able to take these activities for granted, but they are truly luxuries. It may surprise and shock you to learn that, according to the Water Supply and Sanitation Collaborative Council, 2.4 billion of the world's citizens lack basic hygiene and sanitation facilities,

TABLE 12.2 Ten great public health achievements in the United States, 1900–1999[a]

Vaccination[b]
Motor vehicle safety
Safer workplaces
Control of infectious diseases (includes sanitational hygiene, antibiotics, and clean water[b]
Decline in deaths from coronary heart disease and stroke
Safer and healthier foods[b]
Healthier mothers and babies
Family planning
Fluoridation of drinking water
Recognition of tobacco as a health hazard

[a]Not ranked in order of importance. Adapted from CDC, *Morbidity and Mortality Weekly Report* **48**:241–243, 1999.
[b]Focused on reduction of infectious diseases.

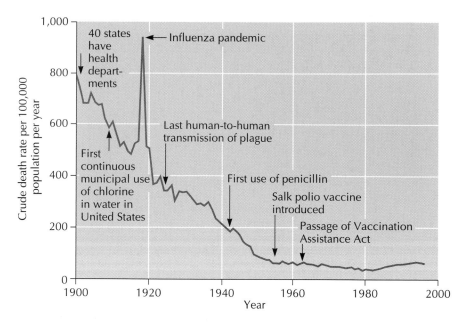

FIGURE 12.1 Major public health events and their influence on crude death rate, 1900 to 1996. Infectious diseases were a major cause of morbidity and mortality in the early 1900s. Public health intervention focused on improvements in sanitation and hygiene, implementation of antibiotics, and immunization programs. These strategies were effective over the generations. (Redrawn from CDC, *Morbidity and Mortality Weekly Report* **48**:621–629, 1999; adapted from G. L. Armstrong, L. A. Conn, and R. W. Pinner, *JAMA* **281**:61–66, 1999. Date of first continuous municipal water chlorination is from American Water Works Association [AWWA], *Water Chlorination Principles and Practices*, AWWA manual M20 [AWWA, Denver, Colo., 1973].)

and more than 1 billion people in the world do not have access to a daily supply of clean water. Every day, about 10,000 people die of diarrheal diseases because they lack the most basic sanitation. World leaders agree that hygienic means of sanitation and a safe supply of drinking water are basic human needs. The safe disposal of human excreta is central to sanitation.

In chapter 1, population growth and the resulting urbanization in the 19th century were cited as major factors contributing to the challenge of infectious disease control. Industrial development and immigration led to an influx into the cities, which, in turn, fueled poor housing, overcrowding, lack of clean water, and lack of facilities for disposal of human waste. John Cairns, a British biologist, termed cities the "graveyards of mankind" (chapter 1). Urbanization is not a new phenomenon but one that dates back millennia to the times when hunter-gatherers first saw a benefit in pooling their meager resources by living together in villages. These villages have evolved into today's megacities resplendent with shopping malls, simultaneously imposing the seemingly impossible burden of establishing public health measures and infrastructure to keep pace with the burgeoning masses (In the News 12.1). The 19th-century outbreaks of cholera in London (chapter 7) illustrate the consequences of inadequate sanitation and hygiene. Plagues have besieged civilization from time immemorial. As civilizations developed, public health measures failed to keep pace; the germ theory and the idea of contagion were yet to be developed. Arno Karlen, in his book *Man and Microbes*, wrote:

IN THE NEWS 12.1

No More Room for Garbage in Japan

People in Congo live with disease and war
Many flee to remote forests, where food is hard to find

One million people crowd into refugee camps

Farmers Found in Contempt of Court in Waste Cleanup Case

A Strain on the Public Health Infrastructure

Conditions in many European capitals matched those of Third World cities today. London's population rose 20% in the 1820s alone. People lived packed in tiny rooms, dripping basements, and airless attics. Coal smoke blackened the air; poor ventilation, at home and at work, aided the spread of airborne diseases. Garbage and refuse were collected irregularly or not at all. The city stank, animal manure filled the street—and the many homes in which people kept pigs and chickens. Houses commonly relied on cesspits that were emptied into streets or open ditches. The poor saved their feces in cellars for sale as night soil. The city's seven sewer systems were uncoordinated and relied on defective pipes. They received tons of human and animal feces, dead animals, waste from abattoirs [slaughter houses], effluvia from hospitals and tanneries, the occasional human corpse, and contaminated ground water from cemeteries.

All of London's refuse ended up in the Thames, which provided most of the city's water. Cholera arrived; there were eight separate water companies and just one experimental filtration system. Water not taken from the Thames came from wells, many as badly polluted as the river. The city drank, cooked, and washed in its own filth. With the crowding and dirt, once a waterborne disease was established, further person to person transmission was virtually assured.

London was not the only city to be so afflicted. Filth and squalor prevailed in Europe and around the world. Cities in the United States were hardly models of cleanliness and sanitation. Sewage disposal systems were few, and outhouses, overflowing cesspools, and garbage-littered streets flourished, as did tuberculosis, diphtheria, scarlet fever, and typhoid fever. Somewhere about the middle of the 19th century, a sanitary reform movement gradually arose from the ashes of human corpses, debris, and human and animal wastes, perhaps fired by the third cholera epidemic to hit London (on the heels of the second epidemic). The combination of disease, filth, and lack of shelter (Fig. 12.2) led to the enactment of laws relating to sanitation, including sewage and water treatment, garbage collection, and other

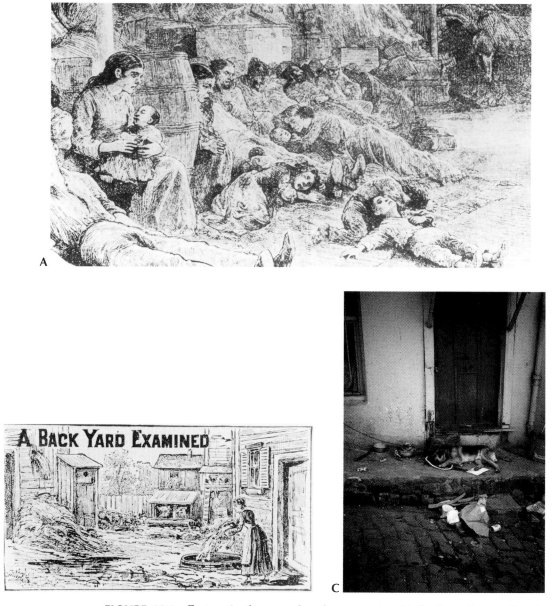

FIGURE 12.2 Factors in the spread and emergence of infectious disease. People living in squalor without adequate shelter and basic sanitation are at increased risk for infectious disease. Despite major advances over the past century, substandard levels of living, such as those shown in these old drawings (**A** and **B**) and photograph taken in 1997 in Turkey (**C**), persist in underdeveloped countries and in pockets of developed countries. (Sources: Public Health Service [**A**]; National Library of Medicine [**B**]; author [**C**].)

public health measures. In the 1880s, Louis Pasteur and Robert Koch triumphed with their discoveries (chapter 6). The germ theory of disease was established, and microbes were the link between sanitation and (microbial) disease, an idea that was embraced in Europe and in the United States.

Sanitation was "in," and sanitary engineers and bacteriologists (a term which preceded "microbiologists") flourished. In 1887, the Marine Hospital Service was established and charged with monitoring cholera in immigrants on ships coming into New York. This facility was the forerunner of today's National Institutes of Health. Other public health laboratories and organizations were established in major cities around the world, and bacteriologists and sanitary engineers worked in concert. Public health statutes promoting sanitation and good hygiene were passed and implemented (Fig. 12.3 and 12.4), and by 1900 40 states had established health departments. Over the 20th century, their focus changed from meeting the urgent and immediate needs of basic sanitation to delivering health services (Table

FIGURE 12.3   At the beginning of the 20th century, the age of sanitation began to emerge. **(A)** Public health statutes were passed and enforced by health officers. (Source: Public Health Service.) **(B)** Quarantine signs were required by law to be placed on homes in which an individual was suffering from a "contagious" disease. (Source: CDC.)

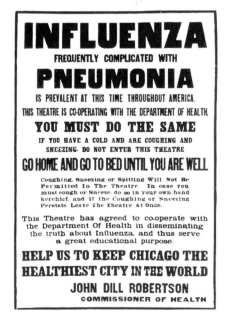

FIGURE 12.4   An early health warning sign. In the early 1900s, departments of health were formed, and they fostered in the population an appreciation of good personal hygiene in an effort to control the dissemination of infectious disease. (Source: Public Health Service.)

TABLE 12.3   The changing role of health departments[a]

Health department services in 1900 (driven by urgent needs)
Sewer construction
Water supply inspection
Sewage disposal
Nuisance and pest control
Privy inspection or removal
Milk supply inspection
Infectious disease control (tuberculosis, diphtheria and croup, scarlet fever, smallpox, and typhoid fever)

Health department services in 1999 (developing an organized approach)
Monitoring community health status to identify potential hazards
Investigating disease outbreaks and safety hazards in the community
Mobilizing community partnerships to solve health problems
Developing policies and plans that support individual and community health efforts
Enforcing laws and regulations that protect health and ensure safety
Linking populations with needed personal health services and ensuring the provision of health care when otherwise unavailable
Ensuring a competent public health and personal health care workforce
Evaluating effectiveness, accessibility, and quality of personal and population-based health services
Researching new ideas and innovative solutions to health problems

[a]Source: CDC.

FIGURE 12.5   Garbage disposal is an important public health measure. Garbage accumulation attracts rats and other rodents and animals that serve as reservoirs and vectors of infectious disease. (Author's photo.)

12.3). The 1920s through the 1950s witnessed great strides in public health strategies to control infectious diseases. Hospitals designed to treat these diseases sprung up in communities across the nation and the industrialized world. Great attention was paid to the construction of water and sewage treatment facilities, chlorination, better housing, control of tuberculosis and venereal diseases (now called sexually transmitted diseases), food production and distribution, animal and pest control measures, and garbage disposal (Fig. 12.5). The public was bombarded with information regarding the evil of "germs" and their transmission from the sick to the healthy by various modes of transmission. The "gospel of germs" was accepted, and rub-a-dub-dub, scrub, dust, and clean-clean-clean were heralded; somewhere along the line, the expression "cleanliness is next only to godliness" became a household dictate (Fig. 12.6); some say this expression is attributed to India's Mahatma Gandhi.

These new efforts paid off. Malaria, plague, and tuberculosis were markedly reduced; the last major outbreak of plague in the United States occurred during 1924 and 1925 in Los Angeles.

The proper disposal of human wastes remains a major problem throughout the world, particularly in developing countries and in poverty pockets in developed countries. "Open defecation" (directly onto the ground) contributes greatly to the presence of microbial diseases. It is estimated that in India, a case in point, over 100 million households have no toilets and 10 million households use buckets for waste disposal. Further,

FIGURE 12.6   "Cleanliness is next to godliness." This expression promotes cleanliness as a means of minimizing the risk of catching a disease by vanquishing "bad spirits." (Source: Public Health Service.)

estimates are that 900 million liters of urine and 135 million kilograms of fecal material need to be disposed of each day. About half a million children die every year in India because of dehydration resulting from diarrheal diseases which are frequently traceable to open defecation. The subject of toilets and defecation is hardly dinnertime conversation, but it is a fact of life and an integral part of the history of human hygiene. You may find this hard to believe, but in New Delhi, India, there is the Museum of Toilets, which has a "rare collection of artifacts, pictures and objects depicting the historical evolution of toilets since the year 2500 B.C." (A replica of the throne of King Louis XIII, on display at the museum, illustrates that he actually had a commode under his throne which he used while giving audience.) Publicity for the museum states that the museum also has "a rare collection of beautiful poems"; figure that one out!

The development of sewage systems in developing countries to provide for safe disposal of human wastes is a top priority for organizations such as the United Nations and the World Bank. While it is true that in developed countries, indoor toilets are pretty much taken for granted, much of the world lacks sanitary disposal methods. Bathrooms and toilets, such as our society is accustomed to, are not necessary goals; certainly, less luxurious and primitive facilities, be they outdoor or indoor, are affordable and effective goals (Fig. 12.7). Programs to improve poor sanitation are in progress in slums and squatter settlements of the world's poorest countries. The Kampung Improvement Program in Indonesia has focused on covering open sanitation drains and on bringing reasonably clean water to families. The Orangi Pilot Project has reached 650,000 people in a poor neighborhood in Karachi, Pakistan. Orangi is the largest squatter settlement in Karachi, with approximately 1 million inhabitants. According to a public health official, "People don't need a flush toilet in every home or a faucet in every room. But with a standpipe for every three units, with adequate pit latrines and other forms of [waste] treatment, the services can be there and the health of the children maintained."

> **AUTHOR'S NOTE** *As a biologist, I am full of useless facts. Here is one for you:* in 1857, Joseph Cayetty invented toilet paper in the United States.

FIGURE 12.7 Human waste disposal. Contamination of food and water by fecal material is a major cause of many infectious diseases. Primitive and simple toilets are relatively inexpensive and efficient if properly constructed. (Author's photos.)

A   B

Developed countries are not spared the aftermath of fires, flood, hurricanes, wars, and other catastrophic events; they, too, witness misery as land and water become fouled with human and animal wastes and carcasses, leading to infectious disease. Hurricane Floyd hit North Carolina with a vengeance in September 1999. According to the state environmental secretary, "The rivers are now swollen with human waste, animal waste, and other pollution." On the other side of the globe, in a poverty-stricken and war-torn part of Indonesia, refugees occupy makeshift shelters in camps and live under miserable conditions with inadequate sanitation, food, and water. A newspaper account tells the plight of one family: "Mr. _____ watched cockroaches crawl over his wife and three children asleep on the dirt floor of a tent they share with three other families. He lamented the lack of food and medicine to treat the diarrhea, measles, eye infections, flu, and other ailments that spread quickly here. There is nothing but misery and sorrow in this place. All we want is to go home, but everyone believes we will be killed if we do."

At the other end of the spectrum, even those able to afford the luxury of a cruise are subject to a breakdown in sanitation and its consequences. Cruise ships can be "ships from hell," a term used by passengers in 1999 when their cruise ship caught fire and floated adrift in rough seas. Water was limited, there was no air-conditioning, and raw sewage ran through the corridors and cabins, posing an immediate threat to the passengers. So much for luxurious vacations in the sun!

The United States and other countries of the world are on a "sanitation kick." Supermarket shelves are loaded with a tremendous variety of sprays, mists, and bubble-producing solutions, all designed to control microbes (Fig. 12.8). One well-known household product claims to "kill bacteria, viruses, molds and mildew." Mattresses, cribs, playpens, and toys now boast that they contain antibacterial materials. Have we gone too far? Stuart Levy of Tufts University Center for Microbial Adaptation thinks that we have; he advocates frequent hand washing with regular soaps, not hand sanitizers. He worries about the production of antibiotic-resistant strains due to overuse of products containing antibacterials, akin to the situation that exists as a result of overexposure of microbes to antibiotics.

FOOD SAFETY

"Give us this day our daily bread" is a familiar line from the Lord's Prayer. If only we could amend it to "Give us this day our daily bread, but, please God, let it be safe and without microbial contaminants." Food safety is considered one of the 10 great public health achievements of the United States during the 20th century (Table 12.2). In January 2000, WHO announced a plan to expand its food safety initiatives in response to new challenges, including health implications of genetically engineered foods. In the United States, the Food Safety Inspection Service of the U.S. Department of Agriculture (USDA) is charged with the safety, labeling, and packaging of the nation's commercial supply of meat, poultry, and eggs.

Leaders of the sanitation movement, initiated in the early 1900s, recognized that typhoid fever, tuberculosis, scarlet fever, botulism, and other diseases were transmitted by food (including milk) and water and advocated safer food-handling procedures, including pasteurization, refrigeration, hand washing, safer food processing, and pesticide application. *The Jungle*, a powerful 1906 novel by Upton Sinclair, describes the misery of Jurgis

FIGURE 12.8 A large variety of cleaning products to control microbes are found on the shelves of markets and other stores. They come in spray, foam, and squirt containers of assorted odors, each with exaggerated claims. (Author's photo.)

Rudkis, a young Lithuanian immigrant, employed in "Packingtown," a filthy Chicago stockyard. An excerpt from the book follows:

> There were some [cattle] with broken legs, and some with gored sides; there were some that had died, from what cause no one could say; and they were all to be disposed of here in darkness and silence. "Downers," the men called them; and the packing houses had a special elevator upon which they were raised to the killing beds, where the gang proceeded to handle them, with an air of business-like nonchalance which said plainer than any words that it was a matter of everyday routine. . . . and in the end Jurgis saw them go into the chilling rooms with the rest of the meat, being carefully scattered here and there so that they could not be identified.

The public reacted so strongly that within a year after publication of Sinclair's book, the Pure Food and Drug Act was passed. In recent years, news shows such as *Dateline* and *20/20* have shocked and sickened the public with episodes about food processing. Despite great progress, food-associated illnesses continue to be a major problem and something we need to chew on (bad pun). The incidence of typhoid fever dramatically decreased in the second half of the 20th century. In the 1940s, *Trichinella spiralis* (chapter 10) was present in 16% of the U.S. population, with 300 to 400 cases of trichinellosis diagnosed and 10 to 20 deaths each year. In 1996, thanks to modern public health measures, only 38 cases and three deaths were reported. Despite these advances in the safety of the food supply, food-borne diseases remain a major cause of morbidity and mortality. Well over 25 microbes, including bacteria, viruses, protozoans, and worms, are causative agents of food-borne diseases, many of which are discussed in chapters 8, 9, and 10. Further, changes in food processing technologies, personal eating habits, and food distribution have contributed to the presence of new emergent pathogens (chapter 1). The CDC estimates the number of cases of food-borne diseases in the United States to be somewhere between 6.5 million and 76 million each year, with 325,000 hospitalizations and 5,000 to 9,000 deaths. More accurate estimates are not possible because of differences between states in disease reporting, data collection, diagnostic procedures, and analytical methods.

How does food become contaminated along the pathway from farm to plate? What are the sources of contamination? The answer to these questions points the way to implementation of food safety practices designed to decrease the incidence of food-borne illnesses. In 1999, the American Academy of Microbiology issued a report on food safety which cited seven practices in the food supply system that potentially contribute to the creation and persistence of situations that favor the likelihood of microbial contamination at levels high enough to cause disease (Table 12.4).

Ultimately, foods are prepared and consumed in the home—the last stage in the journey from farm to plate. Cleanliness in the kitchen has received considerable attention in recent years. A biologist who calls himself the "Sultan of Slime" claims that "our bathrooms may be cleaner than our kitchens"; perhaps a course in "kitchen ecology" is not a bad idea. Dish towels, sponges, counter surfaces, and hands are all culprits involved in the spread of food-borne pathogens. Chapter 8 lists some commonsense tips to minimize the problem.

New measures to increase food safety, some of which are "gimmicky," are in vogue (In the News 12.2). The great (but not too inspiring) debate regarding the use of wooden versus plastic cutting boards still exists. One

TABLE 12.4 Practices significant to food-borne diseases[a]

Preslaughter practices
Preharvest practices
Slaughtering practices
Harvesting practices
Agriculture
Food processing
Food preparation and consumption in the home

[a]Source: S. Doores (preparer), *Food Safety: Current Status and Future Needs* (American Academy of Microbiology, Washington, D.C., 1999).

New Products in the Fight against Food-Borne Infections

entrepreneurial company now advertises Cut and Toss disposable cutting boards. Toxin Alert, a Canadian company, is developing a plastic food wrap that changes color when it is contaminated with salmonellae, campylobacters, *Escherichia coli,* or other bacteria. Pasteurized eggs are now on the scene and are marketed under the brand name Davidson's Pasteurized Eggs; they look and taste like unpasteurized eggs. Maybe "sunny side up," "over easy," and "runny" eggs will soon be okay—an "eggcellent" advance! This new technique, allowing in-shell pasteurization, may "egg on" other producers to make pasteurized eggs more available.

There is some comfort in knowing that city and state health departments, as well as agencies at the national and international levels, including the CDC, the USDA, and WHO, are all working to develop more efficient strategies of surveillance and of standardization of sampling procedures at the international level. In the United States, since 1998, the largest meat and poultry processing plants have been required by the USDA to implement the Hazard Analysis and Critical Control Point program targeted at reducing contamination during food processing. In 1997, PulseNet, a network designed to track food-borne illnesses at a national level, was established. The network allows comparison of DNA fingerprint patterns through electronic communication. In 1998, PulseNet revealed that isolates of *Listeria* responsible for outbreaks of infection involving an estimated 100 cases and 22 deaths were linked to contaminated hot dogs and deli meats from a single processing plant.

It is convenient to adopt a "Big Brother" attitude and to rely on the food industry and governmental agencies as our guardians to protect us from the perils of food-borne diseases. However, as consumers, we share in the responsibility of minimizing the consumption of contaminated foods by exercising common sense in our daily lives. There is no need to become a "food safety fanatic." On the other hand, the specter of mad cow disease, a prion-caused disease which now terrifies the European Union, sounds a word of caution the world over. (Prions and mad cow disease are discussed in chapter 15.)

About 175 years ago, Jean Anthelme Brillat-Savarin, a renowned gastronome, stated, "Tell me what you eat and I will tell you what you are." What do you think of this?

CLEAN WATER

"Water, water everywhere, nor any drop to drink." This famous line from the classic 1798 poem *The Rime of the Ancient Mariner,* by Samuel Taylor Coleridge, depicts the desperate plight of an old mariner surrounded by a sea of undrinkable water. It serves to depict the desperation of one-fourth of the world's population that have only limited access to water that may or may not be safe (Fig. 12.9). Not surprisingly, there is a strong correlation between access to safe drinking water and child health (Fig. 12.10). WHO estimates that at any given time, perhaps one-half of all people in the developing world are suffering from one or more of the six main diseases associated with drinking contaminated water. In 1999, an international panel of health experts reported that the poorest people in the world are paying many times more than their richer compatriots for the water that they need to live and are getting more than their share of deadly diseases because supplies are dangerously contaminated. Control of waterborne diseases is a particularly difficult problem because of its inextricable link to sanitary facilities for excreta disposal, preventing fecal material from gaining access to water sources and food supplies. Anywhere from a 20 to 80% decline in morbidity and mortality is possible with improved water sanitation. A variety of water-related diseases caused by bacteria, viruses, protozoans, and worms are presented in chapters 8, 9, and 10.

Improvement of water quality was, and continues to be, a major public health priority which has resulted in the decline of waterborne and water-associated diseases, including typhoid fever and cholera. Unfortunately, people in developing countries do not have the luxury of clean

FIGURE 12.9   Clean water: a luxury. For much of the world's population, water is not piped into their homes and must be transported in containers. Bodies of water may serve for clothes washing, bathing, water for animals, and drinking water. (Author's photos.)

A

B

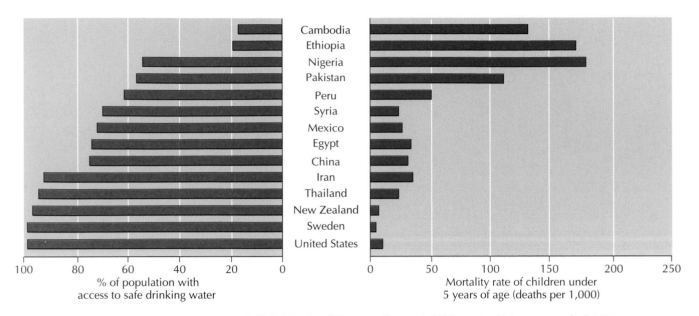

FIGURE 12.10 Water quality and child survival. Access to safe drinking water correlates with low child mortality. Compare the data for Cambodia with those for the United States; the differences in the availability of safe drinking water and in child mortality are striking. (Sources: UNICEF, *The State of the World's Children 1993* [UNICEF, Geneva, Switzerland]; UNICEF statistical databases [http://www.child info.org].)

water for bathing, drinking, and recreational use. In some countries you could end up with guinea worms biting their way through your flesh or with cholera so devastating that in several hours your life is threatened by severe dehydration resulting from massive loss of water through diarrhea. In the village of Chiladi in rural Brazil, you could end up with brown spots on your hands and serious symptoms due to drinking the arsenic-contaminated water that was supposed to be clean (Box 12.1). However, even in developed countries, vigilance needs to be maintained. A safe water supply can never be taken for granted, as the citizens of Milwaukee painfully discovered in 1993 when *Cryptosporidium parvum* caused the largest outbreak of waterborne disease in U.S. history (chapter 10).

Since 1920, data have been available on disease outbreaks associated with dirty water. The CDC and the U.S. Environmental Protection Agency (EPA) have maintained a collaborative surveillance system since 1971 and continue to report on the occurrence of waterborne diseases with the goal of characterizing and identifying the causative agents. Over the past century, legislation has been implemented in an effort to regulate the nation's water supply and, as shown in Fig. 12.11, the initiatives have paid off. The EPA, Food and Drug Administration (FDA), CDC, U.S. Geological Survey, and National Marine Fisheries Service as well as state and municipal health departments have contributed to this effort. In 1999, the American Society for Microbiology released a report that was critical of proposed initiatives to promote safer water supplies in the United States. The criticism was based on several points, including the likelihood that fragmentation of regulatory responsibilities among several federal agencies and between state and local levels would lead to inconsistencies in water quality and treatment standards. The report was also critical of the use of *E. coli* as an indicator of fecal contamination, since it does not indicate the

BOX 12.1   Arsenic in the Well and in the Wood

Sometimes, in an effort to alleviate a problem, well-meaning public health officials initiate interventions that backfire and create a different problem. The building of the Aswan Dam (see chapter 1) is an example. Another example is the current arsenic crisis in Bangladesh and in West Bengal, India. To combat the high incidence of death and disease resulting from drinking surface water contaminated with microorganisms, health officials installed tube wells to capture groundwater, but this led to an unexpected public health problem of a different kind. The number of illnesses resulting from waterborne diseases was markedly reduced, but the price was too high; the groundwater was contaminated with arsenic in concentrations well above the accepted levels.

A wide variety of bacteria, viruses, and protozoans (see chapters 8, 9, and 10) are waterborne; "clean" water is a luxury not universally available. Epidemics of cholera, polio, diarrhea, typhoid fever, and amebiasis are a few examples of waterborne diseases. Until about 30 years ago, millions of people in Bangladesh and West Bengal, which are poor and densely populated regions, drank surface water contaminated with disease-producing microbes from shallow hand-dug wells, streams, and ponds. The result was devastating epidemics of disease. In the 1970s, international agencies, particularly the United Nations Children's Fund (UNICEF), invested huge sums of money in sinking tube wells to tap the clean water of the Ganges. (A tube well is a simple device constructed of steel pipes fitted with a pump handle. You might find one at a roadside picnic area where piped water is not available. The pump is sealed topside to prevent water leaking back down the pipe. Microbes are filtered out as the groundwater trickles through the aquifer, resulting in

microbiologically safe water.) The number of tube wells in Bangladesh and West Bengal is estimated at 3.5 million. Untreated tube well water was expected to be the answer to the epidemics associated with the use of contaminated unsafe surface water. But in the 1980s, evidence of arsenic contamination was reported, and by the mid-1990s, an arsenic crisis existed. The source of the arsenic was geologic; it is released into groundwater under naturally occurring conditions. Millions of people in the area are now drinking water contaminated with arsenic far above acceptable levels. Many have been diagnosed with arsenic poisoning. UNICEF's chief of water and environmental sanitation claims that Bangladesh has become the victim of its own success. Arsenic is usually excreted from the body, but if excess amounts are ingested, it accumulates. According to the Arsenic Crisis Information Centre, "Water with high levels of arsenic leads to health problems such as melanosis, leuko-melanosis, hyperkeratosis, black foot disease, cardiovascular disease, hepatomegaly, neuropathy and cancer. . . .Early symptoms of arsenic poisoning can range from the development of dark spots on the skin to a hardening of the skin into nodules—often on the palms and on the soles." WHO estimates that these symptoms can take 5 to 10 years of constant exposure to arsenic to develop. Over time, these symptoms can become more pronounced, and in some cases, internal organs, including the liver, kidneys, and lungs, can be affected. In the most severe cases, cancer can develop in the skin and internal organs, and limbs can be affected by gangrene. While there is evidence that links arsenic to cancer, it is difficult to say how much exposure, over what period of time, will result in this disease.

It seems that the people of Bangla-

desh and West Bengal have unwittingly gone "from the frying pan into the fire." The choice may be between drinking arsenic-free, microbiologically contaminated water, with its inherent risk of serious illness and possible (relatively quick) death, and drinking arsenic-contaminated, microbiologically safe water, with the risk of dying after several years from arsenic poisoning. Fortunately, new and alternative safe water options based on rendering surface water microbiologically safe and on treating arsenic-contaminated groundwater are increasingly available. The choice is dependent upon community factors, including culture and socio-economic conditions.

Look around your neighborhood, and chances are you will see decks, fences, and playground equipment built with pressure-treated wood touted for its resistance to rot and insect damage. These qualities, which led to the popularity of the product, are the result of lacing the wood with an arsenic-based preservative—the same arsenic that has poisoned millions of people in Bangladesh, West Bengal, and numerous other regions. In February 2002, representatives of the chemical and home improvement industry agreed to phase out arsenic-treated products over the next 2 years and to seek alternatives. Environmental groups are urging companies to stop selling the wood before the end of 2003. According to a senior vice president of the Environmental Working Group, the product (pressure-treated wood impregnated with arsenic) "should never have been put on the market in the first place." Meanwhile, the EPA is studying whether children with repeated exposure to arsenic preservative are at a heightened risk of developing lung, skin, or bladder cancer.

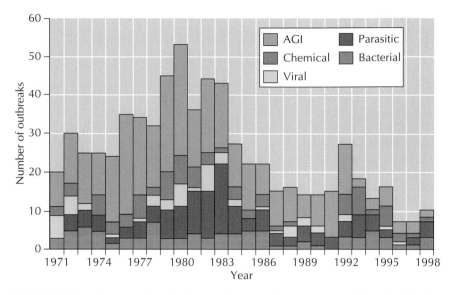

FIGURE 12.11 Number of waterborne disease outbreaks associated with drinking water in the United States, 1971 to 1998. Legislation to regulate the nation's drinking water over the past 3 decades has been successful. National, state, and local organizations and governments have worked together in this achievement. AGI, acute gastrointestinal illness of unknown etiology. [Redrawn from R. S. Barwick, D. A. Levy, G. F. Craun, M. J. Beach, and R. L. Calderon, *Morbidity and Mortality Weekly Report* **49**(SS-4):1–44, 2000.]

presence of pathogenic viruses, protozoans, and nonfecal pathogens; thus, there is a need for the development and implementation of newer, more sophisticated indicator strategies. Further confounding the issue of water safety is the fact that some pathogenic protozoans (*Giardia, Cyclospora,* and *Cryptosporidium* [chapter 10]) are resistant to the usual levels of chlorination. The EPA recently (2001) announced a proposal to use ultraviolet disinfection in New York to treat drinking water. In this procedure, ultraviolet light rays shine through the water and disrupt the DNA of pathogens that resist chlorination as they pass through the rays. Ultimately, the provision of clean water to all residents requires constant surveillance, maintenance, and improvement of the supply system at the community level (In the News 12.3).

IMMUNIZATION

Background

A 1999 CDC report (Table 12.2) lists vaccination (immunization) as one of the 10 great public health achievements of the 20th century. No wonder, considering the millions of lives that have been saved around the world as a result of vigorous vaccination campaigns targeted against bacterial and viral diseases. The lives of 3 million children are saved in the United States alone each year because of routine immunization. For some diseases, the decline over the past century is 100%, or close to 100% (Table 12.5). According to the CDC, in the United States in 1999 there were 900,000 fewer cases of measles than in 1941 and 21,000 fewer cases of polio than in 1951.

A "miracle" occurred in 1980 when WHO announced that smallpox had been eradicated from the face of the earth, thanks to a long process that

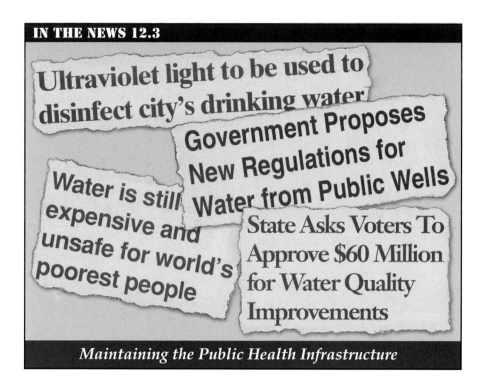

IN THE NEWS 12.3

Ultraviolet light to be used to disinfect city's drinking water

Government Proposes New Regulations for Water from Public Wells

Water is still expensive and unsafe for world's poorest people

State Asks Voters To Approve $60 Million for Water Quality Improvements

Maintaining the Public Health Infrastructure

began with the pioneering work of Edward Jenner in 1796. Smallpox was the first disease to be eradicated, because of the use of vaccination and the fact that the disease has no human reservoirs. In 1881, almost a century after Jenner's work, Louis Pasteur developed a vaccine against anthrax in animals, and only a few years later (1885), he developed a vaccine against human rabies. Before 1900, vaccines against three additional diseases were available, and during the 20th century, 21 additional diseases were added to the list (Table 12.6). Polio is near eradication, and other diseases are on the "hot list" for eradication, including guinea worm and measles. Vaccine development is an area of intensive research. Vaccines against AIDS, tu-

TABLE 12.5   Impact of 20th-century public health achievements on selected diseases

Disease	% decline
Smallpox	100
Diphtheria	100
Pertussis	95.7
Tetanus	97.4
Poliomyelitis (paralytic)	100
Measles	100
Mumps	99.6
Rubella	99.3
Congenital rubella syndrome	99.4
Haemophilus influenzae type b	99.7

[a]Adapted from CDC, *Morbidity and Mortality Weekly Report* **48:**243–248, 1999.

TABLE 12.6 Vaccine-preventable diseases[a]

Disease[b]	Year of vaccine development or U.S. licensure
Smallpox (V)	1798
Rabies (V)	1885
Typhoid (B)	1896
Cholera (B)	1896
Plague (B)	1897
Pertussis (B)	1926
Tetanus (B)	1927
Tuberculosis (B)	1927
Influenza (V)	1945
Yellow fever (V)	1953
Poliomyelitis (V)	1955
Measles (V)	1963
Mumps (V)	1967
German measles (V)	1969
Anthrax (B)	1970
Meningitis (B)	1975
Pneumonia (B)	1977
Adenovirus (V)	1980
Hepatitis B (V)	1981
Haemophilus influenzae type b (B)	1985
Japanese encephalitis (V)	1992
Hepatitis A (V)	1995
Chicken pox (V)	1995
Lyme disease (B)	1998
Rotavirus (V)	1998 (withdrawn 1999)

[a]Adapted from CDC, *Morbidity and Mortality Weekly Report* **48**:243–248, 1999.
[b]B, bacterial disease; V, viral disease.

berculosis, and malaria—diseases that kill more than 5 million people each year, including many children—are badly needed. The availability of vaccination is a story of "good news and bad news." The good news is that much of the world is immunized against a variety of microbial diseases, while the bad news is that underdeveloped countries have not shared in this advance. Tragically, whooping cough, measles, and tetanus are diseases which take the lives of over 3 million children each year, despite the fact that immunization is available. The good news of the bad news is that high-priority efforts are under way to remedy this situation. The Bill and Melinda Gates Foundation's Child Vaccination Program is one example; the program was established in 1998 with a $100 million grant and aims to reduce the approximately 15-year time lag between the availability of a vaccine in developed countries and its availability in developing countries. The Global Fund for Children was launched in 1999 with a $750 million grant and will build on the Gates Foundation's vaccination program. The Global Alliance for Vaccines and Immunization will collaborate in the mission of delivering vaccines to children in 50 or more of the world's poorest countries.

Disease eradication is today's public health buzzword, resulting from the success of vaccination programs, particularly in developed countries around the world. Smallpox is an example of global cooperation involving 33 nations. In 1998, a team of genetic engineers announced that the genetic map of the bacterium that causes syphilis had been decoded, raising hopes for the development of a vaccine against this ancient disease. The impact of vaccination and justification for its inclusion on the list of 10 great public health achievements of the 20th century are summarized by the following statement of the senior officer of health services in Tanzania:

> The ward for measles, if you go there now, is empty! And tetanus—it was killing many of our babies. In Muhimbili alone we were admitting 300 to 400 children per year to the hospital. Now it is down to only two! But while we see lots of progress, we still have our job cut out for us. There are children who don't get immunized at all and only the rich can afford hepatitis B or Hib [*Haemophilus influenzae* type b] vaccine. So we cannot rest, we must strive to bridge these gaps as soon as we can.

Recall from chapter 11 that the body's immune system acts to destroy or get rid of invading microbes by antibody production and cell-mediated immunity. Vaccines act to stimulate the immune system to produce specific antibodies or to mount a cell-mediated response (Fig. 12.12). They "trick" the immune system into responding as though the vaccines were the real thing—that is, the actual microbes. To put it another way, a vaccine does not cause infection but mimics a microbe, or microbial component, eliciting an appropriate immune response. For example, measles vaccines result in anti-measles virus antibodies, flu vaccines result in anti-influenza virus antibodies, and polio vaccines result in antipoliovirus antibodies. This is like a dress rehearsal for future challenges. As discussed in chapter 11, B cells are the antibody-producing cells, but some are held aside, establishing a bank of preprogrammed memory cells against the antigens involved. Hence, although the antibodies resulting initially from the antigens in the vaccine may be long gone, the memory cells are "jump-started," resulting in circumvention of the usual 10- or 12-day lag time. The presence of antibodies within several hours is like achieving (almost) "instant protection." The memory cells are like guard dogs on patrol ready to attack. Some vaccines

FIGURE 12.12   How immunization works.

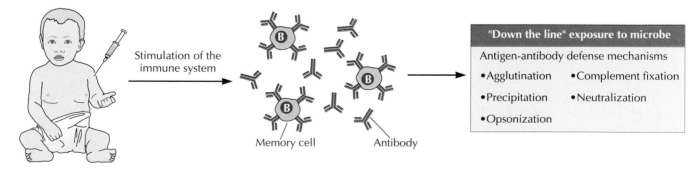

TABLE 12.7 Outline of immunization

Active immunization (individual makes own antibodies)
Natural (subclinical or clinical disease and recovery)
Artificial (vaccines for immunization "shots")
Live, attenuated microbes
Killed microbes
Toxoids and other purified microbial components
New and experimental vaccines
Passive immunization (individual receives preformed antibodies)
Natural (in utero mother-to-infant passage; breast milk)
Artificial (use of immune globulin)

afford lifetime protection; others require "boosters," and the flu shot needs to be administered every year.

Active Immunization

Table 12.7 outlines the categories of immunization. Active immunization is the result of stimulating a person's immune system to produce antibodies and memory cells and generally confers immunity over a relatively long time. Natural active immunity is achieved by the natural process of recovering from a particular disease. Analysis of an individual's blood serum frequently reveals the presence of antibodies against which there is no clinical history of disease, indicating that disease at a subclinical level has occurred. The use of vaccines, however, is artificial in the sense that vaccines are administered into the body to provoke an antibody immune response as a future protective measure.

As indicated in Table 12.7, active immunization can be accomplished in a few ways; the method of choice reflects the best protection for the particular disease. All of the procedures must meet three basic requirements: safety, effectiveness, and stability. Additionally, an ideal vaccine (Table 12.8) would be affordable to developing countries.

In the United States, the FDA is the agency responsible for issuing a license to vaccine manufacturers, allowing the vaccine to be widely distributed. Safety issues are of prime importance; vaccine safety is further addressed in a later section. For a vaccine to be effective, it must stimulate an immune response which affords protection to the recipient of the vaccine. Vaccine preparations need to be stable over time to make them cost-effective. Some vaccines can be stored at room temperature, whereas others need to be refrigerated, presenting a problem in distribution of these vaccines to Third World countries. Unfortunately, on too many occasions, vaccines end up being junked because refrigeration requirements have not been adhered to. Strict attention needs to be paid to expiration dates and to conditions of storage. Recombinant vaccines, including recombinant plant vaccines, are described below and, assuming their effectiveness, will allow easier vaccine distribution to the Third World.

Types of Active Vaccines

The point to be discussed now is how live, disease-producing microbes are turned into non-disease-producing antibody-stimulating agents. This can be accomplished in four ways (Table 12.7).

TABLE 12.8 An ideal vaccine

General vaccine requirements	Ideal vaccine requirements
Safety	Safety
Effectiveness	Effectiveness
Stability	Stability
	Affordability
	Administration as a nasal spray or edible vaccine
	No need for refrigeration; stability at ordinary "tropical" temperatures
	One dose or one shot
	Long shelf life

Live Attenuated Microbes

Vaccines made from live attenuated microbes, the majority of which are viruses, confer long-lasting immunity, frequently lifetime. The word "attenuated" means weakened in terms of virulence. These vaccines produce limited infection but not overt clinical disease; the infection may be manifested by nausea, headache, fatigue, and soreness at the site of injection for about 24 hours; frequently, there are no symptoms. Most live attenuated viral vaccines have been achieved by serial (repeated) transfer in tissue culture (Fig. 12.13), allowing the production of random and unpredictable mutants. These mutants are tested in laboratory animals, and those strains producing no symptoms are selected for further trial. The **bacillus Calmette-Guérin (BCG) vaccine** against tuberculosis utilizes a tuberculosis strain (*Mycobacterium bovis* BCG) attenuated by repeated subculturing over a 10-year period (1908 to 1918) on laboratory medium with no reversion to virulence for over 80 years. Pasteur's vaccine against rabies was the result of serial passage in the spinal cord of dogs and rabbits, a procedure which rendered the live virus nonvirulent but continued to provoke protective antibodies against the virus. The major concern of the use of live and attenuated vaccines is the possibility of reversion to virulent forms.

Killed (Inactivated) Microbe Vaccines

Killed microbe vaccines are used when attenuation has not been accomplished or when reversion to the virulent type is considered to be too risky. Killed microbes present no risk but are not as effective at stimulating antibody production; some require multiple doses to maintain protective antibody levels. The Salk polio vaccine and the vaccines against plague and influenza are examples.

Toxoids and Other Purified-Component Vaccines

Some of the most serious bacterial diseases (diphtheria, tetanus, cholera, and botulism) result from the production of very potent exotoxins (chapter 8). Neutralization of these toxins with antibodies is an effective strategy. Bacterial toxins can be inactivated by heat or formaldehyde, resulting in a loss of toxicity, but they retain their property of stimulating antibody production; these altered toxins are referred to as toxoids. The diphtheria-pertussis-tetanus (DPT) vaccine, commonly administered to children at about the age of 2 months, contains diphtheria and tetanus toxoids.

New and Experimental Vaccines

In recent years, the development of DNA technology has yielded many biological products that have proved to be of considerable value in the prevention and treatment of a variety of diseases, including those caused by microbes. Recombinant DNA vaccines are made by inserting genes for specific antigens into nonvirulent host organisms, including bacteria, yeasts, viruses, and insects. These genes are expressed in the host organism and result in production of the inserted antigens, which are then used as vaccines to stimulate an immune response. A vaccine against hepatitis B was the first recombinant-antigen vaccine to be used in humans. This vaccine was produced by cloning (inserting) a major surface antigen of the virus into yeast cells, which were then grown in huge vats. The yeasts express the hepatitis B virus antigen during their growth. These recombinant antigens are safe and produce protective antibodies.

DNA vaccines are a new and promising approach to immunization. In

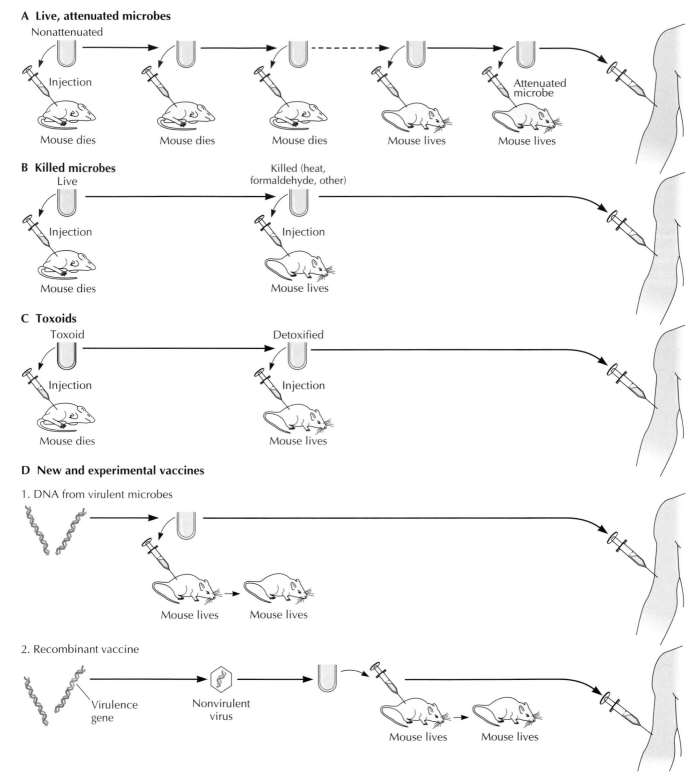

A Live, attenuated microbes

Nonattenuated

Injection

Attenuated microbe

Mouse dies   Mouse dies   Mouse dies   Mouse lives   Mouse lives

B Killed microbes

Live   Killed (heat, formaldehyde, other)

Injection   Injection

Mouse dies   Mouse lives

C Toxoids

Toxoid   Detoxified

Injection   Injection

Mouse dies   Mouse lives

D New and experimental vaccines

1. DNA from virulent microbes

Mouse lives   Mouse lives

2. Recombinant vaccine

Virulence gene   Nonvirulent virus

Mouse lives   Mouse lives

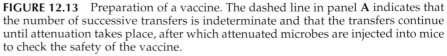

FIGURE 12.13 Preparation of a vaccine. The dashed line in panel **A** indicates that the number of successive transfers is indeterminate and that the transfers continue until attenuation takes place, after which attenuated microbes are injected into mice to check the safety of the vaccine.

this strategy, microbial DNA is inserted into plasmids (chapter 4). These plasmids carrying the microbial DNA are then injected directly into the host, and the microbial DNA is expressed by the host cells as protein. (Recall from chapter 4 that DNA "produces" proteins.) These proteins are recognized as foreign by the host's immune system and stimulate an immune response. DNA vaccine development for a variety of microbial diseases, including influenza, tuberculosis, malaria, Lyme disease, and hepatitis C, is under way.

Vaccine Safety

How safe are the vaccines on the market? No vaccine (or other medication) is 100% safe and without risk. The better question to be asked is, "Do the benefits of the vaccine outweigh the risk?" The FDA, the licensing agent, does not approve a vaccine unless initial trials indicate the benefits clearly outweigh the risks. Vaccines are subject to particularly high safety standards, because, unlike other health treatments, they are given as preventives to healthy people. Vaccines have greatly reduced the burden of infectious diseases around the world, particularly in the richer countries, which is proof of their effectiveness. Licensure is a rigorous process involving three phases of clinical trials and may take 10 or more years.

Vaccines are manufactured by pharmaceutical companies under the watchful eye of the FDA; each batch of vaccine must be approved before it can be released for use by health providers. Issues of safety, effectiveness, sterility, and purity are all evaluated by laboratory procedures, and post-marketing surveillance is conducted to identify undesirable side effects that might occur in large groups of people over long periods. As the incidence of diphtheria, whooping cough, measles, and a variety of other diseases has declined, attention has focused on the risks associated with vaccines. This is ironic in the sense that, if these diseases were still prevalent, little attention would be paid to the rare and unfortunate mishap that might occur. For example, oral polio vaccine, a live vaccine preparation, carries a risk of polio of about 1 in 7 million people. Now that polio is nearly eradicated, it is unfortunate that unwarranted concern is sometimes paid to the one case, and the realization is lost that 7 million people minus one individual were candidates for paralytic polio. While it is true that some adverse reactions have occurred after vaccination, it is also true that vaccines may be falsely blamed because unrelated events may coincidentally occur shortly after vaccine administration.

In the last 2 decades, vaccine safety issues have become prominent, and lawsuits have been filed and damages awarded on behalf of those presumably injured by vaccines, despite the lack of scientific evidence to support these claims. In 1986, in response, Congress passed the National Childhood Vaccine Injury Act, mandating that health care providers furnish a vaccine information sheet to recipients describing the risks and benefits of the vaccine. Providers are also required to report certain side effects following vaccination to the FDA's Vaccine Adverse Event Reporting System. Further, "no-fault" vaccine compensation is provided to those injured by vaccines.

The pertussis (whooping cough) vaccine, formerly made from whole cells, has in recent years been derived from a component of the microbe, resulting in fewer adverse effects. The newer DPT vaccine is called **DaPT vaccine** because of the use of this acellular pertussis (aP) microbe component. Vaccines have side effects and need to be constantly monitored and modified or withdrawn as circumstances dictate. For example, in 1976, Fort Dix, N.J., was threatened by the appearance of a new and deadly strain of swine flu viruses; in response, a vaccine was quickly developed, and 45 million people were vaccinated. Unexpectedly, in some cases, the vaccine triggered a debilitating and potentially fatal neuromuscular disease called **Guillain-Barré syndrome.** The old DPT vaccine has been associated with brain damage, autism, and learning disabilities; measles-mumps-rubella (MMR) vaccine has also been cited as a possible cause of autism, a condition characterized by severe social, communication, and behavioral problems. In May 2001, the American Academy of Pediatrics released a report based on over 1,000 references in the medical literature, concluding that there is no support for the hypothesis that the MMR vaccine causes autism. The new strategy for polio immunization recommends that all four doses consist of the Salk killed virus to eliminate the slight risk of vaccine-associated paralytic polio that might occur with the Sabin live attenuated oral polio vaccine. Rotavirus infection (chapter 9) is a potentially fatal disease in children, and the development of the RotaShield vaccine was heralded as a preventive measure against this disease. About 1 year later, RotaShield was withdrawn because of a strong association with the occurrence of intussusception (twisting and obstruction of the bowel) in 23 infants 1 to 2 weeks after vaccination. Recently, reports have suggested a possible link between Lymerix, a vaccine that protects against Lyme disease, and arthritis. Lymerix has been on the market only since December 1998. In 1995, Vivax, a chicken pox vaccine, was introduced, but it has not been widely accepted by the public because of unwarranted fears. To date, 25 states have made immunization against chicken pox mandatory; Georgia recently joined the list.

Although these examples are worrisome, you should take some comfort in the realization that the FDA attempts to stay on top of the situation. During this new century, new vaccines will be added to the list of those now available. Oral vaccines and vaccines administered by nasal sprays are on the horizon and may make vaccines easier to implement in Third World countries, reducing the burden of infectious diseases around the world. Vaccines have come a long way since the pioneering work of Jenner and Pasteur.

Childhood Immunization

The burden of infectious disease has been reduced throughout the world, most notably in the United States and in other industrialized countries, through the routine practice of childhood immunization. Examples of this have been cited in this chapter and also in chapters 8 and 9. Figure 12.14 illustrates the most current recommended childhood immunization schedule in the United States, according to the CDC Advisory Committee on Immunization Practices, the American Academy of Pediatrics, and the American Academy of Family Physicians. Included are routine immunizations against 11 diseases and immunizations against 2 others (hepatitis A and influenza) for selected populations. Some of the immunizations are against bacterial diseases, but most are against viral diseases; attenuated, killed, subunit, and genetically engineered vaccines are all represented. All of these immunizations need to be started by the age of 2 years.

The relative simplicity and low cost of immunization make a June 2000 report of the Institute of Medicine (a branch of the National Academy of Sciences that advises Congress on matters of health policy) disturbing in its statement that "thousands of inner city kids have not been immunized and millions of adults have never had basic immunizations." The report also states that "about 70,000 adults die every year from illnesses, such as the flu or pneumonia, that could have been prevented by vaccines." What a shame!

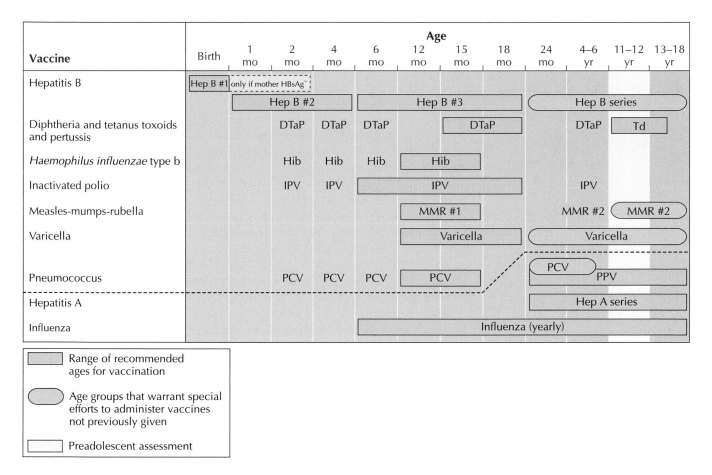

Vaccine	Birth	1 mo	2 mo	4 mo	6 mo	12 mo	15 mo	18 mo	24 mo	4–6 yr	11–12 yr	13–18 yr
Hepatitis B	Hep B #1 only if mother HBsAg⁻		Hep B #2			Hep B #3				Hep B series		
Diphtheria and tetanus toxoids and pertussis			DTaP	DTaP	DTaP		DTaP			DTaP	Td	
Haemophilus influenzae type b			Hib	Hib	Hib	Hib						
Inactivated polio			IPV	IPV		IPV				IPV		
Measles-mumps-rubella						MMR #1				MMR #2	MMR #2	
Varicella						Varicella				Varicella		
Pneumococcus			PCV	PCV	PCV	PCV			PCV / PPV			
Hepatitis A										Hep A series		
Influenza						Influenza (yearly)						

Range of recommended ages for vaccination

Age groups that warrant special efforts to administer vaccines not previously given

Preadolescent assessment

FIGURE 12.14   Childhood immunization. The CDC publishes a schedule for recommended immunizations that is approved by the Advisory Committee on Immunization Practices, the American Academy of Pediatrics, and the American Academy of Family Physicians. Vaccines below the dashed line are for selected populations. Abbreviations: Hep B, hepatitis B; HBsAg⁻, hepatitis B surface antigen negative; DTaP, diphtheria and tetanus toxoids and acellular pertussis vaccines (referred to as DaPT in the text); Td, tetanus and diphtheria toxoids, absorbed, for adult use; Hib, *Haemophilus influenzae* type b; IPV, inactivated polio vaccine; PCV, pneumococcal conjugate vaccine; PPV, pneumococcal polysaccharide vaccine; MMR, measles, mumps, and rubella; Hep A, hepatitis A. (Sources: CDC and American Academy of Pediatrics.)

A measles epidemic occurred in the United States from 1989 to 1991, resulting in 55,000 reported cases and 150 deaths. In response, in 1991, former first lady of the United States Rosalynn Carter and former first lady of Arkansas Betty Bumpers founded the not-for-profit organization Every Child by Two (ECBT), which aims to reduce infant mortality through on-time immunization. Their catchy slogan is "shots by age two if they're important to you."

Passive Immunization

Active immunization is based on stimulating a recipient's immune system to produce antibodies and memory cells. In passive immunization, by contrast, the recipient receives preformed antibodies from human or animal sources. A big advantage to passive immunization is that antibodies are immediately present at the time of infection and can be of lifesaving value, as in cases where exposure or symptoms have already occurred. The immunity gained, however, is relatively short-lived and limited to the presence of the administered antibodies in the recipients; further, there is no immunological memory. Table 12.9 summarizes the distinctions between active and passive immunization.

Before the 1940s and the advent of antibiotics, immune therapy (the use of immune serum) was commonly used, particularly for diphtheria, tetanus, and pneumococcal pneumonia. The first two diseases, as previously described, are toxemias; that is, the diseases are a manifestation of the production of a lethal toxin. Antitoxins (antibodies against toxoids) are produced by the injection of toxoids (inactivated toxins) into horses to produce immune serum. In the treatment of diphtheria, about all that could be done in the days of the "horse-and-buggy" doctors was the administration of antiserum in a desperate attempt to save lives.

Immunotherapy with animal serum is hazardous because of possible complications arising from the fact that the antibodies, along with other components of the blood serum, are considered foreign protein by the recipient's immune system. A condition known as **serum sickness** can result; it is characterized by the formation of antigen-antibody complexes which are deposited in the skin, kidney, and other body sites from immunotherapy. Nevertheless, under certain circumstances, immunotherapy is still used when immediate antibody protection is required.

The use of human immune serum, taken from individuals following vaccination or during their convalescence period from specific diseases, will minimize but not eliminate the risk of serum sickness. This immune serum, known as immune globulin, contains high levels of specific antibodies. For example, tetanus immune globulin is rich in antibodies against tetanus, and varicella-zoster immune globulin has a high concentration of antibody against the virus that causes chicken pox and shingles. Consider a case in which an individual goes to an emergency room having sustained a puncture and is at risk for tetanus. Should that person receive active immunization by a booster shot with tetanus toxoid, passive immunization with tetanus immune globulin, or both? The answer depends on the person's immune history. If the individual has, within the last 10 years, received tetanus toxoid as a booster, all that is necessary is another booster to effectively stimulate those memory cells preprogrammed to produce tetanus antibodies almost immediately. However, if the individual has not received (or has no knowledge of) past immunizations against tetanus, immediate protection against the tetanus toxin is necessary. There is not sufficient time to make antibodies from scratch, in which case tetanus immune globulin should be administered, along with the first dose of toxoid, to be followed by the tetanus series to establish future protection.

Antitoxins against the deadly toxins injected into the body by certain species of snakes and arachnids (spiders, scorpions, etc.) are examples of

TABLE 12.9 Active versus passive immunization

Property	Active	Passive
Protection	Waiting period	Immediate
Duration	Extended memory	Limited (no memory)
Adverse reactions	Possible	Possible

lifesaving passive immunization. If you are bitten by a poisonous snake or scorpion, you need immediate antibody protection. There are a very large number of poisonous scorpions in Mexico.

ANTIBIOTICS

History and Background

The first "wonder" drugs were the sulfonamide (sulfa) drugs, introduced in the pre-antibiotic era. These drugs, although they are antimicrobials, are not antibiotics, since they are synthetic compounds and not products of microbes. The sulfa drugs saved millions of lives in World War II; medics were taught to sprinkle sulfur on wounds sustained on the battlefields. In 1983, Lewis Thomas wrote:

> Then came the explosive news of sulfanilamide, and the start of the real revolution in medicine. I remember with astonishment when the first cases of pneumococcal and streptococcal septicemia were treated in Boston in 1937. The phenomenon was almost beyond belief. Here were moribund patients, who would surely have died without treatment, improving in their appearance in a matter of hours of being given the medicine and feeling entirely well within the next day or so.

The antibiotic era was ushered in with the discovery of penicillin. On February 12, 1941, Police Constable Robert Alexander of Oxford, England, was the first patient in the world to receive penicillin. Alexander was seriously ill with a staphylococcal infection that started with a small sore at the corner of his mouth. Despite treatment with sulfonamides, the staphylococci spread uncontrollably into his bloodstream, resulting in numerous abscesses over his body and the spread of infection to the rest of his face, eyes, and scalp, necessitating removal of his left eye. Death seemed imminent. But 24 hours after receiving penicillin, miraculously, he was much improved: his lesions showed signs of healing, his elevated body temperature dropped toward normal, and within several days his right eye was almost normal. Unfortunately, the small amount of penicillin that was available was insufficient for continued treatment. In a heroic effort to save the patient's life, doctors extracted penicillin from his urine and injected it back into his bloodstream. But the microbes gained the upper hand, and Alexander's condition deteriorated. He died on March 15, 1941.

With this dramatic event the antibiotic era began, contributing to the decline in infectious diseases fostered by improvements in sanitation and hygiene, safer foods, cleaner water, and implementation of vaccines. **Antibiotics** can rightly be considered the single most important discovery for the treatment of diseases in the history of medicine. They serve as testimony that "nature knows best"; the secretion of metabolic products by soil bacteria and fungi that inhibit the growth of other microbes, resulting in less competition, is an example of ecological antagonism at the microbial level.

The story of the discovery of penicillin is a fascinating one and centers on the observations of Alexander Fleming in 1928. It should be noted that others had previously described antibacterial properties of the penicillium mold, but it was Fleming who followed through on a serendipitous (chance) observation. (Serendipity is a significant factor in several important scientific discoveries; the genius is in recognizing the significance of the observation. Louis Pasteur wrote, "In the field of observation, chance favors only the prepared mind." In other words, it is not just sheer luck but

AUTHOR'S NOTE *While in Guatemala, I was about to stick my bare foot into my slipper without first checking it out as I had been advised to do. Suddenly a large scorpion—at least I think it was a scorpion—crawled out. It pays to follow advice! I did see one of the world's most poisonous snakes and captured it on film. That was probably the fastest close-up I have ever taken!*

rather the ability to recognize the unexpected.) Fleming was working with staphylococci and left a petri dish streaked with this organism on his lab bench while he went away on a 2-week vacation. On returning, he noted that the plate was contaminated with a common mold and that the staphylococci failed to grow only in the vicinity of the mold; the mold had produced an inhibitory substance. Fleming isolated an extract from the mold, which he called penicillin. It appears that Fleming was a modest man; he stated, "Nature created penicillin. I only found it."

Penicillin's therapeutic potential was not fully investigated until several years after its discovery, when Ernst Chain, Howard Florey, Edward Abraham, and Norman Heatley purified penicillin and successfully cured mice that had been injected with fatal doses of bacteria. Human trials were initiated and proved to be very successful. World War II triggered the production of penicillin on a scale large enough to be available to the military to treat wounded combatants, resulting in the saving of thousands of lives. By 1944, supplies of penicillin were abundant and were released for the civilian population without prescription. In 1945, Fleming was awarded the Nobel Prize in physiology or medicine, along with Chain and Florey, who helped develop penicillin into a widely available medical product.

Penicillin became a prescription drug in the mid-1950s. Chemists learned how to manipulate and modify the penicillin molecule, giving rise to semisynthetic penicillin derivatives, including methicillin, ampicillin, and penicillin V, each with distinctive and beneficial properties. In the post-World War II period, many other antibiotics were discovered, and their use brought about a rapid decline in deaths due to diseases caused by bacteria.

There is no such thing as a universal antibiotic any more than there is a universal disinfectant. Bacteria vary in their antibiotic susceptibility, and each antibiotic has a spectrum of activity against certain bacteria. Some antibiotics are more effective against gram-positive organisms, while others exhibit greater activity against gram-negative bacteria (chapter 4), but there are exceptions. A broad-spectrum antibiotic is one that is inhibitory to a large variety of gram-positive and gram-negative bacteria, while a narrow-spectrum antibiotic is inhibitory to a limited range of bacteria. Some antibiotics are extremely effective but, unfortunately, exhibit marked toxicity, rendering them not useful. Some antibiotics are very expensive, while others are not, and some are more prone to result in antibiotic-resistant strains than others. In prescribing an antibiotic from among the many that are available, cost and antibiotic resistance are considered, but effectiveness and toxicity are the central factors. A broad-spectrum antibiotic is generally prescribed when an individual is seriously ill and the causative bacteria have not been identified, so as to target a broad range of suspects. The downside of a broad-spectrum antibiotic is that a large number of bacterial species of the normal flora are killed, causing ecological disruption and allowing non-antibiotic-susceptible organisms to flourish. Some people receiving antibiotics develop oral thrush, a painful yeast infection, resulting from disruption of the normal bacterial flora that allows yeasts to outgrow the normal flora. The tongue has a whitish appearance as a result of the colonies of yeast that have colonized it; fortunately, thrush responds well to mouth rinses with antiyeast drugs. In females receiving a course of antibiotics, vaginal thrush may develop and cause pain, burning, and itching, which is treatable by vaginal rinses. Narrow-spectrum antibiotics cause less disruption in the ecological balance of microbes and also minimize the likelihood of antibiotic resistance.

Mechanisms of Antimicrobial Activity

Antibiotics act by interfering with or disrupting the metabolic pathways of the bacterial cell. (It may be useful to review chapter 4 here.) Each antibiotic has a specific mechanism of action, although there is some overlap. For example, penicillin interferes with the ability of bacterial cells to synthesize cell walls, rendering these disabled cells subject to lysis, whereas erythromycin inhibits protein synthesis. Some antibiotics are bactericidal (they kill directly), whereas others are bacteriostatic (they keep the population from growing, thus allowing the body's defense mechanisms to get rid of the invaders). Selective toxicity is exhibited by penicillin because this drug interferes with cell wall production, and human cells do not have cell walls. On the other hand, amphotericin B is quite toxic, since it acts upon the cell membrane, a structure present in human cells.

Following are descriptions of five mechanisms of action for antibiotics (Table 12.10 and Fig. 12.15).

Interference with Cell Wall Synthesis

As described above, bacterial cells have rigid cell walls (peptidoglycan) that afford protection against lysis when bacteria are exposed to the low osmotic pressure of body fluids. The beta-lactam antibiotics penicillin and cephalosporins contain structures (beta-lactam rings) which interfere with enzymes responsible for cell wall synthesis. Vancomycin, sometimes considered the "last antibiotic stronghold" because it is used when other antibiotics fail, blocks a crucial reaction necessary for cell wall synthesis.

Interference with Protein Synthesis

Protein synthesis is an integral part of a cell's activity and is the culmination of expression of cellular DNA. After DNA is decoded into mRNA, amino acids, the building blocks of protein, are assembled on structures in the cytoplasm called ribosomes; the process is similar in all cells. Bacterial ribosomes differ in their size and structure from human ribosomes, and thus they can be used as targets for some antibiotics. (Bacterial cells are procaryotic and their ribosomes are 70S, whereas human cells are eucaryotic and their ribosomes are 80S. S represents Svedberg units, a measurement of sedimentation rates.) Streptomycin is a powerful antibiotic that was discovered in 1944; its use is now usually reserved for treatment of tuberculosis. The tetracyclines, chloramphenicol, and erythromycin are commonly used antibiotics that, like streptomycin, interfere with protein synthesis by binding with procaryotic ribosomes. Chloramphenicol can cause a potentially fatal condition called aplastic anemia in which the bone marrow ceases to

TABLE 12.10 Mechanisms of antimicrobial activity

Mechanism	Example(s) of antimicrobial agent(s)
Interference with cell wall synthesis	Penicillin, cephalosporin, vancomycin
Interference with protein synthesis	Streptomycin, erythromycin, chloramphenicol
Interference with cell membrane function	Polymyxin B
Interference with nucleic acid synthesis	Rifampin, nalidixic acid
Interference with metabolic activity	Sulfa drugs

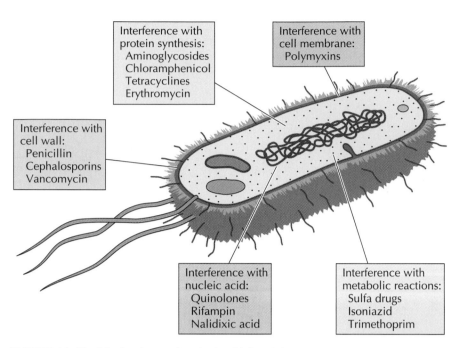

Interference with protein synthesis:
Aminoglycosides
Chloramphenicol
Tetracyclines
Erythromycin

Interference with cell membrane:
Polymyxins

Interference with cell wall:
Penicillin
Cephalosporins
Vancomycin

Interference with nucleic acid:
Quinolones
Rifampin
Nalidixic acid

Interference with metabolic reactions:
Sulfa drugs
Isoniazid
Trimethoprim

FIGURE 12.15 Mechanisms of antimicrobial activity.

produce red blood cells. Nevertheless, it is used for treating typhoid fever and certain life-threatening infections caused by anaerobic bacteria.

Interference with Cell Membrane Function

Cell membranes (chapter 4) function in a vital capacity as "gatekeepers." On the basis of their chemical and physical makeup, they control what goes into and out of the cell. Polymyxin B is an antibiotic that binds to and distorts the bacterial cell membrane, resulting in increased permeability and leakage of important molecules out of the cell.

Interference with Nucleic Acid Synthesis

The synthesis of the nucleic acids, RNA and DNA, is a necessary step in the expression of the cell's DNA; the process is a long and complicated series of chemical reactions that can be targeted for antibiotic activity. Rifampin and nalidixic acid are antibiotics that block RNA synthesis. Quinolones are a large family of synthetic drugs that act by inhibiting the action of an enzyme called DNA gyrase. DNA gyrase is responsible for the supercoiling of bacterial DNA, enabling the cell to pack DNA. Mammalian cells use different enzymes for this activity and hence are not affected by this antibiotic—another example of selective toxicity.

Interference with Metabolic Activity

Metabolism, the ability to carry out energy-generating reactions, is a key characteristic in the distinction between life and nonlife (chapter 2). Antimetabolites are drugs that are structurally similar to natural compounds involved in metabolism and that competitively bind with these enzymes, rendering them inactive; this is called molecular mimicry. The sulfa drugs work in this fashion. They mimic *p*-aminobenzoic acid, a component of the microbial cell that is necessary for synthesis of a necessary compound called folic acid. In so doing, they interfere with folic acid synthesis, preventing bacterial multiplication. Mammalian cells do not make folic acid but obtain this compound from their diet; hence, sulfa drugs can be used in human therapy.

Acquisition of Antibiotic Resistance

The experience of the past 50 years has revealed that bacteria are "smart" and develop mechanisms of resistance to antibiotics that are designed to bring about their death. (You really can't blame them!) But they are not really smart; the development of antibiotic resistance is a manifestation of the Darwinian process of natural selection—"survival of the fittest"—resulting from the widespread misuse of antibiotics. In chapter 1, the development of antibiotic resistance was cited as a major public health problem contributing to the threat of emerging infections. Antibiotic resistance is a crisis at the international level demanding attention (In the News 12.4).

How do cells develop resistance to antibiotics? What are the biological factors involved? The development of antibiotic resistance is based on genetic changes. Recall from chapter 3 that the bacterial cell, like all cells, has DNA and that the expression of its DNA confers the cell's properties. In bacterial cells, the DNA is present in a single circular chromosome; additionally, some cells possess plasmids, which are extra bits of DNA independent from the DNA in the chromosome. Some plasmids are R (resistance) plasmids, which are transferred from one cell to another and are referred to as infectious agents. Antibiotic resistance can result from mutations in the chromosomal DNA or plasmid DNA. Mutations in DNA occur spontaneously and randomly in populations of growing cells at a rate higher than 1 in 10 million; they are not caused by selective pressure, but rather they are selected for survival by selective pressure. Only the survivors multiply, and, in so doing, pass on their new "survival genes" along with all the other genes. The outcome is that the next generation carries these new survival genes. This is the basis for Charles Darwin's concepts of

IN THE NEWS 12.4

Scientists Warn of Dangers of Antibiotic-Resistant Bacteria

Antibiotic Use for Colds Can Lead to Drug Resistance

Peptide molecules show promise against drug-resistant bacteria

Gonorrhea Bacteria Becoming More Resistant to Antibiotics

Antibiotic Resistance: an International Crisis

survival of the fittest and of evolution, which also apply at the microbial level. As an analogy, consider a hypothetical population of trees in a geographical area that has suffered a drought for several years. Gradually, most of the trees die, except for a few survivors that are "lucky" enough to have preexisting mutations in their genes that allow them to survive with less water. In the years to come, the forest will be populated by drought-resistant trees. The drought did not cause the mutations but selected those spontaneous, random, and preexisting mutations that allowed the trees to survive—hence, survival of the fittest. In the same fashion, antibiotics do not cause mutations but select preexisting mutations which conferred antibiotic resistance; the genes with those mutations are passed on to successive generations during cell division.

The genes for antibiotic resistance result either from chromosomal mutations or from transfer of R plasmids from antibiotic-resistant strains to antibiotic-sensitive ones. Chromosomal mutations usually confer resistance to only a single antibiotic; the presence of R plasmids can confer resistance to several antibiotics at one time, a phenomenon which was first reported in Japan in 1959 when lab personnel noted the emergence of *Shigella* bacteria that were resistant to several antibiotics. The origin of these R plasmids is not known. There is a third mechanism of antibiotic resistance, and that is **transposons,** or "jumping genes," which may carry antibiotic resistance. Transposons are versatile and can integrate into chromosomes or into or between plasmids, allowing rapid dissemination of antibiotic resistance.

The exchange of genetic material within bacterial populations, whether it occurs on chromosomes or plasmids, leads to the development of antibiotic resistance and is a major public health concern.

The mechanisms by which antibiotics exert their bactericidal or bacteriostatic effects are described above, as are the mechanisms by which antibiotic resistance emerges and spreads throughout bacterial populations. Antibiotic-resistant microbes "fight back" by countering the effects of the antibiotic (Fig. 12.16) in several ways. In some cases, antibiotic resistance is the result of the production of enzymes that bring about inactivation of

FIGURE 12.16   Microbes fight back.

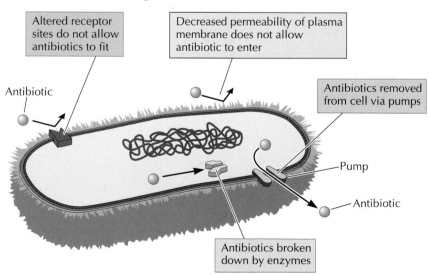

Altered receptor sites do not allow antibiotics to fit

Decreased permeability of plasma membrane does not allow antibiotic to enter

Antibiotics removed from cell via pumps

Antibiotic

Pump

Antibiotic

Antibiotics broken down by enzymes

the antibiotic, e.g., penicillin or a cephalosporin, rendering it ineffective. Some bacteria alter the uptake of the antibiotic; they possess pumps that actively transport antibiotics in and out of the cell. A variety of gram-positive and gram-negative bacteria are resistant to the antibiotic tetracycline by this mechanism. Drug uptake may be altered by a decrease in the permeability of the cell membrane to certain antibiotics. Another mechanism of countering the effects of antibiotic activity is by modification of the drug receptor site or of an essential metabolic pathway, preventing binding of the antibiotic. Penicillin resistance in streptococci and methicillin resistance in staphylococci are associated with alteration in the antibiotic receptor sites. Finally, as pointed out above, sulfonamides and some antibiotics act by interfering with essential metabolic pathways, but some bacteria develop an alternative pathway, rendering the antibiotic ineffective. It appears that evolutionary forces are constantly at play in humans' attempts to rein in bacteria; they fight back by adaptation. Darwin was right. Antibiotic resistance was not caused by the discovery and use of antibiotics; the genes for antibiotic resistance were present long before. Pathogenic (and other) bacteria acquired these preexisting genes from other bacteria through **horizontal gene transfer,** a mechanism by which genes are passed from one mature bacterial cell to another. (Vertical gene transfer is the passage of genes from parent to offspring.) The widespread use of antibiotics has fostered the emergence of antibiotic-resistant strains. Nobel laureate Joshua Lederberg, in an article in a 1996 issue of the *Journal of the American Medical Association*, stated, "Pitted against microbial genes we have mainly our wits."

Antibiotic Misuse

According to the CDC, approximately 150 million prescriptions for antibiotics are written in the United States each year; of these, nearly 50 million are unnecessary. Many of these prescriptions are written for colds and flu, despite the fact that antibiotics are not effective against these viral illnesses. The justification is "just in case." Too frequently, physicians are pressured by demanding patients or by overly anxious parents whose child has "the sniffles." Studies have indicated that patients who walk out of their physician's office without a prescription for an antibiotic will often complain that their physician "billed them for nothing." So the problem boils down to a misuse of antibiotics; this has led us to a critical state, with many biologists warning that we may be forced back to the pre-antibiotic era.

Here are a few alarming facts resulting from the misuse of antibiotics.

- Gonorrhea is increasingly more difficult to treat, and the organisms are resistant to two antibiotics. In Hawaii, resistance to the quinolone antibiotics jumped from 1.4% in 1997 to 9.5% in 2000. The resistance of gonorrhea to a new antibiotic, azithromycin, has been on the increase.

- More than 90% of the strains of *Staphylococcus aureus* are resistant to penicillin and other antibiotics.

- Ear infections (otitis media) are the second leading cause of office visits to physicians and account for over 40% of all outpatient antimicrobial use in children. An alarming number of bacterial strains that cause this condition are now antibiotic resistant.

- Antibiotic resistance is a global problem. Resistance to vancomycin, once considered the "last stronghold," has been reported to occur in staphylococci and enterococci.
- According to a 2000 WHO report, drug-resistant strains of the tuberculosis bacterium are increasing worldwide.
- Approximately half of all the antibiotics produced are used for disease control and promotion of growth in animals destined for the table, a practice which fosters the selection of antibiotic-resistant pathogens in animals and a potential risk of possible transmission to humans.
- The marked increase in domestic and international travel allows exposure to antibiotic-resistant pathogens that can be spread within a country and between countries. Antibiotic-resistant strains of the gonorrhea bacterium that originated in Africa and in Asia are now prevalent throughout the world.

AUTHOR'S NOTE *Now that you have learned that viruses are subcellular and do not have the target sites for antibacterial activity, you know that you should not expect or ask for an antibiotic when you have a viral infection. Frequently, students complain to me about the health services on campus and complain that "the doctor didn't even give me an antibiotic!" You now know better.*

Working toward the Solution

The seriousness of antibiotic resistance as an impending global crisis has been established. What is the solution? The answer lies in the hands of physicians and patients, both of whom share the responsibility for the misuse and overuse of antibiotics that have resulted in the emergence of "superbugs," and, therefore, both are obligated to work toward the solution. Perhaps you as a patient have unintentionally misused antibiotics. When you are ill with what seems to be a cold, you might expect, or demand, that your physician prescribe an antibiotic. You might not be content when advised that your cold is progressing "normally" and you will be fine. You may be guilty of not following instructions to take the full dose, because after a few days you feel better. In so doing, you contribute to the emergence of antibiotic-resistant bacteria. Another too common scenario is that you do not feel well and pull out of the medicine cabinet some leftover antibiotic or ask your roommate if he or she has any antibiotics on hand. You, of course, have no idea of the identity of the microbe. Physicians share in the responsibility for the antibiotic crisis and in its control; too often, they succumb to patient pressure and fail to spend the time explaining to patients why they do not need an antibiotic. Young physicians trying to build a practice cannot afford the reputation passed on from one patient to a potential patient that "my doctor didn't know enough to prescribe an antibiotic." Also, physicians fear being sued by a patient who claims that his or her illness is a result of not being "put on an antibiotic." As stated in chapter 1, a paradox regarding the misuse of antibiotics exists between the "haves" and the "have-nots" of the world. In the Third World, antibiotic resistance is fostered by not completing the course of antibiotics in order to save them "for a rainy day," while in developed countries misuse of antibiotics is based on their ready availability. Further, in some countries, antibiotics can be purchased over the counter in markets and pharmacies without a prescription. The judicious use of antibiotics can stave off and contain an impending antibiotic resistance crisis, allowing society to continue to enjoy the benefit of Fleming's serendipitous observation of the antagonism between a mold and a bacterium, both residents of soil that are fighting for their lives.

ANTIVIRAL AGENTS

Effective antiviral agents are few and far between. As pointed out above, antibiotics are not effective against viruses because they lack the target components against which antibiotics are directed; further, their intracellular existence is a problem in terms of achieving selective toxicity. Many of the drugs that can penetrate a cell and destroy viruses are likely to damage cells and to cause serious side effects, rendering them unusable. In certain circumstances, the use of antibiotics for individuals with viral infections is justified when secondary bacterial infection is a potential threat. For example, senior citizens with influenza are at risk for bacterial pneumonia, a potentially fatal disease, and are frequently put on antibiotics as a preventive measure.

Several antiviral agents are available, and a race is on to develop new ones. At some point in the not too distant future, antiviral chemotherapy may parallel antibiotic chemotherapy. In 1999, two new antiflu drugs, zanamivir (Relenza) and oseltamivir (Tamiflu), effective against influenza A and B viruses, were approved by the FDA. They do not prevent or cure the flu, but if taken early, they decrease duration of the illness by a few days. This may not sound like much, but if you have ever had the misfortune of having the flu, a few days is a lot!

The AIDS pandemic drives the search for antiviral drugs. The drug zidovudine (also called azidothymidine or **AZT**), an inhibitor of reverse transcriptase (chapter 5), and a group of **protease inhibitors** have achieved some success in AIDS therapy, but not surprisingly, drug resistance is an emerging problem. Antiviral agents are designed to interfere with some aspect of viral replication; recall from chapter 5 that viral replication involves the steps of adsorption, penetration, replication, assembly, and release, offering a variety of targets for effective antiviral therapy. We have a long way to go!

OVERVIEW The 20th century witnessed an increase in life expectancy in many nations of the world. U.S. residents live 30 years longer, on the average, than they did in 1900. At least 25 of those gained years are attributable to public health achievements in sanitation and hygiene, food and water safety, immunization, and antibiotics—the topics of this chapter.

Other industrialized countries of the world share in these successes. The sad part of this story is that a tremendous disparity exists between the "haves" and the "have-nots." The disparity is so great that in Japan, people live on the average to age 80, as contrasted to a life span of 26 years in Sierra Leone. ■

SELF-EVALUATION

PART I Choose the *single* best answer.

1. The MMR vaccine uses
   a. live viruses
   b. only live bacteria
   c. a mixture of live and dead viruses
   d. a combination of toxoids and viruses

2. The relationship between safe drinking water and child survival is
   a. inverse
   b. direct
   c. country related
   d. without correlation

3. The Kampung Improvement Program in Indonesia focused on
   a. immunization programs
   b. providing antibiotics
   c. improving the sewage system
   d. food safety

4. What agency is responsible for licensing vaccines?
   a. WHO
   b. FDA
   c. USDA
   d. CDC

5. Serum sickness can result from
   a. attenuated vaccines
   b. live vaccines
   c. toxoid administration
   d. passive immunization

6. All of the following are antibiotics with the exception of
   a. chloramphenicol
   b. penicillin
   c. sulfa drugs
   d. erythromycin

PART II Fill in the following.

1. What is the mechanism of activity of penicillin?
   _____

2. Which organism is commonly used as an indicator of clean water? _____

3. *The Jungle,* Upton Sinclair's 1906 book, had to do with _____.

4. The DPT vaccine protects against three diseases. Name them. _____

5. Name a microbial disease characterized by exotoxin production for which a toxoid is available for treatment. _____

6. Why is penicillin not toxic for humans? _____

PART III Answer the following.

1. Sanitation was "in" during the 20th century. Explain (give examples) of this statement.

2. Discuss the merits of "live versus dead" vaccines. Discuss three mechanisms by which live microbes are rendered safe for immunization purposes.

3. Describe the work of local health departments as partners in microbial disease control.

4. From a Darwinian point of view, describe the emergence of antibiotic-resistant strains.

13

PARTNERSHIPS IN THE CONTROL OF INFECTIOUS DISEASES

I am my brother, and my brother is me.

RALPH WALDO EMERSON

PREVIEW Chapter 12 emphasized the enormous strides made in the 20th century in decreasing the burden of infectious diseases worldwide, particularly in developed countries. The successes are the result of collaborative partnerships in pooling funds, talents, and resources toward common goals. Continued progress depends on partnerships ranging from the local to the national and international levels and requires cooperation between the public and private sectors. Measures need to be implemented to correct the disparity in providing "health for all" so that the poorer nations of the world receive their share of the benefits of progress.

BACKGROUND AND GOALS

"The attainment for all people of the world by the year 2000 of a level of health that will permit them to lead a socially and economically productive life." This was the stated goal that emerged from the International Conference on Primary Health Care held in Alma-Ata, USSR (now Almaty, Kazakhstan), from September 6 to 12, 1978; it was further declared:

> The Conference strongly reaffirms that health, which is a state of complete physical, mental, and social wellbeing, and not merely the absence of disease and infirmity, is a fundamental human right and that the attainment of the highest possible level of health is a most important world-wide social goal whose realization requires the action of many other social and economic sectors in addition to the health sector.

Over 2 decades have passed since the Alma-Ata conference. Where do we stand? Has there been progress toward the attainment of these ambitious goals? As discussed in the chapter 12, life expectancy has dramatically increased since the beginning of the 20th century, and Alma-Ata and other international alliances have built on these gains. The World Health

Assembly adopted the slogan "Health for all by the year 2000." That year has come and gone, and the goal has yet to be realized. Nevertheless, progress has been made, partially as a result of Alma-Ata and its emphasis on primary health care. "Health for all" is an elusive and moving target and may not be realistic; but striving toward it can only have positive consequences. The achievements of public health and advances in medical science have not been equally distributed across the board, resulting in the poorer nations' not receiving their share of the benefits of progress; this disparity remains a challenge.

The attainment of health in all its dimensions, including reducing the burden of infectious diseases, is a matter of public health concern at several levels. These levels range from the individual to the community, to state departments of health, to national agencies such as the Centers for Disease Control and Prevention (CDC) and the Food and Drug Administration (FDA), and to international agencies such as the World Health Organization (WHO). Barry Bloom, dean of the Harvard School of Public Health, states, "One of the myths of the modern world is that health is determined largely by individual choice and is therefore a matter of individual responsibility." Realistically, individuals are limited in their efforts to achieve and maintain good health, necessitating public health strategies aimed at populations to reduce the burden of infectious and other diseases.

The eradication of infectious diseases has been a goal since the establishment of Koch's postulates (chapter 6) and the germ theory of disease and, to date, has been achieved only in the case of smallpox. Thomas Jefferson, a few years after the introduction of smallpox vaccination in 1796, commented, "One evil more [smallpox] is withdrawn from the condition of man." In 1892, a contagious pleuropneumonia of cattle (which was imported into the United States in 1847) was declared eradicated from the country as the result of a 5-year, $2 million campaign to identify and slaughter infected animals. The Rockefeller Foundation ambitiously campaigned to eradicate yellow fever (chapter 9) and hookworm disease (chapter 10) in the early years of the 20th century but failed because of the complexity of eradication programs. Malaria eradication seemed plausible in the period of 1955 to 1965 but was unsuccessful primarily because of the emergence of drug-resistant parasites and mosquitoes.

The International Task Force for Disease Eradication (ITFDE), composed of a group of scientists, convened six times between 1989 and 1992 at the Carter Center and evaluated 94 diseases in terms of their potential eradication, using smallpox as the yardstick. The task force targeted three viral diseases—mumps, polio, and rubella (chapter 9)—and three worm diseases—guinea worm disease, lymphatic filariasis, and cysticercosis (chapter 10)—for eradication. The ITFDE defines eradication as "reduction of the worldwide incidence of a disease to zero as a result of deliberate efforts, obviating the necessity for further control measures." Elimination, according to the ITFDE, is "control of the manifestations of a disease so that the disease is no longer considered to be a public health problem," for example, blindness resulting from onchocerciasis or trachoma (a chlamydial disease). Further, elimination generally refers to a limited geographical area (a single country or continent), whereas eradication is used in a global sense.

The eradication of smallpox in 1980 stands as a public health triumph of the 20th century. Generations to come will never know the horrors of this disease, but images will serve as a reminder (Fig. 13.1). What were the unique characteristics of smallpox and of the smallpox virus that led to success in

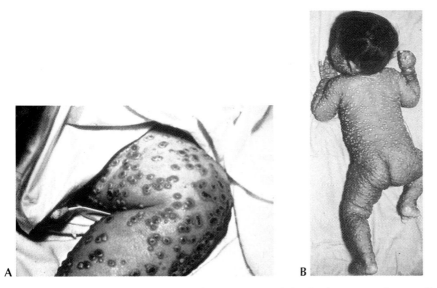

FIGURE 13.1   Smallpox: a disease of the past (maybe). The last case of naturally acquired smallpox occurred in Somalia in 1977; in 1980, WHO announced that smallpox was eradicated. This was a major public health victory of the 20th century. The disease is characterized by the appearance of pustules on the body. The use of smallpox by terrorists is a potential worldwide threat. (Source: WHO.)

smallpox eradication and its establishment as the criterion by which to evaluate other diseases as targets for eradication? They are as follows:

- It is a disease only of humans; there are no natural reservoirs or biological vectors.
- The infection is easily diagnosed because of a characteristic rash.
- The duration and intensity of infectiousness is limited.
- Recovery establishes permanent immunity.
- A safe, effective, inexpensive, easily administered, stable (even in tropical climates) one-dose vaccine is available.
- Vaccination confers long-lasting, possibly lifetime immunity.
- Vaccination results in a permanent and recognizable scar, allowing for detection of immune versus nonimmune individuals in a population.

The degree to which other diseases mimic smallpox reflects their potential for eradication, but these are not absolute criteria (Box 13.1). For example, a biological vector is a part of the guinea worm life cycle, polio immunization requires four doses, and neither disease produces visible early manifestations. Despite these considerations, both of these diseases are on the "hot list" for eradication, and considerable progress has been made toward that achievement. The last case of polio in the Americas occurred in Peru in 1975. Not all diseases (Table 13.1) reviewed by the ITFDE are considered candidates for eradication for a variety of reasons, highlighting the complexity of eradication programs.

PARTNERSHIPS IN INFECTIOUS DISEASE CONTROL

Reducing the incidence of infectious diseases is a tremendous challenge. The successes that have been accomplished to date are largely the result of

BOX 13.1 Criteria for Assessing Eradicability of Diseases and Conditions

Scientific Feasibility
- Epidemiologic vulnerability (e.g., existence of a nonhuman reservoir, ease of spread, natural cyclical decline in prevalence, naturally induced immunity, ease of diagnosis, and duration of any relapse potential)
- Effective, practical intervention available (e.g., a vaccine or other primary preventive, a curative treatment, and a means of eliminating the vector). Ideally, intervention should be effective, safe, long lasting, and easily deployed.
- Demonstrated feasibility of elimination (e.g., documented elimination from an island or other geographic unit)

Political Will and Popular Support
- Perceived burden of the disease (e.g., extent, deaths, or other effects; true burden may not be perceived; the reverse of benefits expected to accrue from eradication; relevance to rich and poor countries)
- Expected cost of eradication (especially in relation to perceived burden from the disease)

Source: CDC, *Morbidity and Mortality Weekly Report* **42**(RR-16):1–25, 1993.

collaborative partnerships resulting in the sharing of funds, talents, and resources toward common goals. Partnerships are not unique to public health; the concept is embodied in business, law, and medicine. Continued successes in public health will depend upon partnerships within and between the public and private sectors. In preceding chapters, numerous agencies, both public and private, have been cited for their leadership in the fight against infectious diseases, and this chapter describes some of them.

Today's crowded societies and sharing of resources are far removed from hunter-gatherer societies, where family units were relatively isolated and depended only on their own efforts and ingenuity to stay alive. As populations grew and urbanization developed, a sharing of community responsibilities emerged in all aspects of life, including health. A negative aspect of all this "togetherness" is that individuals are limited in measures that can be taken to minimize exposure to pathogens. In a collective effort, the early 1900s saw the establishment of community and state departments of health, which evolved into a complex network from the state level to the national and international levels and their partnerships in the public and private sectors.

At the Local Level

Community, City, and State Health Departments
Every state has a department of health, and in the larger states, subordinate health departments at the community and city levels also exist. Although the organizational charts vary from state to state, they are to a large extent a reflection of the size of the population covered. Health departments focus on the prevention of disease and the promotion of health and safety of the people within their jurisdiction by minimizing environments that potentially spread disease. Reference has been made in several previous chapters to the hazards of food-borne and waterborne diseases. Chapters 8, 9, and 10, in particular, discuss bacterial, viral, and worm diseases that are associated with food and water. A major responsibility of health departments is to establish and implement safety regulations pertaining to food and water sanitation (In the News 13.1). For example, there is rigorous control of the milk industry in each state, involving the health and the environment of dairy cows as well as the submission of milk and milk products on a strict schedule to state departments of public health laboratories to ensure the safety of the product before delivery to markets. Food workers are also re-

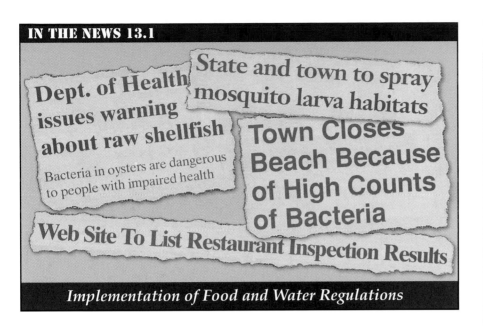

IN THE NEWS 13.1

Dept. of Health issues warning about raw shellfish

Bacteria in oysters are dangerous to people with impaired health

State and town to spray mosquito larva habitats

Town Closes Beach Because of High Counts of Bacteria

Web Site To List Restaurant Inspection Results

Implementation of Food and Water Regulations

TABLE 13.1   ITFDE classification of diseases by potential for eradication[a]
Diseases targeted for eradication
Dracunculiasis (W)[b]
Poliomyelitis (V)[b]
Diseases that may be eradicable
Lymphatic filariasis (W)[b]
Diseases of which some aspects could be eliminated
Hepatitis B (V)[b]
Neonatal tetanus (B)[b]
Onchocerciasis (W)[b]
Rabies (V)[b]
Trachoma (B)
Yaws (B)
Diseases that are not eradicable now
Ascariasis (W)[b]
Cholera (B)[b]
Diphtheria (B)[b]
Hookworm (W)[b]
Leprosy (B)[b]
Measles (V)[b]
Pertussis (B)[b]
Rotaviral enteritis (V)[b]
Schistosomiasis (W)[b]
Tuberculosis (B)[b]
Yellow fever (V)[b]
Diseases that are not eradicable
Amebiasis (P)[b]
Bartonellosis (B)
Clonorchiasis (W)
Enterobius (pinworm) disease (W)[b]
American trypanosomiasis (P)[b]
Varicella and zoster (V)[b]

[a]P, protozoan; B, bacterial; V, virus; W, worm. Source: Modified from CDC, *Morbidity and Mortality Weekly Report* **42**(RR-16):1–25, 1993.
[b]Discussed in this book.

quired to submit stool specimens in order to avoid a repeat of "Typhoid Mary" (chapter 8). Public health restaurant inspectors make spot visits to eating establishments to ensure adherence to proper temperatures for the cooking and storing of foods; the absence of mice, rats, roaches, and other vermin; and the practice of appropriate measures of sanitation and hygiene by food workers. The use of disposable gloves by food workers (Fig. 13.2) is an increasingly mandated practice, as is the use of protective shields over salad bars. No longer do you need to witness food workers in college cafeterias and in restaurants mixing huge quantities of salad or hamburger with their bare hands. Some communities, in addition to imposing fines and closing noncompliant establishments, make restaurant inspection reports available to the public on a website and have implemented other methods to alert the public. Los Angeles requires restaurants to post inspection grades in the windows, and the local newspapers in central Florida and other areas publish the detailed results of restaurant inspections and the fines imposed on those failing to achieve a clean bill of health. Boston has a new website (http://www.mayorsfoodcourt.com) which lists the most recent "report card" for Boston's 2,370 eateries. In some cases, the results are hard to swallow (pun intended). Violations include potentially hazardous, uncooked food held at unsafe temperatures; food handlers preparing foods with their bare hands; raw chicken stored over raw meat in the cooler; feta cheese, ranch dressing, and cheddar cheese held at improper temperatures; and roaches crawling over counters and food. Additionally, some areas require that food workers, in those restaurants failing to pass inspection, attend a hospitality education program.

On March 11, 2000, student health services at a university in the District of Columbia notified the city health department of acute gastroenteritis in a large number of students. The health department initiated an investigation, which identified food-transmitted rotavirus as the cause. This account illustrates the reporting mechanisms and the detective work employed in investigating food-borne outbreaks. Investigation of such outbreaks is a major function of health departments at the local, state (Fig. 13.3), and national (CDC) levels.

Surveillance and control of infectious diseases are major functions of

FIGURE 13.2 Protection of the public. Local departments of health in most areas of the country require food handlers to wear disposable gloves, as shown in this photo, and require eating establishments to use protective shields over salad bars. (Author's photo.)

Foodborne Illness Complaint Worksheet
STATE DEPARTMENT OF HEALTH

Date: _03/11/97_
#: _97-076_

PERSON COMPLETING INFORMATION
Name: _Xavier Onassis_   ☎ : (_512_) _555_ - _1234_
Affiliation: ☐ Local BOH *(town)*: _____ ☒ State DPH *(division)*: _Epi_   ☐
Other: _____

REPORTER / COMPLAINANT
Name: _Refused to provide_   ☎ : ( ) _____ - _____

Affiliation: ☒ Consumer   *specify:*
                 ☐ Laboratory   division,
                 ☐ Local BOH   *facility,*   _____
                 ☐ Medical Provider address,   _____
                 ☐ State DPH   town, etc.
                 ☐ Other   _____

ILLNESS INFORMATION
 # Persons ill: _2_
Symptoms: (% reporting)

☒ Diarrhea *(both)*   ☒ Vomiting *(both)*   ☒ Nausea *(both)*   ☐ Abdominal cramps
☒ Fever *(only one)*   ☐ Bloody stool   ☒ Headache *(only one)* ☐ Muscle aches
☐ Chills   ☐ Loss of appetite   ☒ Fatigue *(both)*   ☐ Dizziness
☐ Burning in mouth   ☐ Other symptoms: _None_

Onset: Earliest   Date: _03/10/97_   Time: _11:30_   ☐ AM ☒PM
              Latest **(if ≥ 2 ill)** Date: _03/11/97_   Time: _2:30_   ☒ AM ☐ PM

Duration: ☐ Less than 24 Hours ☐ 24-48 Hours ☐ More than 48 Hours ☒ Ongoing ☐ Unknown

Ill Persons:
	Name	Address/Town	☎	Age (yrs)	Occupation	Med. Provider/ ☎
1	☒ same as reporter	University X	refused	18	student	none
2	refused	University X	refused	19	student	none
3						
4						

Medical attention received *(by anyone)*? ☐ Yes ☒ No   ☐ Unknown → *If Yes, specify above:*
Stool specimens submitted *(by anyone)*? ☐ Yes ☒ No   ☐ Unknown → **To SLI** [1]? ☐ Yes ☐ No ☐ Unknown
Medical diagnosis reported?

FIGURE 13.3 A reporting form that is filled out in the event of a food-borne illness and sent to a state health department.

---

local health departments. For example, an outbreak of meningitis occurred in Rhode Island in 1998 and was a cause of alarm and near panic. Health officials responded with control strategies, including vaccination of thousands of schoolchildren against meningitis; the CDC was called in as a consultant. State health departments are required to notify the CDC of 52 specific diseases (Table 7.1) so that a national network of surveillance and communication can be maintained.

Vaccine campaigns and implementation of regulations requiring immunization against infectious diseases are another function of state health departments. Local departments of health are involved in numerous other endeavors to foster the health and welfare of citizens, including toxicology, vital statistics, public awareness health programs, prevention and treatment of drug abuse, cancer surveillance, and lead paint screening. The mission statement of the Rhode Island Department of Health is typical of the missions of local and state departments of health and is presented in Box 13.2 along with those of other agencies.

BOX 13.2 Missions and Objectives of Partners in Infectious Disease Control

Public Sector: Governmental and International Agencies

STATE AND LOCAL LEVELS

Health Departments (e.g., Rhode Island Department of Health)

The primary mission of the Rhode Island Department of Health is to prevent disease and to protect and promote the health and safety of the people of Rhode Island. All people in Rhode Island will have the opportunity to live a safe and healthy life in a safe and healthy community.

NATIONAL LEVEL

Centers for Disease Control and Prevention

The CDC's mission is to promote health and quality of life by preventing and controlling disease, injury, and disability. The CDC pledges to the American people:

- To be a diligent steward of the funds entrusted to it.
- To provide an environment for intellectual and personal growth and integrity.
- To base all public health decisions on the highest quality scientific data, openly and objectively derived.
- To place the benefits of society above the benefits to the institution.
- To treat all persons with dignity, honesty, and respect.

Public Health Service

The mission of the PHS is to improve the health of every individual by conducting research, engineering systems for safe delivery of water and disposal of waste, overseeing food and drugs, studying and developing means to contain or eliminate disease, and promoting a safe and healthful environment at work or home.

Federal Emergency Management Agency

To provide leadership and support to reduce the loss of life and property and protect our nation's institutions from all types of hazards through a comprehensive, risk-based, all-hazards emergency management program of mitigation, preparedness, response, and recovery.

Food and Drug Administration

The FDA ensures the safety of food and cosmetics and the safety and efficacy of pharmaceuticals, biological products, and medical devices. The FDA is charged with protecting American consumers by enforcing the Federal Food, Drug, and Cosmetic Act and several related laws.

National Institutes of Health

The NIH mission is to uncover new knowledge that will lead to better health for everyone. NIH works towards that mission by conducting research in its own laboratories; supporting the research of non-Federal scientists in universities, medical schools, hospitals, and research institutions throughout the country and abroad; helping the training of research investigators; and fostering communication of medical information.

INTERNATIONAL LEVEL

World Health Organization

The objective of WHO is the attainment by all peoples of the highest possible level of health. Health, as defined in the WHO Constitution, is a state of complete physical, mental and social well-being and not merely the absence of disease or infirmity. In support of its main objective, the organization has a wide range of functions, including the following:

- To act as the directing and coordinating authority on international health work.
- To assist governments, upon request, in strengthening health services.
- To establish and maintain such administrative and technical services as may be required, including epidemiological and statistical services.
- To provide information, counsel, and assistance in the field of health; to stimulate the eradication of epidemic, endemic, and other diseases.
- To promote improved nutrition, housing, sanitation, working conditions, and other aspects of environmental hygiene.

- To promote and coordinate biomedical and health services research.
- To develop international standards for food, biological and pharmaceutical products, and to standardize diagnostic procedures.
- To assist in developing an informed public opinion among all peoples on matters of health.

Pan American Health Organization

The essential mission of PAHO is to strengthen national and local health systems and improve the health of the peoples of the Americas, in collaboration with ministries of health, other governmental and international agencies, nongovernmental organizations, universities, social security agencies, community groups, and many others. The fundamental purposes of PAHO are to promote and coordinate the efforts of the countries of the Americas to combat disease, lengthen life, and promote the physical and mental health of their people.

Private Sector: Nongovernmental Organizations

Carter Center

The Carter Center, in partnership with Emory University, is guided by a fundamental commitment to human rights and the alleviation of human suffering; it seeks to prevent and resolve conflicts, enhance freedom and democracy, and improve health.

Rotary International

In 1985, Rotary International launched Polio Plus, a 20-year commitment to eradicate polio. Polio Plus is one of the most ambitious humanitarian undertakings ever made by a private entity. It will serve as a paradigm for private-public collaborations in the fight against disease well into the 21st century.

Bill and Melinda Gates Foundation

This foundation will support programs in global health and learning, with the hope that as we move into the 21st century advances in these critical areas will be available for all people.

At the National Level

The Centers for Disease Control and Prevention

The CDC has been cited repeatedly throughout this book. It is the nation's premier public health facility, and its impact is global. The agency was founded in 1946 and employs about 8,500 people. It operates on a budget (in fiscal year 2002) of about $4.3 billion, of which $345 million is earmarked for control of infectious diseases, $628 million is earmarked for immunization, and over $1.1 billion is to be used for prevention of AIDS, other sexually transmitted diseases, and tuberculosis. The CDC is a vital member of partnerships with local health departments and national and international organizations (Box 13.2). The CDC is like a large microbial zoo; within its locked freezers are every known microbe on Earth (including smallpox virus) caged in small vials under the watchful eye of a microbe keeper or in the live bodies of rabbits, mice, rats, and monkeys in locked corridors. The "deadliest of the deadly" viruses, such as those causing acquired immune deficiency syndrome (AIDS), rabies, Ebola hemorrhagic fever, and hantavirus pulmonary syndrome, are a part of the CDC's microbial zoo population. A CDC scientist studying the transmission of malaria uses himself as a "living laboratory"; he donates his own blood once a week to breed his 10,000 "mosquito pets" per day.

The CDC has come to the rescue on numerous occasions in just about all parts of the world, frequently in conjunction with its public health partners; examples include the smallpox eradication program, the 1996 hantavirus outbreak in the Four Corners region of the southwestern United States, identification of the cause of the *Legionella* outbreak in Philadelphia in 1976, and the Ebola hemorrhagic fever outbreak in Kikwit, Zaire (now the Democratic Republic of the Congo), in 1995. In addition to these exotic plagues are the less publicized, less dramatic, and mundane day-to-day scenarios that need to be dealt with, including food-borne and waterborne outbreaks, "superbugs" (antibiotic-resistant bacteria), and nosocomial infections, all topics discussed in other chapters of this book.

The CDC (Fig. 13.4) is based in Atlanta, Ga., and has seven centers, one of which is the National Center for Infectious Diseases, regarded as the nation's main line of defense against threatened epidemics and plagues. The agency fields about 1,000 calls for help each year, many of which can be handled by over-the-phone consultation. As deemed necessary, its "SWAT" teams of epidemiologists head into the fields, frequently in collaboration with partnership agencies, equipped with ready-to-go containers stocked with syringes, needles, vaccines, intravenous fluids, examination gloves, refrigerators to store samples of blood and other tissues, generators, stacks of questionnaires, and other items to conduct guerrilla warfare against the microbes. These elite teams have been called "virus jocks," "disease cowboys," "Green

FIGURE 13.4   The CDC headquarters in Atlanta, Ga. The CDC is the nation's premier agency in infectious disease surveillance and works in conjunction with state and local health departments and with WHO and other agencies. (Source: Public Health Image Library, CDC.)

Berets," "disease detectives," and "the Sherlock Holmeses of disease," all terms describing their mission to identify, seek out, and destroy the microbial enemy. Intriguing and heroic tales of their battles against an invisible enemy have been recounted in movies such as *Outbreak* and *Epidemic* and in books such as *The Coming Plague* and *The Hot Zone*.

Frequently, the specific disease agent cannot be definitively isolated and identified under primitive field conditions, so the disease detectives must ship tissue samples from sick and dead victims to the CDC labs in Atlanta for identification. The labs are designed to work with deadly microbes, and their locked doors bear large biohazard signs (Fig. 13.5). Many of the labs maintain negative pressure so that air is actively whisked out of the rooms into a series of filters to trap microbes, preventing their escape to the outside world. The facility has four levels of biohazard containment, and the assigned level is a function of the virulence of the organism, the mechanism of transmission, and the availability of a cure or a vaccine (Fig. 13.6). AIDS is considered level 3; biocontainment level 4, CDC's famous Building 15, is the "hot zone" prepared to handle the deadliest of microbes, including Ebola virus and hantaviruses, for which there is neither cure nor vaccine. Air leaving Building 15 is passed through a series of filters, water is boiled before entering sewer lines, and the fortresslike building has a camera trained on its one entrance. Workers strip naked, shower, and don biohazard suits, known as blue suits, and as many as three pairs of gloves before entering the level 4 lab; air is supplied to them through a tube attached to the blue suit. Their protection against deadly pathogens such as Ebola virus and hantaviruses is limited to a layer of fabric which can be penetrated by the jagged glass of a broken test tube or by a syringe needle. As much as possible, researchers work in pairs and monitor each other for fatigue and for tears in their gloves and blue suits. The U.S. Army Medical Research Institute of Infectious Diseases and the Southwest Foundation for Biomedical Research (San Antonio, Tex.) also have level 4 laboratory facilities. It takes a person of great courage to work with a deadly microbial enemy that cannot be seen!

In its role as the nation's watchdog, the CDC has developed a four-

FIGURE 13.5 The biohazard symbol. Laboratories working with highly virulent microbes post biohazard signs on their doors. Potentially infectious materials from hospitals, laboratories, and other facilities are disposed of in clearly marked biohazard bags, which are then incinerated. (Author's photo.)

FIGURE 13.6 Level 4 biocontainment. Ebola virus, hantaviruses, and other microbes are potentially deadly and must be handled in specially designed biohazard hoods to protect laboratory personnel and the environment from the risks of contamination. (Source: PHS.)

point strategy for the 21st century (Box 13.3) to counter the threat of new, emerging, and reemerging infections. According to the CDC,

> . . . the fulfillment of CDC's vision will require the sustained and coordinated efforts of many individuals and organizations. As CDC carries out this plan, it will coordinate with state and local health departments (e.g., on surveillance of infectious diseases), academic centers and other federal agencies (e.g., on research agendas), health care providers and health care networks (e.g., on development and dissemination of guidelines), international organizations (e.g., on outbreak responses overseas), and other partners.

Do you want to go on a trip, possibly a cruise, to some far-off destination? If so, you would be well advised to consult the "green sheet," CDC's *Summary of Sanitation Inspections of International Cruise Ships*, available to travel services around the world. Because of major disease outbreaks, CDC established the Vessel Sanitation Program in the early 1970s in an attempt to minimize the risk for diarrheal diseases among passengers. All vessels with a foreign itinerary that carry more than 13 passengers and call on U.S. ports are subject to unannounced twice-yearly inspections by Vessel Sanitation Program staff and to reinspections when necessary. To pass, the ship must score a minimum of 86 points on a 100-point scale. The program is similar to inspections of eating establishments conducted by local and state health officials and focuses on the water supply, spas and pools, and food-handling practices. Further, the general cleanliness, personal hygiene, and physical condition of the crew, along with training programs in environmental and public health practices, are evaluated.

BOX 13.3   Preventing Emerging Infectious Diseases

The CDC's strategy for combating emerging infections has four goals.

Goal I: Surveillance and Response

Objective I-A.   Strengthen infectious disease surveillance and response.

Objective I-B.   Improve methods for gathering and evaluating surveillance data.

Objective I-C.   Ensure the use of surveillance data to improve public health practice and medical treatment.

Objective I-D.   Strengthen global capacity to monitor and respond to emerging infectious diseases.

Goal II: Applied Research

Objective II-A.   Develop, evaluate, and disseminate tools for identifying and understanding emerging infectious diseases.

Objective II-B.   Identify the behaviors, environments, and host factors that put people at increased risk for infectious diseases and their sequelae.

Objective II-C.   Conduct research to develop and evaluate prevention and control strategies in the nine target areas.

Goal III: Infrastructure and Training

Objective III-A.   Enhance epidemiologic and laboratory capacity.

Objective III-B.   Improve CDC's ability to communicate electronically with state and local health departments, U.S. quarantine stations, health care professionals, and others.

Objective III-C.   Enhance the nation's capacity to respond to complex infectious disease threats in the United States and internationally, including outbreaks that may result from bioterrorism.

Objective III-D.   Provide training opportunities in infectious disease epidemiology and diagnosis in the United States and throughout the world.

Goal IV: Prevention and Control

Objective IV-A.   Implement, support, and evaluate programs for the prevention and control of emerging infectious diseases.

Objective IV-B.   Develop, evaluate, and promote strategies to help health care providers and other individuals change behaviors that facilitate disease transmission.

Objective IV-C.   Support and promote disease control and prevention internationally.

Source: CDC, *Preventing Emerging Infectious Diseases: a Strategy for the 21st Century* (CDC, Atlanta, Ga., 1998).

Which shots do you need before traveling to the Amazon, Mozambique, or Tahiti? These exotic foreign destinations are potential sources of disease that can make you very ill and even "do you in." Before embarking, consult the CDC's website (http://www.cdc.gov) for up-to-date traveler's health information, including vaccinations, safe food and water, and disease outbreaks in the land of your dreams.

The Public Health Service

The Public Health Service (PHS), headed by the U.S. surgeon general, consists of a 6,000-member Commissioned Corps and support staff and is a component of the U.S. Department of Health and Human Services (Box 13.2). It originated as the Marine Hospital Service in 1889, and in 1912 the name was changed to the PHS. Initially, the PHS focused on sailors and their medical care in an attempt to alleviate the burden on public hospitals in caring for merchant seamen. In 1891, the service was charged with being the nation's medical gatekeeper by providing medical inspection of arriving immigrants in order to weed out "idiots, insane persons, persons likely to become a public charge and persons suffering from a loathsome or a dangerous contagious disease." The commissioner general of immunization clearly stated in 1902 that America should not become "the hospital of the nations of the earth." It is for this reason that immigrants were screened by PHS officers at Ellis Island for "germ diseases," including cholera, typhus, plague, smallpox, yellow fever, and trachoma (which can cause blindness). Ellis Island was dubbed "The Island of Hope" and also "The Island of Tears"—"hope" for the new and promising way of life for those who made it through the inspection line, but "tears" for those who were separated from their families and returned to their place of origin.

The PHS has taken a leading role, particularly during wars, in campaigns against syphilis, gonorrhea, and other venereal diseases, now called sexually transmitted diseases (Fig. 13.7); in immunization campaigns (Fig. 13.8); in vector control programs (Fig. 13.9); and in public health awareness programs (Fig. 13.10).

The Federal Emergency Management Agency

The Federal Emergency Management Agency (FEMA), founded in 1979, is an independent agency of the federal government and reports directly to the president. Its slogan is "Helping People, before, during, and after Disaster." Microbial diseases frequently follow in the wake of natural disasters (e.g., hurricanes, floods, tornadoes, and cyclones) as a result of the destruction of the public health infrastructure that leads to polluted waters and compromised sanitation. It is common to hear in such instances that the president has declared a community to be eligible for FEMA funds and assistance in coping with the disaster. FEMA works in partnership with local, national, and international agencies in performing its role (Box 13.2).

The Food and Drug Administration

The FDA is part of the PHS. It focuses on safety in truth and labeling of foods, vaccines, antibiotics and other medicinal (biologic) products, and medical devices, all of which play a role in the prevention and control of infectious disease. This agency has "the last word" before these products are approved for release (chapter 12). The safety of the nation's blood supply system is under the umbrella of the FDA, whose inspectors routinely test blood and blood products for contamination. The agency has the authority

A   B   NO is the best tactic; the next, PROphylactic!

FIGURE 13.7 The wartime fight against sexually transmitted diseases. These diseases remain a very significant public health problem in the United States and around the world. Today, AIDS overshadows the presence of other sexually transmitted diseases. (Source: PHS.)

to direct withdrawal of products, either voluntarily or legally, as in the withdrawal of the RotaShield vaccine (chapter 9) against gastroenteritis shortly after its approval. Products not directly related to infectious diseases also fall under the scrutiny of the FDA. The agency is charged with protecting the consumer by enforcing the Federal Food, Drug, and Cosmetic Act and related public health laws (Box 13.2).

The National Institutes of Health

The National Institutes of Health (NIH) is one of the most distinguished medical research centers in the world and the federal focal point of medical research in the United States. It has come a long way since its founding in 1887 as the one-room Laboratory of Hygiene to 75 buildings spread over a 300-acre campus in Bethesda, Md. (Fig. 13.11). Its annual budget was $300 in 1887 and has grown to more than $15.6 billion in 1999. Its organizational chart shows 25 separate institutes and centers, one of which is the National Institute of Allergy and Infectious Diseases. The primary function of NIH (Box 13.2) is to administer and support biomedical research at over 2,000 sites in the United States and abroad; many of the research findings have direct applicability to public health measures.

At the International Level

The World Health Organization

The days are long gone when a continent's, or a country's, public health problems were unique and limited to that geographical area. The planet's microbes know no boundaries and travel freely, without passport, from

FIGURE 13.8 A person being immunized in a PHS immunization campaign. The PHS has played a major role in immunization campaigns around the world to halt epidemics by immunizing the population. (Source: PHS.)

FIGURE 13.9   A PHS vector control program. Insect vectors, including fleas, the vectors of typhus fever, are highly significant in the transmission of many infectious diseases. (Source: PHS.)

hemisphere to hemisphere in a matter of hours (Table 1.7). WHO headquarters in Geneva, Switzerland, functions as a command post in its extensive partnerships and employs sophisticated systems of surveillance and communication to keep track of microbial diseases on a global level. WHO's activities are multifold and not limited to microbial disease surveillance and control (Box 13.2).

FIGURE 13.10   PHS public health awareness programs. Informing and advising the public are important functions of public health agencies. All households were sent a copy of the pamphlet "Understanding AIDS" several years ago by the PHS. (Source: PHS.)

FIGURE 13.11 Building 1 at NIH. This research-oriented organization is located in Bethesda, Md., and is vital to the nation's health. It works with the CDC and other agencies.

WHO partnerships date back to the early years of the organization; its influenza surveillance network is responsible for identifying each year the strains of influenza to be used in the next year's vaccine, thereby serving as a global watchdog for surveillance of influenza. WHO has entered into many partnerships in an effort to combat rabies, malaria, leprosy, sleeping sickness, filariasis, guinea worm disease, and practically every other microbial disease on the planet.

The growth of information technology provides increased opportunities for disease surveillance and response requiring rapid assessment to initiate control efforts with minimal delay and to screen out unsubstantiated reports. In early 1997, WHO established an innovative approach to global disease surveillance aimed at improving epidemic disease control by rapid communication of potentially significant outbreaks to health professionals around the globe (Fig. 13.12). WHO, in its alliance with 191 member countries, is in an ideal position to monitor infectious disease surveillance and control. During the verification process, WHO offers assistance in investigation and disease control, drawing on its own resources and that of its partners.

The Pan American Health Organization

The Pan American Health Organization (PAHO) has 100 years of experience in striving to improve the health and living standards of the people of the Americas. It is a component of the United Nations and serves as the Regional Office for the Americas of WHO. Its member states include all 35 countries in the Americas; Puerto Rico is an associate member, and several European states collaborate. Its mission (Box 13.2) is achieved in association with other governmental and nongovernmental agencies, universities, and community groups.

PAHO promotes primary health strategies and assists countries in com-

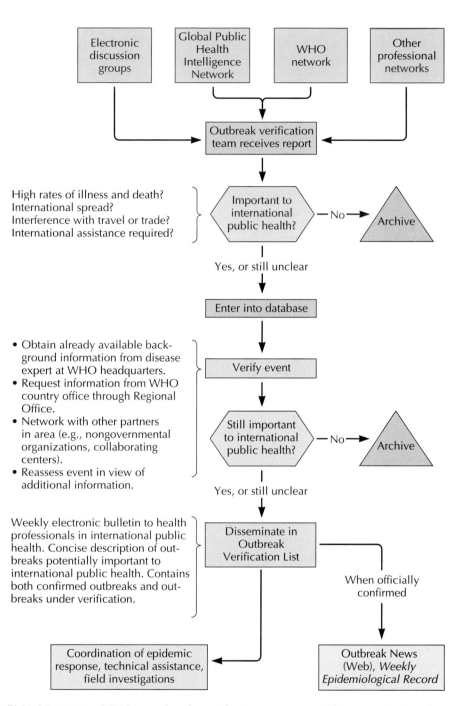

FIGURE 13.12 WHO's outbreak verification program. This organization has teams ready to go in the event of verification of an impending outbreak of disease. (Redrawn from T. W. Grein, K.-B. O. Kamara, G. Rodier, A. J. Plant, P. Bovier, M. J. Ryan, T. Ohyama, and D. L. Heymann, *Emerging Infectious Diseases* **6:**97–102, 2000.)

bating cholera, dengue, tuberculosis, and the spreading AIDS epidemic and is committed to ensuring that blood for transfusion is safe and not a source of disease. Further, the organization aims to eliminate all vaccine-preventable diseases and embarked on polio eradication efforts in 1985, leading to a declaration of polio-free Americas in 1994. The last case of polio in the Americas was identified on August 23, 1991, in Luis Fermin Tenorio Cortez, a young boy in Junin, Peru. Improvement of drinking water, sanitation, and health care for the poor remain top priorities for PAHO, with a focus on equity. In collaboration with WHO and other organizations, PAHO has proposed Healthy Children: Goal 2002, an initiative aimed at preventing 100,000 deaths in children under 5 years of age in the Americas, through the International Management of Childhood Immunization strategy. This approach promotes three main components: improving the case management skills of health professionals, improving the quality of the health care system, and improving family and community practices.

The Private Sector

The role of government agencies as partners in the control of infectious diseases has been outlined, but the success achieved over the past century is, in no small measure, also attributable to the role of the private sector, the nongovernmental organizations. They are equal partners in the ongoing battle to lessen the world's burden of infectious diseases. Some of these private sector participants and their objectives are indicated in Box 13.2 and are discussed in other chapters. Rotary International has been a key player in the polio eradication program since 1985, when it launched Polio Plus, a 20-year commitment which will cost nearly $500 million by the polio eradication target date of 2005. The Carter Center has committed to a variety of projects, including partnerships with both governmental and nongovernmental organizations, for the eradication of guinea worm disease, onchocerciasis, and filariasis. In January 2001, Microsoft billionaire Bill Gates pledged over $100 million to the search for an AIDS vaccine and challenged others to pitch in. The Bill and Melinda Gates Foundation has awarded millions of dollars over the years to numerous projects for the elimination of lymphatic filariasis, for the treatment of African sleeping sickness and leishmaniasis, and to PAHO for the improved safety of blood transfusion in the Americas. Yahoo!, an Internet search engine, also promised a $5 million grant over 3 years toward the International AIDS Vaccine Initiative and is a valuable corporate sponsor. Pharmaceutical companies, too, have been significant partners. Ivermectin, an important drug in the onchocerciasis control program, has been donated free of charge by Merck & Co. to countries where this disease is prevalent; SmithKline Beecham (now part of GlaxoSmithKline) has supplied the drug albendazole, an orally administered broad-spectrum anthelmintic drug, to countries as needed; other drug companies have also made generous contributions.

PARTNERSHIPS: THE WAY TO GO

Chapter 12 emphasized the enormous public health strides made over the last century that have led to an increased life span for people in many countries and to better containment of microbial diseases around the globe. These achievements are the results of collaborative partnerships from the community level to the international level; the players on the team are described in this chapter. The potential burden of infectious diseases is too enormous to be handled without teamwork, particularly given the explosive nature of microbial populations, the ease of transmission resulting from globalization, and the increased threat of new, emerging, and reemerging infections.

Partnerships are the key to further success in reducing the burden of microbial diseases over the coming years and in responding without delay to populations endangered by the ravages of disease as a consequence of floods, earthquakes, and other natural disasters. History and current events provide many examples of the misery and deaths that occur in the aftermath of catastrophic events because of the collapse of the public health infrastructure. The lack of safe drinking water and foods and the lack of sanitation and personal hygiene, coupled with overcrowded and makeshift living quarters, create an environment that is conducive to outbreaks of diarrheal, respiratory, and a multitude of other diseases. The circumstances are more than can be handled at the local level and require the cooperation of disaster relief and public health agencies at the global level. While the primary motive (one would like to think) of these partnerships is humanitarianism, there is also a selfish motivation, and understandably so, considering the rapid transmission of microbes from country to country.

Examples of partnerships in action in response to catastrophic events are abundant. A massive earthquake, the worst in 50 years, rumbled through western India on January 28, 2001, causing thousands of houses to collapse in Bhuj, a city of 150,000 close to the quake's epicenter, and spreading damage in its path. The earthquake was quickly followed by the arrival of international disaster teams and pledges of financial aid from Britain's International Rescue Corps, the Swiss Red Cross, the International Federation of the Red Cross, the United Nations, the European Union, Germany, Turkey, Norway, and the Netherlands. Pakistan, India's arch rival, was among the first to offer condolences and demonstrated that human tragedy transcends politics.

Another example of partnership was the response to the outbreak of cholera that hit South Africa in August and September 2000; more than 150,000 people became ill, about 500 were hospitalized, and 59 died over the next 5 months. According to a health official, "it's a typical South African outbreak, where you have large numbers residing in poor circumstances, in the rural areas, and these are the people who are infected." This scenario triggered South Africa to enlist the help of WHO epidemiologists to work with local health officials in developing control strategies against waterborne diarrheal and other diseases.

A final example of public health collaboration in the control of infectious disease outbreaks was the outbreak of leptospirosis, a spirochetal disease (chapter 8), among nine athletes participating in the 12-day EcoChallenge in Sabah, a state in the Malaysian part of Borneo, in September 2000. The event included jungle trekking, open-water swimming, river and ocean paddling, mountain hiking, canoeing, scuba diving, sea paddling, and spelunking in the jungles and along the coast of Sabah. GEOSentinel, established in 1995, is a network of 26 tropical medicine clinics (15 of which are in the United States) that monitor global trends in disease among travelers, immigrants, and refugees. The network has the capacity to facilitate rapid response by electronic dissemination to surveillance sites, including the 1,500-partner International Society of Travel Medicine, the CDC, and health care providers around the world. Three GEOSentinel sites, along with the Idaho Department of Health and Welfare, the Los Angeles County Department of Health, and the San Diego County Health Department, were among the first to notice and report clusters of suspected leptospirosis at the EcoChallenge, alerting all 300 athlete-participants and public health officials in each country represented to watch for possible exposure to the spirochete that causes leptospirosis.

The earthquake in India, the cholera outbreak in South Africa, and the

EcoChallenge leptospirosis outbreak are excellent examples of partnerships in action at the local and international levels, involving both public and private sectors, working together to monitor and control infectious disease outbreaks. There is no doubt that everything works better when everyone works together. Partnerships are the way to go!

OVERVIEW The 20th century witnessed tremendous accomplishments in public health that led to a worldwide decrease in the burden of infectious diseases, particularly in developed countries. The attainment of "health for all" is a goal yet to be realized. Further successes require teamwork at the local, national, and international levels and between the public and private health sectors. The players on the team are described in this chapter.

Surveillance and control of microbial disease at the community level are the responsibility of local and state departments of health. They are responsible for alerting the CDC to the occurrence of specified communicable diseases so that a national network of surveillance and communication can be maintained. At the national level, a number of agencies, most notably the CDC, cooperate to bring about the containment of infectious diseases. Microbes spread readily and quickly from continent to continent, creating the need for international agencies, such as WHO, that have the capacity to respond to outbreaks of disease at the global level. ∎

SELF-EVALUATION

PART I Choose the *single* best answer.

1. The Rockefeller Foundation was involved in attempts to eradicate
   a. smallpox   **c.** leptospirosis
   b. hookworm disease   **d.** malaria

2. Which characteristic of smallpox was significant in contributing to its eradication?
   a. natural reservoir relatively easy to control
   b. limited duration of infection
   c. requires only two doses of vaccine
   d. can be grown on petri dishes in the lab

3. The expression "microbial zoo" best applies to
   a. WHO   **c.** a dense population of people
   b. NIH   **d.** CDC

4. The safety of the nation's blood supply is monitored by
   a. Red Cross   **c.** FDA
   b. George Bush   **d.** FEMA

PART II Fill in the following.

1. What does ITFDE stand for? _____

2. "Hot" microbes, such as Ebola virus, are dangerous to work with and require special facilities known as biosafety level _____.

3. Rotary International is a key player in the campaign to eradicate _____.

4. Athletes in the EcoChallenge in Borneo were threatened by an outbreak of _____.

PART III Answer the following.

1. Smallpox is a model chosen by ITFDE for evaluating eradication of other microbial diseases. Discuss four or five characteristics of smallpox.

2. Describe the work of local health departments as partners in microbial disease control. Cite a few specific examples.

3. You, as director of WHO, are well aware of the disparity between the developing and developed countries in matters of public health. Outline your approach to address this issue.

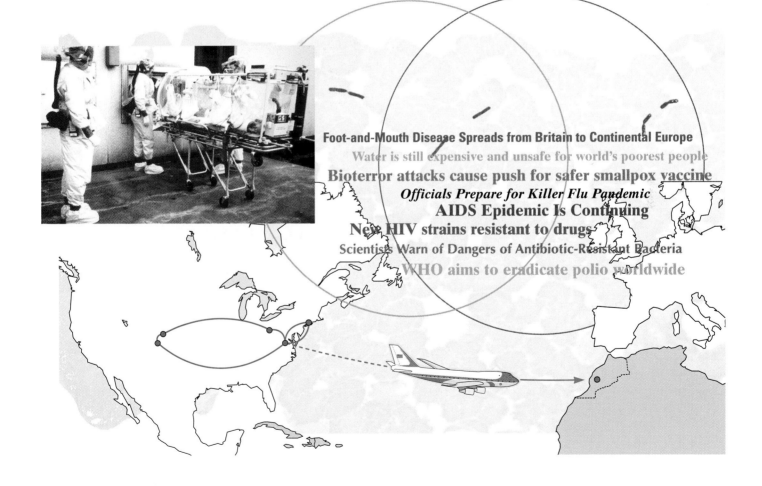

PART III
CURRENT CHALLENGES

Foot-and-Mouth Disease Spreads from Britain to Continental Europe

Water is still expensive and unsafe for world's poorest people

Bioterror attacks cause push for safer smallpox vaccine

Officials Prepare for Killer Flu Pandemic

AIDS Epidemic Is Continuing

New HIV strains resistant to drugs

Scientists Warn of Dangers of Antibiotic-Resistant Bacteria

WHO aims to eradicate polio worldwide

BIOLOGICAL WEAPONS*

They shall beat their swords into plowshares and their spears into pruning hooks. Nation will not lift sword against nation; there will be no more training for war.

ISAIAH 2:4

PREVIEW Perhaps you recognize the popular biblical quote above. In the 1960s, a time of reaction to the Vietnam War, society was in turmoil, and antiwar protesters played a major role through their political activities and seemingly outlandish behavior. A song of the times was "I Ain't Gonna Study War No More." It is, unfortunately, all too obvious that society has still not been able to achieve such a lofty goal. Rather, through the centuries, the conventional weapons of warfare have evolved from the days of hand-to-hand combat with spears and knives to highly sophisticated weaponry, from carrying warriors and supplies on horses and mules to using aircraft carriers, nuclear powered submarines, and jet aircraft that can deploy thousands of military personnel and support equipment from one corner of the world to another in less than 24 hours.

Current events bear witness to this all too well. Ironically, the progress in microbiology, geared to alleviate humankind's suffering by minimizing and controlling infectious diseases, has been subverted by biological weapons designed to bring about mass destruction. These weapons are deadly and inexpensive and can be deployed inside intercontinental missiles, in aerosol cans, or by crop dusters. Those who make light of the threat of biological warfare and terrorism are, like the proverbial ostrich, hiding their heads in the sand.

Biological warfare, the sanctioned employment of biological weapons by governments, has a long history. Armies learned that the natural outbreak of disease can be deadly to the opposition. Despite international treaties, biological warfare remains a threat which escalated during the last

*Credit is given to *Textbook of Military Medicine, Part I, Medical Aspects of Chemical and Biological Warfare* (Office of the Surgeon General, United States Army, 1997), as a major reference for the history of biological warfare in this chapter.

decade of the 20th century and persists because of political unrest in the Middle East.

Biological terrorism, the use of biological weapons by non-state government groups and crazed individuals, surfaced in the middle of the 20th century and is a threat to the world community. Biological weapons used for warfare or for acts of bioterrorism are described as the "poor man's weapons of mass destruction" because of the relative ease and low cost of their deployment.

Anthrax is the number 1 biological threat, resulting in the authorization of the military to vaccinate all U.S. service members. This decision has caused considerable controversy. The employment of biological weapons, whether for warfare or for terrorism, is a real threat. The Department of Health and Human Services, of which the Centers for Disease Control and Prevention (CDC) is an agency, is charged by presidential directive with the responsibility of preparing a national response to emergencies resulting from the use of biological weapons.

BIOLOGICAL WARFARE AND TERRORISM

Biological weapons are the tools of **biological warfare, bioterrorism,** and **biocrime;** they are microbes deployed to grow in (or on) their target host, or microbial products, that are deliberately employed to produce a state of clinical disease with the intent of incapacitating or killing individuals or producing mass casualties. Biological warfare is the sanctioned employment of biological weapons by nations in the conduct of war. Biological terrorism is the employment of biological weapons by non-state government groups, including religious cults, militants, and crazed individuals. Biocrime is a bioterrorist act that targets a specific individual or group rather than the masses.

Records of biological warfare date back to antiquity. Since the late 1900s, the threat has escalated as a result of ongoing aggression sanctioned by Iraqi president Saddam Hussein in the Persian Gulf, which served as a wake-up call to nations. Evidence of biological terrorism surfaced in the middle of the 20th century and has gained momentum in about the last 12 years. On April 19, 1995, what was then the worst attack in American history occurred when a truck bomb caused a massive explosion that took down the north face of the Alfred P. Murrah Federal Building in Oklahoma City. One hundred sixty-eight people died, including 19 children, and hundreds were wounded. The terrorist was Timothy McVeigh, who was seeking revenge for "crimes" committed by the U.S. government. McVeigh was sentenced to death on June 13, 1997, and was executed on the morning of June 12, 2001, by lethal injection. This shocking episode shows the depth that terrorists resort to in order to further their cause. It was an eye-opener in terms of our vulnerability to terrorist attacks, whether chemical or biological, and indicates the need for readiness.

The study of history reveals the influential role played by microbes in battles and wars, perhaps even more influential than artillery, tanks, and other weapons of conventional warfare. The natural outbreak of disease may be more devastating to military troops than the forces brought to bear by the opposition. Battlefield conditions are conducive to the spread of infectious diseases because of unsanitary conditions, particularly resulting from the problem of disposal of large amounts of human fecal material, the unavailability of uncontaminated drinking water, poor personal hygiene, crowding, inadequate supplies of safe foods, and overwhelmed and inadequate medical facilities. A variety of diarrheal diseases can quickly bring soldiers to their knees, as in the Battle of Crecy in 1346 (chapter 8). Epidemic typhus fever and relapsing fever (chapter 8), transmitted from person to person by the feces of body lice, prevail under crowded conditions and the inability to practice good personal hygiene. Napoleon's army was defeated by typhus fever. During the Civil War, bacterial, viral, and protozoan diseases, including typhoid fever, staphylococcal and streptococcal infections, battlefield gangrene, smallpox, chicken pox, measles, malaria, and amebic dysentery, killed and incapacitated more soldiers than did rifles, cannons, and other weapons. Undoubtedly, in all wars, armies have had to defend against microbes within their own troops as a natural consequence of battlefield conditions, as well as against their opposition. In stark contrast, however, the deliberate release of microbes to kill is a heinous crime and challenges the rules of warfare.

Historically, circumstances which herd together large groups of people living in proximity under unsanitary and nonhygienic conditions (e.g., refugees, prisoners of war, and inhabitants of concentration camps) promote outbreaks of microbial diseases. International relief agencies, including the Red Cross and the United Nations (UN), play a vital role in attempting to minimize human misery. In the wake of disasters caused by natural catastrophic events (e.g., hurricanes, floods, tornadoes), communities need to implement disaster relief plans to minimize and cope with infectious diseases, particularly those due to contaminated water and food supplies.

No country and no community are safe from warmongers or from terrorists. Secretary of State Madeleine Albright, in 1998, remarked that "biological weapons know no boundaries." In 1999, President Bill Clinton stated that the possibility of germ attack kept him awake at night and, further, "a chemical attack would be horrible, but it would be finite . . . but a biological attack could spread . . . kind of like the gift that keeps giving." In the presidential motorcade, you may notice a black van; it is outfitted for decontamination in the event that the president is exposed to biological or chemical weapons. In Israel, citizens are encouraged to have "shelter" rooms in their homes in case of biological warfare.

Keep in mind that animals, plants, water supplies, and food supplies are also likely targets for biological weapons. During World War I and previous wars, horses and mules played a significant role in the transport of soldiers and supplies in the conduct of war and were potential targets that would seriously cripple an army if they were to become ill or die.

Documentation of the employment of biological weapons is difficult because of several factors, including confirmation of allegations of employment, the lack of reliable microbiological and epidemiological data, propaganda, the secrecy inherent in the use of biological weapons, scare tactics, and the occurrence of naturally caused diseases, particularly during war. Nevertheless, there is ample evidence that biological weapons have been employed since ancient times.

HISTORY OF BIOLOGICAL WEAPONRY

The history of the use of biological weapons will be considered over three time frames: (i) early history to World War II, (ii) World War II to the 1972 Biological Weapons Convention, and (iii) 1972 to September 11, 2001.

Early History to World War II

Ancient warriors recognized the devastating impact of infection and poor sanitation on armies and utilized their observations in the crude use of filth, human excrement, and human and animal carcasses to produce epidemics of infection and to pollute wells and other water sources of enemy forces (Table 14.1). These kinds of atrocities continued into the 20th century.

In preparation for naval battle against King Eumenes II of Pergamum in 184 B.C., Hannibal, the famous Carthaginian leader, instructed his army to fill earthen pots with "serpents of every kind" which were then hurled onto enemy ships to terrorize warriors. Hannibal's forces were victorious as the opposition, crazed with fear, now had to battle two forces. During the siege of Kaffa (now Feodossia, Ukraine) in 1346, Tatar military leaders recognized that diseased corpses could be used as weapons and catapulted their own dead soldiers into the midst of the enemy. According to one account:

> The Tatars, fatigued by such a plague and pestiferous disease, stupefied and amazed, observing themselves dying without hope of health ordered cadavers placed on their hurling machines and thrown into the city of Kaffa, so that by means of these intolerable passengers, the defenders died widely.

During a plague outbreak, bodies of dead soldiers along with large quantities of excrement were hurled into the ranks of the enemy at Karlstein, Bohemia, in 1422. In 1710, Russian troops battling Swedish forces in Reval (now Tallinn, Estonia) threw their plague victims over the walls into the midst of the opposition. The agent of smallpox, long before immunization and knowledge of its viral nature, was a potent weapon. In the 16th century, Spanish conquistadors were reported to have given smallpox-contaminated clothing to natives of South American natives. During the French and Indian War (1754 to 1763), the English, feigning humane intent, supplied Indians loyal to the French with smallpox-contaminated blankets. General W. T. Sherman in his *Memoirs* complained that the Confederate troops were deliberately shooting farm animals in ponds so that their "stinking carcasses" would contaminate the water supplies of the Union forces, resulting in troops weakened and demoralized by the onset of gastrointestinal disease.

The end of the 19th century and the beginning of the 20th century marked the emergence of modern microbiology and are frequently referred

TABLE 14.1 Examples of biological warfare in early history

Event	Description
Hannibal's battle with King Eumenes II, 184 B.C.	The great Carthaginian leader had his army hurl serpents onto enemy ships
Siege of Kaffa, 1346	The Tatars, fatigued by a plague, catapulted cadavers into the city of Kaffa
Siege of Karlstein, 1422	Excrement and bodies of dead soldiers thrown into enemy ranks
Use of smallpox	French and Indian War, 1754–1763; Civil War
U.S. Civil War	General Sherman accused Confederate soldiers of shooting farm animals in ponds so that their "stinking carcasses" would contaminate the water supplies of Union forces

to as the "golden era of microbiology." The pioneering work of Louis Pasteur, Robert Koch, and others (chapter 2) established that infectious diseases are caused by microbes and not by miasmas—"bad air" or "swamp gas." Hence, the capability for the production of selected pathogens which could be used for biological warfare was born as the result of, and on the heels of, advances designed to control microbial infections. What an irony! During World War I, Germany was accused of conducting biological warfare by shipping infected horses and cattle to the United States and to other countries. Further, it was alleged that in 1915, German forces attempted to spread cholera into Italy and Russia and air-dropped bacterially contaminated fruit, chocolate, and children's toys into Romania. However, it is important to realize that these events are allegations and were never proved. Germany was exonerated by the League of Nations because of the lack of "hard evidence," but suspicion remains; Germany was not exonerated from the charge of conducting chemical warfare.

The horror experienced from the use of gas in World War I is depicted in Fig. 14.1 and in the following excerpt from a poem by the British poet Wilfred Owen.

> Gas! GAS! Quick, boys!—An ecstasy of fumbling,
> Fitting the clumsy helmets just in time,
> But someone still was yelling out and stumbling
> And flound'ring like a man in fire or lime.—
> Dim through the misty panes and thick green light,
> As under a green sea, I saw him drowning.
> In all my dreams, before my helpless sight
> He plunges at me, guttering, choking, drowning.

> WILFRED OWEN, "DULCE ET DECORUM EST"

After the war, the Protocol for the Prohibition of the Use in War of Asphyxiating, Poisonous or other Gases and of Bacteriological Methods of Warfare, more frequently referred to as the Geneva Protocol of 1925, was signed by 108 nations, including Iraq (Table 14.2). This was a landmark event; it represented the first multilateral agreement prohibiting the use of biological weapons. The term "bacteriological" was subsequently accepted to be synonymous with the term "biological" to include viruses and fungi. There were no provisions prohibiting the use of microbes for basic research or mandating inspection for purposes of monitoring compliance. These

FIGURE 14.1 *Gassed*, a painting by John Singer Sargent. Poison gas added to the horror of World War I. (Source: Imperial War Museum, London, United Kingdom.)

TABLE 14.2   Early landmark treaties against biological warfare

Treaty	Description
Geneva Protocol, 1925	Protocol for the prohibition of the use in war of asphyxiating, poisonous, or other gases and of bacteriological methods of warfare
Biological Weapons Convention, 1972	Convention on the prohibition of the development, production, and stockpiling of bacteriological (biological) and toxin weapons and on their destruction

limitations lie at the heart of the current threat of bioweapons produced by Iraq and its defiance of inspection. The Geneva Protocol allows the right of retaliatory use of chemical or biological weapons. Any treaty, not just those restricted to matters of warfare, is only as good as the intent of those who sign it.

World War II to 1972

The occasions on which biological weapons were or may have been used during the next period of approximately 50 years (to the 1972 Biological Weapons Convention) are numerous, and the evidence is confounded with allegations, secrecy, intrigue, charges, and countercharges. Japan employed biological warfare agents in Manchuria, China, from 1932 until the end of World War II. The infamous Japanese Army Units 731 and 100 were biological warfare units and conducted unforgivable experiments on prisoners in China. Unit 731 carried out experimental studies in Ping Fan, Hancheng, Changchun, and other sites on prisoners from many countries, including Manchuria, the Soviet Union, the United States, Great Britain, and Australia. The prisoners were injected with pathogens, including *Bacillus anthracis, Neisseria meningitidis, Vibrio cholerae, Shigella* species, and *Yersinia pestis.* At least 3,000 prisoners died at Ping Fan. Further allegations against Japan, in at least 11 Chinese cities, included contaminating water and food supplies with a variety of pathogens, throwing cultures into homes, or spraying infective aerosols from aircraft over cities. In Japan's attack on Changteh in 1941, utilizing cholera as a weapon, a backlash occurred, resulting in 1,700 deaths among Japanese troops. This illustrates a potential pitfall to the aggressor employing biological weapons. While some of these incidents have not been conclusively substantiated, the Japanese government later stated that its conduct was "most regrettable from the viewpoint of humanity."

In the events leading up to World War II and during the war, there is no documentation that Adolf Hitler's Nazi Germany conducted biological warfare. Winston Churchill is said to have considered anthrax as a retaliatory measure if Germany used biological agents against Britain. During the years after World War II, allegations regarding the employment of biological agents were carelessly thrown about. It is true that epidemics of disease occurred, but were they the result of biological weapons? Certainly, the devastation of war infringes strongly on a country's infrastructure, including a subsequent decline in public health measures and a resulting increase in disease. To some extent, these allegations were purposely employed as blatant propaganda and scare techniques, exploiting the fear of biological warfare.

In response to the international threat of the employment of biological weapons, the United States embarked on an offensive biological weapons program in 1942. The Soviet Union also began such a program during World War II. The U.S. initiative was under the direction of the War Reserve Service, a civilian agency, headquartered at Camp Detrick (later renamed Fort Detrick) in western Maryland. At its peak in 1945, almost 2,000 scientific and military personnel worked at this maximum-security campus, which had the dubious distinction of being the world's most advanced biological warfare unit. After the Korean War, the U.S. program focused on retaliatory capability and was centered at Fort Detrick (Fig. 14.2) under the

FIGURE 14.2   Fort Detrick, the location of the U.S. Army's main facility for research on biological warfare defense. **(A)** An old photo of the main gate, probably taken in the 1950s; **(B)** technicians working with highly virulent microbes in biological containment hoods; **(C)** a huge vat used to culture microbes on a large scale. (Source: Public Health Service.)

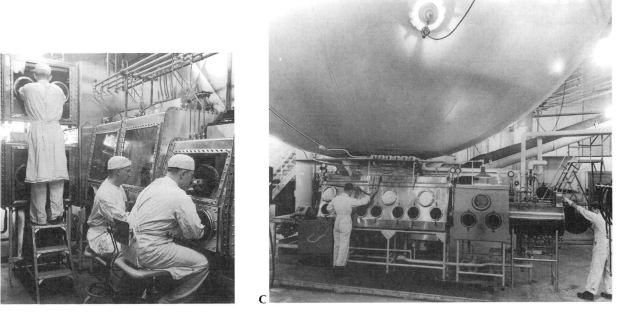

jurisdiction of the U.S. Army. Weapons systems disseminating a variety of pathogens were tested at Fort Detrick, major cities in the United States, and remote desert and Pacific sites. New York and San Francisco were unknowingly used as models to test aerosolization and dispersal methods using "harmless bacteria." In 1950, an outbreak of nosocomial urinary tract infections caused by *Serratia marcescens* occurred at Stanford University Hospital, resulting in 11 cases and one death, following the secret use of this supposed nonpathogen. A panel of public health experts concluded that the outbreak was unrelated to the aerosolization experiment, but who really knows? In 1966, *Bacillus globigii*, a nonpathogenic sporeformer related to *B. anthracis*, was released into the subway system of New York City to determine the effectiveness of aerosolization and the city's vulnerability. Results indicated that the entire tunnel system could be infected by release of spores into a single station because of the air currents generated by the trains. More recent research suggests that the bacteria are carried on the trains rather than on the air currents in the tunnels. They get sucked into the trains' ventilation systems and then emitted onto station platforms when the doors open.

During the 1960s, international concern peaked regarding the risk, unpredictability, lack of control measures, and the failure of the 1925 Geneva Protocol for preventing biological weapons proliferation. Nevertheless, in a surprise move in November 1969, President Richard Nixon terminated the U.S. offensive biological weapons program, stating, "We'll never use the damn germs. . . . If someone uses germs on us we'll nuke 'em." The 1972 Convention on the Prohibition of the Development, Production, and Stockpiling of Bacteriological (Biological) and Toxin Weapons and on their Destruction, commonly known as the Biological Weapons Convention, was convened, and a treaty (Table 14.2) was signed by 103 nations, including Iraq. One point of the treaty prohibited the development of delivery systems intended to dispense biological agents and, further, required signatory nations to destroy their stocks of biological agents, delivery systems, and equipment within 9 months of signing. The treaty was ratified in April 1972 and went into effect in March 1975.

1972 to September 11, 2001

As a result of the Biological Weapons Convention of 1972 and the termination of the offensive biological warfare program, the biological arsenal at Fort Detrick was destroyed, including the causative agents of the potentially lethal diseases anthrax, tularemia, and Venezuelan equine encephalitis, the toxins botulinum toxin and staphylococcal enterotoxin B, and several anticrop biological agents. The focus at Fort Detrick shifted from offensive to defensive strategies and was placed under the auspices of the U.S. Army Medical Research Institute of Infectious Diseases (USAMRIID). The availability of high-level containment facilities at the Fort Detrick facility still allows for the study of diseases caused by highly virulent pathogens.

As was the case following the Geneva Protocol of 1925, the 1972 Biological Weapons Convention has proved to be ineffective. As an example, on Monday, April 2, 1979, an epidemic of anthrax occurred in Sverdlovsk (now Yekaterinburg), a city 4 kilometers downwind of a Soviet military microbiology facility. Inhabitants of the community reported to hospitals with symptoms of anthrax. At least 77 cases and 66 autopsy-confirmed deaths resulted, the largest epidemic of inhalation anthrax in history. In addition to its impact on humans, the outbreak caused the death of livestock. The Soviet Union continued an offensive biological weapons program through the 1980s under the control of the Ministry of Defense. Western intelligence suspected that the Sverdlovsk facility was a biological warfare research center and that the epidemic was due to the accidental airborne release of anthrax spores. The Soviets denied the allegation and insisted that the incident was due to ingestion of contaminated meat. However, in 1992, almost 15 years later, Russian president Boris Yeltsin admitted to a cover-up; he acknowledged that the microbiology facility was part of an offensive biological weapons program and that failure to activate air filters resulted in the accidental release of anthrax spores. He proclaimed that he would terminate biological warfare initiatives. To what extent this happened is not clear; it is believed that a large number of workers are still employed in the biological weapons business (In the News 14.1).

Iraq continues to be a major actor on the world scene in terms of the threat of biological warfare. Between 1986 and the end of the Gulf War in April 1991, Iraqi scientists investigated the potential of a large number of biological weapons, including bacteria, fungi, and viruses, as well as botulinum toxin. At least 200 bombs and 25 ballistic missiles laden with biological agents were available at the time of the Gulf War but were never utilized. As preposterous as it may sound, the *Clostridium botulinum* culture used for the mass production of botulinum toxin was obtained from the United States. Despite the Gulf War defeat, the Iraqi biological warfare threat remains. In 1990, the UN imposed sanctions on Iraq because of its invasion of Kuwait and the threatened use of biological weapons. Iraq's failure to allow inspection of its biological weapons by the UN Special Commission led to air strikes by U.S. and British forces in December 1998, the results of which have not been fully evaluated. Concern remains that large quantities of anthrax spores are still present. Iraq is required to be certified free of weapons of mass destruction for the UN sanctions to be lifted. A stalemate now exists.

IN THE NEWS 14.1

Iran tries to hire ex-Soviet germ warfare researchers

Soviet Biological Warfare Program of 1980s Not Detected by West

Biological weapons: a relic of the Soviet era or part of Russia's future plans?

Russia's Biological Warfare Program

EMERGENCE OF BIOLOGICAL TERRORISM

In the past 15 years, the threat of biological weaponry has taken on a new face: that of biological terrorism, a strategy developed by nongovernment users, including religious cults, terrorists, and crazed individuals, to further their own personal or political agendas by employing biological weaponry. In some respects, biological terrorism is an even greater threat than biological warfare; it is more difficult to detect and control and can strike without warning (Table 14.3). In 1981, members of a religious commune, followers of Bhagwan Shree Rajneesh, intentionally contaminated 10 restaurant salad bars in an Oregon community with *Salmonella enterica* serovar Typhimurium, leading to a total of over 750 cases of salmonella gastroenteritis in a community in which fewer than five cases per year are usually reported. This was intended as a "rehearsal" for a plan to cause an epidemic on election day in order to influence the outcome of the county elections. The terrorist threat posed by radical groups was again demonstrated in 1995 by the Aum Shinrikyo ("Supreme Truth") cult, possibly the world's most infamous apocalyptic sect, when they released sarin gas into a Tokyo subway, killing at least 11 people and making thousands of others ill. Perhaps even more scary is the fact that these terrorists previously attempted at least nine biological attacks with the intent of causing mass murder. Their lack of sophistication, not any lack of determination, resulted in the failure of these activities. The cult possessed pathogenic microbes as well as sarin. In still another incident, on May 5, 1995, Larry Harris, a laboratory technician in Ohio, received three vials of *Y. pestis*, the causative agent of plague, from the American Type Culture Collection. His intent is not clear, but the fact that he needed only a credit card and a false letterhead is frightening. Another incident at first thought to be a possible case of bioterrorism occurred on March 25, 1999, when a 38-year-old woman was admitted to a hospital in New Hampshire with the possibility of brucellosis. This bacterial disease is transmitted to humans from infected animals (cattle and pigs) or animal products. Initially, it was thought that the infection might have resulted from exposure to "cultures" kept in her apartment by her boyfriend, a foreign national. "Good news and bad news" concludes this scenario: the good news is that brucellosis and bioterrorism were ruled out; the bad news is that the woman died from other causes. This case illustrates an effective public health response as a

TABLE 14.3 Examples of biological terrorism

Year	Perpetrator	Description
1981	Bhagwan Shree Rajneesh cult	Contaminated 10 salad bars in Oregon community with *Salmonella enterica* serovar Typhimurium
1995	Aum Shinrikyo ("Supreme Truth") cult	Sarin gas attack in Tokyo subway; previously attempted at least nine biological attacks
1995	Larry Harris	Laboratory technician in Ohio ordered and received three vials of *Yersinia pestis* (plague bacillus) from the American Type Culture Collection
2001	Unknown	Anthrax-contaminated letters mailed to media organizations and Senate offices (see Box 14.3)

result of the coordination of clinical, public health, and law enforcement activities to deal with suspected bioterrorism. These are only a few examples of the many reported incidents involving pathogens in the hands of terrorists acting singly or in groups (In the News 14.2).

The entire world is vulnerable to bioterrorism. The United States and other countries can no longer focus on biological warfare alone but also must heed the threat of bioterrorism from non-state actors. In this regard, three major statutes (Table 14.4) have been passed in recent years by the U.S. Congress: (i) the Biological Weapons Act of 1989, (ii) the Chemical and Biological Weapons Control and Warfare Elimination Act of 1991, and (iii) the Anti-Terrorism and Effective Death Penalty Act of 1996. In 1997, the CDC was charged with the regulation and transfer of 24 infectious agents and 12 toxins that pose a significant public health risk (Table 14.5).

ASSESSMENT OF THE THREAT OF BIOLOGICAL WEAPONRY

If this account "makes your skin crawl," it is with good reason, because the threats of biological warfare and biological terrorism are real and should not be underemphasized. The numerous events employing biological weaponry from antiquity to current times make this point all too obvious;

IN THE NEWS 14.2

Bioterrorism: a Threat to Public Health

Two Employees of Brentwood Mail Facility Die of Anthrax
Two Others in Hospital

Bioterror attacks cause push for safer smallpox vaccine

New Anthrax Case Reported
NBC Worker in New York Has Positive Test

Anthrax spores found in four buildings in D.C.

Army Research Facility Failed To Keep Track of Anthrax Specimens

Postal Investigators, FBI Step Up Search for Sender of Anthrax-Tainted Mail

Biological Terrorism

TABLE 14.4   U.S. statutes and directions against biological terrorism

Law	Description
Biological Weapons Act, 1989	The knowing development, manufacture, transfer, or possession of any biological agents, toxin, or delivery for use as a weapon is to be considered a federal crime
Chemical and Biological Weapons Control and Warfare Elimination Act, 1991	Stipulates a system of economic sanctions and export controls by Congress to curb the proliferation of biological weapons; to be imposed on international companies in countries designated by the United States as terrorist states
Presidential Decision Directive 39, 1995	In response to the threat of biological weapons, President Clinton ordered the Department of Health and Human Services to develop countermeasures
Anti-Terrorism and Effective Death Penalty Act, 1996	Extended the Biological Weapons Act of 1989 by broadening criminal provisions to include threats or attempts to develop or to use biological weapons, including DNA, or other technological advances to create new and more virulent pathogens
CDC charged with regulation and transfer of biological agents, 1997	24 infectious agents and 12 toxins identified as biological hazards to be regulated by CDC (see Table 14.5)

TABLE 14.5   CDC-regulated biological hazards[a]

Viruses (13)
Eastern equine encephalitis virus
Ebola virus
Smallpox virus
Hantavirus
Yellow fever virus
Bacteria (7)
Anthrax bacillus
Botulism bacterium
Plague bacterium
Rickettsiae (3)
Typhus fever rickettsia
Rocky Mountain spotted fever rickettsia
Fungi (1)
Toxins (12)

[a]A partial list including only those named in chapter 8 or chapter 9. The fungi and toxins are not discussed in the text.

it is within the capability and resources of many countries to employ microbes as tools of mass murder (Table 14.6). Biological agents are readily accessible because of their use in legitimate research activities throughout the world. The equipment and facilities for their mass production are found in the pharmaceutical, agricultural, and food industries, whereas both nuclear and chemical warfare require special facilities that would make their detection relatively easy. Biological weaponry is referred to as the "poor man's weapons of mass destruction." The weapons used to deliver biological agents are proliferating and are cost-effective. A group of technical experts estimated that "for a large-scale operation against a civilian population, casualties might cost about $2,000 per square kilometer with conventional weapons, $800 with nuclear weapons, $600 with nerve gas weapons, and $1.00 with biological weapons."

The employment of biological weapons, whether by governments, groups, or individuals, has inherent advantages and disadvantages (Table 14.7). The advantages of biological weapons include deadly or incapacitating effects on the target population, low cost, continued microbial proliferation, difficulty of immediate detection, and lack of physical damage to the area. The disadvantages include danger to the health of the aggressors, the effects of physical factors such as weather conditions on the success of an attack, public aversion to use, and the environmental persistence of some agents.

A large number of microbes and biological toxins are potential agents of biological weaponry directed against humans; the list consists of bacteria (including rickettsiae), viruses, fungi, toxins, and genetically modified forms of these agents, some of which are listed in Table 14.5. Further, some of these agents can be employed against animals and crops. Considering the numerous factors necessary for effective deployment, relatively few biological weapons meet the criteria; anthrax, smallpox, plague, and botulinum toxin are front-runners.

The psychosocial responses following even the threat of a biological attack add another dimension to the horror of biological weaponry. The fear of "bugs" and infection is terrifying; the resulting panic would produce staggering numbers of psychiatric casualties that could overwhelm health care facilities. The threat of the release of biological agents in a particular community, transportation system, school, or shopping mall causes a crippling effect. For example, in March 1992, an individual perpetrated a hoax by spraying his roommates with a fluid that he claimed contained anthrax bacteria. The house was placed under quarantine; 20 people, including a pregnant woman, were treated for possible exposure to anthrax while awaiting test results. On December 26, 1998, 750 people were quarantined after police received a call claiming that the anthrax-causing bacteria had been released into a popular nightclub in Pomona, Calif. This incident was reported to be the sixth anthrax hoax in the area in 2 weeks.

TABLE 14.6 International biological weapons programs[a]

Known
Iraq
Former Soviet Union
Probable
China
Iran
North Korea
Libya
Syria
Taiwan
Possible
Cuba
Egypt
Israel

[a]Source: U.S. Congress, House Committee on Armed Services, Special Inquiry into the Chemical and Biological Threat, *Countering the Chemical and Biological Weapons Threat in the Post-Soviet World* (Government Printing Office, Washington, D.C., 1993).

TABLE 14.7 Advantages and disadvantages of biological weapons[a]

Advantages
The potential deadly or incapacitating effects on a susceptible population
The self-replicating capacity of some biological agents to continue proliferating in the affected individual and, potentially, in the local population and surroundings
The relatively low cost of producing many biological weapons
The insidious symptoms that can mimic endemic diseases
The difficulty of immediately detecting the use of a biological agent, owing to the current limitations in fielding a multiagent sensor system on the battlefield, as well as to the prolonged incubation period preceding onset of illness (or the slow onset of symptoms) with some biological agents
The sparing of property and physical surroundings (compared with conventional or nuclear weapons)
Disadvantages
The danger that biological agents can also affect the health of the aggressor forces
The dependence of effective dispersion on prevailing winds and other weather conditions
The effects of temperature, sunlight, and desiccation on the survivability of some infectious organisms
The environmental persistence of some agents, such as spore-forming anthrax bacteria, which can make an area uninhabitable for long periods
The possibility that secondary aerosols of the agent will be generated as the aggressor moves through an area already attacked
The unpredictability of morbidity secondary to a biological attack, since casualites (including civilians) will be related to the quantity and the manner of exposure
The relatively long incubation period for many agents, a factor that may limit their tactical usefulness
The public's aversion to the use of biological warfare agents

[a]From the *Textbook of Military Medicine, Part I, Medical Aspects of Chemical and Biological Warfare*, Office of the Surgeon General, United States Army, 1997.

CATEGORY A BIOLOGICAL THREATS

Biological agents and the diseases they cause have been categorized into three groups based on their risk to the national security (Box 14.1). Anthrax, botulism, plague, smallpox, tularemia, and viral hemorrhagic fevers make up category A. Although these diseases have been discussed in earlier chapters (anthrax, botulism, plague, and tularemia in chapter 8; viral hemorrhagic fevers in chapter 9; and smallpox [with regard to its eradication] in chapter 13) and should now be reviewed, they are summarized in Table 14.8. More needs to be said about anthrax and smallpox here.

Anthrax

Anthrax is discussed briefly in chapter 8. It has the dubious distinction of being the most likely biological weapon to be employed and was designated by the chairman of the Joint Chiefs of Staff as the number 1 biological threat. The anthrax microbe has been identified as being in the arsenal, or in the process of being developed, in at least 10 countries, including Iraq. There are also reports that scientists in at least one of these countries have genetically engineered strains of anthrax to produce new toxins.

What is anthrax, and why is it at the top of the list of biological weaponry? Anthrax is a rapidly progressing acute infectious disease caused by the organism *B. anthracis* (Fig. 14.3), a rod-shaped spore-forming bacterium. It is a disease of antiquity and thought to have caused the fifth Egyptian plague around 500 B.C.; during the Middle Ages it was known as the Black Bain and was responsible for nearly destroying the cattle herds of Europe. The ability of *B. anthracis* to form spores makes anthrax a choice weapon of mass destruction, since these spores can be disseminated by bombs, artillery shells, aerial sprayers, and other methods of dispersal.

BOX 14.1 Categories of Biological Diseases and Agents

Category A Agents

The U.S. public health system and primary health care providers must be prepared to address varied biological agents, including pathogens that are rarely seen in the United States. High-priority (category A) agents include organisms that pose a risk to national security because they can be easily disseminated or transmitted from person to person, cause high rates of mortality with the potential for major public health impact, might cause public panic and social disruption, and require special action for public health preparedness.

Category A agents include variola virus, which causes smallpox; *B. anthracis*, which causes anthrax; *Y. pestis*, which causes plague; *C. botulinum* toxin, which causes botulism; *Francisella tularensis*, which causes tularemia; and filoviruses (Ebola and Marburg viruses) and arenaviruses (e.g., Lassa and Junin viruses), which cause viral hemorrhagic fevers.

Category B Agents

The second-highest-priority agents include those that are moderately easy to disseminate, cause moderate morbidity and low mortality, and require specific enhancements of the CDC's diagnostic capacity and enhanced disease surveillance.

Category B agents include *Coxiella burnetii*, which causes Q fever; *Brucella* species, which cause brucellosis; *Burkholderia mallei*, which causes glanders; ricin toxin from *Ricinus communis* (castor beans); epsilon toxin of *Clostridium perfringens*; and *Staphylococcus* enterotoxin B.

Category C Agents

The third-highest-priority agents include emerging pathogens that could be engineered for mass dissemination in the future because of availability, ease of production and dissemination, and potential for high morbidity and mortality and major health impact.

Category C agents include Nipah virus, hantaviruses, tick-borne hemorrhagic fever viruses, tick-borne encephalitis viruses, yellow fever virus, and multidrug-resistant *Mycobacterium tuberculosis*.

Source: CDC, *Morbidity and Mortality Weekly Report* **49**(RR-4):1–26, 2000.

TABLE 14.8 Category A agents and diseases

Disease(s)	Agent(s)	Method of transmission	Incubation time[a]	Symptoms	Treatment
Anthrax	*Bacillus anthracis*	Inhalation	1–7 days	Flu-like symptoms, fatigue, breathing difficulty, possible death	Antibiotics (early administration), prevention by vaccination
Botulism	*Clostridium botulinum* toxin	Food borne	12–36 hours	Double vision, blurred vision, slurred speech, difficulty swallowing, dry mouth, muscle weakness, paralysis of breathing muscles leading to death	Antitoxin
Plague (pneumonic)	*Yersinia pestis*	Inhalation	1–6 days	Fever, headache, cough, weakness	Antibiotics
Smallpox	Variola virus	Inhalation	7–17 days	Characteristic rash, fever, fatigue	Vaccination (within 4 days after exposure)
Tularemia	*Francisella tularensis*	Inhalation	1–14 days	Fever, ulcerated skin lesions, chills and shaking, headache, weakness; can progress to respiratory fever and shock	Antibiotics
Viral hemorrhagic fevers	Ebola virus, Marburg virus, Lassa virus	Close contact with an infected person or his or her body fluids	7–14 days (Ebola fever), 5–10 days (Marburg fever), 6–21 days (Lassa fever)	Fever, fatigue, dizziness, weakness, internal bleeding, bleeding from body orifices, shock, delirium	Supportive therapy, ribavirin (depending on circumstances)

[a]Considerable variation.

Spores are condensed bits of the organism's protoplasm, including the genetic material, surrounded by a complex layer of coats rendering them resistant to the usual lethal effects of sterilization by heat and chemical disinfection; in other words, they are hard to kill and reside in soils worldwide for many years. A case in point is that in 1942, the British used Gruinard Island, a site 3 miles off the northwest coast of Scotland, as a site for anthrax experiments. The soil remained contaminated with anthrax spores for over 50 years, making the area unsafe for human entry. Spores are like seeds (chapter 4) in that under appropriate conditions, as are met on and in the body, they germinate into actively multiplying and toxin-producing bacterial forms. The reasons why anthrax is the number 1 choice as a biological weapon are summarized in Table 14.9.

There are three varieties of anthrax disease varying by the route of transmission and subsequent symptoms: (i) Cutaneous (skin) anthrax is a hazard to those employed in occupations requiring frequent close contact with the hides, hair, wool, bones, or bone products of infected animals. The bacterial spores enter the body through a break in the skin. About 20% of untreated cases result in death, but death is rare with appropriate antibiotic therapy. (ii) Gastrointestinal anthrax is rare but can be acquired by the ingestion of inadequately cooked contaminated meat. Death occurs in 25 to 60% of untreated cases. (iii) Inhalation anthrax poses the greatest threat

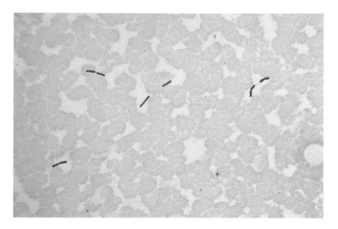

FIGURE 14.3 Micrograph showing anthrax bacteria (the purple, rod-shaped objects). (Photographer unknown.)

with the most deadly consequences. Initially, symptoms resemble a common cold, which progresses to severe breathing problems. The incubation period is 1 to 7 days; death usually results in 1 to 2 days after the occurrence of acute symptoms. All it takes is a millionth of a gram (8,000 to 10,000 spores) of *B. anthracis* to constitute a lethal dose; 1 kilogram could kill billions of people. Human-to-human transmission does not occur. The death rate is 99% in nonimmunized and untreated individuals. Antibiotic therapy can suppress the disease but only if administered within the first 24 to 48 hours after exposure; if antibiotic treatment is not begun within that period, the death rate is about 95%.

During the early 1990s, UN-sanctioned inspection teams reported that Iraq had produced 8,000 liters of anthrax spores, an amount believed capable of killing every man, woman, and child on Earth. In 1995, Iraq admitted to the UN that it had loaded anthrax spores into warheads during the Gulf War. More recently, from October 30 to December 23, 1998, a series of bioterrorist threats of exposure to anthrax were reported by the CDC (Box 14.2). In those incidents, letters alleged to be impregnated with anthrax spores were sent to health clinics in Indiana, Kentucky, and Tennessee and to a private business in California. Additionally, three telephone threats warning of anthrax contamination of ventilation systems in private and public buildings were made. All of the threats turned out to be hoaxes, but they indicate the need for a state of readiness. Further, these incidents pointed out the psychological stress that accompanies threats of biological weapons.

After a 3-year study, the secretary of defense, William Cohen, on May 18, 1998, approved the Pentagon's plan to vaccinate all U.S. military service members for anthrax, a decision causing considerable controversy. As a show of confidence, Secretary Cohen was immunized. Full immunization requires six injections administered over an 18-month period. Implementation of the immunization program in the military has been designed as a three-phase strategy that is already in place, with a completion date of 2003; it has a price tag of about $60 to $80 per person for the full six-dose regimen and a total cost in excess of $200 million for the estimated 2.4 million personnel.

The anthrax vaccine is a cell-free filtrate, meaning that it has no whole bacteria, alive or dead, and thus is incapable of causing infection. Hence, one of the arguments against use of the vaccine, namely, that it might cause

TABLE 14.9 Anthrax: the number 1 choice as a biological weapon

High fatality rate if untreated
Treatment only partially successful and only if administered within 24–48 hours
Organism easy to grow
Produces spores which remain viable for years
Can be easily dispersed by missiles, rockets, aerial bombs, crop dusters, and sprayers
No indication of exposure

BOX 14.2 Anthrax Threats Reported by the CDC, October 30 to December 23, 1998

Indiana

A threatening letter was sent to a health clinic. Thirty-one adults considered to be possibly exposed to spores were detained for 3 hours and decontaminated. They placed their clothing and personal effects in plastic bags and showered, using soap and water plus a dilute bleach solution. The desktop where the letter was found was washed with household bleach (50% hypochlorite solution). All 31 persons were transported to local emergency departments to receive treatment. Some underwent additional decontamination.

The letter was found to be negative for *B. anthracis* by the Indiana State Department of Health and by USAMRIID in Fort Detrick, Md.

Kentucky

A threatening letter was sent to a health clinic. The circumstances and decontamination procedure were similar to those in Indiana. The letter was found to be negative for *B. anthracis* by the University of Louisville Hospital Clinical Microbiology Laboratory and by USAMRIID.

Tennessee

A threatening letter was sent to a health clinic under circumstances similar to those in Indiana and Kentucky. The letter was found to be negative for *B. anthracis* by USAMRIID.

California

Four threats of *B. anthracis* contamination were received by businesses or government offices from December 17 to 23, 1998; all premises were found to be negative for *B. anthracis*.

The first threat was made in a letter mailed to a private business; all 28 "at-risk" adults were decontaminated and given antibiotics.

The second threat was made by a telephone caller to a government building claiming to have contaminated the building's air-handling system. The decision was made not to decontaminate approximately 95 adults, but all received chemoprophylaxis.

The third threat was made by a telephone caller to 911 claiming to have contaminated the air-handling system of a federal building; 1,200 to 1,500 persons (including at least one pregnant woman and two children) were at risk. Decontamination and chemoprophylaxis were not practiced, but all were advised to wash their vehicles with bleach, place clothing in plastic bags pending laboratory results, and shower.

The fourth threat was made by a telephone caller to 911 claiming to have contaminated an air-handling system; approximately 200 persons were at risk. Decontamination and chemoprophylaxis were not practiced.

---

anthrax, has no basis. The vaccine was first approved by the Food and Drug Administration in 1970 as a measure to protect at-risk workers with high exposure to anthrax against the cutaneous form of the disease.

The Pentagon's anthrax program has been faced with severe opposition by Citizen Soldier (an advocacy group for veterans) in defense of service members who have refused to take the shots because of the reports of adverse effects in a small number of individuals. Congress is calling for voluntary, not mandatory, immunization or suspension of the program (In the News 14.3) pending further study. Meanwhile, the Pentagon continues to strongly defend its decision and to vaccinate military personnel, resulting in courts-martial and discharges for those refusing to be immunized. On February 16, 2000, the court-martial of Air Force major Sonnie G. Bates, the Air Force's first court-martial involving refusal of anthrax vaccination, was avoided when Major Bates agreed to face a lesser administrative review, resulting in a general discharge under honorable circumstances.

Criticism of the vaccine is based on the following major points: (i) the difficult immunization schedule (six shots over a period of 18 months); (ii) the fact that full immunity is not conferred until after 18 months; (iii) the need for an annual booster; (iv) the lack of proven effectiveness against a large dose of aerosolized spores; and (v) the use of a vaccine approved by the Food and Drug Administration in 1970 for prevention of cutaneous anthrax, which opponents claim is insufficient grounds to prove effectiveness against inhalation anthrax.

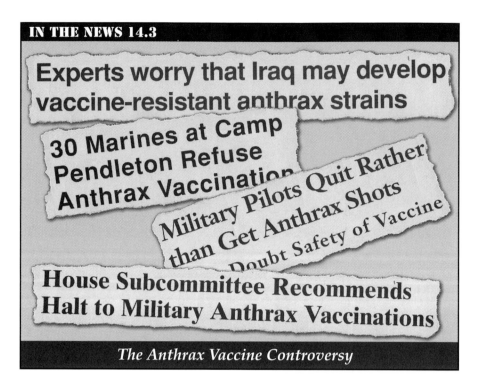

IN THE NEWS 14.3

Experts worry that Iraq may develop vaccine-resistant anthrax strains

30 Marines at Camp Pendleton Refuse Anthrax Vaccination

Military Pilots Quit Rather than Get Anthrax Shots

Doubt Safety of Vaccine

House Subcommittee Recommends Halt to Military Anthrax Vaccinations

The Anthrax Vaccine Controversy

Smallpox

Smallpox, the first disease to be eradicated, is now a top biological candidate weapon. The unique properties of the disease that made its eradication possible are discussed in chapter 13. Routine immunization against smallpox ended in 1972 in the United States and, within a few years, ended around the world. Ironically, the science that led to the eradication of smallpox resulted in a world population under the age of 25 that is susceptible to smallpox (including about half the population of the United States). Those who were immunized have a waning immunity. The significance is that if smallpox is introduced into the population, it has the potential to spread like wildfire through the (nonimmunized) population.

If smallpox has been eradicated from the face of the earth, why is it a threat? The answer is that the disease has been eradicated, but not the virus, which exists in freezers under lock and key at a designated facility in Russia and at the CDC in Atlanta, Ga. However, it is believed that covert stocks of the virus are in the hands of potential terrorists or unreliable foreign governments.

The early signs of smallpox include high fever, fatigue, headaches, and backaches, followed in a few days by an extensive rash over the body (Fig. 13.1). The rash is characterized by flat red lesions that become filled with pus and crust over in the second week; the lesions break out at the same time. (The rash of chicken pox appears in waves.) The crusts dry up and fall off, leaving deeply pitted scars, particularly on the face. The case fatality rate is about 30%.

Smallpox is transmitted directly from person to person by virus-infected saliva droplets expelled from an infected individual onto a mucosal surface of another. Smallpox virus-contaminated clothing and bedding can also spread the virus. The first week of illness is the most infectious time because of the high numbers of viruses or viral particles in the saliva.

COUNTERMEASURES TO BIOLOGICAL WEAPONRY

Perhaps you found the foregoing account of biological warfare and terrorist activities unsettling and fearsome. If you did, it is with good reason. Unfortunately, the potential exists and is no longer limited to science fiction and to James Bond movies. Adding to your anxiety is the fact that, since the dissolution of the Soviet Union, Iraq is luring Russian scientists with expertise in the field of biological weapons based on their past employment in Russia's very extensive biological warfare program. These scientists are attracted by "big bucks," far more than they can possibly earn in Russia's troubled economy.

The use of biological weapons presents a unique set of challenges to the public health infrastructure of countries worldwide, as compared with the use of conventional weapons or even chemical or nuclear warfare, including the following:

1. An attack might not be immediately discernible and would remain unnoticed until individuals became seriously ill; incubation periods vary as a function of the biological agent involved and also from person to person.
2. Imagine if an attack were to occur in a subway station, a train terminal, or, even worse, in an airport. Victims would be widely dispersed and, within 24 hours, could spread the disease worldwide before even being aware of their own illness.
3. If the agent is communicable from person to person, each infected individual would seed the agent into an ever-increasing circle of disease in successive waves.
4. Studies have shown that bioterrorist events trigger chaos, panic, and civil disorder to an extent greater than that resulting from other forms of terrorism.
5. Bioterrorist attacks are directly targeted at human populations.
6. Dispersal by aerosolization can be inexpensively accomplished with crop dusters, spray cans, and other simple devices.
7. The first responders would most likely be health care personnel, who would succumb to infection, thwarting further efforts of response.
8. The pathogen selected would be one that is not routinely seen in a population. This would mean that the population would be particularly susceptible because of a lack of immunity, e.g., in the case of smallpox, since vaccination is no longer carried out.
9. Diagnosis may be delayed because the disease would most likely be one with which medical personnel have little or no familiarity or one in which the symptoms are vague and may resemble the symptoms of a variety of microbial infections. For example, smallpox has not been seen for over 25 years, and anthrax initially has cold-like symptoms.

Given these factors, what is the nation's response, particularly to the threat posed by terrorists? How prepared are we, and what precautions are being taken (In the News 14.4)? Today's public health officials are challenged not only by the problem of emerging and reemerging infection but also by the deliberate release of biological weapons in the conduct of warfare and by terrorists intent on furthering their own agendas at any cost. Presidential Decision Directive 39, issued by President Clinton in 1995, designated the Department of Health and Human Services (DHHS) as "the

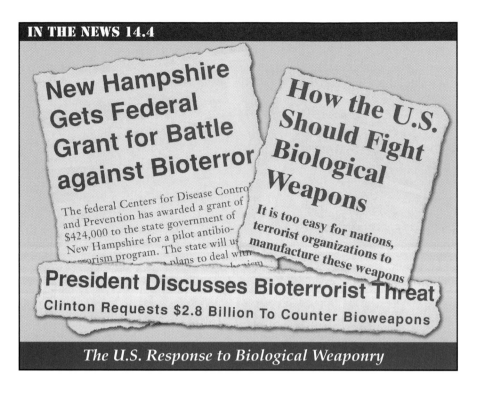

IN THE NEWS 14.4

New Hampshire Gets Federal Grant for Battle against Bioterror

The federal Centers for Disease Contro_ and Prevention has awarded a grant of $424,000 to the state government of New Hampshire for a pilot antibio-_ _orism program. The state will u_ _plans to deal wi___

How the U.S. Should Fight Biological Weapons

It is too easy for nations, terrorist organizations to manufacture these weapons

President Discusses Bioterrorist Threat

Clinton Requests $2.8 Billion To Counter Bioweapons

The U.S. Response to Biological Weaponry

lead federal agency to plan and prepare for a national response to medical emergencies arising from terrorist use of weapons of mass destruction." (The CDC is an agency within DHHS.) The total allocation for fiscal year (FY) 2000 for spending on bioterrorism preparedness was $230 million, with an increase to $1.4 billion in the FY 2002 budget.

On March 3, 1995, an assistant secretary of the DHHS testified before a public health subcommittee of the United States Senate and outlined the department's initiative and strategic plan for coping with potential bioterrorism as follows:

1. Deterrence by controlling access to and handling of dangerous pathogens, including monitoring of the facilities and procedures currently in use with these agents. The CDC is charged with this responsibility.
2. Surveillance and rapid detection, based on the number of individuals becoming ill and their symptoms, of whether a biologic agent has been released. The public health surveillance network will be improved.
3. Medical and public health response aimed at strengthening the response at the local level, including the capability of mass immunization or prophylactic management, safe disposition of the deceased, infection control, and assessment of extent of the problem.
4. Creation of a stockpile of pharmaceuticals as a national resource, since it would be beyond the resource of local governments to develop and maintain such a facility. Appropriate pharmaceuticals will be deployed to reach victims within 24 hours. CDC has the responsibility for developing the stockpile; $51 million was appointed for FY 1999 and again for FY 2000.
5. Expansion of support for research and development related to pathogens that might be used as biological weapons, particularly the

development of rapid diagnostic methods since "time is of the essence." New vaccines and new or improved antiviral and antibiological agents are critical.

It appears that a plan of readiness is under way as the nation recognizes the use of biological weapons as a real threat. Hopefully, no nation will ever need to meet the challenge of biological warfare, and all acts of terrorists will be contained with minimal consequences. There is no absolute deterrent to the crazed minds of those who would employ biological weapons. The best deterrent is an appeal to human reason and to the sanity of all individuals. Being prepared to intelligently meet a threat is the best rational approach. The use of biological weapons is a clear and dramatic example of the interfaces of biology, public health, international law, and ethics and is a topic that needs to be addressed by society at large. Some might argue that this is a scare tactic, but hiding one's head in the sand is not a viable alternative.

IN THE AFTERMATH OF SEPTEMBER 11, 2001

September 11, 2001, was a second date of infamy for America. (The first was December 7, 1941, the date of the Japanese attack on Pearl Harbor, which President Franklin D. Roosevelt called "a date which will live in infamy.") Nineteen terrorists of Arab origin commandeered four jets, loaded with jet fuel and bound from the East Coast to the West Coast, and turned them into deadly flying missiles. (A Boeing 767 carries 240,000 gallons of fuel.)

The devastation was indescribable. Television coverage over the next several hours showed bodies in the air falling from the towers (including a man and a woman holding hands on their way down), hundreds of people running in the streets pursued by billowing clouds of smoke, choking fumes and swirling dust, people blackened by soot, blood-stained streets, and blood-soaked people. More than 40,000 people worked in the twin towers of the World Trade Center, and more than 10,000 others visited each day.

Earlier in this chapter the bombing of the Federal Building in Oklahoma City was described as the worst terrorist attack on the United States in history. That description now applies to the events of September 11.

By the next day, President George W. Bush declared, "We're at war—a war against terrorism." The United States began to build a coalition of countries against terrorism and to prepare for military action in Afghanistan against the Taliban, a reactionary Islamic organization that controlled most of that country and harbored Osama bin Laden, leader of the terrorist al Qaeda network. Within a few days, CNN labeled its news coverage "America's New War."

In the days following September 11, Americans were gravely concerned and anxiety stricken about what terror might follow and might be even worse. In a *Time*/CNN poll, 53% of those surveyed feared a chemical or biological attack, while 23% feared a nuclear attack would follow.

Anthrax hoaxes are described above (Box 14.2), but the reality of employing *B. anthracis* spores as biological weapons surfaced only 2 weeks later when, on September 25, a 38-year-old assistant to NBC's Tom Brokaw developed the first case of cutaneous anthrax (Box 14.3), presumably as the result of opening a letter, postmarked September 18, containing a powder. Over the next few weeks, cases of inhalation anthrax and cutaneous anthrax targeting primarily those in the media emerged; New York City was particularly hard hit, with seven cases of cutaneous anthrax (two at NBC,

AUTHOR'S NOTE *The morning of Tuesday, September 11, heralded a perfect day in my hometown; the temperature was 70°F, and the sky was cloudless. I was at home watching the news and working on this book. I could not believe my eyes when the TV screen clearly showed a Boeing 767 jet about to crash into the 110-floor north tower of the World Trade Center. At 8:45 a.m., the plane slammed into the tower, resulting in a huge burst of flames and smoke. My immediate thought was that the plane was out of control, but about 20 minutes later, just after 9 a.m., a second jet crashed into the south tower of the World Trade Center. About 40 minutes later, a third jet slammed into the Pentagon. A fourth jet, presumably headed for the White House, crashed in a field close to Pittsburgh less than half an hour after the Pentagon crash. A series of terrorist attacks had occurred. I knew that this chapter would need to be updated.*

BOX 14.3 The Anthrax Attacks after September 11, 2001

September 18, 2001

A 38-year-old assistant to NBC News anchorman Tom Brokaw handles a letter containing powder which was postmarked on this date. On September 25, she notices a raised skin lesion on her chest, and over the next 3 days she experiences progressive erythema and edema. On September 29, she develops malaise and a headache. By October 1, the lesion has developed into a 5-centimeter oval with raised borders and satellite vesicles. There is left cervical lymphadenopathy, and a black eschar soon develops. This turns out to be the first case of cutaneous anthrax. The patient recovers with antibiotics.

September 28, 2001

A 7-month-old son of an ABC producer is taken to his mother's workplace at ABC in New York. On September 29, a large weeping lesion is noted on his left arm. Over the course of days, the lesion develops into an ulcer with a black eschar. The child is hospitalized and develops hemolytic anemia and thrombocytopenia. He is diagnosed as having cutaneous anthrax and recovers uneventfully with antibiotics. Subsequently, cases of cutaneous anthrax turn up at CBS, the *New York Times,* and the *New York Post.* All of the patients recover.

October 2, 2001

Robert Stevens, 63, is admitted to a Lake Worth, Fla., hospital gravely ill with the presumptive diagnosis of meningitis. The diagnosis of inhalation anthrax is made after testing. On October 5, he dies from inhalation anthrax, the first bioterrorism casualty of this millennium. Anthrax bacteria are found at his workplace at American Media Inc. A coworker, Ernest Bianco, is admitted to a Miami hospital with the diagnosis of pneumonia, which is later changed to inhalation anthrax. He eventually recovers.

October 14, 2001

An aide to Senate Majority Leader Tom Daschle opens a letter to the senator postmarked Trenton, N.J. The letter is loaded with high-grade, light, fine-textured anthrax spores. Three days later, 28 persons test positive for exposure. This letter was handled at the Brentwood postal facility in Washington, D.C., where two workers soon die from inhalation anthrax.

October 16, 2001

Joseph Curseen, Jr., 47, a Washington, D.C., postal worker, falls ill with flu-like symptoms but is not hospitalized. He returns to the hospital on October 22 and dies of inhalation anthrax. More than 2,200 Washington postal workers are placed on a 10-day supply of ciprofloxacin.

October 19, 2001

Thomas Morris, Jr., 53, an employee of the Brentwood post office, which handles all mail delivered into Washington (including the mail sent to Senator Daschle's office), is admitted from the emergency room at Inova Fairfax Hospital in Fairfax, Va. He is subsequently diagnosed as having contracted inhalation anthrax. He is the third person in the current outbreak to have contracted this form of anthrax. As a result, many federal office buildings (e.g., the Supreme Court, the Central Intelligence Agency) shut down for various times. On October 21, Morris dies of inhalation anthrax.

October 23, 2001

Attorney General John Ashcroft releases the text from the anthrax letters sent to Daschle, Brokaw, and the *New York Post.* The letters dated September 11 contain the phrases "Take the penicillin now," "Death to America," "Death to Israel," and "Allah is Great." Meanwhile, the CDC faces public criticism at a bioterrorism hearing at the Capitol. Defenders cite the CDC's lack of adequate resources. Two cases in point: the CDC operates in World War II-era buildings, and bad wiring caused a power outage that delayed by 15 hours the CDC's ability to identify the anthrax case at NBC News.

October 24, 2001

A 59-year-old Washington, D.C., postal worker is admitted to an emergency room with a fever of 100.8°F, sweats, myalgia, chest pain, nausea, vomiting, diarrhea, and abdominal pain. Blood cultures lead to a diagnosis of anthrax.

October 25, 2001

Homeland Security Director Tom Ridge reports that the anthrax powder sent to Daschle's office was highly concentrated and designed to be disseminated and inhaled easily. The Postal Service decides to begin environmental testing at 200 postal facilities along the East Coast. The number of confirmed anthrax cases rises to 13, of which 7 are inhalation and cutaneous anthrax. Most of the cases are linked to mail passing through New Jersey, New York City, or Washington, D.C. An estimated 10,000 people have been placed on prophylactic antibiotics.

The same day, Kathy Nguyen, a 61-year-old woman who works in the stockroom at Manhattan Eye, Ear and Throat Hospital, falls ill with myalgia and malaise. She is hospitalized on October 28, and over the course of the next few days she develops shortness of breath, chest discomfort, and a cough with blood-tinged sputum. Cultures of her blood grow *B. anthracis,* the anthrax bacillus. On October 31, Nguyen dies of inhalation anthrax.

October 28, 2001

Officials confirm a new case of inhalation anthrax in a New Jersey postal worker. He works at the facility that processed the three anthrax-laden letters going to New York City and Washington, D.C.

October 29, 2001

The CDC reports two new cases of anthrax in New Jersey. One is inhalation anthrax in a postal worker, and the other is cutaneous anthrax in a private citizen who may have contracted the disease from mail.

BOX 14.3 The Anthrax Attacks after September 11, 2001 (continued)

November 16, 2001
A 94-year-old woman in Connecticut is diagnosed with inhalation anthrax. On November 21, she dies of this disease. How she contracted anthrax remains a mystery.

January 9, 2002
The CDC reports that a total of 23 cases of anthrax have occurred: 11 cases of confirmed inhalation anthrax and 12

cases of cutaneous anthrax (7 confirmed and 5 suspected).

March 4, 2002
Cutaneous anthrax is diagnosed in a laboratory worker.

Sources: P. Rega, *Biological Terrorism Response Manual*, Maumee, Ohio, 2001 (http://www.bioterry.com); B. Mina, J. P. Dym, F. Kuepper, R. Tso, C. Arrastia, I. Kaplounova, H. Faraj, A. Kwapniewski, C.

M. Krol, M. Grosser, J. Glick, S. Fochios, A. Remolina, L. Vasovic, J. Moses, T. Robin, M. DeVita, and M. L. Tapper, *JAMA* **287:**858–862, 2002; and L. A. Barakat, H. L. Quentzel, J. A. Jernigan, D. L. Kirschke, K. Griffith, S. M. Spear, K. Kelley, D. Barden, D. Mayo, D. S. Stephens, T. Popovic, C. Marston, S. R. Zaki, J. Guarner, W. J. Shieh, H. W. Carver II, R. F. Meyer, D. L. Swerdlow, E. E. Mast, J. L. Hadler, and the Anthrax Bioterrorism Investigation Team, *JAMA* **287:**863–868, 2002.

one at CBS, three at the *New York Post,* and one at ABC) and one case of inhalation anthrax. The buildings housing these offices tested positive for anthrax spores. The offices of Senate Majority Leader Tom Daschle and Senator Patrick Leahy were contaminated on October 14 by letters postmarked Trenton, N.J., and containing a finely milled version of "anthrax powder." Dozens of personnel were contaminated. Mail facilities in the Washington, D.C., area became centers for cross-contamination of mail delivered into the city by mail trucks serving as vectors of disease. Two postal workers died of inhalation anthrax, and over 2,000 postal workers in the nation's capital were placed on a 10-day supply of the antibiotic ciprofloxacin (Cipro). In the aftermath, government bodies, including the Senate, the Central Intelligence Agency, and the Supreme Court, closed for various times; Americans were asked not to hoard ciprofloxacin following a buying frenzy of the antibiotic; Army-Navy stores unloaded their stocks of gas masks; and antidotes, water filters, disaster kits, and protective garb became hot shopping items to a panicked nation. The Postal Service instituted new security measures and distributed information about how to recognize and report suspicious mail (Fig. 14.4). The nation's airlines were threatened with bankruptcy, and security measures in and around airports were heightened to protect the flying public. (Northwest Airlines even went as far as to ban the sweetener Sweet 'n Low because of its resemblance to the powdery substance associated with anthrax spores.) Fire and police departments and ambulance services were bombarded with calls to check out "suspicious powders," and emergency room facilities were overwhelmed with people fearful that they had been exposed to anthrax spores.

After several weeks, the pandemonium died down, but many people claimed that "America would never be the same." To date, anthrax has not been linked to Osama bin Laden or his al Qaeda network. There is speculation that anthrax bacteria might have been stolen from Fort Detrick, the former army biological warfare station. The anthrax powder has been described as "weaponized," a new term meaning that the microbes have been highly milled to a particle size that is aerosolized and can reach the lungs. This suggests that the organisms are not of the common or "garden variety" but are in the hands of a trained microbiologist. In the fall of 2001, the Federal Bureau of Investigation asked the American Society for Microbiology (ASM) to make its membership list available to assist in the criminal investigation of transmission through the mail of the bacterium that causes

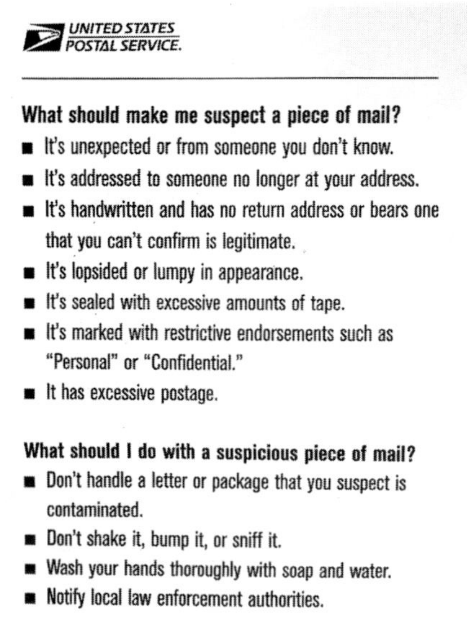

FIGURE 14.4 Postal Service instructions for dealing with suspicious mail. (Author's photo.)

UNITED STATES POSTAL SERVICE.

What should make me suspect a piece of mail?
- It's unexpected or from someone you don't know.
- It's addressed to someone no longer at your address.
- It's handwritten and has no return address or bears one that you can't confirm is legitimate.
- It's lopsided or lumpy in appearance.
- It's sealed with excessive amounts of tape.
- It's marked with restrictive endorsements such as "Personal" or "Confidential."
- It has excessive postage.

What should I do with a suspicious piece of mail?
- Don't handle a letter or package that you suspect is contaminated.
- Don't shake it, bump it, or sniff it.
- Wash your hands thoroughly with soap and water.
- Notify local law enforcement authorities.

anthrax. The request was considered by elected officers of and legal counsel to ASM, with the decision that it was appropriate to cooperate; anthrax had already claimed the lives of five individuals.

Countermeasures to the employment of biological weapons are described in an earlier section and centered on a state of readiness "just in case." But bioterrorism is no longer just a threat; it is a harsh reality that has caused individuals to die from inhalation anthrax and others to come down with cutaneous anthrax. Postal workers and other government employees by the thousands were placed on protective antibiotic therapy; all have been terrified, and many have suffered from the adverse effects of antibiotics. The nation has moved from a state of readiness to a state of alert.

Hospitals, fire and police departments, and other agencies charged with protecting the public have developed strategic plans to cope with disaster. Images of those on the front line wearing protective suits dramatized the public's anxiety (Fig. 14.5). A new government agency, the Office of Homeland Security, has been created by President Bush to develop, coordinate, and implement measures to protect the nation.

Smallpox, again, rears its ugly head; history proves that it is an effective biological weapon. There is great concern that the amount of smallpox vaccine necessary to immunize and to treat the public is woefully deficient, but two recent developments are cause for optimism in this regard. Researchers have demonstrated that the nation's smallpox vaccine stockpile can be diluted at least 5-fold (maybe 10-fold) and still be effective; this would result in about 77 million doses. In addition, a drug company recently reported the availability of 85 million doses left over from the past that might also be effective when diluted. Further, the government has ordered 200 million new doses from drug companies with deliveries scheduled by the end of 2002 or shortly thereafter. Thus, the country seems to be in good shape in that its 280 million residents are not totally defenseless against the scourge of smallpox. The secretary of DHHS, Tommy Thompson, stated, "We hope our smallpox vaccine stockpile will serve as a deterrent to those who may consider using smallpox as a weapon. We will have the necessary medicine to save and protect every American should there be an outbreak."

The CDC's plan advocates that the vaccine should not be used until after an attack occurs and then should be rapidly deployed. Others call for

FIGURE 14.5   Emergency workers in biosafety suits. (Photographer unknown.)

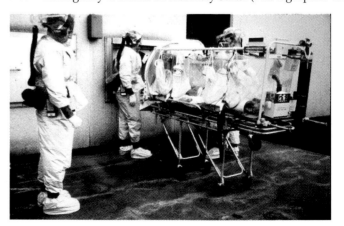

mass inoculations, fearing that an epidemic may develop too rapidly to be contained, particularly if an attack were to occur simultaneously in several cities. But Anthony Fauci, director of the National Institute of Allergy and Infectious Diseases, warns against this. According to Fauci, "if we could vaccinate people with virtually no incidence of any serious toxicity…we tomorrow could eliminate the threat of a smallpox bioterrorist attack. Unfortunately, that is not the case." Estimates are that if all Americans were vaccinated, 180 to 400 people could die just from the vaccine's side effects.

So what do we do? What is the take-home lesson? If only there were an easy answer. As individuals we must avoid panic, but we must always be on the alert. As stated in an earlier chapter, we must avoid an epidemic of fear. Perhaps we can look forward to better times and heed the admonition of the song title cited in the opening of this chapter, "I Ain't Gonna Study War No More."

OVERVIEW Biological warfare has been deployed for centuries as armies realized the devastating effect of causing disease among the opposition. Human excrement, dead human and animal bodies, and contaminated drinking water are crude methods to cause outbreaks of disease that continued to be practiced into the 20th century.

Acts of biological terrorism are a more recent phenomenon, having emerged in the mid-20th century, and pose a threat more difficult to detect than biological warfare.

Anthrax is considered the number 1 threat for several reasons, particularly because *B. anthracis,* the causative microbe, is a spore-forming bacterium. Microbes and their toxins are considered "the poor man's weapons of mass destruction," based on their cost-effectiveness, putting them within reach of many nations. The military has been authorized to immunize all personnel against anthrax, a decision which has provoked considerable controversy. The CDC, an agency within the DHHS, has been designated as the lead agency responsible for a plan of readiness against the potential use of biological weaponry in warfare and in terrorism. ■

SELF-EVALUATION

PART I Choose the *single* best answer.

1. Napoleon's army was defeated by
   a. typhoid fever
   c. diarrhea
   b. typhus fever
   d. battlefield gangrene

2. When was the Geneva Protocol signed?
   a. 1920
   c. 1948
   b. 1925
   d. 1972

3. In 184 B.C., Hannibal used a crude form of biological warfare. His act involved
   a. poisoning wells
   c. serpents
   b. poisoning horses
   d. catapulting human cadavers

4. In 1981, a religious commune purposely contaminated salad bars in an Oregon community. Which microbe was employed?
   a. *Salmonella*
   c. influenza virus
   b. *B. anthracis*
   d. *Clostridium botulinum*

5. Which form of anthrax poses the greatest threat?
   a. cutaneous anthrax
   c. gastrointestinal anthrax
   b. inhalation anthrax
   d. waterborne anthrax

6. The anthrax vaccine consists of
   a. dead bacteria
   c. attenuated bacteria
   b. live bacteria
   d. cell-free filtrate

PART II Fill in the following.

1. Biological weapons are the least expensive method of mass destruction to employ. What are the most expensive weapons? _____

(continues)

SELF-EVALUATION *(continued)*

2. Anthrax is the number 1 choice as a biological weapon. Name two properties that make anthrax such an ideal weapon. _____

3. An epidemic of a disease occurred in Sverdlovsk, Soviet Union, in 1979. Name the disease. _____

PART III Answer the following.

1. Cite three landmark treaties associated with biological warfare. Describe each, and give an approximate date for each.

2. Briefly describe the three types of anthrax and how they are acquired. Why is anthrax on the list of biological weapons?

3. How do you feel about the Pentagon's decision to make the anthrax vaccine mandatory for those in the military? Defend your position.

CURRENT "PLAGUES"

Soldiers have rarely won wars. They more often mop up after the barrage of epidemics. And typhus, with its brothers and sisters—plague, cholera, typhoid, dysentery—has decided more campaigns than Caesar, Hannibal, Napoleon, and the inspectors general of history. The epidemics get the blame for defeat, the generals get the credit for victory. It ought to be the other way around.

HANS ZINSSER

PREVIEW The occurrence of large-scale outbreaks of disease, commonly referred to as plagues, is the result of the lack of adaptation between humans and microbes. A variety of microbe-caused plagues have decimated populations and influenced the course of civilizations since ancient times. The Black Death and plagues of smallpox are infamous. Plagues are not relegated to history; new diseases and new epidemics continue to emerge as humans and microbes match wits in the ongoing process of adaptation. Mad cow disease, acquired immune deficiency syndrome (AIDS), and tuberculosis (TB) are modern-day plagues that will be discussed in this chapter.

THE NATURE OF PLAGUES

"A pox on you" and "a dose of the pox" are expressions found in the early literature used to wish evil on one's enemy. Pox probably refers to the lesions associated with the smallpox virus when the disease reached epidemic proportions in 18th-century Europe or, possibly, in earlier smallpox epidemics. Gradually, the term pox came to mean all forms of grotesque diseases manifested by horrendous sores on the skin. It serves as a reminder that epidemics date back to the evolution and coexistence of humans and microbes. The occurrence of disease is the result of a failure of adaptation; the more serious and fatal diseases are those with the least adaptation. In fact, today's normal flora may well have been yesterday's pathogens, and today's current pathogens may be tomorrow's normal flora. It is a matter of adaptation as each species battles to survive and reproduce to guarantee the next generation; the battle of genetic wits leads to survival of the fittest.

On many occasions throughout history, circumstances cause humans and microbes to lock horns, and a battle of epidemic or, possibly, pandemic proportions ensues. The term "plague," strictly speaking, refers to the plague of the Black Death, but the word is used in a popular and somewhat loose sense to describe outbreaks of disease. In time, plagues run their course and die out as the population of susceptible (nonimmune) individuals decreases. Implementation of public health measures and treatment of disease during the 20th century were highly significant in interrupting the cycle of disease. Epidemics and pandemics of the Black Death, smallpox, malaria, typhus fever, yellow fever, and influenza are examples of plagues that have sporadically wiped out populations and influenced the course of civilization; examples are presented in the chapters on bacterial diseases (chapter 8) and viral diseases (chapter 9). Smallpox has been eradicated and other diseases have been controlled, while the threat of others reaching epidemic and pandemic levels remains. The threat of new diseases and new plagues—including AIDS, tuberculosis, West Nile virus, and **Creutzfeldt-Jakob disease** (CJD; associated with **mad cow disease**)—continues.

Some microbiologists have described the outbreak of an epidemic as being like an outbreak of a forest fire. The human-microbe conflict is the result of the combination of environmental changes, population increase and urbanization, technological advances, and human behavior, all of which are discussed in chapter 1. These factors serve to kindle an outbreak; all it takes is a spark to ignite the flames. Once ignited, a disease, like a fire, can spread to pandemic proportions if the susceptible population is large enough to sustain and foster its transmission.

The outbreak of diphtheria in the newly independent states of the former Soviet Union in the 1990s was kindled by the decline in the public health infrastructure resulting from internal wars and a devastated economy, accompanied by a decline in hygiene and in immunization of the population. Hence, soldiers returning from distant battlefields may have been the spark to fuel diphtheria in the nonimmune population.

The Bible speaks of pestilence, and many of its descriptions appear to be of epidemics. Medical historians trace the natural history of infectious disease and epidemics to the days when nomadic hunter-gatherer societies began to form community partnerships to share their meager resources. But, in so doing, they also shared and fostered the spread of their microbes from one individual to another, a process which magnified as villages merged into cities and, ultimately, into urban sprawl. A variety of microbe-caused plagues ravaged areas of the world and spread terror. Wars through the ages left epidemics in their wake, resulting in devastation greater than that wrought by the war itself. It seems that conflict between humans and microbes is a natural part of the biosphere. As the human species evolves, accompanied by technological advances, expectations increase for a healthier and longer life. Adaptation between humans and microbes is like being on a merry-go-round and trying to capture the elusive brass ring. Along the ride, new diseases and new epidemics are spawned as progress is made against older diseases. The proof of this is in the pudding, as evidenced by daily news headlines (In the News 15.1) and news broadcasts informing us of the continued pandemic of AIDS, the outbreaks in Europe of mad cow disease, and the reemergence of TB. These three diseases, all current public health problems, will be discussed in this chapter; they show that plagues are not

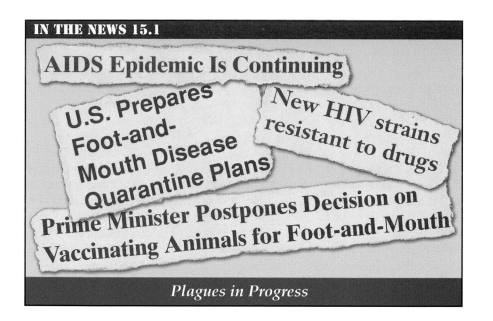

AIDS Epidemic Is Continuing

U.S. Prepares Foot-and-Mouth Disease Quarantine Plans

New HIV strains resistant to drugs

Prime Minister Postpones Decision on Vaccinating Animals for Foot-and-Mouth

Plagues in Progress

events to be dismissed as ancient history but are a part of today's and tomorrow's history. Mad cow disease is caused by **prions,** AIDS is caused by a virus, and TB is caused by a bacterium. This illustrates the diversity of infectious agents responsible for large-scale or threatened large-scale outbreaks.

PRIONS: MAD COW DISEASE AND VARIANT CREUTZFELDT-JAKOB DISEASE

Biology of Prions: New Infectious Particles

Name a particle smaller than a virus that lacks both DNA and RNA, is responsible for panic in Europe, and poses a potential threat to the United States and other countries of the world because it causes mad cow disease in animals and **variant Creutzfeldt-Jakob disease (vCJD)**, a brain disease, in humans. The answer is **prions,** a shorthand term for "proteinaceous infectious particles."

Prions are proteins that have been found in the brains of all mammals that have been studied. Like all proteins, they are composed of specific sequences of amino acids, and their presence is the result of gene expression in the same way that eye color is a manifestation of protein pigments conferred by eye color genes. In their misfolded state, prions are highly stable and resist freezing, drying, and heating at usual cooking temperatures; they are resistant to pasteurization and conventional sterilizing temperatures. Table 15.1 describes the biological distinctions between bacteria, viruses, and prions. The gene which encodes the prion protein is the **PrP gene,** located on chromosome 20.

According to the prion theory, **bovine spongiform encephalopathy (BSE)**, CJD, vCJD, and other **transmissible spongiform encephalopathies (TSEs)** are the result of abnormally folded prions that are able to latch onto normal prions and convert them into an altered, misfolded form, a property that reveals their infectious nature. The expression "one bad apple spoils the barrel" is applicable (Fig. 15.1). Another analogy is from Robert Louis Stevenson's novel *Dr. Jekyll and Mr. Hyde.* Dr. Jekyll is the "good guy,"

TABLE 15.1   A comparison of bacteria, viruses, and prions

Organism	Characteristics	Infectious	Genetic material	Immunogenic	Resistance
Bacteria	Prokaryotic cell	Yes	DNA and RNA	Yes	Moderate[a]
Viruses	Protein coat and nucleic acid, noncellular	Yes	DNA or RNA	Yes	Moderate[a]
Prions	Abnormally folded protein, noncellular	Yes	No genetic material	No	Marked

[a]Depends on the particular test condition and the specific bacterial or viral species.

and Mr. Hyde is the "evil beast" living within Jekyll. The consequence of abnormally folded prions in the brain is that the victim loses motor coordination, exhibits dementia and other neurological symptoms, and dies. Autopsy reveals a Swiss cheese-like or spongelike brain full of holes. There is no treatment and no cure.

The biology of prions is summarized in a press release of the Karolinska Institute's announcement of the 1997 Nobel Prize in physiology or medicine to Stanley Prusiner for his discovery of prions: "Prions exist normally as innocuous cellular proteins. However, prions possess an innate capacity to convert their structures . . . that ultimately result[s] in the formation of harmful particles, the causative agents of several deadly brain diseases of the dementia type in humans and animals."

The BSE-vCJD puzzle is not complete; not all the pieces of the puzzle are in place, nor have they been identified. The idea that abnormal prions—infectious proteinaceous particles without DNA or RNA—are able to convert other molecules to an altered form is a revolutionary hypothesis. Nucleic acids are the building units of genes, the hereditary material found in all

FIGURE 15.1   Conversion of normal prions to infectious prions.

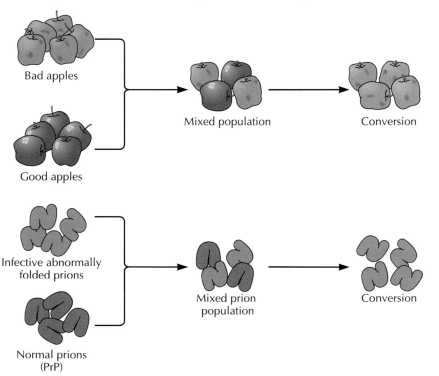

microbes that is passed on from generation to generation. Genes encode proteins, which are essential for life's activities. Prions lack genes but are, nevertheless, infectious, representing a new agent of infection. Some skeptics of the prion hypothesis have been won over but continue to argue that definitive proof that a slow virus is not involved is lacking. (Slow viruses cause diseases in which symptoms take years to develop.) Perhaps a virus is the trigger necessary for initiation of conversion of proteins to a mutant infectious agent, particularly since there is no evidence that prions, free of viruses, synthesized in a test tube will produce disease. But earlier, in the 1960s, it was already suggested that an infectious agent lacking genetic material might be responsible for disease. One landmark study indicated that brain tissue removed from sheep with scrapie (a disease now known to be a TSE) remained infectious, even after radiation that would destroy DNA or RNA; another study suggested that a misfolded protein could initiate other proteins to misfold. The biggest missing piece in the prion puzzle has to do with the trigger that induces the normal protein to misfold. Many scientists believe that the prion protein alone is an infectious agent. Prusiner acknowledges the need for further experiments and suggests the possibility of a "missing factor" that might chaperone PrP into its abnormal shape. Perhaps Prusiner's missing factor will turn out to be a virus yet to be discovered! The deputy chair of the Nobel Committee stated that the committee was not bothered by the unanswered questions but that the prize was awarded for the discovery of the prion and its role in the disease process. He stated, "The details have to be solved in the future. But no one can object to the essential role of the prion protein in those brain diseases." Further, prions are an exciting discovery that may eventually be linked to Alzheimer's and Parkinson's diseases as well as other neurological conditions.

Mad Cow Disease

It all began in 1986 when an outbreak of BSE, better known as mad cow disease, in the United Kingdom captured and frightened the attention of meat eaters around the world and raised the questions, "Can it happen here? Is it transmissible to humans?" In the United States the mad cow disease scare was fueled by an Oprah Winfrey show about the disease and Oprah's statement, "I will never eat another hamburger." Television news shows wasted no time in showing pastures populated by mad cows drooling, stumbling, and unable to rise to their feet.

BSE is one of a group of diseases known as TSEs, characterized by spongy degeneration of the brain accompanied by severe and fatal neurological damage. Since the 1986 initial reports of BSE, according to the Centers for Disease Control and Prevention (CDC), approximately 180,000 cases have been confirmed in the United Kingdom as of December 2000. The disease has spread to cattle in other European countries, primarily France, Switzerland, Portugal, and Ireland. There are other forms of TSEs, in addition to BSE, also characterized by spongy deterioration of the brain. Scrapie is a TSE of sheep and is manifested by animals scraping against trees, fencing, and whatever else might be available. TSEs that resemble BSE are also found in household cats, deer, elk, and mink, as well as in ruminants.

Human TSEs

CJD, the model for the human TSEs, is transmitted in three ways (Table 15.2). Sporadic cases occur throughout the world at a rate about 1 case per 1 million people, accounting for 85 to 90% of CJD cases. Another 5 to 10%

TABLE 15.2 Comparison of CJD and vCJD

Disease	Average age of infection (years)	Duration of illness (months)	Transmission
CJD	65	4.5	Sporadic (cause unknown), 85-90%; genetic mutation, 5-10%; iatrogenic, <5%
vCJD	29	14	Meats (100%)

of CJD cases are due to hereditary predisposition associated with gene mutation. Fewer than 5% of CJD cases are **iatrogenic,** meaning that they are transmitted via contaminated surgical equipment, corneal transplants, or natural human growth hormone. The prototype human TSE was **kuru,** which was epidemic in Fore natives of New Guinea. Kuru was spread by an ancient cannibalistic practice in which, as a sign of respect, Fore people, particularly females, ate the brains of deceased relatives, thereby infecting themselves with prions. This ritual has been abolished, leading to the disappearance of kuru.

In March 1996, a new form of CJD appeared. CJD develops at an average age of 65 years and has a mean duration of illness of about 4.5 months, whereas the new form, vCJD, affects individuals at an average age of 29 years and has a duration of illness of about 14 months (Table 15.2). Further, vCJD is transmissible through meats. By November 2001, 106 cases had been identified in the United Kingdom, 4 cases had occurred in France, and 1 case had occurred in Ireland.

The BSE-vCJD Link

The most plausible explanation for vCJD is the contamination of beef products by central nervous system (brain and spinal cord) tissue. The cluster of vCJD cases is most likely due to the same agent that causes BSE. Support for this hypothesis is based on several lines of evidence, including an association between these two TSEs in time and place, a resemblance of pathological features in the brains of monkeys inoculated with either vCJD or BSE, and the nearly identical and distinctive distribution and pathology of the infectious agent in the brains of mice injected with infected cow BSE tissues or with human vCJD tissue.

Assuming that BSE and vCJD are caused by the same prion strain, how did all of this come about? The natural history of the events that brought us to this stage is illustrated in Fig. 15.2. Livestock farmers figure into the scenario as potential victims of vCJD.

The widely held view is that the outbreak of BSE in cattle in 1986 in the United Kingdom was the result of the leap of prions from sheep to cattle as the result of prion-infected animal proteins' being used as cattle feed to "beef up" milk production. (There are many examples of the "species leap," AIDS being another case in point.)

The term "animal protein" sounds innocuous, but, in fact, animal protein is a concoction of ground-up carcasses of sheep and cattle, including their brains, spleens, thymus glands, tonsils, and intestines, a mixture called **offal.** In essence, cattle were eating the remains of other cattle, including "downers" (sick and dead animals found in the fields). Chapter 12 contains excerpts from Upton Sinclair's 1906 novel *The Jungle* describing conditions in "Packingtown" and the inclusion of downers in meats

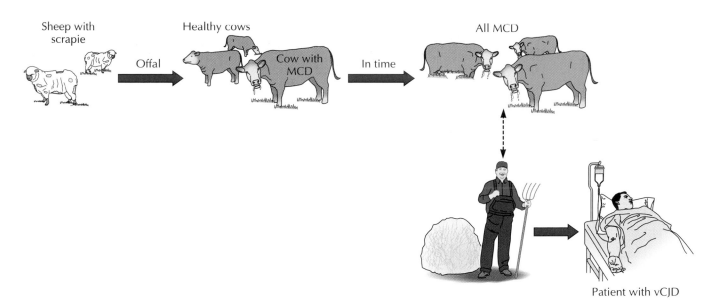

Sheep with scrapie

Offal

Healthy cows

Cow with MCD

In time

All MCD

Patient with vCJD

FIGURE 15.2 Transmission of spongiform encephalopathies.

packaged for human consumption, a practice that continued for almost another 100 years. In 1989, the use of offal was banned in the United Kingdom. Cattle are natural herbivores, and feeding them parts of other animals does not make sense, other than economically. Science writer Joseph Duggan describes the practice as

> . . . high tech cannibalism, a cannibalistic practice being extended and implemented in chickens, pigs, mink, and even pets in the high tech world of profit-driven factory farms. Further, this new doctrine of cannibalism is a violation of "inherent" safety measures (in Nature) which the animal kingdom follows instinctually. For instance, predators very seldom eat the head of their prey. Cats often leave the head of a bird or mouse uneaten. Larger prey have skulls so hard that predators cannot break them open. Humans, not being limited by the strength of their jaws, crack open the skull of any animal including other humans to eat the contents. . . . Nothing is wasted as chicken feed is spiked with protein derived from the feathers, blood, and even the feces of other chickens. Pigs are fed the recycled remains of other pigs; cows the rendered remains of downer cattle; and mink the remains of any animal. Pet foods may even contain the remains of other pets put down by veterinarians.

But why did this outbreak of BSE suddenly occur in 1986? What changed on the farm or en route from farm to table? The most plausible explanation has to do with a change that occurred in the **rendering process** used in the production of food products for cattle. (The rendering industry also produces cosmetics and pet foods as by-products). Rendering is a process somewhat like the making of stew, in which all of the ingredients are put into a pot and boiled. In the late 1970s, the rendering process was altered by the elimination of the solvent-steam treatment phase. Solvents and steam reduce the infectivity of prions and, hence, the transmissibility of TSEs. The upshot, allegedly, is that the scrapie prion, in the absence of the solvent-steam treatment, survived the rendering process and contaminated the dietary supplement, which was then distributed throughout the United Kingdom, resulting in infected herds.

Experiments to study the rendering process and its effect on the inactivation of prions confuse the situation, since none of the different rendering processes completely inactivate prion infectivity. Paul Brown, a career investigator of TSEs at the National Institute of Neurological Disorders and Stroke, suggests that even though the number of infectious scrapie particles may have been only modestly reduced by the solvent-steam treatment, the reduction may have been sufficient to make the difference if, in fact, the level of infectivity was at the borderline of the number of prions necessary to cause BSE in cattle that are fed offal. This might explain why the United States, and other countries that similarly modified the rendering process at about the same time as the United Kingdom, did not experience an outbreak of BSE. But why the outbreak in the United Kingdom? Since the ratio of sheep to cattle is lower in the United States than in the United Kingdom, so is the number of prions (from scrapie-infected sheep), resulting in less than the number of prions required for infectivity. The explanation is plausible!

Control Measures: Breaking the Cycle of Transmission

The cycle of disease transmission, discussed in chapter 7, can be applied to counter the threat of BSE and vCJD. The measures to be implemented are direct and are targeted at interrupting transmission from sheep to cattle and from cattle to humans. Control of TSEs requires coordinated surveillance, reporting, and policing at the local, national, and international levels of partnerships.

Israel has the distinction of being the first country in the world to guard against mad cow disease. In 1988 it banned imports of live cattle and meat and bone meal supplements that included the ground-up remains of dead cattle, some of which were later shown to be infected with BSE. As a result of bans against the inclusion of animal products in feed, the incidence of new cases is on the decline in many countries, while in other countries BSE continues to exist or is making its first appearance, possibly as the result of increased surveillance. Meanwhile, the livestock and spin-off industries have been brought to their knees as the result of the slaughter of 4 million cattle in a containment effort, including some that were possibly not diseased. In June 2000, the European Union's Council on Food Safety and Animal Welfare imposed a restriction on its member states to remove "specified risk materials" from animal and human food chains effective as of October 1, 2000, a practice that had already been instituted in most of the union's member countries. Bovine milk and milk products are not considered to pose any risk for transmission of TSEs.

During the summer of 2000, the U.S. Department of Agriculture (USDA) exercised its authority in deciding to kill about 400 East Friesian sheep in two flocks in a rural Vermont community. This breed, imported from Belgium, is known to produce about 10 times more milk than American sheep. Some of the animals fell ill and were suspected of harboring BSE prions. The incident provoked strong emotions in the press, extending beyond Vermont. It was a heyday for the press, which reported the projected slaughter of sheep such as Mrs. Friendly (who liked to be petted), her lamb Buttercup, Pumpkin, Olga, and Sweetpea. Two-year-old Moe was so beloved that the slogan "Hell, no, don't kill Moe!" became popular. One community resident likened the government's action to that of seizing (at gunpoint) Elian Gonzalez, a young Cuban boy, from his Miami relatives and shipping him back to Cuba.

For officials of the USDA, their victory was a hollow one. In this rare instance of a pastoral community pitted against the federal government, the government held fast to its guns and vowed to "continue our efforts to protect American agriculture, and ultimately, the health of the American people." The animals were confiscated in March 2001 to be disposed of.

The USDA's Animal and Plant Health Inspection Service has conducted active surveillance for BSE in the United States since 1990. As of November 2001, it had examined over 15,000 bovine brain specimens without finding any evidence of BSE. Severe restrictions are in place regarding importation of ruminants (hoofed mammals that chew their cud) and ruminant products from all European countries. The USDA instituted a ban on use of ruminant remains as livestock feed effective as of October 1997. It is, therefore, extremely unlikely that BSE is a food-borne hazard in the United States. Further, the CDC monitors the incidence of CJD by examination of death certificates and other practices of ongoing surveillance.

Since 1991, the World Health Organization (WHO) has convened nine scientific gatherings on issues related to animal and human TSEs. Since 1997, it has conducted training courses emphasizing surveillance procedures, particularly in developing countries. The organization has also made a number of recommendations to protect against BSE (Box 15.1) and vCJD.

BSE provokes serious and far-reaching political, economic, and health consequences (In the News 15.2). For example, although there is no evidence that vCJD spreads through blood, the Food and Drug Administration prohibits blood donations from those who spent a total of at least 6 months in Great Britain between 1980 and 1996 and plans to extend the ban to individuals who have lived in or traveled to France or Portugal for over 10 years since 1980. The American Red Cross calls for an even stricter ban. The danger is that these restrictions will threaten the nation's supply of blood and blood products. Even the sport of bullfighting, the *corrida de toros*, integral to Spain's culture, is feeling the brunt of BSE. Thousands of bulls are slaughtered each year in the rings amid the sound of blaring trumpets and wildly cheering crowds, generating millions of dollars for the meat industry. But recently, under the authority of the European Union, fighting bulls killed in the rings are required to be tested for BSE before being released for consumption. Testing is an expensive process associated with practical problems which interfere with the animals' being sold to slaughterhouses.

AUTHOR'S NOTE *Do you think the USDA was right in slaughtering the sheep? Were the rights of the farmers violated? The enforcement roles of government agencies in health matters pose serious ethical issues. The tale of Typhoid Mary, presented in chapter 7, is another example. How about people who knowingly have AIDS and continue to have sex with others without advising them? The list goes on. These questions are great fodder for class discussion.*

BOX 15.1   WHO Recommendations for BSE Control

To protect human health, WHO has recommended the following:

- No part or product of any animal which has shown signs of a TSE should enter any (human or animal) food chain.
- Countries should not permit tissues that are likely to contain the BSE agent to enter any (human or animal) food chain.

- All countries should ban the use of ruminant tissues in ruminant feed.
- Human and veterinary vaccines prepared from bovine materials may carry the risk of transmission of animal TSE agents. The pharmaceutical industry should ideally avoid the use of bovine materials and materials from other animal species in which TSEs naturally occur. If absolutely

necessary, bovine materials should be obtained from countries which have a surveillance system for BSE in place and which report either zero or only sporadic cases of BSE. These precautions apply to the manufacture of cosmetics as well.

Source: WHO.

IN THE NEWS 15.2

Mad cow deaths in English village linked to butchering technique...

Bullfighters Worry about Mad Cow Disease

McDonald's increases mad cow monitoring

Following Canada's Lead, U.S. Bans Brazilian Beef

More people consider vegetarianism

The Impact of BSE

McDonald's Corporation suffered its first decline in profits in almost 3 years because of Europe's mad cow disease.

AIDS

Today's college students are living in the shadow of a pandemic of **AIDS,** a dread disease caused by the **human immunodeficiency virus (HIV),** which surfaced in 1981 and has since spread havoc and desperation throughout the world, particularly in Africa. The story of Mark Gardner Hoyle is symbolic; Mark died of AIDS on October 26, 1986, at the age of 14. He was a student at

| AUTHOR'S NOTE | *If all this were not enough, foot-and-mouth disease surfaced in the United Kingdom in February 2001 (In the News 15.3). This disease, not to be confused with BSE and other TSEs, is a highly infectious viral disease of cloven-hoofed livestock (cattle, pigs, sheep, and goats). Close to 1 million animals, primarily in Europe but also in the United States and other countries, have been slaughtered as a containment measure. Infected animals have painful sores on their mouths and feet, high fevers, and excessive salivating. The virus can be carried by the wind, on boots and clothing, on tires, and in hay, water, and manure. TV, magazines, and newspapers display terrible scenes of cranes lifting dead animals and dumping them onto massive pyres and of the thick, black smoke as the animals burn.* |

IN THE NEWS 15.3

Officials admit British foot-and-mouth epidemic is out of control

Foot-and-Mouth Disease Spreads from Britain to Continental Europe

Healthy sheep to be slaughtered to prevent foot-and-mouth spread

Tougher Inspections for Travelers Returning from Britain, Europe

Foot-and-Mouth Disease

Case Junior High School in Swansea, Mass., and his name is inscribed on home plate at the school's playing field. Ryan White was an Indiana teenager who died of AIDS on April 8, 1989, at the age of 19. You may have heard of the Ryan White CARE (Comprehensive AIDS Resources Emergency) Act, which is designed to assist the growing number of Americans living with AIDS but without adequate health care insurance. Both boys and their families struggled with an unsympathetic, hostile, and uninformed public intent on ostracizing them. They helped to teach the nation about compassion and the needs of people unfortunate enough to have contracted this inevitably fatal disease. It is ironic that both boys contracted and died from this so-called "gay" disease when, in fact, both suffered from a bleeding disorder known as hemophilia and required blood products in their treatment. Society has come a long way since the 1980s in acceptance of gays and people with AIDS, but not far enough. The tragic death of Matthew Shepard, a young gay student at the University of Wyoming, shocked the nation and the world in 1999. Matthew was severely beaten by local boys, tied to a fence, and left to die. Society still has a long way to go!

In the United States, the AIDS story began in the late 1970s when physicians in Los Angeles, San Francisco, and New York City observed clusters of patients with symptoms that included severe fungal pneumonia, swollen lymph nodes, sudden weight loss, a history of recent and new infections, and blue-purplish spots on their skin (Fig. 15.3). These spots were later diagnosed as a rare vascular cancer called **Kaposi's sarcoma,** first seen in the late 1800s by Moritz Kaposi, a Viennese dermatologist. Oddly, or so it seemed at the time, all of the patients were young, male, and homosexual.

Society became increasingly ruthless (and still is) as it became obvious that AIDS was the "kiss of death" and was associated with gay men, despite the occurrence of cases in nonhomosexual patients who had been transfused with blood or blood products. AIDS quickly became known as the "gay man's disease"; many people expressed the opinion that "it's their own fault," "God is punishing gays for their lifestyle," and "they deserve it." Gay bathhouses, places of assemblage for gay men, promoted promiscuity and oral and anal sex, and the disease took off.

The term AIDS first appeared in 1982 in CDC's *Morbidity and Mortality Weekly Report* and was described as "a disease, at least moderately predictive of a defect in cell-mediated immunity, occurring with no known cause with diminished resistance to that disease." The updated 1993 description was expanded to include 23 diseases (Box 15.2). Further, the case definition

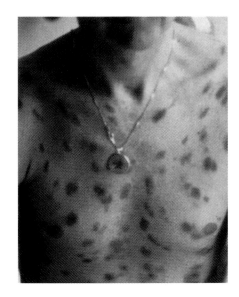

FIGURE 15.3 Kaposi's sarcoma in a young man infected with HIV. (Source: Paul Volberding. Reprinted from J. A. Levy, *HIV and the Pathogenesis of AIDS,* 2nd ed., ASM Press, Washington, D.C., 1998.)

BOX 15.2 The CDC's Revised Definition of AIDS (January 1993)

Expanded U.S. AIDS Surveillance Case Definitions

- All HIV-infected adolescents and adults with fewer than 200 CD4[+] T lymphocytes per microliter
- Or HIV positive and having one or more of the following:
  Cryptosporidiosis, cytomegalovirus, isosporiasis, Kaposi's sarcoma, lymphoma, lymphoid pneumonia (hyperplasia), *Pneumocystis carinii* pneumonia, progressive multifocal leukoencephalopathy, toxoplasmosis, candidiasis, coccidioidomycosis, cryptococcosis, herpes simplex virus, histoplasmosis, extrapulmonary tuberculosis, other mycobacteriosis, salmonellosis, other bacterial infections, HIV encephalopathy (dementia), HIV wasting syndrome, pulmonary tuberculosis, recurrent pneumonia, and invasive cervical cancer.

Source: CDC.

now includes HIV infection in an individual with a CD4$^+$ T-cell count less than 200 cells per microliter of blood. Note that the number of CD4$^+$ T cells is critical in the revised definition of AIDS. (The critical role of these cells in specific immunity is highlighted in chapter 11.) But what about the definition of AIDS in developing countries where diagnostic facilities do not have the capability to do CD4$^+$ T-cell counts? In these cases, epidemiologists employ a case definition based on the presentation of clinical symptoms and the exclusion of other known causes of immunosuppression, such as cancer and malnutrition, resulting in underreporting of the incidence of AIDS.

Biology of AIDS

Viruses are discussed in chapter 5. As in all viruses, the replication of HIV consists of the five stages of adsorption, penetration, replication, maturation, and release (Table 5.3 and Fig. 5.4). Figure 15.4 illustrates the virus "budding" out of host cells. The biology of this virus and other retroviruses is unique in that the flow of genetic information starts with RNA, not with DNA. As discussed in chapter 5, DNA is the usual starting point. (A review of chapter 5 is recommended.)

There are at least two types of HIV, the best known of which is HIV type 1 (HIV-1), the most common cause of AIDS worldwide. In West Africa, HIV-2 is most common. Within each group are a number of subcategories. This chapter focuses on HIV-1.

Origins of AIDS

Where did AIDS come from, and why did it emerge in the latter half of the 20th century and explode in the population during the last 2 decades of that century? There is considerable speculation regarding this question, but there is no definitive answer. Genetic studies link HIV to simian (monkey) viruses and indicate that HIV-1 and HIV-2 arose from different evolutionary lines. It is postulated that somewhere along the line, as presumably happened with vCJD and other diseases, an interspecies leap from monkeys to humans occurred, possibly by humans eating monkey meat or being bitten by a monkey. This association with monkeys might have been the spark that was fueled by changing sexual and social patterns, including urbanization, travel, increasing immigration, and the rise of prostitution. Other explanations, all of which are false, have been advanced: that HIV is

FIGURE 15.4 HIV "budding" out of host cells. (Source: WHO.)

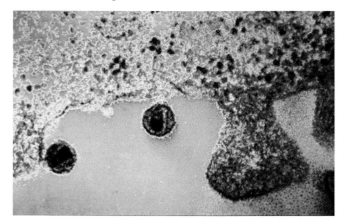

the result of warfare research in the United States, the result of sexual relations between humans and sheep or humans and monkeys, or the result of a scheme between Israel and South Africa. A controversial theory that existed for years centered on contamination of an experimental polio vaccine under trial in Africa in the 1950s with the AIDS virus. Several studies have conclusively proved that the polio vaccine theory is no more than a myth.

The first well-documented case of AIDS occurred in an African man in 1959, although the diagnosis of AIDS was not made until decades later. Speculation is that HIV was first carried to the United States and Europe by way of the Caribbean, by a promiscuous airline steward.

Before the identification of HIV as the cause of AIDS, sperm antibodies, fungi, and recreational drugs were, at various times, named as the cause of AIDS. The discovery of the AIDS virus in 1982 is credited to two research teams, one headed by Luc Montagnier of the Pasteur Institute in Paris and the other by Robert Gallo of the National Institutes of Health (Fig. 15.5). A battle ensued as to which of these two investigators should be regarded as the discoverer of the virus, and legal and ethical sparring went on between the United States and France, involving scientific integrity, ethics, and royalties. Some 20 years later, there is still debate among scientists, although most of the scientific community acknowledges codiscovery by Montagnier and Gallo. The evidence that HIV causes AIDS is abundant, but, nevertheless, there are skeptics. The evidence for and against HIV as the causative role is presented in a later section.

FIGURE 15.5 Robert Gallo, codiscoverer of HIV. (Source: Public Health Service.)

Transmission of AIDS

Near panic ensued in the early 1980s because of the severity and prognosis of the disease, its association with gay sex, and the stigmatization of those infected with AIDS. It seems that everyone had an opinion on the mechanisms of transmission. The myths abounded, leading to further stigmatization and harassment of HIV-infected individuals. Some of these individuals had medical conditions that necessitated their receiving blood or blood products prior to the development of a screening test for the virus. Insects, food, terrorist attacks, casual contact, and "God's curse" were all considered possible vehicles of transmission.

It is important to understand how AIDS is transmitted (Box 15.3) in order to break the epidemic cycle (chapter 7) and to dispel the myths, particularly the "I can't get it" myth that persists in those of high school age and older. HIV can be transmitted in four ways (Fig. 15.6):

1. Sexual contact with an infected partner, whether it be male to male, male to female, female to male, or female to female. The virus can penetrate the lining of the vagina, vulva, penis, rectum, or mouth during sexual activity. Some sexual behaviors are considered more risky and unsafe than others. Anal sex is the most dangerous because the lining of the anus is more subject to tears and injury than is the lining of the vagina, allowing the AIDS virus (and other microbes) easier passage into the blood.
2. Contact with infected blood or blood products. Prior to identification of the cause of AIDS and the development of an HIV screening test in April 1985, persons needing transfusions of blood or blood products, including those with hemophilia, were hard hit. Many recipients of blood or blood products died as a result of AIDS, includ-

BOX 15.3 How You Can and Cannot Become Infected with HIV

How do you get HIV from sexual intercourse?

- HIV can be spread through unprotected sexual intercourse from male to female, female to male, or male to male. Female-to-female sexual transmission is possible, but rare. Unprotected sexual intercourse means sexual intercourse without correct and consistent condom use.
- HIV may be in an infected person's blood, semen, or vaginal secretions. It is thought that it can enter the body through cuts or sores—some so small you do not know they are there—on tissue in the vagina, penis, or rectum, and possibly the mouth.
- HIV is transmitted by anal, vaginal, or oral intercourse with a person who is infected with HIV.
- Since many infected people have no apparent symptoms of the condition, it is hard to be sure who is or is not infected with HIV. So, the more sex partners you have, the greater are your chances of encountering one who is infected and becoming infected yourself.

How do you get HIV from using needles?

- Sharing needles or syringes, even once, is an easy way to be infected with HIV and other germs. Sharing needles to inject drugs is the most dangerous form of needle sharing. Blood from an infected person can remain in or on a needle or syringe and then be transferred directly into the next person who uses it.
- Sharing other types of needles also may transmit HIV and other germs. These types of needles include those used to inject steroids and those used for tattooing or piercing.
- If you plan to get a piercing or a tattoo, make sure you go to a qualified individual.

If somebody in my class at school has AIDS, am I likely to get it too?

- No. HIV is transmitted by unprotected sexual intercourse, needle sharing, or infected blood. It can also be given by an infected mother to her baby during pregnancy, birth, or breast-feeding.
- People infected with HIV cannot pass the virus to others through ordinary activities of young people in school.
- You will not become infected with HIV just by attending school with someone who is infected or who has AIDS.

Can I become infected with HIV from "French" kissing?

- Not likely. HIV occasionally can be found in saliva, but in very low concentrations—so low that scientists believe it is virtually impossible to transmit infection by deep kissing.
- However, the possibility exists that cuts or sores in the mouth provide direct access for HIV to enter the bloodstream during prolonged deep kissing.
- There has never been a single case documented in which HIV was transmitted by kissing.
- Scientists cannot absolutely rule out the possibility of transmission during prolonged, deep kissing because of possible blood contact.

Can I become infected with HIV from oral sex?

- It is possible.
- Oral sex often involves semen, vaginal secretions, or blood—fluids that contain HIV.
- HIV is transmitted by the introduction of infected semen, vaginal secretions, or blood into another person's body.
- During oral intercourse, the virus could enter the body through tiny cuts or sores in the mouth.

AUTHOR'S NOTE *How comforting is the 0.5% statistic to someone who has accidentally stuck a finger? Not very, as my experience tells; it happened to me almost 20 years ago when drawing blood from a patient at Hadassah Hospital in Jerusalem. Fortunately, I did not pick up any infections from the needle. In 1991, a dentist with AIDS was responsible for infecting six patients. Improperly sterilized needles used to deliver local anesthetic were found to be contaminated with the dentist's blood.*

ing tennis star Arthur Ashe, who died in 1995 as a result of having received contaminated blood 10 years earlier during heart surgery.

3. Sharing blood-contaminated needles and syringes, as is practiced by drug users when shooting up, runs a high risk of HIV transmission. All it takes is a minute amount of contaminated blood on a needle. Accidental needle sticks and glove tears are a nightmare to all health care workers, although the estimated risk of acquiring AIDS from such an incident is less than 0.5%.

4. Transmission from mother to unborn child through the passage of HIV across the placenta carries about a 20% risk. The virus can also be transmitted at the time of birth by infected vaginal secretions or, after birth, through infected breast milk.

Transmission of HIV can have legal implications. Recently, a 20-year-old man was sentenced to 7 to 10 years in prison for having sex with a 14-year-old; he knew he was infected with HIV. Since 1985, a ban has prohibited men who have had gay sex (even once) since 1977 from donating

How You Can and Cannot Become Infected with HIV *(continued)*

As long as I use a latex condom during sexual intercourse, I won't get HIV infection, right?

- Latex condoms have been shown to prevent HIV infection and other sexually transmitted diseases.
- You have to use them properly. And you have to use them every time you have sex—vaginal, anal, and oral—with a male.
- The only sure way to avoid infection through sex is to abstain from sexual intercourse or to engage in sexual intercourse only with someone is not infected.

My friend has anal intercourse with her boyfried so that she won't get pregnant. She can't get AIDS from doing that, right?

- Wrong. Anal intercourse with an infected partner is one of the ways in which HIV has been transmitted.
- Whether you are male or female, anal intercourse with an infected person is very risky.

If I have never injected drugs and have had sexual intercourse only with a person of the opposite sex, could I have become infected with HIV?

- Yes. HIV does not discriminate.

You do not have to be homosexual or use drugs to become infected.

- Both males and females can become infected and transmit the infection to another person through intercourse.
- If a previous sex partner was infected, you may be infected as well.

Is it possible to become infected with HIV by donating blood?

- No. There is absolutely no risk of HIV infection from donating blood.
- Blood donation centers use a new, sterile needle for each donation.

I had a blood transfusion. Is it likely that I am infected with HIV?

- Is it highly unlikely. All donated blood has been tested for HIV since 1985.
- Donors are asked if they have practiced behaviors that place them at increased risk for HIV. If they have, they are not allowed to donate blood.
- Today the American blood supply is extremely safe.
- If you are still concerned about the remote possibility of HIV infection from a transfusion, you should see your doctor or seek counseling about getting an HIV antibody test.

Call the CDC National AIDS Hotline (1-800-342-AIDS) or your local health department to find out about counseling and testing facilities in your area.

Can I become infected with HIV from a toilet seat or other objects I routinely use?

- No. HIV does not live on toilet seats or other everyday objects, even those on which body fluids may sometimes be found. Other examples of everyday objects are door-knobs, phones, money, and drinking fountains.

Can I become infected with HIV from a mosquito or other insects?

- You cannot get HIV from a mosquito bite. The AIDS virus does not live in a mosquito, and it is not transmitted through a mosquito's salivary glands like other diseases such as malaria or yellow fever. You cannot get it from bed bugs, lice, flies, or other insects either.

Source: CDC, AIDS Prevention Guide.

blood. This ruling has recently been challenged as being discriminatory because today's blood-testing procedures can detect HIV-infected blood, but the ban has been upheld.

AIDS: the Disease

A diagnosis of AIDS is terrifying because the best that an infected person can hope for is a long life without too much misery. HIV does not directly cause death, as other microbes do by their secretions of toxins or by tissue damage. Rather, HIV depletes the number of T-helper cells, resulting in the individual's becoming immunocompromised (chapter 11) and vulnerable to opportunistic diseases caused by an array of microbes (Table 15.3).

Infection with HIV does not initially constitute AIDS, but does, with rare exceptions, progress to clinical AIDS over an incubation period that can vary from a few years to as many as 15 years. Those few HIV-infected individuals who continue to survive after 10 years live in a state of fear and uncertainty, not knowing when full-blown AIDS will develop. The CDC

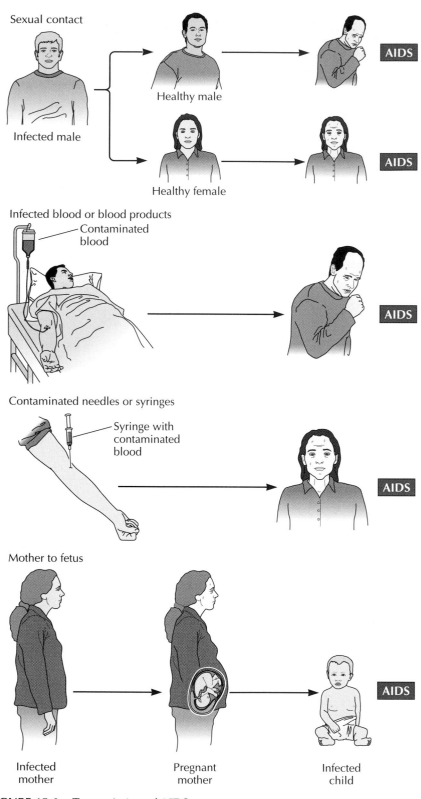

Sexual contact

Infected male

Healthy male → AIDS

Healthy female → AIDS

Infected blood or blood products
Contaminated blood

→ AIDS

Contaminated needles or syringes
Syringe with contaminated blood

→ AIDS

Mother to fetus

Infected mother

Pregnant mother

Infected child

AIDS

FIGURE 15.6   Transmission of AIDS.

TABLE 15.3   Opportunistic infections associated with AIDS

Microbe	Disease	Characteristics
Bacteria		
Mycobacterium tuberculosis	Tuberculosis	Primarily lungs, but dissemination possible
Mycobacterium avium, M. intracellulare	Extrapulmonary tuberculosis	Lungs and other organs
Salmonella spp.	Salmonellosis	Gastrointestinal symptoms
Legionella pneumophila	Legionnaires' disease	Pneumonia-like respiratory symptoms
Viruses		
Herpes simplex virus	Herpes	Lesions on skin and mucous membranes
Cytomegalovirus	Cytomegalovirus	Encephalitis, blindness
Varicella-zoster virus	Shingles	Painful, itchy lesions on skin
Epstein-Barr virus	Hairy leukoplakia	Lesions in mouth
Protozoans		
Pneumocystis carinii	Pneumonia	Pneumonia-like respiratory symptoms
Cryptosporidium parvum	Cryptosporidiosis	Diarrhea
Toxoplasma gondii	Toxoplasmosis	Brain and nervous system
Isospora belli	Isosporiasis	Diarrhea
Fungi		
Candida albicans	Candidiasis	Lesions on throat, pharynx, lungs
Cryptococcus neoformans	Cryptococcosis	Meningitis and other disseminated symptoms
Histoplasma capsulatum	Histoplasmosis	Pneumonia and other disseminated symptoms

has classified HIV infection into four stages. It should be emphasized that no two cases are alike, and there might be considerable variation in the time and course of infection. Further, these stages progress from one to another without sharp lines of demarcation.

Stage 1, the **prodromal stage,** is characterized by fever, diarrhea, rash, aches, headaches, lymphadenopathy (enlarged lymph glands), and fatigue. The symptoms resemble mononucleosis and usually last only a few weeks to a few months. Laboratory evaluation indicates an initial drop in the number of CD4$^+$ T cells from its normal value of 800 to 1,000 per microliter of blood, but the count remains sufficiently high for the immune system to function. The viral load—the number of HIV particles in the blood—can also be tracked in the laboratory.

Stage 2 is the **latency period,** which can last from 2 or 3 years to as many as 15 years, with an average of about 10 years. During this time, most infected persons are free of symptoms since the virus is in hiding, and the number of CD4$^+$ T cells keeps ahead of the viral load, enabling destruction of virus-infected cells. An AIDS blood test, based on the detection of antibodies against HIV, is the only diagnostic measure and may result in a false-negative report, since the number of antibodies may be insufficient to be detected. Hence, persons may or may not be aware that they are infected, but they can infect others.

Stage 3, persistent generalized **lymphadenopathy,** is characterized by the appearance of swollen lymph nodes, recurrent fevers, night sweats, persistent diarrhea, persistent cough, extreme fatigue, and, in some cases, neurological impairment, including memory loss, confusion, and depression. Opportunistic infections, frequently caused by yeasts, result in the production of painful sores in the mouth and on the tongue (oral thrush) or in the vagina (vaginal thrush); herpesvirus infection resulting in painful eruptions on the skin around the genital area and around the mouth may also occur. This stage usually signals onset of stage 4 within a short time.

Stage 4 is manifested as full-blown AIDS, and death usually occurs within a few years. This stage is one of misery, since the patient is tormented by one debilitating and painful opportunistic infection after another. About the best that can be done medically is to treat the symptoms of each infection as it develops; frequently, this means hospitalization. Since the advent of AIDS, much has been learned about management of opportunistic infections, affording some measure of relief and extended life to those with AIDS. Opportunistic infections occur because the immune system is severely compromised (weakened) as a result of the CD4$^+$ T-cell count's dropping below 200 per microliter of blood and an accompanying degeneration of the lymph glands.

The Cause of AIDS

Virtually all AIDS researchers and other scientists agree that AIDS is caused by HIV, yet there is a dissident point of view advanced by a handful of scientists. In the interest of scientific objectivity, the accepted and the dissident points of view are presented in Box 15.4.

Evidence that HIV Causes AIDS

The fulfillment of Koch's postulates, which were developed in the late 19th century (chapter 6), is generally accepted as the criterion establishing the link between a pathogen and a disease. These postulates have been somewhat modified over the years, particularly with regard to viruses, but the basic principles have remained the same for more than a century and continue to serve as the litmus test of causality. The National Institute of Allergy and Infectious Diseases (NIAID) presents the following criteria based on Koch's postulates:

1. Epidemiological association: the suspected cause must be strongly associated with the disease.
2. Isolation: the suspected pathogen can be isolated—and propagated—outside the host.
3. Transmission pathogenesis: transfer of the suspected pathogen to an uninfected host, human or animal, which results in disease in the host.

Numerous studies around the world since 1981 indicate that virtually all individuals with AIDS are **HIV seropositive** (i.e., have antibodies against HIV), satisfying postulate 1. In fulfillment of postulate 2, the virus has been isolated in virtually all AIDS patients and in almost all HIV-seropositive individuals. Further, the presence of HIV genes in virtually all persons with AIDS and in those in the earlier stages of HIV disease has been

BOX 15.4 The HIV-AIDS "Controversy"

The Evidence that HIV Causes AIDS

- HIV fulfills Koch's postulates as the cause of AIDS.
- AIDS and HIV infections are invariably linked in time, place, and population group.
- Severe immunosuppression and AIDS-defining illnesses occur almost exclusively in individuals who are HIV infected.
- Before the appearance of HIV, AIDS-related diseases were rare in developed countries; today, they are common in HIV-infected individuals.
- HIV can be detected in virtually everyone with AIDS.
- Nearly everyone with AIDS has antibodies to HIV.
- The specific immunological profile that typifies AIDS (a persistently low CD4+ T-cell count) is extraordinarily rare in the absence of HIV infection or other known causes of immunosuppression.
- Among HIV-infected patients who receive anti-HIV therapy, those whose viral loads are driven to low levels are much less likely to develop AIDS or die than patients who do not respond to therapy. Such an effect would not be seen if HIV did not have a central role in causing AIDS.

An Argument that HIV Does Not Cause AIDS

- HIV antibody testing is unreliable.
- HIV cannot be the cause of AIDS because researchers are unable to explain precisely how HIV destroys the immune system.
- AZT and other antiretroviral drugs, not HIV, cause AIDS.
- Behavioral factors such as recreational drug use and multiple sexual partners account for AIDS.
- AIDS among transfusion recipients is due to underlying diseases that necessitated the transfusion, rather than to HIV.
- High usage of clotting factor concentrate, not HIV, leads to CD4+ T-cell depletion and AIDS in hemophiliacs
- The spectrum of AIDS-related infections seen in different populations proves that AIDS is actually many diseases not caused by HIV.
- HIV cannot be the cause of AIDS because the body develops a vigorous antibody response to the virus.

Source: Adapted from *NIAID Fact Sheet* (updated November 29, 2000), *The Evidence that HIV Causes AIDS* (NIAID, Bethesda, Md.).

determined by employing techniques of molecular biology. Postulate 3 has been satisfied in tragic incidents, including the cases of three laboratory technicians who developed HIV disease after accidental exposure to the virus and the case of transmission of HIV from an infected dentist to six patients. The development of AIDS has also been documented in health care workers after accidental exposure to the virus, in blood transfusion cases, in mother-to-child transmission, and in injection by drug users. Koch's postulates have also been met in experiments with animals, including chimpanzees, monkeys, and certain strains of mice.

A Dissenting View about the Cause of AIDS

It should be clear, from the preceding discussion, that HIV causes AIDS. However, the dissenting point of view is championed by Peter Duesberg (University of California, Berkeley), the most outspoken opponent of the HIV-AIDS hypothesis. He is no crackpot. He has top scientific credentials and is acknowledged as an authority on retroviruses. Duesberg and his associates, including people with Ph.D.'s and M.D.'s, three of whom are Nobel Prize winners, are convinced that AIDS is not an infectious disease but is caused by lifestyle, environmental drugs, recreational drugs, and anti-HIV drugs used in the treatment and prophylaxis of AIDS. They believe that these are factors which depress the immune system (Box 15.4).

More recently, South African president Thabo Mbeki has also expressed skepticism about HIV causing AIDS and about the treatment of AIDS. He expressed his dissenting opinion at the 13th International AIDS Conference on July 10, 2000, causing many delegates to walk out in protest during his speech. Mbeki's critics claim he is wasting his country's time and resources and promoting confusion among AIDS patients who say, "The president [Mbeki] says that HIV doesn't cause AIDS. So why are you telling me to wear a condom?"

AIDS Treatment and Prevention

Treatment

The bottom line is that there is no cure for AIDS, and there is no preventive vaccine. From time to time, over the past 20 years, "breakthroughs" have been announced, but they always fall short of meeting expectations. Nevertheless, considerable progress has been made: the virus has been isolated, diagnostic laboratory tests have been developed, the dynamics of HIV are better understood, treatment of opportunistic infections has improved, and mother-to-child transmission has been drastically reduced thanks to the antiviral drug zidovudine (also called azidothymidine [AZT]). Further, society has become more tolerant toward individuals with AIDS, be they homosexual or heterosexual. The world awaits a cure to deal with the millions of infected individuals and, even more important, a vaccine that will prevent AIDS. Realistically, despite the millions of dollars dedicated to AIDS research, the ultimate goals of prevention and cure will not be met in the immediate future.

The life cycles of all viruses are complex; this is particularly true of HIV, which can remain in a latent state (in hiding) for more than 10 years before returning to a replicative cycle. Like all retroviruses, HIV defies the central dogma of biology by storing its genetic information in RNA and using the reverse transcriptase enzyme to get it back on the track to DNA. (Recall from chapter 9 that the flow of information is from RNA to DNA to RNA to protein.) To prevent or cure AIDS, a vaccine and drugs must somehow interfere with the virus's life cycle. Potential targets are adsorption, penetration, inhibition of reverse transcriptase, and assembly of new viral particles (Fig. 15.7). Drugs currently in use are directed against some of these targets and, to varying degrees, inhibit viral replication but only on a temporary basis.

An early treatment that ended in failure was to flood the body with soluble free-floating CD4 molecules which, presumably, would act as decoys and tie up the "docking" **gp120** molecules on the surface of HIV (Fig. 15.8). Recall that CD4 molecules are on T-helper and other cells that "fit" the gp120 molecules on HIV. However, flooding the body with decoy gp120 molecules to tie up the CD4 sites was unsuccessful.

AZT was the first clinically safe and effective drug used to treat HIV infection. AZT and other anti-HIV drugs are inhibitors of the reverse transcriptase enzyme and act at an early stage by preventing synthesis of DNA from RNA. The AZT molecule resembles an important building block in viral DNA synthesis, and when it is mistakenly utilized by the virus, DNA synthesis is brought to a halt. AZT and related drugs slow the spread of HIV in the body and temporarily forestall opportunistic infections. They are not cures.

A second class of drugs called **protease inhibitors** have been developed and approved for treating HIV infection. These drugs act by interfering with the final assembly and maturation of the virus; whereas AZT slows viral reproduction, protease inhibitors act to shut down the assembly line. Because HIV develops resistance to AZT and to protease inhibitors, combination treatment is now considered to be the best approach. The so-called **AIDS cocktail** includes two reverse transcriptase inhibitors

AUTHOR'S NOTE *Fifteen years ago I lectured to my microbiology class that within the next several years, there would be a vaccine for AIDS. Several years later, I all but promised that by the end of the century AIDS would be defeated. So, what do I say now?*

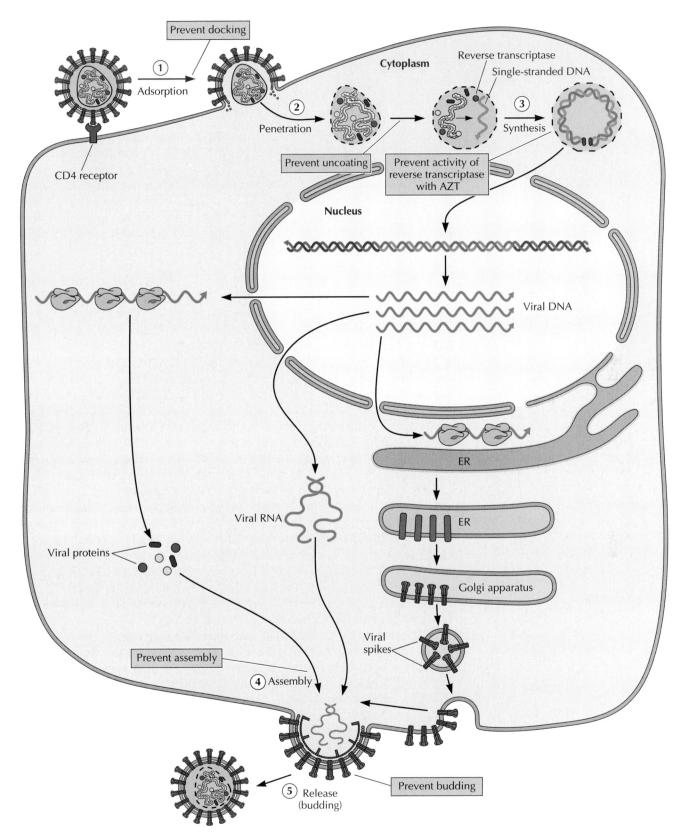

FIGURE 15.7 HIV replication cycle and potential drug targets. ER, endoplasmic reticulum.

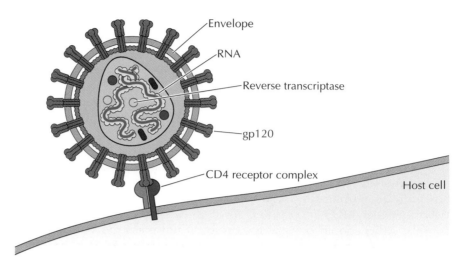

FIGURE 15.8   "Docking": CD4-gp120 complementarity.

and one protease inhibitor—a "triple whammy"—in an effort to block viral replication at two target sites. HIV-infected patients undergoing this combination treatment often become ecstatic because their T-cell counts rise, their viral loads decline, and their health improves remarkably. However, the improvement is not permanent.

In 1996, researcher David Ho announced a new strategy of AIDS treatment, namely, to administer the cocktail to patients during the first few weeks of infection. "Hit early, hit hard" was the strategy. Studies of a small group of patients indicated that this new strategy might actually eliminate—not just decrease or slow down—HIV from the body. The use of this strategy resulted in a decline of AIDS in the United States and elsewhere. David Ho was named *Time*'s 1996 "Man of the Year" for his work. AIDS began to be cautiously viewed for the first time as "curable." However, only 4 or 5 years later, the dark side of the "hit early, hit hard" approach emerged. The appearance of drug-resistant HIV strains increased, as did the toxic side effects, necessitating reconsideration about the wisdom of early and intensive therapy. Some think that the "big guns" should be saved for later in the battle. The situation is complex and is one of those "damned if you do, damned if you don't" situations, forcing physicians and patients to choose between the consequences of early treatment and the consequences of HIV when treatment is postponed. None of the currently available drugs cure HIV infection or AIDS; they all have side effects, some of which are very serious, they conflict with other drugs, they require strong motivation and self-discipline because of the large number of pills to be taken daily on a strict time schedule, and they cost over $10,000 a year. All of these factors contribute to the frustration of Third World countries as they witness the onslaught of AIDS. One major breakthrough is that the risk of HIV transmission from a pregnant woman to her baby is markedly reduced if she takes AZT during pregnancy, labor, and delivery; AZT is also given to the infant during the first 6 weeks of life.

Vaccines

Development of a safe and effective vaccine for AIDS is a priority; most scientists believe that the current AIDS pandemic will be controlled and future ones will be prevented only when a vaccine is available. A sense of urgency exists among AIDS researchers involved in vaccine development. Ongoing studies of monkeys, chimpanzees, and mice are designed to ultimately come up with a suitable vaccine. Over 25 vaccines are in the pipeline, and several have reached the stage of being candidates. Vaccines are being evaluated in the United States and in other countries, including an oral AIDS vaccine, a subunit vaccine utilizing gp120 and other single HIV components, and genetically engineered viruses. One study that appeared to have promise for monkeys was based on a live attenuated HIV virus; this would be an unlikely candidate for humans because of the possibility of reversion to a mutant virulent form. So, what is the conclusion regarding prevention? The conclusion is that your best protection against AIDS is a matter of your own personal conduct. Theoretically, AIDS should be easy to control, if not eradicate, because there are no vectors or animal reservoirs. On the other hand, matters of behavior, particularly sexual behavior, are sensitive and difficult to deal with. Mechanisms of transmission of HIV are discussed above; they point the way for preventive measures centered on abstinence or safe sex and avoidance of drug use. The AIDS cocktail and the implementation of early treatment dramatically reduced AIDS deaths, but recent CDC reports indicate that the decline is slowing. From 1996 to 1997, deaths dropped by 42%, but they dropped by only 20% from 1997 to 1998. A false sense of complacency has emerged from the myth that the cocktail and early intervention are cures for AIDS, resulting in an increase in high-risk sexual behavior. The CDC, along with other agencies, plays a vital role in educating the public about AIDS and AIDS transmission (Fig. 15.9). AIDS conferences are frequently held (Fig. 15.10) to evaluate, improve, and develop strategies to counter AIDS.

The sex industry flourishes in many countries, and many females are forced into a life of prostitution in order to earn a living. Prostitution is an important factor in considering the spread of AIDS throughout Africa and on other continents. Sex stops are abundant along the major truck

FIGURE 15.9   AIDS education posters. (Source: CDC.)

FIGURE 15.10 An AIDS conference in the late 1980s. Among those in the photo are then-vice president George Bush and leading AIDS researcher Anthony Fauci, the director of the National Institute of Allergy and Infectious Diseases, a unit of the National Institutes of Health. (Source: Public Health Service.)

routes; buying sex buys HIV and selling sex sells HIV. The drivers may be gone from home for weeks or months and return home with money and with HIV. HIV (but not the money) is quickly transmitted to their wives and girlfriends. The use of condoms is, to a large extent, not acceptable in many societies, and for the female sex partner to even suggest that her male partner use one is an insult to the man's "macho" character, dignity, and manhood. Females are frequently considered second-class citizens, and such suggestions are likely to be dealt with by vicious beatings and ostracism. The horror stories that have surfaced are beyond your imagination. Hopefully, time and education will lead to gentler, kinder, and wiser societies.

The Consequences of AIDS

You are witnessing the AIDS pandemic and are constantly reminded of its consequences in the media. Perhaps you even know someone with HIV infection or AIDS, or perhaps you know someone who has died of the disease. The virus knows no discrimination in terms of race, sex, or age. It attacks the rich and the famous, the young and the old. All countries, particularly Third World countries, have been hit, and all countries, rich or poor, bear the scars of AIDS and its repercussions in politics, in the economy, and in the well-being of citizens. AIDS and its consequences have been called "crimes against humanity," and all countries are its victims. The statistics are appalling, as indicated in a recent NIAID report (Box 15.5).

AIDS in Africa

The news out of Africa is particularly catastrophic. Africa, presumably the birthplace of AIDS, is plunged into despair (In the News 15.4). AIDS adds a heavy weight to the existing burden of other microbial and nonmicrobial diseases. It is the top killer in some parts of the African continent and has left 6 million children orphaned, accounting for 70% of the world's

BOX 15.5 NIAID Fact Sheet on HIV-AIDS Statistics

HIV and AIDS Worldwide

- As of the end of 2000, an estimated 36.1 million people worldwide (34.7 million adults and 1.4 million children under 15 years old) were living with HIV-AIDS. More than 70% of these people (25.3 million) live in sub-Saharan Africa; another 16% (5.8 million) live in South or Southeast Asia.
- Worldwide, approximately 1 in every 100 adults aged 15 to 49 is HIV infected. In sub-Saharan Africa, about 8.8% of all adults in this age group are HIV infected. In 16 African countries, the prevalence of HIV infection among adults aged 15 to 49 exceeds 10%.
- Worldwide, approximately 47% of the adults with HIV-AIDS are women.
- An estimated 5.3 million new HIV infections occurred worldwide during 2000; that is, about 15,000 infections each day. More than 95% of these new infections occurred in developing countries.

- Worldwide, more than 80% of all adult HIV infections have resulted from heterosexual intercourse.
- Mother-to-child (vertical) transmission has accounted for more than 90% of all HIV infections worldwide in infants and children

HIV and AIDS in the United States

- The CDC estimates that 800,000 to 900,000 U.S. residents are living with HIV infection and that one-third are unaware of their infection.
- For new infections among men in the United States, the CDC estimates that approximately 60% of men were infected through homosexual sex, 25% were infected through injection drug use, and 15% were infected through heterosexual sex. Of newly infected men, approximately 50% are black, 30% are white, 20% are Hispanic, and a small percentage are members of other racial or ethnic groups.
- For new infections among women in the United States, the CDC estimates

that approximately 75% of women were infected through heterosexual sex and 25% were infected through injection drug use. Of newly infected women, approximately 64% are black, 18% are white, and a small percentage are members of other racial or ethnic groups.
- The rate of new AIDS cases reported in the United States in 1999 (per 100,000 population) was 66.0 among blacks, 25.6 among Hispanics, 7.6 among whites, 8.8 among American Indians and Alaskan natives, and 3.4 among Asians and Pacific Islanders.
- As of December 31, 1999, 430,441 deaths among people with AIDS had been reported to the CDC. AIDS is now the fifth leading cause of death in the United States among people aged 25 to 44, behind unintentional injuries, cancer, heart disease, and suicide.

Source: NIAID.

IN THE NEWS 15.4

UN Warns That AIDS May Destabilize Southern Africa

AIDS Epidemic Reduces Life Expectancy in Southern Africa

More Africans Die of AIDS than in War

AIDS is leading cause of death in some parts of Africa

Out of Africa

orphans. There are now 25 million people infected in Africa, about three times higher than what demographers predicted in 1991. Life expectancy in some African countries is less than 40 years, compared with 76 years in the United States, and is heading downward in some nations. Hospital wards are overtaxed, and the beds are filled with bodies wasting away as flesh disappears from the bones of AIDS victims. Corpses are piled high in morgues to the point where individual identity is obliterated, and the landscape is littered with unmarked graves. Shame, stigma, poverty, prostitution, sexual violence, and ignorance enter into the cycle, contributing to the paralysis of the already weak public health infrastructure. The AIDS pandemic has also caused grave political and economic problems in Africa. Many hospital victims are told to "go home and die," because there are no drugs to help them. Further, many choose not to be tested since they know that treatment is expensive and only temporary, so why bother? The situation is desperate. What is to be done about it?

Brazil might have at least a partial answer and may serve as a role model for other developing countries. In 1992, Brazil embarked on a controversial program of producing its own AIDS drugs and distributing them free of charge; its government laboratories now produce at least five AIDS drugs. In the past 5 years, the number of AIDS-related deaths in Rio de Janeiro and São Paulo, the country's hot spots of AIDS, has plummeted by 40% and 53%, respectively. Developing countries in Africa and elsewhere are studying Brazil's example. The strategy flies in the face of politics, patent and copyright laws, and big (pharmaceutical) business, but it is working.

AIDS: the Future

Tarot cards, astrology, and gazing into crystal balls are of little value in trying to determine when the AIDS epidemic will end or when definitive AIDS treatment and prevention will be found. Will it be within this decade? The best answer to this question is "hopefully so." Meanwhile, active research in all phases of HIV infection and AIDS is under way under the sponsorship of both the private and public sectors—another example of partnerships (chapter 13).

In early 2001, the CDC announced a new campaign to "break the back" of the epidemic by identifying healthy Americans who are not aware that they are carrying HIV. Hopefully, these individuals will exercise appropriate precautions to protect others and take AIDS medication in an effort to interrupt the cycle of transmission.

TB

History

"Consumption," a 19th-century term for tuberculosis (TB), is an ancient bacterial disease and was so called because its victims appeared to be consumed by their illness. It was as though they had a demon inside which caused them to become excessively thin, pale, and weak. The disease is caused by *Mycobacterium tuberculosis*, also referred to as the tubercle bacillus, and was briefly introduced in chapter 8. TB is included in this chapter because it exists at epidemic levels. Globally, TB is the leading cause of death resulting from a single infectious disease. TB may be an ancient disease, but it is not, by any means, a disease of the past. In the United States and other industrialized countries, it is a reemerging disease, yet only 20 years ago it was on the brink of extinction. The disease may be better considered as an "emerged" disease that has returned despite the years of work toward its control.

Paintings from tombs in Egypt and examination of mummies dating back to 4000 B.C. suggest the antiquity of TB. The disease emerged in Neolithic times as human populations increased, settled down, and domesticated cattle. Presumably, human TB may have arisen from bovine TB. TB is as much a social disease as it is a microbial disease; in the 18th and 19th centuries, its development as an urban plague was associated with poverty, poor housing, crowding, inadequate nutrition, and unemployment, all spawned in the wake of the Industrial Revolution. It is ironic that progress fueled the social conditions that allowed TB and other diseases to flourish. Before the discovery of the tubercle bacillus by Robert Koch in 1882, the environment of crowded tenements was associated with the cause of the disease. Subsequently, one public health worker described the plight of the poor urban worker:

> It is in these rooms, they do everything, They cook, they eat, they sleep. This is where our patients cough, where they spit, where they waste away, and where they die. . . . the consumptive is left alone all day: he coughs, he spits on the ground; it is easy, then, to understand the danger faced by children coming home from school or workers coming home to rest. This is the time when they pretend to clean the room. They sweep, and from the dried sputum, the microbe is lifted and suspended in the air.

The impact of TB was so profound in the 19th century that some feared it would bring about the end of European civilization. In many cities of America and Europe, TB was the leading cause of death, accounting for as many as 15 to 20% of fatalities.

As social conditions improved during the first few decades of the 20th century, accompanied by the development of immunization in the 1920s and antibiotics in the 1940s, TB declined. The treatment of TB centered on fresh air, bed rest, and good nutrition with plenty of fresh eggs, milk, and cream, all of which were a part of the treatment in the TB sanatoria (as hospitals for the care of people with TB were called) around the country. (Cholesterol levels were not considered in those days!) Porches and decks were prominent in these facilities to allow the patients to be in fresh-air environments, even during the winter. In some cases, surgical intervention to bring about a collapsed lung was practiced. The rationale was that in a "rested" state an infected lung would heal more quickly. (A tube was inserted into the chest cavity to allow air to enter and collapse the lung.) Streptomycin, an antibiotic effective against the tubercle bacillus, was discovered in the 1940s, followed by the introduction in 1952 of isoniazid, a drug which is still widely used to combat TB.

Current Status

TB kills 2 million people per year; if the disease is left unchecked, that figure may increase by as much as 10-fold. In 1993, WHO declared TB a global emergency and estimated that between 2000 and 2020, nearly 1 billion people will be newly infected, 200 million people will get sick, and 35 million people will die from TB. Here are some more statistics that will knock you off your feet: one-third of the world's population is currently infected with TB; 5 to 10% of people who are infected with TB develop active TB during their lifetime; someone in the world is newly infected with TB every second; over 1.5 million TB cases per year occur in sub-Saharan Africa, largely because of the AIDS epidemic; nearly 3 million TB cases per year occur in southeast Asia; over a quarter of a million TB cases per year occur in eastern Europe.

TB: the Disease

The causative agent of TB is *M. tuberculosis* (chapter 4). TB is an infectious disease of the lower respiratory tract. The bacilli are acquired by the inhalation of infected droplets sprayed into the air by TB-infected individuals during coughing, sneezing, singing, talking, or laughing. Large numbers of TB bacilli are coughed up by infected individuals, and, since the waxy outer coat protects the bacilli from drying, transmission is efficient. Hence, people at the greatest risk are those who spend relatively long periods with an infected person, including family members, friends, and coworkers. Before pasteurization, milk was significant in the transmission of TB. Less commonly, the bacilli can be acquired through the skin, and cases have been reported in laboratory personnel who handled specimens containing the bacilli and in people who have received tattoos. An unusual case occurred in which an embalmer contracted TB from an infected corpse while preparing the body for burial. In most cases (Fig. 15.11), the primary infection is in the lungs, but infection can occur in the brain, spinal cord, kidney, bone, or cutaneous (skin) tissue (Fig. 15.12).

The first exposure to tubercle bacilli results in primary infection; in most cases there are no symptoms, and the individual is not even aware that infection has occurred. Cell-mediated immunity walls off the bacilli in lesions known as **granulomas.** This is the usual response in about 90% of

FIGURE 15.11   Common sites of TB infection.

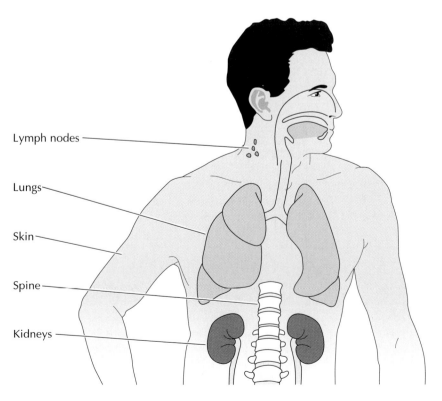

Lymph nodes

Lungs

Skin

Spine

Kidneys

individuals with primary infection. In the other 10%, the immune response is not adequate, resulting in the escape of bacilli from the granulomas; the bacilli cause symptomatic primary TB manifested by fevers, night sweats, weight loss, fatigue, and the coughing up of blood-tinged sputum. A classical sign of active TB is a blood-stained handkerchief. Symptomatic primary infection is more likely to occur in children, in the elderly, and in the immunocompromised, especially those with HIV.

Although most cases of TB infection do not progress to acute disease, reactivation TB, or secondary TB, can result because the bacilli can remain in a dormant state for years and be reactivated decades later. If this happens, any of the anatomical sites seeded during the primary infection may be affected, including primarily the lungs but also the bones and joints. Perhaps Quasimodo, the hunchback in Victor Hugo's classic tale *The Hunchback of Notre Dame,* was a victim of spinal TB. The course of TB infection is illustrated in Fig. 15.13.

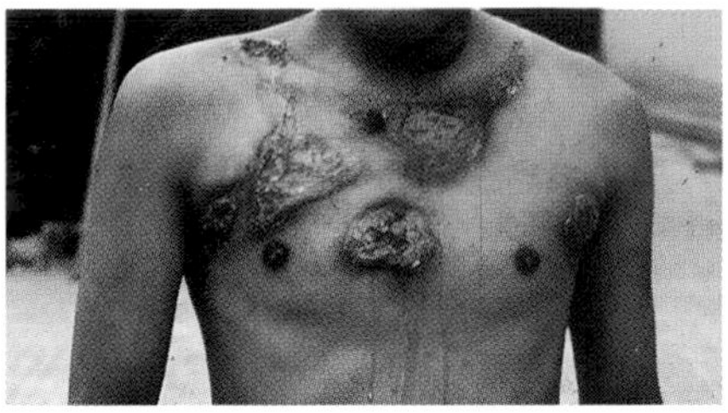

Diagnosis

The early symptoms of pulmonary TB resemble those found in a plethora of microbial diseases, including some caused by fungi. The presumptive diagnosis is based on the appearance of clinical signs and symptoms supported by a positive tuberculin skin test (Fig. 15.14), X-ray examination (Fig. 15.15), microscopic examination of sputum for the presence of tubercle bacilli (Fig. 15.16), and culture of sputum specimens for the demonstration of *M. tuberculosis* (Table 15.4).

FIGURE 15.12 Cutaneous TB. (Author's photo.)

FIGURE 15.13 The course of TB infection and disease.

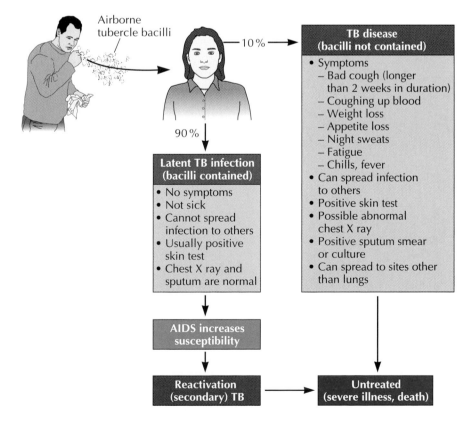

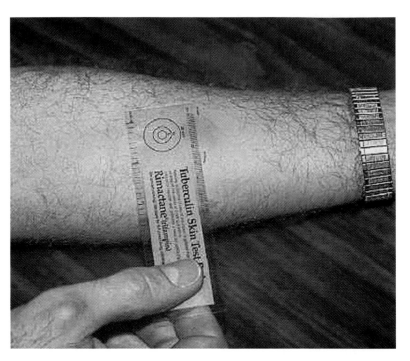

FIGURE 15.14   A positive tuberculin skin test. (Source: New Jersey Medical School National Tuberculosis Center.)

The **tuberculin skin test (Mantoux test)** is performed by the intradermal injection of a minute amount of **purified protein derivative (PPD)** from *M. tuberculosis*; this test is harmless, since it does not use whole organisms. After 48 to 72 hours, the injection site is visually examined for the presence of an induration (a red, raised lesion), which, if measuring 5 millimeters or more in diameter, constitutes a positive skin test. A positive skin test does not indicate active TB disease but demonstrates previous exposure to tubercle bacilli—a condition referred to as TB infection, not TB disease. A positive skin test is an indicator for further testing. It is important to emphasize the distinction between TB infection and TB disease. In infection, the bacilli are walled off in the granulomas, and there are no symptoms,

FIGURE 15.15   **(A)** A normal chest X ray; **(B)** a chest X ray showing TB infection in the left upper lobe of one lung. (Source: New Jersey Medical School National Tuberculosis Center.)

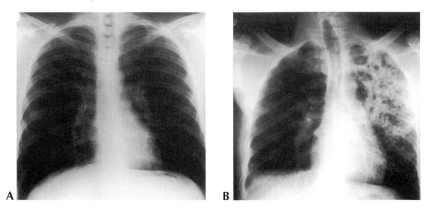

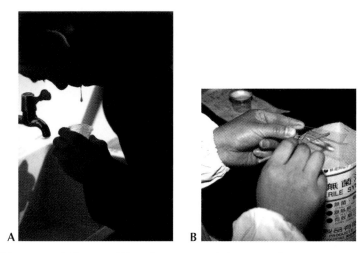

FIGURE 15.16   Collecting **(A)** and preparing **(B)** sputum for examination. (Source: WHO.)

whereas active symptomatic disease is the result of bacilli escaping from granulomas. TB scares have occurred on many college campuses and in other environments, causing those who have close contact with a TB-infected person to be skin tested. Those testing positive may falsely jump to the conclusion that they have TB. Remember that a positive skin test only indicates previous exposure to the organism. It may be that an individual had TB and recovered; such individuals test skin positive for many years. Further, foreign students make a large contribution to the population of colleges and universities scattered across the United States. Some of these people may have come from countries with a high rate of TB; they may have completely recovered from the disease but still manifest a positive skin test. Further, students from those countries may have received a vaccine (described below) as a preventive measure; these individuals also test positive.

Antibiotic Therapy

Antibiotic therapy is available for both TB infection and active TB disease. For those with TB infection (not disease), the antibiotic isoniazid is completely effective if taken for 6 months as a preventive measure. Treatment for active TB disease involves taking a combination of antibiotics (usually three) for 6 to 9 months along with supportive measures of adequate rest, a good diet, and oxygen therapy (Fig. 15.17). Periodic examinations are necessary to check for reexposure, but keep in mind that once a skin test is positive, it will always be positive. Five antibiotics are particularly effective against the tubercle bacilli: isoniazid, ethambutol, rifampin, streptomycin, and pyrazinamide. Toxicity and antibiotic resistance need to be taken into consideration when the appropriate combination of antibiotics is chosen. The use of multiple antibiotics greatly reduces the chances of the emergence of antibiotic-resistant strains.

   Although antibiotics that are effective in curing an individual with TB are available, these antibiotics need to be taken for at least 6 months, possibly creating a serious problem of infected persons' noncompliance with the treatment; further, they are expensive. Both of these factors severely limit the treatment of TB both in developing countries and in developed countries. This is particularly true in the population of homeless people—

TABLE 15.4   Bases for diagnosis of TB

Clinical signs and symptoms: weight loss, cough, fatigue, night sweats, low-grade fever
X-ray examination
Positive skin test (indurated lesion 5 mm or greater)
Demonstration of bacilli in sputum (presence of acid-fast bacilli on microscopic examination of sputum)
Culture of *M. tuberculosis* (sputum cultured on agar plate and observed for 6 to 8 weeks for colonies of *M. tuberculosis*)

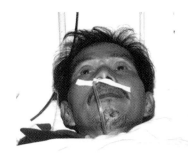

FIGURE 15.17   A TB patient receiving antibiotic and oxygen therapy. (Source: WHO.)

many of whom have TB or AIDS as well as drug or alcohol addiction—who live on the streets, in subways, in parks, and even in sewer pipes in large cities. Even if the antibiotics are provided free of charge and made available at gas stations, police stations, and local restaurants, noncompliance remains a major issue. Many people are more intent on satisfying their addiction to alcohol than on getting clean; many have given up on life. According to a WHO report, in 1995 approximately 30% of San Francisco's homeless population and 25% of London's homeless population were infected with TB. The infection rates in prisons and homeless shelters are staggering.

DOTS

Direct observational therapy short course (**DOTS**) is simple, and its strategy is to make certain that those infected with TB regularly swallow the right medicines. The strategy has proved to be successful since its adoption by WHO in 1992. Thousands of lives have been saved and the potential emergence of drug-resistant strains has been minimized as a result of treatment's being carried out without interruption. Essentially, DOTS is an attempt to deal with complacency and noncompliance (Fig. 15.18).

The DOTS strategy combines the five elements of political commitment, microscopy services, drug supplies, surveillance and monitoring, and direct observation. Once an individual's sputum shows tubercle bacilli, health care workers watch the patient swallow the full course of the prescribed anti-TB drugs on a daily basis. The sputum-smear testing is repeated after 2 months and at the end of the 6- to 8-month treatment schedule.

The drugs used in the DOTS system are not new, and when used correctly, they have a cure rate approaching 100%. The problem is one of compliance because of the length and cost of treatment. DOTS does not require hospitalization for monitoring treatment and progress; hospitalization is a luxury not available for the countries hardest hit by TB. Within the first 2 to 4 weeks of treatment, patients become noninfective. Previous attempts to control TB by simply giving infected individuals take-home medicines have been unsuccessful; the medications were taken in a haphazard fashion and actually added to the problem by fostering the development of multiple-drug-resistant strains of the tubercle bacillus. The DOTS program is designed for large, poor populations (rural or urban) and, when properly implemented, results in permanent cures. The program is cost-effective and varies for each country, but $100 to $200 per patient serves as a standard; the total cost of the drugs to effect a cure does not exceed $20 per patient. The World Bank has rated DOTS "one of the most cost-effective of all health interventions."

The DOTS success stories are numerous and almost too good to be true, but they are true; DOTS implementation revolutionized TB control in those countries of the world with massive rates of TB. In China, the old TB proverb, "ten get it, nine die," has been replaced by "ten detect it, nine cured." In Bangladesh, a poverty-stricken country with minimal public health infrastructure, rural areas have witnessed an 85% cure rate, with DOTS being delivered by village women. In Peru, DOTS has resulted in successful treatment of about 90% of the TB cases over its 5 years of use. The DOTS strategy has been used in New York City and in other cities of the developed world with impressive cure rates. Even President Clinton got into the act in India when he marked World Tuberculosis Day on March

FIGURE 15.18   A patient in a DOTS program taking her TB medicine. (Source: WHO.)

24, 2000, by administering the final dose of medicine to patients undergoing DOTS therapy. He noted that

> Today is World Tuberculosis Day. It marks the day the bacteria [*sic*] which causes TB was discovered 118 years ago. And, yet, even though this is 118-year-old knowledge, in the year 2000, TB kills more people than ever before, including one almost every minute here in India. These are human tragedies, economic calamities, and far more than crises for you, they are crises for the world. The spread of disease is the one global problem from which. . .no nation is immune.

By the end of 1998, the 22 countries with the greatest burdens of TB had adopted DOTS, and in that same year, 21% of the estimated number of TB patients received treatment under DOTS—double the percentage reported in 1995.

Factors Contributing to Reemergence

The statistics presented earlier are frightening, particularly since TB was on the brink of elimination 20 years ago in the United States and in other industrialized countries (Table 15.5). The public health infrastructure was caught flat-footed, and TB skyrocketed, with one-third of the world's population becoming infected. What happened? You can probably guess: factors discussed in chapter 1 were responsible. In the United States, the increase was first noted in 1986 in New York City; the nation experienced a 20% increase between 1985 and 1992 as a consequence of TB and HIV coinfection, multiple-drug resistance, complacency, and travel and immigration.

TB-and-HIV Coinfection

TB and HIV are a dynamic and deadly duo; each accelerates the other's progress. HIV weakens the immune system, fostering an 800-times-higher probability that an HIV-infected individual will develop active TB (Fig. 15.19). TB is the leading cause of death in HIV-infected populations, accounting for about 15% of deaths. It stands to reason that those countries hardest hit by AIDS are those with an increased incidence of TB.

TABLE 15.5 Reported tuberculosis cases, United States, 1985 to 2000[a]

Year	No. of cases
1985	22,201
1986	22,768
1987	22,517
1988	22,436
1989	23,495
1990	25,701
1991	26,283
1992	26,673
1993	25,287
1994	24,361
1995	22,860
1996	21,337
1997	19,851
1998	18,361
1999	17,531
2000[b]	16,372

[a]Source: CDC.
[b]Provisional 2000 data.

FIGURE 15.19 An HIV-infected patient with TB. (Source: WHO.)

Multiple-Drug-Resistant TB

Chapter 12 cites the development of antibiotics as a major factor responsible for the increase in life expectancy and quality of life during the 20th century. The antibiotic success story was tarnished by the emergence of antibiotic-resistant microbes, including the **multiple-drug-resistant** tubercle bacillus. Misuse of antibiotics (chapter 12) is the major cause. The regimen of TB therapy is long and expensive, frequently resulting in noncompliance and fostering the development of antibiotic resistance.

Complacency

Human progress frequently suffers because of a "why bother" or "it's too much trouble" attitude. Complacency was a significant factor in the 1980s, when the industrialized world relaxed its TB control programs.

Travel and Immigration

As with many microbial diseases, the period of incubation for TB is longer than the time it takes to travel from one part of the world to another. Hence, tourists may return from countries that have a high incidence of TB laden with an armful of gifts and souvenirs but also with a chest full of tubercle bacilli. Health experts warn that immigrants from countries with high rates of TB negate the gains made in reducing the disease in the United States. The Institute of Medicine recommends that immigrants from Mexico, the Philippines, Vietnam, and other countries with high TB rates be tested for the latent as well as the active form of the disease. If positive, these immigrants would not be denied entry but would be issued special visas requiring them to be treated before becoming permanent residents. Between 1994 and 1998, in addition to approximately 3.9 million legal immigrants, 2.7 million persons from Mexico and Central America resided in the United States illegally, contributing to the burden of TB, particularly considering Mexico's higher TB rate. Children adopted from some countries are a potential source of infection; since 1986, more than 125,000 children have been adopted by U.S. families. In 1999, a boy from the Marshall Islands in the western Pacific spread TB to 56 people, including many of his playmates, after moving to rural North Dakota.

Concern exists about the risk of transmission of TB during long airplane flights. The CDC reported an investigation in 1995 involving transmission of the tubercle bacillus from a passenger with active TB to others on the plane. The report revealed that four passengers were infected during an 8½-hour flight, but none of them had developed active disease 1 year later. The CDC concluded that the risk of transmission on airplanes is no greater than that in any other confined space, including other forms of transportation. Further, the agency issued recommendations calling for passengers and flight crews to be advised when they have been exposed to TB.

Prevention

Vaccine Development

Public health strategies focus on prevention of disease; the implementation of vaccines was a major factor in disease control in the 20th century. The **bacillus Calmette-Guérin (BCG) vaccine** is available, but its use is controversial. The protection rate is about 80% in children and less than 50% in adults, with the duration of protection ranging from 5 to 15 years. The BCG vaccine is not used in the United States because the incidence of TB does not justify its use, except among high-risk groups, including health professionals charged with the care of patients with TB and military personnel in areas with a high rate of TB. Further, those receiving the BCG vaccine will have a positive tuberculin reaction for life, necessitating repeated chest X rays for monitoring purposes.

The high worldwide incidence of TB affords top priority to the development of new and effective vaccines against the disease. New approaches utilizing the tools of molecular biology, including plasmid DNA vector-based vaccines, recombinant DNA vaccines, mutant BCG vaccines, and subunit vaccines, are encouraging.

DOTS Implementation

More countries need to adopt the DOTS strategy or other measures to treat those infected with TB on an uninterrupted basis with the right combination of antibiotics.

Improved Social Conditions

Many of the world's citizens live in abject poverty and suffer from a lack of clean water, malnutrition, and inadequate housing conditions, particularly in the developing world. These factors favor the development and spread of all microbial diseases. TB is as much a social disease as it is a microbial disease, and alleviation of poor social conditions will alleviate the incidence of TB.

The number of TB cases in the United States has decreased in recent years (Table 15.5), and the time is ripe to eliminate TB in this country, an attainable goal. Care must be taken to ensure that complacency and neglect, the price of success, will not once again increase the nation's vulnerability. In other nations, particularly in the developing world, WHO and its partners can only continue to chip away at the problem and to assist in DOTS implementation or alternative strategies.

OVERVIEW Plagues continue to occur throughout history as circumstances dictate humans and microbes locking horns in a battle for survival. Mad cow disease has reached epidemic proportions, resulting in a cluster of cases of vCJD disease in humans. Mad cow disease and vCJD are TSEs caused by prions.

AIDS, a plague which surfaced in the last 2 decades of the 20th century and continues into the 21st century, has taken a terrible toll throughout the world, particularly in Africa. Treatment and prevention of this disease remain elusive, although considerable progress has been made in improving the quality and life span of those infected.

TB is an ancient disease, but it is not a disease of the past. It is a disease which has reemerged with a vengeance and has infected one-third of the world's population. Treatment of the disease is seriously complicated by the development of multiple-drug-resistant strains and its association with AIDS. ■

SELF-EVALUATION

PART I Choose the *single* best answer.

1. Identification of HIV as the cause of AIDS is credited to
   - **a.** Duesberg
   - **b.** Montagnier
   - **c.** Gallo
   - **d.** more than one of the above

2. Prions contain
   - **a.** DNA
   - **b.** RNA
   - **c.** both DNA and RNA
   - **d.** no nucleic acid

3. Which variety of HIV is most common in West Africa?
   - **a.** HIV-1
   - **b.** HIV-2
   - **c.** HIV-3
   - **d.** about the same for HIV-1 and HIV-2

4. Which is correct regarding the BCG vaccine?
   - **a.** killed microbes
   - **b.** subunit vaccine
   - **c.** DNA vaccine
   - **d.** attenuated microbes

5. Which term is associated with TB?
   - **a.** Kaposi's sarcoma
   - **b.** granuloma
   - **c.** iatrogenic
   - **d.** kuru

6. Prions cause all of the following with the exception of
   - **a.** Foot-and-mouth disease
   - **b.** CJD
   - **c.** TSE
   - **d.** Mad cow disease

PART II Fill in the following.

1. The gene which encodes prion production is called the _____ gene.

2. TSE is an abbreviation for _____.

3. Some individuals with AIDS develop blue-purplish lesions on their skin known as _____.

4. The first stage of HIV infection is called _____.

5. What is DOTS? _____

6. TB is frequently found as a coinfection with _____.

PART III Answer the following.

1. What are the biological distinctions between bacteria, viruses, and prions?

2. "One bad apple spoils the barrel" is a popular expression. Discuss this expression as applied to CJD.

3. The course of AIDS can be followed by measuring the patient's CD4$^+$ T-cell level. Explain this.

THE UNFINISHED AGENDA

TEXT REVIEW

AN ONGOING BATTLE

CHRONIC DISEASES

RESEARCH

HEALTH FOR ALL

Article 25. (1) Everyone has the right to a standard of living adequate for the health and well-being of himself and his family, including food, clothing, housing and medical care and necessary social services, and the right to security in the event of unemployment, sickness, disability, widowhood, old age or other lack of livelihood in circumstances beyond his control.

UNIVERSAL DECLARATION OF HUMAN RIGHTS, 1948

PREVIEW This is the concluding chapter of *The Microbial Challenge,* and it is time to reflect on the book's theme. As quoted in the opening chapter, in 1967 Surgeon General William H. Stewart stated, with good reason, that it was "time to close the book on infectious diseases." Sanitation, vaccines, and antibiotics had made enormous strides against pathogenic microbes, and there was little evidence that the tide would turn. But the forces of natural selection and the ability of microbes to adapt, along with urbanization, complacency, technological advances, ecological disruption, and social evolution reflected in human behavior proved otherwise, and the tide did turn. More recently, one scientist remarked, "The war has been won, but, by the other side"; this facetious statement reflects the theme of this book, namely, the challenge of new, emerging, and reemerging infections. Outbreaks of microbial diseases are frequent subjects of current news (In the News 16.1) and indicate the need to avoid complacency. Prions, the cause of mad cow disease and Creutzfeldt-Jakob disease, appear to be a new category of infectious agent that defies fundamental biological concepts.

TEXT REVIEW

Part I (chapters 1 to 10) presents the challenges and the challengers. It is about the biology of microbes and the diseases they cause. Chapter 3 stands as an exception; it describes the beneficial aspects of microbes (the other side of the coin)—their role in the cycles of nature, without which life could not exist, and their exploitation in the production of medicinals, foods, and industrial products as agents for bioremediation. Chapter 3 is included

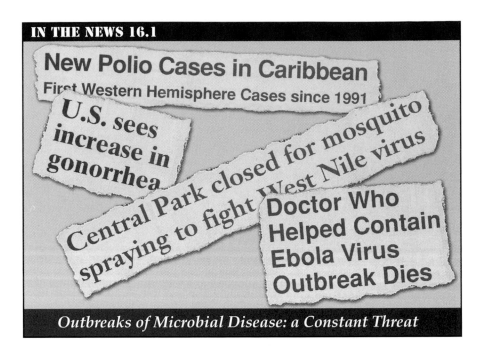

IN THE NEWS 16.1

New Polio Cases in Caribbean
First Western Hemisphere Cases since 1991

U.S. sees increase in gonorrhea

Central Park closed for mosquito spraying to fight West Nile virus

Doctor Who Helped Contain Ebola Virus Outbreak Dies

Outbreaks of Microbial Disease: a Constant Threat

early in the text to counter the notion that all microbes are disease producers and are "out to get us"; the truth is that the majority of microbes are not pathogens. In a positive sense, the further harnessing of microbes for beneficial purposes is part of the microbial challenge.

Part II (chapters 11 to 13) focuses on meeting the microbial challenge. It describes the human immune system, a system adept at recognizing foreignness (inherent in pathogens), and the body's response to eliminate pathogens by countering with antibodies and cell-mediated immunity. Public health measures and partnerships to meet the challenge are considered, with an emphasis on sanitation, vaccines, and antibiotics.

In part III, chapters 14 and 15 emphasize the current status of the challenge posed by microbes. Biological warfare continues to pose a threat to the world and is within the capability of many nations; biological terrorism (real or threatened) strikes fear into populations. The use of biological weapons indicates the need for research to develop rapid methods of microbial identification. The world is still plagued by plagues, and the spectrum varies from nation to nation and is in a constant state of flux. Tuberculosis and AIDS are rampant throughout the world. Mad cow disease is epidemic in Europe and threatens cattle herds throughout the world. Its human counterpart, variant Creutzfeldt-Jakob disease, has killed about 120 people. A worrisome aspect of this disease is that it has an extended incubation period which may be up to 15 years or longer, making it impossible to estimate the number of people harboring prions in their brains because of exposure that predated recent control measures.

AN ONGOING BATTLE

Microbes have challenged their hosts since the time of their appearance and will continue to do so. Are microbes really deliberately challenging humans? Is it in a microbe's best interest to annihilate its human (or other) host, and is it in humans' best interest to annihilate microbes? The answer to each of these questions is a definite "no." The point is that many different life forms and species make up the biotic component of ecosystems. A successful ecosystem is one in which all members of the biotic community live harmoniously with one another as a result of adaptation in their continuing co-evolution. It is a matter of adaptation, not annihilation.

Host-parasite interactions are dynamic. There are no scientific equations that can predict the outcome of specific interactions; the outcome depends on a variety of circumstances reflecting the time and place when microbes and humans are on a collision course. History reveals that epidemics come and go as a result of shifts in the seesaw of host-parasite relationships. On the surface, it would seem that microbes have the upper hand. They reproduce asexually and quickly without being hampered by requiring suitable partners (sexual reproduction) with which to share their genetic material. There are two advantages to not having a sex life; one is that when large numbers of the population are killed, asexual reproduction, coupled with generation times that can, in most cases, be expressed in minutes, leads to a rapid recovery of the population. Second, rapid reproduction results in enormous genetic variability, which, in turn, provides for genetic adaptation. A case in point is the rapid emergence of antibiotic-resistant strains of microbes. Sometimes, microbial adaptation is accompanied by an increase in virulence, as in toxigenic *Escherichia coli* and toxic shock-producing streptococci and the emergence of an unusually contagious and rapidly growing strain of *Mycobacterium tuberculosis*.

So why have humans (and other hosts) been able to survive the onslaught of microbes which periodically occurs at epidemic and even pandemic levels? Outbreaks of plague, smallpox, influenza, tuberculosis, and acquired immune deficiency syndrome make it clear that microbial threats are real. Nobel laureate Joshua Lederberg, in a 1999 article in *ASM News*, stated, in reference to the 1918 influenza epidemic, "If the mortality rate had been another order of magnitude higher, our species might not have survived." This statement, and other evidence through the ages, makes it clear that the battle between microbes and humans has been a close call at times. It is our wits which will continue to give the human species the upper hand.

A notable point to be made is the ability of microbes, based on their genetic variability, to make the leap from animal hosts to humans. Many biologists believe that tuberculosis and plague jumped from animal reservoirs to human hosts centuries ago. More recently, the emergence of AIDS and mad cow disease and its human variant supports this fact. In 1999, an outbreak of Nipah virus occurred in pigs in Malaysia and Singapore and took its toll on human lives. According to the director of the Centers for Disease Control and Prevention's Division of Viral and Rickettsial Diseases, "All we know is that some events caused the virus to jump species from its natural host, which is probably fruit bats, into pigs and then into humans." This outbreak is a further reminder and a warning that microbial diseases can be transmitted from wild and domestic animals to human populations.

The ongoing saga of microbial diseases is a reflection of the ongoing evolutionary dance; there is no reason to expect the band to play after the ball is over. On the other hand, and as a part of the unfinished agenda, the continued practices of personal and public health (Boxes 16.1 and 16.2) can be brought to bear on the prevention and control of outbreaks. The take-home message is that since humans and microbes coexist, individuals and public health officials need to be guided by the mistakes and successes of the past and to be prepared to nip new threats in the bud as they emerge.

BOX 16.1   Good Personal Health Practices

- Wash hands frequently (but not obsessively).
- Cook foods properly and refrigerate them promptly.
- Stay up to date on immunizations.

- Use antibiotics prudently.
- Practice abstinence or safe sex before marriage.
- Check with the Centers for Disease Control and Prevention or a travel

clinic before traveling, particularly if traveling to a Third World country.
- Avoid complacency.
- Use common sense in health matters.

CHRONIC DISEASES

The scientific and medical community found it difficult to accept Barry Marshall's theory, presented in 1983, that stomach and internal ulcers are caused by the bacterium *Helicobacter pylori*, an organism present in the acidic environment of the stomach. Marshall even drank a vial of *H. pylori* to induce ulcers, which were subsequently treated by antibiotics, during his 9-year battle to prove his point. This work set the precedent that microbes may play a role in a variety of chronic diseases that are not usually thought to be caused by infectious agents. The establishment of microbes as causative agents would be great news for those infected with a variety of diseases, because antibiotic treatment, antiviral agents, and vaccines could prevent, cure, or at least relieve their symptoms. For instance, the time-honored treatment for ulcers was for patients to drink a lot of milk and to take prescribed medicines that stocked the shelves of drug stores. In some cases, surgery was performed. Now, antibiotics are the treatment of choice, and researchers report that an antiulcer vaccine is on the way.

Research is important to uncover causal relationships between microbes and a variety of chronic illnesses. Scientists are finding links that indicate that microbes may play a role in diseases in which no one suspected their involvement. Microbes may directly act as causative agents, as cofactors, or as triggers that cause damage through autoimmune reactions. Microbes are suspected of playing a role in a variety of chronic diseases (Table 16.1), including stomach cancer, lung cancer, cervical cancer, strokes, asthma, heart disease, and juvenile diabetes. The list of diseases is a formidable one and involves a multitude of bacteria and viruses. In fact, some biologists believe that the majority of chronic illnesses are infectious in nature.

Confirmation of causality between infectious agents and chronic diseases would relieve the burden of disease in individuals and in public health, since drug therapy and immunization would be (or become) available.

BOX 16.2   Important Strategies To Improve Public Health

- Develop new partnerships between the public and private sectors.
- Continue the war on poverty in the developing and developed world.
- Maintain and improve vector control programs.
- Continue and increase monitoring

and surveillance.
- Target the Third World to reduce disparities in standards of living.
- Increase research in matters related to public health.
- Educate the public in matters related to public health (e.g., antibiotics and

vaccinations).
- Develop and maintain the public health infrastructure.
- Be prepared to handle outbreaks.
- Avoid the danger of complacency.

RESEARCH

Continued scientific research leads to continued advancement in society's understanding of its environment and establishes directions for future research for the betterment of humankind. The advancements in human welfare that have been made over the centuries, and in particular, in the past century, are the result of rigorous research. Microbiological research is directed toward an understanding of the microbial world and its interactions with humans and other species. Many microbes are essential for the maintenance of life on the planet, and others have been harnessed for a better quality of life. The American Academy of Microbiology hosted a colloquium in May 1996 "to identify those research areas most clearly deserving future attention, those most likely to provide optimal returns on scientific and monetary investment, and those offering the greatest promise for solving critical problems over the next decade." The colloquium participants identified six broad areas in the microbiological sciences that were deemed to offer the greatest potential for understanding microbes (Table 16.2). Additionally, microbial genome sequencing projects are an active area of research; a number of microbes have already been sequenced. This research will have an enormous impact on our understanding of virulence factors, vaccine design, countermeasures to antibiotic resistance, and microbial diversity.

HEALTH FOR ALL

Half a century has passed since the proclamation of the Universal Declaration of Human Rights, and almost a quarter of a century has transpired since the 1978 Alma-Ata Conference (see chapter 13). Article 25 of the declaration speaks of the highest attainment of health for all people as a fundamental right, and Alma-Ata defined its goal as "the attainment for all people of the world by the year 2000 of a level of health that will permit them to lead a socially and economically productive life." Substantial progress has been made toward these goals, as witnessed by impressive gains in healthy life expectancy for citizens fortunate enough to live in the industrialized world. But 85% of the world's population now live in the Third World and will not reap the full benefits of modern health care. To a large extent, their lives are spent in abject poverty with little access to decent housing, clean water, sanitary waste disposal systems, appropriate nutrition, or education. The reality is that the Declaration of Human Rights and the goal of Alma-Ata are only empty words with little promise. The future is limited to tomorrow, and people in the Third World live life on the edge (Fig. 16.1). Consider that there is about a 40-year disparity in healthy life expectancy between industrialized countries and Sierra Leone and Botswana, where the healthy life expectancy is under 40 years. It is true that the successes achieved in the 20th century were remarkable, but the biggest failure is that the scales are badly tipped (Fig. 16.2), as are the scales of justice, at the expense of developing countries, which bear the brunt of the burden of disease and malnutrition. Balancing the scale and reducing the disparity between the "haves" and the "have-nots" need to be priorities in international public health, and the old must give way to the new (Fig. 16.3). There is no mistaking the fact that this is the single most significant item on the unfinished agenda.

Poverty is frequently considered to be the root of all evils; poverty lies at the heart of the burden of inequality and needs to be the target of social, economic, and public health intervention. Poverty, not microbes, is the fun-

TABLE 16.1 Chronic diseases possibly caused by microbes

Juvenile diabetes
Atherosclerosis
Strokes
Heart disease
Stomach cancer
Lung cancer
Cervical cancer
Arthritis
Asthma
Chronic lung disease
Hypertensive renal disease
Crohn's disease

TABLE 16.2 Most important research areas in microbiology[a]

Microbial diversity and versatility
Microbial ecology and physiology
Pathogenic microbiology and immune responses
Microbial control of environmental pollution and recycling
Biotechnology
Education

[a]From W. J. Payne (preparer), *Basic Research for the Future: Opportunities in Microbiology for the Coming Decade* (American Academy of Microbiology, Washington, D.C., 1997).

FIGURE 16.1   Life on the edge. (Author's photo.)

damental cause and the challenge that allows disadvantaged populations to suffer more than their fair share of microbial (and some other) diseases. Microbes are the agents of disease that are allowed to flourish in the midst of poverty and its attendant consequences. The war against malaria and tuberculosis, as well as numerous other diseases, is a war against poverty; malaria is endemic in some of the most impoverished areas of the world. The battle against microbial diseases is a social and economic developmental issue as well as a public health issue.

The poor exist in pockets of poverty in developed countries as well as in developing countries. In response to the established association between poverty and disease, the World Health Organization (WHO) launched new initiatives for health that could save millions of lives in the 21st century. Gro Harlem Brundtland, director general of WHO, stated, "The world could end the first decade of the 21st century with notable accomplishments. Most of the world's poor people would no longer suffer today's burden of premature death and excessive disability, and poverty itself would thereby be much reduced. Healthy life expectancy would increase for all. . . . The fi-

FIGURE 16.2   The scales of (in)justice.

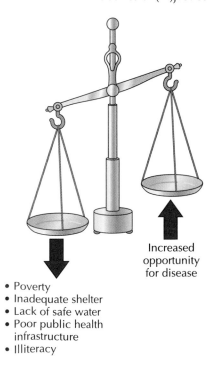

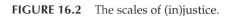

Increased opportunity for disease

• Poverty
• Inadequate shelter
• Lack of safe water
• Poor public health infrastructure
• Illiteracy

FIGURE 16.3   The old must give way to the new to improve the public health infrastructure in Third World countries. (Author's photo.)

nancial burdens of medical needs would be more fairly shared, leaving no household without access to care or exposed to economic ruin as a result of health expenditure. And health systems would respond with greater compassion, quality and efficiency to the increasingly diverse demands they face." But the responsibility of addressing poverty does not rest only with WHO; partnerships at the international level are vital. Recently, the United Nations published a report outlining several remedies aimed at helping poor nations develop.

From Brundtland's inspiring words, it is obvious that new initiatives are targeted at addressing the disparity that exists in health care delivery, and, in doing so, attempt to bring to fruition the intent of the Universal Declaration of Human Rights and the Alma-Ata Conference. The cycle of disease is complex and decidedly multifaceted (Fig. 16.4).

The Harvard University School of Public Health bears the following inscription, repeated in several languages, on an outside wall: "The highest attainable standard of health is one of the fundamental rights of every human being."

AUTHOR'S NOTE *I have walked by this inscription at least a few hundred times, but each time is like the first time; my eyes, my mind, and my emotions focus on these words; they continue to inspire me.*

FIGURE 16.4 Relationships among determinants of emerging infections. (Redrawn from D. B. Louria, p. 255, *in* W. M. Scheld, D. Armstrong, and J. M. Hughes [ed.], *Emerging Infections 1* [ASM Press, Washington, D.C., 1998], with permission.)

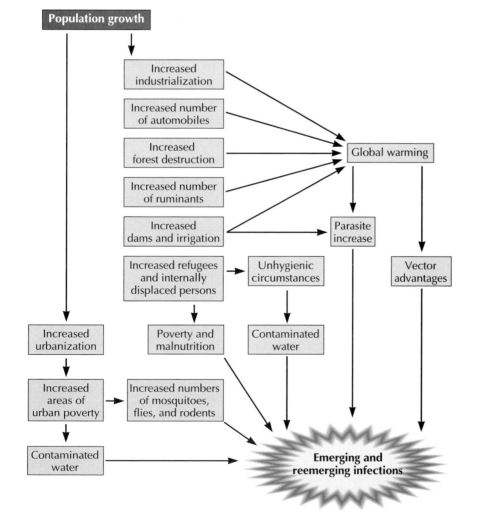

Glossary

AB model An explanation of the mode of toxin activity involving A and B toxin fragments.

Abiotic Nonliving.

Acquired immune deficiency syndrome *See* AIDS.

Active carriers Individuals who have a microbial disease that can be transmitted to others.

Acute leukemia One of two classifications of leukemia, in which the symptoms appear suddenly and progress rapidly, and death occurs within a few months.

Adenoids Tissues located in the back of the throat that are associated with the secondary immune structures; their lymphocytes play a role in protection from microbes entering through the nose and mouth.

Adenosine triphosphate *See* ATP.

Adsorption The first stage in the replication cycle of viruses during which viruses attach to the surface of host cells.

Aedes aegypti A species of mosquito that is a vector for dengue fever.

Aedes albopictus A species of mosquito that is a vector for dengue fever; these mosquitoes are commonly known as Asian tiger mosquitoes.

Aerobes Organisms that require oxygen for their metabolic activities.

Aerosols Suspensions of airborne particles ranging from 1 to 5 micrometers that are a means of transmission of microbes.

African trypanosomiasis *See* Sleeping sickness.

AIDS Acquired immune deficiency syndrome, a disease caused by the human immunodeficiency virus in which the immune system becomes severely compromised.

AIDS cocktail A combination of drugs used to block the replication of human immunodeficiency virus, including two reverse transcriptase inhibitors and one protease inhibitor.

Airport malaria Malaria that is the result of the survival of malaria-infected mosquitoes from other countries and is transmitted to people within the vicinity of an airport.

Albendazole A single-dose drug used in the treatment of filariasis.

Algae Organisms, some of which are unicellular and considered microbes.

Allergy An adverse immune response to molecules associated with pollen, dust, foods, mites, antibiotics, or bee stings.

Alpha interferon A drug used in the treatment of hepatitis C.

Amebiasis A disease caused by the protozoan *Entamoeba histolytica;* it is prevalent in areas with poor sanitation.

Anaerobes Organisms that do not require oxygen for their metabolic activities.

Anthrax A potentially fatal disease caused by *Bacillus anthracis;* the organism has been used in bioterrorism and is a potential agent of biological warfare.

Antibiotics Metabolic products produced by bacteria and fungi that inhibit the growth of other microbes.

Antibodies Protein molecules produced in response to antigens; they react specifically with the antigen that triggered their production and are important defense mechanisms.

Antibody-mediated (humoral) immunity One of two categories of specific immunity; it is the body's defense strategy for dealing with extracellular microbes.

Antigen-presenting cells   Cells that phagocytize microbes and present antigenic components to other cells of the immune system.

Antigenic drift   A minor change in the H and N spikes of the influenza virus occurring over a period of years.

Antigenic shift   A major and abrupt antigenic change in the H or N spikes of the influenza virus resulting in a new strain of the virus.

Antigens   Components of microbes, usually protein structures, that are recognized as foreign by the immune system and are targeted for destruction.

APCs   *See* Antigen-presenting cells.

Arboviruses   Viruses that are borne by arthropods.

Archaea   One of three domains in the Woese system of classification, to which archaebacteria are assigned.

Arthropoda   A biological phylum characterized by jointed appendages and divided into three classes; fleas, ticks, mosquitoes, and lobsters are examples.

Asexual stage   Part of the malaria parasite's life cycle that occurs in the liver and red blood cells of infected individuals in which sporozoites multiply without sexual reproduction taking place.

Asian tiger mosquito   *See Aedes albopictus.*

Assembly   A stage in the life cycle of viruses in which the viral components are assembled.

ATP   Adenosine triphosphate, a high-energy molecule that drives most cellular processes.

Autoimmune disease   A disease in which the immune system fails to distinguish "self" from "nonself."

Autoinfection   A process in which infection is perpetuated within the body; pinworm and strongyloidiasis are examples.

Autotrophs   Microorganisms and plants that are capable of utilizing the energy of the sun or derive energy from the metabolism of inorganic compounds.

Azidothymidine   *See* AZT.

AZT   Azidothymidine, the first clinically safe and effective drug used for the treatment of AIDS; it acts as an inhibitor of the reverse transcriptase enzyme. Also called zidovudine.

B lymphocytes   White blood cells that produce antibodies.

Babesiosis   A tick-borne protozoan disease caused by species of *Babesia*.

Bacilli   Rod-shaped bacteria.

Bacillus Calmette-Guérin vaccine   A vaccine against tuberculosis with limited effectiveness.

Bacteria   Unicellular microbes with distinct properties; one of the five distinct types of microbes.

Bacteria   One of three domains in the Woese system of classification.

Basal bodies   The structures by which flagella are anchored to the cell wall and cell membrane.

Basophils   White blood cells that are rich in granules of histamine.

BCG   *See* Bacillus Calmette-Guérin vaccine.

Beaver fever   *See* Giardiasis.

Binary fission   An asexual mode of reproduction in which a cell splits into two new cells.

Binomial system   A system of nomenclature established by Carolus Linnaeus in 1735 in which the genus and species are identified.

Bioaugmentation   Spraying of nutrients on beaches or on other microbe-contaminated sites to foster the growth of microbes indigenous to the area in order to accelerate degradation of pollutants.

Biocrime   A bioterrorist act that targets a specific individual or group rather than the masses.

Biogeochemical cycles   Processes such as the carbon and nitrogen cycles in which bacteria play a critical role.

Biological amplification   The mechanism of activity of the complement system that promotes immune system function.

Biological vectors   Organisms, including mosquitoes, ticks, lice, and flies, that transmit microbial disease.

Biological warfare   The use of microbes, such as the anthrax bacillus and smallpox virus, as weapons in the conduct of war.

Biological weapons   Microbes that are deployed with the intention of producing clinical disease in an attempt to incapacitate or kill large numbers of individuals.

Bioremediation   A process utilizing the enzymatic activity of microbes to break down pollutants such as oil, paper, and concrete.

Biosphere   All the organisms and environments found on Earth.

Bioterrorism   Employment of biological weapons by non-state governments, religious cults, militants, or crazed individuals.

Biotic   Of or relating to life; refers to the living components of an ecosystem.

Biting plates   Mouth parts found in some worms that allow them to attach to and suck blood from their host.

Black Death   *See* Plague.

Blender   An organism that facilitates the creation of a new hybrid strain of a disease; an example is that of a pig that becomes infected with a human strain and a bird strain of the flu, resulting in a new hybrid form of the virus.

Blood flukes   Common name for *Schistosoma* species and some other worms.

Boils   Localized skin infections frequently caused by staphylococci.

Bone marrow   Source of all blood cells.

Botox Minute doses of botulinum toxin used to reduce wrinkles and to treat common disorders associated with muscle overactivity.

Botulism A form of food poisoning caused by *Clostridium botulinum;* a neurotoxin causes flaccid paralysis.

Bovine spongiform encephalopathy A neurological condition in cattle resulting from abnormally folded prions that convert normal prions to the defective form.

Breakbone fever *See* Dengue fever.

Breath test A means of diagnosing peptic ulcers that is based on the production of urease by *Helicobacter pylori.*

Bronchitis Inflammation of the bronchioles.

BSE *See* Bovine spongiform encephalopathy.

Buboes Enlarged lymph nodes that occur when bacteria localize in the lymph nodes, such as in bubonic plague.

Bubonic plague A form of *Yersinia pestis* infection in which the bacteria localize in lymph nodes, causing them to swell to the size of eggs.

Budding A mechanism by which some viruses are released from the host cell.

Bulbar poliomyelitis An extremely serious form of polio in which individuals have difficulty swallowing and breathing because of muscle paralysis.

Burkitt's lymphoma A malignant cancer of the jaw and abdomen which occurs often in children in central and western Africa.

Campylobacter A bacterium that is an important cause of intestinal disease in humans and is associated with abortion and enteritis in sheep and cattle.

Capsid The protein coat surrounding viruses.

Capsomeres Protein units which make up the capsid; they confer helical, polyhedral, or complex shapes.

Capsule A component of the bacterial envelope that may contribute to virulence; it is not present in all species of bacteria.

Carbohydrates Organic molecules that function as an energy source and are found in some cellular structures.

Carbuncles Localized skin infections that are larger and deeper than boils and can reach baseball size.

Cauterization An early method of treating wounds inflicted by rabid animals in which long, sharp, hot needles were inserted deeply into the wounds.

CD *See* Cluster-of-differentiation molecules.

CD4 receptor molecules Receptor molecules found on some T cells that identify them as T-helper cells.

CD8 receptor molecules Receptor molecules found on some T cells that identify them as cytotoxic T cells.

Cell The basic unit of life.

Cell culture A technique for growing cells that can be used to culture viruses.

Cell-mediated immunity A type of specific immunity that results in the production of sensitized T lymphocytes directed against a particular antigen.

Cell membrane A component of the bacterial envelope that regulates the passage of molecules between the bacterial cell and its external environment.

Cell wall A component of the bacterial envelope that gives the cell its characteristic shape and confers structural integrity.

Cercariae Freely swimming immature forms of *Schistosoma* that are responsible for "swimmer's itch."

Cercarial dermatitis A condition characterized by itching and a rash caused by penetration of the skin by cercariae.

Cerebral malaria A deadly form of malaria that affects the brain.

Cervix The opening to the uterus.

Chagas' disease A protozoan disease caused by *Trypanosoma cruzi* that leads to widespread tissue damage, particularly to the heart, causing it to enlarge and impairing its function.

Chancres Sores on the penis or on the cervix that are characteristic of the primary stage of syphilis.

Chemosynthetic autotrophs Organisms that derive energy from the metabolism of inorganic compounds.

Chemotaxis The process of moving toward or away from a chemical stimulus.

Chicken pox A disease characterized by blisterlike lesions on the body and typically occurring in childhood.

Chlamydiae Coccoid bacteria that are obligate intracellular parasites; they are responsible for a variety of diseases, including urethritis, trachoma, and chlamydia.

Chlorella A photosynthetic alga found on the surface of ocean water.

Cholera A disease caused by *Vibrio cholerae* that is characterized by severe diarrhea and dehydration.

Chromosome The structure into which DNA is organized.

Chronic carriers Individuals who harbor a pathogen for long periods without becoming ill with the disease but who may spread the disease to others.

Chronic fatigue syndrome A disease that was incorrectly believed to have been caused by Epstein-Barr virus.

Chronic granulomatous disease An inherited disorder of phagocytes characterized by their inability to kill bacteria.

Chronic leukemia A form of leukemia, cancer of the white blood cells, that can remain for many months.

Cilia Hairlike projections that line certain areas of the respiratory system and assist in the removal of bacteria.

Ciliata A group of protozoans characterized by the presence of cilia.

Clonal selection theory   An explanation of antibody production: a population of antibody-producing B cells exists for every possible antigen and is triggered to clone and produce antibodies upon contact with a specific antigen.

Cluster-of-differentiation molecules   Distinctive molecules found on T cells that identify their specific role.

Coagulase   An enzyme that forms a network of fibers around bacteria, affording protection against phagocytosis.

Cocci   Spherical bacteria.

Collagen   A protein found in connective tissues.

Collagenase   An enzyme that breaks down collagen.

Colonies   Visible masses of bacterial cells growing on an agar surface, each presumably derived from a single cell.

Colonization   Growth and reproduction of a microbe in a particular niche, resulting in large numbers of cells.

Colorado tick fever   A tick-borne viral fever caused by an arbovirus.

Commensalism   A symbiotic relationship between two species in which one benefits and the other is neither harmed nor benefited.

Common cold   A mild illness caused by a variety of viral groups.

Common-source epidemics   Outbreaks of disease arising from contact with a single contaminated source, typically associated with fecally contaminated food or water.

Common warts   Benign, painless, elevated growths caused by human papillomaviruses that occur most frequently on the fingers.

Complement   A series of blood proteins that constitute a significant defense mechanism against disease-causing microbes.

Complex   An arrangement of capsomeres in some viruses.

Congenital syphilis   Syphilis resulting from the passage of spirochetes across the placenta from mother to baby.

Conjugation   A primitive form of sexual reproduction that permits the exchange of genetic material during physical contact of mating pairs.

Consumers   Organisms that take in oxygen and release carbon dioxide.

Convalescence stage   The time in which recovery from an illness takes place, strength is regained, repair of damaged tissue occurs, and rashes disappear.

Copepods   Water fleas that can harbor infective larvae of guinea worms.

Coronaviruses   A group of viruses that are a major cause of the common cold.

CPE   *See* Cytopathic effect.

Creutzfeldt-Jakob disease   A human transmissible spongiform encephalopathy.

Cryptosporidiosis   A protozoan disease transmitted by drinking fecally contaminated water.

Culture   Growth of an organism in a laboratory for the purposes of propagation and study.

Cutaneous anthrax   A form of anthrax which is acquired by contact with *Bacillus anthracis* or its spores via wool, hides, leather, or hair products.

Cutaneous leishmaniasis   A form of leishmaniasis that results in skin lesions.

Cyanobacteria   A group of bacteria that are photosynthetic.

Cyclosporiasis   A protozoan disease.

Cyst   A small sac, frequently filled with pus.

Cysticerci   Tapeworm larvae that are enclosed within a membranous sac.

Cysticercosis   A disease resulting from the ingestion of pork tapeworm eggs.

Cytokines   Products that are released by lymphocytes in response to stimuli and that trigger responses in other cells.

Cytopathic effect   A pathological change that occurs in cells as a result of viral replication.

Cytoplasm   The area of the cell enclosed by the cell membrane containing organelles that function in cell metabolism and multiplication.

Cytotoxic T cells   Members of the T-cell subset designated CD8.

Cytotoxins   Exotoxins that damage or kill host cells.

$D = nV/R$ An equation representing the struggle between disease-producing microbes and host resistance. D represents the severity of infection; n is the number of organisms; V stands for virulence factors; R stands for resistance factors.

DALYs   *See* Disability-adjusted life years.

DaPT vaccine   A newer DPT vaccine that utilizes acellular pertussis.

Debridement   Removal of necrotic (dead) tissue in an attempt to halt an infection.

Decomposers   Microbes that break down compounds into simpler constituents.

Defensive strategies   Adaptations that allow microbes to escape destruction by the host immune system.

Dengue fever   A mosquito-borne disease that is typically self-limiting, with recovery occurring within about 10 days.

Dengue hemorrhagic fever   A potentially fatal disease that can result from infection with a dengue virus strain different from the one causing the initial infection.

Denitrifying   Returning nitrogen to the atmosphere.

Deoxyribonucleic acid   *See* DNA.

Dermotropic viruses   Viruses that have an affinity for and cause diseases in the skin and subcutaneous tissues.

Diatom   A type of unicellular alga.

Differential count   A reflection of the ratio of the white blood cell categories.

Diffusion   The passive movement of a substance from an area of high concentration to an area of low concentration.

DiGeorge syndrome   An immune disorder resulting from abnormal development of the thymus gland.

Dinoflagellate   A type of unicellular alga classified as a microbe; dinoflagellates are the primary source of food in the oceans.

Dioecious   Having distinct male and female forms in a species.

Diphtheria   An upper respiratory tract infection caused by *Corynebacterium diphtheriae*.

Diplococci   Groups of cocci which occur in pairs.

Direct transmission   Person-to-person contact in which the infectious agent is directly transferred from a portal of exit to a portal of entry.

Disability-adjusted life years   A statistic based on healthy years of life (i.e., not including years lost to disability).

Disease   A possible outcome of infection in which health is impaired in some fashion.

Distilled spirits   Alcoholic beverages resulting from bacterial fermentation and having a high alcoholic content, e.g., brandy, rum, and whiskey.

DNA   Molecules which store the genetic information.

Dose   The number of microorganisms to which a host has been exposed.

DOTS   Direct observational therapy short course, a strategy to ensure that individuals infected with tuberculosis take their prescribed medicines.

DPT vaccine   A triple vaccine against diphtheria, pertussis, and tetanus.

Dracunculiasis   A disease caused by the parasitic guinea worm, *Dracunculus medinensis*.

Eastern equine encephalitis virus   A common arbovirus found in the United States.

Ebola hemorrhagic fever   A severe viral infection characterized by extreme hemorrhaging with a high fatality rate.

Ecosystem   A population of organisms in a particular physical and chemical environment.

Ehrlichiosis   A bacterial tick-borne infection similar to Lyme disease.

Elephantiasis   A disease caused by filarial worms that results in blocked lymphatic vessels and the accumulation of large amounts of lymph fluid in the tissues.

Embryonated (fertile) chicken eggs   Chicken eggs containing live embryos; they are used to grow viruses.

Encystation   A process that allows protozoans to survive outside a host; the parasite is surrounded by a thick capsule.

Endemic disease   A disease that is continually present at a steady level in a population and poses little public threat.

Endemic relapsing fever   A tick-borne bacterial disease caused by *Borrelia recurrentis*.

Endemic typhus   A disease caused by *Rickettsia typhi*; it is transmitted to humans from the bite of rat fleas.

Endocytosis   A process of engulfment of material, including viruses, displayed by some cells.

Endotoxin   A toxin produced by some gram-negative bacteria; it is usually released not during cell growth but upon death and disintegration of the microbe.

Enology   The science of wine making.

Enterotest   A method to test for the presence of *Giardia* trophozoites; the patient swallows a gelatin capsule attached to a string, and after 4 hours, the capsule is withdrawn and examined for the presence of trophozoites.

Enterotoxigenic *Escherichia coli* strains   Strains of *E. coli* that are a common cause of traveler's diarrhea.

Enterotoxin   A toxin that affects the intestinal tract.

Envelope   A bacterial structure consisting of a capsule, cell wall, and cell membrane.

Enveloped viruses   A category of viruses characterized by a membrane surrounding the capsid.

Enzymes   Substances that act as catalysts on specific substrates and influence the rate of a chemical reaction; some bacterial enzymes are virulence factors.

Eosinophils   White blood cells associated with helminth infections and allergies.

Epidemic   A disease that has a sudden increase in morbidity and mortality in a particular population.

Epidemic relapsing fever   A spirochete-caused disease transmitted from person to person by the bite of infected body lice.

Epidemic typhus   A disease caused by *Rickettsia typhi*; it is transmitted directly from human to human by the bites of body lice.

Epidemiology   The study of the sources, causes, and distribution of diseases and disorders that produce illness and death in humans.

Epstein-Barr virus   A DNA virus that is the primary cause of infectious mononucleosis.

Erythrocytes   Red blood cells.

Erythrogenic toxin   A secretion, produced by some strains of streptococci, that causes the red rash of scarlet fever.

Espundia   *See* Mucocutaneous leishmaniasis.

ETEC   *See* Enterotoxigenic *E. coli* strains.

Eucarya   One of three domains in the Woese system of classification.

Eucaryotic cells   Cells possessing a nuclear membrane and other membrane-bound organelles.

Excystation   The process by which a cyst comes out of dormancy, resulting in an active stage.

Exfoliative toxin   A substance produced by staphylococci that causes the skin to become blistery and peel away.

Exotoxins   Protein molecules that are released by microbes during their growth and metabolism.

Extracellular microbes   Microbes that colonize on cell surfaces.

Extremophiles   Microbes that grow under harsh environmental conditions.

Extrusion   A process whereby mature virus particles are released from the host cell; also called budding.

Facultative anaerobes   Organisms that grow best in the presence of oxygen but are capable of survival in its absence.

Fermentation   A metabolic reaction regulated by enzymes that break down sugars to acids and carbon dioxide.

Filarial worms   Tiny adult threadlike worms that can block the lymphatic vessels, possibly resulting in elephantiasis.

Filariasis   *See* Elephantiasis.

Fixation   The process through which atmospheric nitrogen is converted to a product (typically ammonia) that plants are capable of using in metabolic processes.

Flagella   Structures composed of the protein flagellin that provide motility to certain species of bacilli and cocci.

Flagellin   The protein of which flagella are composed.

Flatworms   A morphological category of parasitic worms.

Flesh-eating strep   A streptococcus causing a severe form of streptococcal infection, necrotizing fasciitis.

Fomites   Inanimate objects that serve as means of transmission of infectious material.

Food infection   The result of ingestion of bacteria in contaminated foods and their subsequent growth in the intestinal tract accompanied by their secretion of toxin.

Food intoxication   The result of ingestion of bacterial toxins.

Foreign   Not normally present in the body and capable of triggering an immune response.

Fungi   A category of eucaryotic organisms including mushrooms and yeasts. Many fungi are microbes.

Fusion   A mechanism of penetration employed by some viruses in which there is contact between the viral envelope and the host cell membrane.

Gametes   Male or female sex cells.

Gamma globulin   A fraction of the globulin component of blood plasma in which most of the antibodies are present.

Ganglia   Collections of nerve cells.

Gas gangrene   A condition brought about through contamination of a wound with particles of soil containing *Clostridium perfringens* or its spores.

Gastrointestinal anthrax   A form of anthrax resulting from the ingestion of inadequately cooked meat contaminated with *Bacillus anthracis*.

Generation time   The length of time between rounds of binary fission.

Genes   Segments of the DNA molecule, involved in heredity, which encode specific polypeptide, protein, or RNA molecules.

Genetic engineering   A means of manipulating genes to bring about a desired function.

Genital warts   One of the most common sexually transmitted diseases, caused by human papillomaviruses.

Genus   A category in the binomial system of nomenclature above the species level.

Giardiasis   An intestinal disease caused by the protozoan *Giardia lamblia* and acquired by drinking contaminated water.

Glomerulonephritis   A kidney disease that is a potential result of streptococcal infection.

Gonorrhea   A sexually transmitted disease caused by *Neisseria gonorrhoeae*.

gp120   A molecule found on the surface of HIV that "docks" with CD4 receptor molecules on T lymphocytes.

Gram negative   A cell wall-staining property of some bacteria that causes them not to retain the crystal violet stain after decolorizing with alcohol; they subsequently are stained with safranin and appear red.

Gram positive   A cell wall-staining property of some bacteria that causes them to retain the crystal violet stain after decolorizing with alcohol; they appear purple.

Granuloma   A pocket-like arrangement of cells associated with chronic inflammation that walls off the inflammatory agent.

Growth   An increase in the size of individual cells or an increase in a population of cells.

Guillain-Barré syndrome   A potentially fatal and rare nervous system disease of unknown cause; it sometimes occurs after administration of a vaccine.

Gummas   Tumorlike lesions that characterize tertiary syphilis.

H spike *See* Hemagglutinin.

Halophiles Bacteria that live in environments containing extreme salt concentrations.

Hanging drop A method of observing bacterial motility under a microscope by suspending a drop of culture on a slide.

Hansen's disease A disease caused by *Mycobacterium leprae;* formerly known as leprosy.

Hantavirus A virus that causes hantavirus pulmonary syndrome.

Hantavirus pulmonary syndrome A disease caused by a hantavirus that produces severe influenzalike respiratory problems and can result in death.

HBV *See* Hepatitis B virus.

HCV *See* Hepatitis C virus.

Healthy carriers Individuals who have no symptoms of a particular microbial disease but harbor the microbes and may unwittingly pass the disease on to others.

Helical Arranged in a continuous tube containing a virus's nucleic acid. Refers to arrangement of capsomeres.

Hemagglutinin An enzyme on the spikes of influenza viruses, aiding the penetration of cells by the virus.

Hemolysins Secretions produced by some microbes that destroy red blood cells through the destruction of cell membranes.

Hemolytic uremic syndrome Kidney damage occurring primarily in young children as a result of bacterial toxins.

Hepatitis Inflammation of the liver.

Hepatitis A virus A virus found in feces and transmitted through contaminated drinking water and food. It is the most common hepatitis virus and usually causes only mild disease.

Hepatitis B virus A virus transmitted in body fluids. It causes subclinical to severe disease.

Hepatitis C virus A virus transmitted mainly by intravenous drug use; a major reason for liver transplants in the United States.

Hepatitis D virus An incomplete virus that requires hepatitis B virus in order to replicate. It causes severe disease with a high mortality rate.

Hepatitis E virus A virus transmitted by the fecal-oral route. It causes a disease that is usually of moderate severity.

Herbivorous Feeding on plants.

Herd immunity Immunity of enough members of a population to protect against an epidemic.

Hermaphroditic Producing both sperm and eggs.

Herpes simplex virus A virus that causes the highly infectious disease herpes. Herpes simplex virus type 1 causes painful sores around the mouth and lips, frequently referred to as cold sores or fever blisters. Herpes simplex virus type 2 causes painful sores on the penis in males and on the labia, vagina, or cervix in females, typically referred to as genital herpes.

Heterotrophs Organisms that require organic compounds as an energy source.

HGE *See* Human granulocytic ehrlichiosis.

Hiker's diarrhea *See* Giardiasis.

HIV *See* Human immunodeficiency virus.

HIV seropositive Having antibodies against human immunodeficiency virus.

HME *See* Human monocytic ehrlichiosis.

Hookworm disease A disease that is caused by the roundworms *Ancylostoma duodenale* and *Necator americanus.*

Horizontal gene transfer A process by which genes are transferred from one organism to another.

Horizontal transmission A means by which a disease spreads from one person to another.

HPV *See* Human papillomaviruses.

HSV *See* Herpes simplex virus.

Human Genome Project The mapping of the genes located on the 23 pairs of human chromosomes.

Human granulocytic ehrlichiosis A tick-borne bacterial infection caused by an unknown species of *Ehrlichia.*

Human growth hormone A hormone that is used for the treatment of dwarfism and is now produced by genetic engineering.

Human immunodeficiency virus The virus that causes AIDS. Human immunodeficiency virus type 1 is the most common cause of AIDS worldwide, and human immunodeficiency virus type 2 is the most common cause of AIDS in West Africa.

Human insulin A hormone that regulates the amount of glucose in the bloodstream. Much of the insulin now used to treat diabetes is a product of genetic engineering.

Human monocytic ehrlichiosis A tick-borne bacterial infection caused by *Ehrlichia chaffeensis.*

Human papillomaviruses Viruses that cause warts.

Hyaluronidase An enzyme produced by some bacteria that breaks down hyaluronic acid found in connective tissue.

Hydrophobia Fear of water; an old term for rabies.

Hyperthermophiles Bacteria that live in extremely hot environments.

Hyphae Intertwined filaments characteristic of molds.

Iatrogenic infection Infection induced in a patient by a medical procedure.

Icosahedrons Three-dimensional, 20-sided structures with triangular sides; they confer a geodesic shape to a virus.

ID *See* Infectious dose.

Ig   *See* Immunoglobulins.

Illness stage   The time in which disease develops to the most severe stage, as evidenced by typical signs and symptoms.

Immunocompromised   Characterized by a weakened immune system as a result of AIDS or other conditions.

Immunoglobulins   Categories of antibodies (immunoglobulins A, D, E, G, and M), each possessing specific properties.

Impetigo   A superficial infection of the skin that is typically manifested by blisters around the mouth and is usually caused by staphylococci.

Incubation stage   The time between a pathogen's access into the body and the appearance of signs and symptoms.

Indirect transmission   The passage of infectious material from a reservoir to an intermediate host and then to a final host.

Infantile paralysis   A form of poliomyelitis in infants and young children, characterized by muscle paralysis.

Infection   The presence of microbes in the body without definitive symptoms.

Infectious agents   Plasmids capable of exchanging genetic material from one microbe to another; some confer antibiotic resistance.

Infectious dose   The minimum amount of bacteria needed to establish an infection.

Infectious mononucleosis   A disease caused by Epstein-Barr virus in which the salivary glands are infected by the virus; frequently referred to as "mono."

Inflammation   Swelling, pain, redness, and heat in an area of tissue.

Influenza   Exaggerated coldlike symptoms caused by influenza virus.

Influenza A virus   A strain of influenza virus that causes epidemics and occasionally pandemics in which there are animal reservoirs.

Influenza B virus   A strain of influenza virus that causes epidemics and does not have an animal reservoir.

Influenza C virus   A strain of influenza virus that does not produce epidemics and causes mild respiratory illness.

Inhalation anthrax   The most severe form of anthrax, resulting from the intake of anthrax spores.

Inorganic compound   Generally, a chemical compound that does not contain carbon (although carbon dioxide and several other carbon-containing compounds are considered inorganic).

Insects   A large class of arthropods characterized by three body segments and six legs.

Interferon   A component of blood that interferes with viral replication.

Intracellular bacteria   Bacteria that penetrate and grow inside host cells.

Iron lung   A barrel-like device formerly used for persons who had difficulty breathing on their own because of muscle paralysis as a result of polio.

Ivermectin   A drug used to treat onchocerciasis.

Kala-azar   A severe and usually fatal form of leishmaniasis in which the parasites invade the liver and other organs.

Kaposi's sarcoma   A type of vascular cancer that occurs in people suffering from AIDS.

Kinases   Enzymes that break down clots in the blood.

Kissing bug   A triatomid insect that is the vector for Chagas' disease.

Koch's postulates   A set of rules used to establish that a particular organism is the cause of a particular disease.

Koplik's spots   Spots that appear in the mouth in the early stage of measles.

Kuru   A transmissible spongiform disease formerly found among members of the Fore people of New Guinea.

Larva   A stage in arthropod development that occurs after hatching and before maturity is reached.

Laryngitis   Inflammation of the larynx.

Latency   The period in which a microbe is in the host without displaying any visible symptoms.

LD$_{50}$ 50% lethal dose; a laboratory measurement of virulence to determine the dose that kills 50% of the test animals in a given time.

Legionnaires' disease   An airborne pneumonialike disease caused by *Legionella pneumophila*.

Leguminous plant   A type of plant with swellings or nodules along the root system containing *Rhizobium* and other nitrogen-fixing bacteria.

Leishmaniasis   A disease caused by parasites and transmitted by the bite of female sand flies.

Lepromas   Tumorlike skin lesions seen in leprosy.

Leprosy   *See* Hansen's disease.

Leptospirosis   A disease caused by the spirochete *Leptospira interrogans*.

Leukemia   A type of cancer characterized by uncontrolled reproduction of white blood cells.

Leukocidins   Bacterial enzymes that destroy white blood cells.

Leukocytes   White blood cells.

Lipopolysaccharide   A molecule with both a lipid and a polysaccharide component found in the outer wall of gram-negative bacteria; it acts as an endotoxin.

Lockjaw   A symptom of tetanus characterized by contraction of the muscles in the jaw; also an alternative name for tetanus.

Lower respiratory tract infections   Infections that occur in the trachea, larynx, bronchi, bronchioles, or lungs.

Lymph   A tissue fluid derived from blood that is returned to the blood by lymphatic vessels.

Lymph nodes   Structures along lymphatic vessels that act as microbe filters.

Lymphadenopathy   Swelling of lymph nodes that occurs in AIDS and other infections.

Lymphatic system   A system in which lymph is transported in lymphatic vessels.

Lymphatic vessels   Structures that transport lymph fluid.

Lymphocytes   A category of white blood cells that play a key role in immunity.

Lymphocytic leukemia   A form of leukemia characterized by overproduction of lymphocytes, leading to abnormally large numbers of immature and nonfunctional lymphocytes.

Lymphoid path   The development of blood cells that leads to the production of lymphocytes.

Lysis   Bursting of cells.

Lysogenized   Containing bacteriophage nucleic acid; said of a bacterium.

Lysozyme   An enzyme present in tears that contributes to the nonspecific immune system by disrupting bacterial cell walls.

M protein   A protein found in the cell walls of streptococci which confers resistance to phagocytosis.

Macrophages   Phagocytic cells that function in the immune system.

Macroscopic   Visible without the aid of a microscope.

Mad cow disease   A disease of cattle caused by prions that results in spongy degeneration of the brain accompanied by severe and fatal neurological damage.

Major histocompatibility molecules   Molecules associated with presentation of antigenic material to receptor molecules on cytotoxic T cells.

Malaria   A tropical disease caused by the bite of female *Anopheles* mosquitoes infected with the *Plasmodium* protozoan parasite.

Mastigophora   One of the four groups of protozoans, characterized by the presence of flagella.

Mechanical vectors   Arthropod vectors that transmit microbes passively on their body parts; the microbes do not invade, multiply, or develop in the vector.

Memory cells   B cells that do not progress to antibody-producing plasma cells but are retained for a quick immune response in the event of future exposure to the same antigen.

Meninges   Membranes around the spinal cord and the brain.

Meningitis   Inflammation of the meninges that can be caused by microbes.

Merozoites   Forms in the malaria life cycle resulting from the asexual multiplication of sporozoites.

Metastasis   The spreading of cancer cells from one region or tissue to another.

Miasma   A term, used before microbes were identified as agents of disease, meaning "bad air" or "swamp air."

Microfilariae   Mature filarial worms in the blood that can be taken up by mosquitoes, thereby infecting other people with elephantiasis.

Micrometer   A unit of measurement equal to one millionth of a meter.

Microscopic   Visible only with the aid of a microscope.

MMR vaccine   Measles, mumps, and rubella vaccine.

Monocytes   A category of phagocytic white blood cells.

Mononucleosis   *See* Infectious mononucleosis.

Morbidity   A measure of the rate of illness of a particular disease.

Mortality   A measure of the rate of death resulting from a particular disease.

Mucocutaneous leishmaniasis   A form of leishmaniasis characterized by invasion of the parasite into the skin and mucous membranes, causing destruction of the nose, mouth, and pharynx; also called espundia.

Multicellular   Having more than one cell.

Multiple drug resistant   Resistant to more than one antibiotic.

Multiplication   A process resulting in an increase in the total number of cells in an individual.

Murine typhus   A variety of typhus fever caused by *Rickettsia typhi* that is transmitted by fleas.

Mutualism   A symbiotic relationship in which both organisms benefit from the association.

Mycoplasmas   Bacteria that have no cell walls.

Myeloid leukemia   A type of cancer resulting from overproduction of monocytes and granular leukocytes.

Myeloid path   The lineage of blood cell maturation that leads to the production of platelets, red blood cells, monocytes, neutrophils, eosinophils, and basophils.

N spike   *See* Neuraminidase.

Naked viruses   Viruses that are not enclosed by an envelope.

Nanometer   A unit of measurement equal to one billionth of a meter.

Necrotic   Dead; used in reference to tissue.

Necrotizing fasciitis   A condition caused by highly invasive streptococci in which the subcutaneous tissue is infected; the streptococci are sometimes referred to as "flesh-eating."

Negri bodies   Cytopathic structures found in rabies virus-infected cells.

Neonatal (newborn) tetanus   A manifestation of tetanus in newborn children that results from unsanitary conditions during delivery.

Neuraminidase   An enzyme on the spikes of some influenza viruses, aiding in the release of new virions.

Neurocysticercosis   The presence of cysticerci in the brain or spinal cord resulting in seizures, headaches, and possibly death.

Neurosyphilis   A late stage of syphilis involving neurological damage possibly characterized by paralysis and insanity.

Neurotoxins   Toxins released by microbes that interfere with the transmission of neural impulses.

Neutrophils   White blood cells that are phagocytic.

Nitrification   A stage in the nitrogen cycle in which ammonia is converted into nitrogen.

Nitrogen cycle   A cycle in nature in which atmospheric nitrogen is recycled through a pathway involving bacteria.

Nonenveloped viruses   Viruses that do not have an envelope around them; also referred to as "naked."

Nonspecific immunity   Physiological defenses that prevent microbes from gaining access into the body or eliminate those that have penetrated the body.

Norwalk virus   A virus that is a frequent cause of gastroenteritis in older children and adults.

Norwalk-like viruses   A group of viruses, related to Norwalk virus, that frequently cause gastroenteritis in older children and adults.

Nosocomial infections   Infections acquired by patients during hospitalization or during stays in long-term health care facilities.

Nuclein   An early term for DNA.

Nucleocapsid   A viral structure consisting of the viral nucleic acid and the protein coat.

Nucleoid   The DNA-rich area in procaryotic cells; it is not surrounded by a membrane.

Nutrient agar   A semisolid growth medium used to grow bacteria in the laboratory.

Nymph   A preadult stage in tick development.

Obligate intracellular parasites   Microbes that can only replicate within cells.

Offal   A ground-up mixture of organs and trimmings of dead animals used to feed other animals.

Offensive strategies   Adaptations by microbes that result in their ability to damage the host and establish disease.

Onchocerciasis   A parasitic disease caused by *Onchocerca volvulus* that frequently results in blindness.

Oocyst   Infectious form of *Cryptosporidium parvum*.

Ophthalmia neonatorum   A condition in which the corneas are damaged as a result of the transmission of *Neisseria* into the eyes of newborns during delivery.

Opisthotonos   A body position in which the back is severely arched; characteristic of tetanus.

Opportunistic pathogens   Organisms that cause disease when the host's immune system is weakened, as in AIDS.

Oral rehydration therapy   A method of cholera treatment designed to replace lost body fluids; an alternative to intravenous rehydration.

Orchitis   Inflammation of the testes that sometimes occurs in males infected with mumps virus.

Organ system   A collection of organs that contribute to an overall function.

Organic compound   Generally, a chemical compound that contains carbon (although carbon dioxide and several other carbon-containing compounds are considered inorganic).

Organisms   Living entities that are composed of cells.

Organs   Structures composed of more than one tissue type.

Oriental sore   A form of leishmaniasis that results in skin lesions primarily on exposed parts of the body; also known as cutaneous leishmaniasis.

Oseltamivir   A relatively new drug that shortens the duration of influenza caused by influenza A and B viruses. The trade name of this drug is Tamiflu.

Osmosis   The movement of a solvent from an area of low solute (high solvent) concentration to an area of high solute (low solvent) concentration.

Outer membrane   An additional layer, external to the peptidoglycan layer in gram-negative bacteria, that is associated with virulence.

Pandemic   A worldwide outbreak of disease.

Paralytic poliomyelitis   A form of polio caused by replication of poliovirus in nerve cells, sometimes resulting in severely deformed limbs and paralysis.

Parasitic cycle   A chain of events, sometimes quite complex, by which a parasite exits from a host and gains access to a new host.

Parasitism   A form of symbiosis in which one biological agent lives at the expense of another.

Parotid glands   One of the three pairs of salivary glands; these glands may become infected and cause swelling on one or both sides of the face, a characteristic of mumps.

Parthenogenesis   A process by which females produce eggs without fertilization by males.

Pathogens   Microbes capable of producing disease.

Pellicle   A protective, rigid cover outside the cell membrane, found in some protozoans.

Pelvic inflammatory disease   A condition that occurs in about 50% of untreated female gonorrhea patients, characterized by abdominal pain and sometimes sterility.

Penetration   A stage in the replication of viruses in which the virus enters the host cell.

Peptidoglycan   A compound found in gram-positive and gram-negative bacterial cell walls that confers rigidity and tensile strength.

Perforin   A membrane-penetrating protein released by cytotoxic T cells that leads to lysis of cells harboring viruses.

Peritonitis   A potentially fatal condition caused by leakage of intestinal fluids into the abdomen.

Petri dish   A round, dishlike container in which bacteria are grown on agar in the laboratory.

Phage conversion   A process by which bacteriophage DNA is incorporated into the bacterial chromosome and confers new properties on the bacterial host.

Phagocytic cells   Cells that engulf microbes and bring about their destruction by enzymatic activity; these cells play a vital role in the nonspecific immune system.

Phagocytosis   A nonspecific body defense mechanism by which bacteria are ingested and killed by cells.

Photosynthetic autotrophs   Organisms that utilize the energy of the sun and that use carbon dioxide as a carbon source.

PID   *See* Pelvic inflammatory disease.

Pili   Bacterial appendages, found in gram-negative bacteria, that act as adhesins; some serve as a bridge allowing genetic exchange.

Pilin   The protein that composes pili.

Pinworm   A common helminthic disease caused by the worm *Enterobius vermicularis.*

Plague   A bacterial disease caused by *Yersinia pestis,* a highly virulent bacterium; also known as the Black Death.

Plankton   The primary food source for many aquatic organisms, consisting mainly of protozoans.

Plantar warts   Deep, painful warts found on the soles of the feet; caused by human papillomavirus.

Plasma cells   Antibody-producing cells derived from B cells.

Plasmids   Small molecules of nonchromosomal DNA found in some bacteria.

Platelets   Blood cells that initiate the clotting of blood.

Pleurisy   Inflammation of the pleural lining of the lungs.

Plug drugs   Drugs that act to both prevent and treat influenza by interfering with the N spikes of the virus.

Pneumonia   Inflammation of the lungs; can be caused by a variety of microbes.

Pneumonic plague   A form of bubonic plague that develops into pneumonia.

Pneumotropic viruses   Viruses that have an affinity for the respiratory tract.

Polio   Poliomyelitis, a highly infectious viral disease that can lead to muscle paralysis.

Polyhedral   Consisting of icosahedra; a term used to describe an arrangement of viral capsomeres.

Pontiac fever   A pneumonialike illness caused by *Legionella pneumophila.*

Portal of entry   The site at which microbes enter a host.

Portal of exit   The site from which microbes leave a host and may infect another host.

PPD   *See* Purified protein derivative.

Primary immune structures   The bone marrow and the thymus gland, in which maturation of B and T lymphocytes takes place.

Primary producers   Photosynthetic organisms that utilize the energy of the sun and produce organic compounds and oxygen.

Primary syphilis   An early stage of syphilis characterized by the formation of chancres on the genitals.

Prions   Infectious, highly stable, misfolded proteins that can cause neurological disease; they do not contain DNA or RNA.

Procaryotic   Not having the genetic material contained within a membrane in the cell.

Prodromal stage   An early stage in a microbial disease characterized by headache, tiredness, and muscle aches. Also the first stage of human immunodeficiency virus infection characterized by fever, diarrhea, rash, aches, fatigue, and lymphadenopathy.

Proglottids   Compartmentlike segments of tapeworms containing both testes and ovaries.

Propagated epidemics   Epidemics resulting from direct person-to-person transmission.

Prophage   Genetic material of a bacteriophage incorporated into the bacterial chromosome.

Protease inhibitors   A class of drugs used in AIDS therapy that prevent viral replication.

Protozoans   A category of unicellular eucaryotic microbes.

PrP gene   Gene located on chromosome 20 which is responsible for prion proteins.

Psychrophiles   Cold-loving bacteria that grow best at about 15° Celsius.

Purified protein derivative   A product derived from *Mycobacterium tuberculosis* and used in the tuberculin skin test.

R (resistance) factors   Genes carried on plasmids that confer antibiotic resistance.

Rabies   A viral disease transmitted by the bite of a rabid animal, resulting in damage to the nervous system and eventual death if not treated promptly by vaccination; formerly called hydrophobia.

Recombinant DNA technology   *See* Genetic engineering.

Red blood cells   Cells of the circulatory system that carry oxygen throughout the body; also called erythrocytes.

Release   The last stage in the viral replication cycle, during which mature viruses are released from host cells.

Relenza   *See* Zanamivir.

Rendering process   The process by which animal parts are turned into animal feed.

Replication   The process by which viruses multiply in host cells.

Reservoir   A site in nature where microbes survive and multiply and from which they may be transmitted.

Resistance   A defensive function of the immune system affording protection against microbial invasion; also, the ability of microbes to counter the effects of antimicrobial agents.

Respiratory syncytial virus   A prevalent cause of respiratory illness, most commonly found in infants, that produces nonspecific symptoms, including fever, runny nose, ear infection, and pharyngitis.

Rheumatic fever   A condition involving the heart and joints resulting from repeated bouts of streptococcal infection.

Rhinoviruses   A large group of viruses that are responsible for many common colds.

Ribavirin   A drug used in the treatment of hepatitis C, with limited success, and in the treatment of pneumonialike illnesses caused by respiratory syncytial virus.

Ribosomal ribonucleic acid   A type of ribonucleic acid (RNA) that is associated with ribosomes.

Rickettsiae   Rod-shaped bacteria that are (with a single exception) transmitted through the bite of arthropods; they are intracellular obligate parasites.

River blindness   A common type of blindness resulting from the migration of larval forms of the worm *Onchocerca volvulus* into the eye; also called onchocerciasis.

Rocky Mountain spotted fever   A tick-borne disease common in the southeastern United States that is caused by *Rickettsia rickettsii*.

Ropy milk   A condition that occurs when *Alcaligenes viscolactis* sheds its slime layer into milk.

RotaShield   An oral vaccine against rotavirus; the vaccine was withdrawn about a year after its approval because of adverse reactions.

Rotaviruses   A group of viruses that are a common cause of viral gastroenteritis in children under the age of 5 years.

Roundworms   A morphological category of worms.

Rubella   A viral disease characterized by a rash on the face which spreads to the trunk and the extremities.

Saber shins   A condition sometimes seen in syphilis patients in which the shinbone develops abnormally.

Sabin vaccine   A polio vaccine, developed by Albert Sabin, consisting of attenuated polioviruses.

Salk vaccine   The first vaccine against polio, developed by Jonas Salk, containing inactivated poliovirus.

Salmonellosis   A condition caused by ingestion of salmonella bacteria and resulting in gastroenteritis manifested by nausea, vomiting, abdominal cramps, and diarrhea.

Sand flies   Insects that transmit *Leishmania* species.

Sarcodina   A group of protozoans that exhibit a "creeping" movement.

Scalded skin syndrome   A condition caused by a staphylococcal toxin which results in the skin's becoming blistery with a tendency to peel.

Scarlet fever   A disease caused by a strain of streptococcus that produces an erythrogenic toxin leading to the development of a red rash and a strawberry-colored tongue.

Scavengers   A synonym for decomposers.

Schistosomiasis   A parasitic disease caused by blood flukes that enter the body through the skin.

SCID   *See* Severe combined immunodeficiency.

Scolex   The head of a tapeworm; it attaches to the host intestinal wall by suckerlike projections.

Secondary syphilis   A stage of syphilis characterized by a rash on the palms and soles; during this stage, the spirochetes multiply and spread throughout the body.

Secondary bacterial infections   Infections that result from bacteria in individuals suffering from the flu or other conditions that lower immune resistance.

Secondary immune structures   The spleen, tonsils, adenoids, lymph nodes, and patches of tissue associated with the intestinal tract that are seeded with mature B and T cells and phagocytic cells.

Selectively permeable   Able to be permeated by some but not all types of molecules; a property of cell membranes.

Septicemic plague   A variety of plague resulting from the spread of infection from the lungs to other parts of the body.

Serum sickness   A condition characterized by the formation of antigen-antibody complexes that are deposited in the skin, kidney, and other sites, as might occur in treatment with immunoglobulin.

Severe combined immunodeficiency A disease in which individuals lack both functional T and B cells and cannot mount either an antibody- or a cell-mediated immune response.

Sex pili Structures that function in exchange of DNA between bacteria by forming a bridge between cells.

Sexual stage A stage of malaria in which sporozoites are produced from *Plasmodium* parasites in the blood.

Sexually transmitted diseases Diseases caused by transmission of microbes from the warm, moist mucous membranes of one individual to the mucous membranes of another individual during sexual contact.

Shigellosis A bacterial gastrointestinal illness caused by the ingestion of *Shigella*-contaminated foods and water, resulting in symptoms of diarrhea, abdominal cramping, and, in some cases, dysentery.

Shingles A disease caused by varicella-zoster virus that occurs in some individuals with a history of chicken pox; the virus infects nerve fibers, creating intense pain.

Sleeping sickness A protozoan disease caused by the bite of a tsetse fly carrying the parasite. Two types (West and East African trypanosomiasis) are known.

Slime layer A heavy mucuslike material that accumulates around some bacteria.

Small animalcules The name for microbes first described by Antony van Leeuwenhoek during his examination of tooth scrapings with a primitive microscope.

Species The fundamental rank in the binomial system of classification of organisms; in the name *Escherichia coli*, "*coli*" refers to the species.

Specific immunity Immunity that is acquired after birth and responds to the presence of foreign and potentially harmful microbes that have breached the external and nonspecific defense mechanisms.

Spikes Projections that extrude through the surface of some viruses.

Spirilla Spiral-shaped bacteria.

Spirochetes Flexible, corkscrew-shaped, motile bacteria.

Spleen An organ that is part of the secondary immune system; it contains phagocytic cells and both mature B and T cells.

Spontaneous generation A false but once popular theory that nonliving structures could give rise to living organisms. Louis Pasteur played a major role in disproving this theory.

Sporadic Occurring occasionally and at irregular intervals in a random and unpredictable fashion.

Spores Structures contained within dormant bacterial cells that are highly resistant to heat, drying, radiation, and a variety of chemical compounds. The genera *Bacillus* and *Clostridium* are sporeformers.

Sporozoites Forms of *Cryptosporidium parvum* that penetrate the intestinal cells, where multiplication results in more oocysts; also, infective forms of malaria parasites.

St. Louis encephalitis virus A common mosquito-borne arbovirus found in the United States that causes encephalitis.

Stage of decline The time following an illness in which symptoms begin to disappear and the body returns to normal.

Staph food poisoning An illness caused by the secretion of an enterotoxin from *Staphylococcus aureus*; characteristic symptoms are abdominal cramps, nausea, vomiting, and diarrhea.

Staphylococci Bacteria of the genus *Staphylococcus*, which group together in clusters resembling bunches of grapes.

Stem cells Undifferentiated cells capable of differentiating into specialized cells.

Stomach flu A common but meaningless term that is associated with gastroenteritis.

Strep throat A mild airborne infection caused by *Streptococcus pyogenes* with characteristic symptoms of red or sore throat, fever, and headache.

Streptococci Bacteria of the genus *Streptococcus*, which group together in chains resembling strings of pearls.

Streptokinase A product of streptococci that dissolves blood clots.

Stromatolites Fossilized mats formed from microorganisms and dating as far back as 3.5 billion to 3.8 billion years.

Strongyloidiasis Illness caused by the nematode *Strongyloides stercoralis*; symptoms are nausea, vomiting, anemia, weight loss, and chronic bloody diarrhea.

Subclinical Asymptomatic and not diagnosed.

Sucking disks Structures by which *Giardia lamblia* trophozoites attach to the lining of the intestine.

Swamp fever *See* Leptospirosis.

Swimmer's itch *See* Cercarial dermatitis.

Symbiosis A relationship between two or more organisms that live together; mutualism, commensalism, and parasitism are all forms of symbiosis.

Syphilis A sexually transmitted disease caused by *Treponema pallidum*; the symptoms are sores on the penis or cervix, rash, and degeneration of organs and tissues.

Systemic infections Blood-borne infections that spread throughout the body.

T-cell subset A collective term for the categories of T cells differentiated by clusters of differentiation.

T-helper cells Members of the T-cell subset designated CD4.

T lymphocytes A subcategory of lymphocytes that function in the immune system in several ways.

Tamiflu *See* Oseltamivir.

Taxonomy The science of classification of organisms.

Tertiary syphilis The third stage of syphilis during which organs and tissues undergo degenerative changes.

Tetanospasmin A neurotoxin produced by *Clostridium tetani* that results in rigid paralysis.

Tetanus A bacterial disease acquired by exposure to *Clostridium tetani* or its spores. Symptoms include stiffness in the jaw and contraction of muscles in the limbs, stomach, and neck. Also called lockjaw.

Tetrads Groupings of cocci in clusters of four.

Thymus gland The organ in which T-cell maturation is completed; it is located behind the sternum and just above the heart.

Tissue A group or collection of cells that are all of the same type.

Tonsils Structures of the secondary immune system located at the back of the throat that aid in protection from microbes entering through the nose and throat.

Toxemia An illness resulting from the presence of an exotoxin in the body.

Toxic shock syndrome A condition caused by certain strains of toxin-producing staphylococci; primarily associated with the use of highly absorbent tampons.

Toxigenicity The ability of microbes to produce toxins.

Toxins Major virulence factors that are harmful to the body and are produced by many pathogenic microbes; tetanus, botulinum, and erythrogenic toxins are examples.

Toxoplasmosis A protozoan disease caused by *Toxoplasma gondii* and acquired by the ingestion of oocysts present in cat feces. Symptoms include sore throat, low-grade fever, and lymph node enlargement.

Tracheitis Inflammation of the trachea.

Tracheotomy The cutting of a hole in the throat to facilitate breathing.

Transforming principle A term used by Frederick Griffith to describe the transfer of virulence from virulent to nonvirulent bacteria; the active component was later discovered to be DNA.

Transmissible spongiform encephalopathies Degenerative brain diseases that are thought to be the result of abnormally folded prions that latch onto normal prions and convert them into an altered, defective form.

Transmission A link in the cycle of microbial disease between reservoir and portal of entry.

Transovarial transmission The passage of microbes from adult ticks to their eggs.

Transposons Genes (sometimes conferring antibiotic resistance) that can be integrated into plasmids or chromosomes.

Traveler's diarrhea An illness caused by enterotoxigenic *Escherichia coli* strains.

Treponemes Spirochetes released by chancres in individuals suffering from syphilis.

Triatomid insect The vector for Chagas' disease, commonly referred to as the kissing bug.

Trichinellosis A roundworm disease caused by *Trichinella spiralis* that is transmitted in undercooked pork. Symptoms include nausea, diarrhea, vomiting, fatigue, fever, headaches, chills, aching joints, and itchy skin.

Trichomoniasis A sexually transmitted disease caused by *Trichomonas vaginalis*. Symptoms include intense itching, urinary frequency, pain during urination, and vaginal discharge in females. In males, symptoms include pain during urination, inflammation of the urethra, and a thin, milky discharge.

Trophozoite The reproductive and feeding stage of parasitic amoebae and other protozoan parasites.

Tuberculin skin test (Mantoux test) A method for diagnosis of tuberculosis; purified protein derived from *Mycobacterium tuberculosis* is injected into the patient, and the presence of an induration between 48 and 72 hours after injection indicates past or present exposure to the tubercle bacillus.

Tuberculosis A contagious lower respiratory tract disease caused by *Mycobacterium tuberculosis*. Symptoms include fever, night sweats, weight loss, fatigue, and coughing up of blood-tinged sputum.

Typhoid fever A disease caused by *Salmonella typhi* that is transmitted by flies and fomites.

Typhoid Mary A nickname for Mary Malone (1869–1938), a notorious carrier of typhoid fever.

Ulcers Lesions that occur in the mucous membrane lining the stomach, usually caused by *Helicobacter pylori* bacteria.

Unicellular Having only one cell.

Upper respiratory tract infections Infections occurring in the tonsils or pharynx.

Urea A metabolic product of the body resulting from the breakdown of the enzyme urease.

Urease The enzyme produced by *Helicobacter pylori* that breaks down urea.

Variant Creutzfeldt-Jakob disease A fatal human transmissible spongiform encephalopathic disease caused by prions. It is similar to mad cow disease in that it causes degeneration of brain tissue and severe neurological damage.

Varicella-zoster virus The DNA virus that causes shingles and chicken pox.

VD *See* Venereal diseases.

Vector borne Carried by a particular organism.

Vegetative cells Spore-forming bacteria without spores.

Venereal diseases The earlier name for sexually transmitted diseases, commonly referred to as VD.

Vertical transmission A method of transmission characterized by passage of pathogens from parent to offspring across the placenta, in breast milk, or in the birth canal.

Vesicle A membrane-bound structure in the cytoplasm; also, a fluid-filled lesion that appears on the skin.

Vibrios Spirillar bacteria in the shape of comma-curved rods; *Vibrio cholerae,* the causative agent of cholera, is an example.

Virion A complete viral particle.

Virulence The capacity of microbes to produce disease as a result of defensive and offensive strategies.

Viruses One of the categories of microbes; characterized as subcellular obligate intracellular parasites.

Visceral leishmaniasis The most severe form of leishmaniasis; symptoms are fever, weakness, weight loss, anemia, and protrusion of the abdomen. Also known as kala-azar.

Viscerotropic viruses Viruses that affect internal organs (viscera) such as the liver, spleen, and intestines.

Wandering macrophages Macrophages derived from monocytes of the blood that are phagocytic and move freely about the tissues.

West Nile virus An arbovirus transmitted by the bite of a mosquito that emerged in New York in 1999; it causes encephalitis.

Western equine encephalitis virus A common arbovirus in the United States.

White blood cells Cells that play a critical role in body defense mechanisms; also called leukocytes. Lymphocytes, neutrophils, monocytes, basophils, and eosinophils are the five types of white blood cells.

Whoop The characteristic sound of whooping cough; deep and rapid inspirations throughout the partially obstructed passages are responsible for the sound.

Whooping cough A highly infectious disease caused by *Bordetella pertussis*. Symptoms include spasms of violent hacking and persistent, recurrent coughing with a whooplike noise. Also called pertussis.

Xenodiagnosis A diagnostic method used in Chagas' disease. "Kissing bugs" free of trypanosomes are allowed to feed on individuals suspected of having the disease. A few weeks later, the insects are examined for the presence of the parasite.

Yellow fever A mosquito-borne viral disease also known as yellow jack; symptoms include fever, bloody nose, headache, nausea, muscle pain, vomiting, and jaundice.

Zanamivir An antiflu drug that is effective against both A and B influenza viruses. It reduces the severity of the flu if taken within 48 hours of the appearance of symptoms. The trade name of this drug is Relenza.

Zidovudine *See* AZT.

Zoonoses Diseases for which domestic and/or wild animals are the reservoirs and which can be transmitted to humans.

Index